HANDBOOK OF SENSOR NETWORKS

WILEY SERIES ON PARALLEL AND DISTRIBUTED COMPUTING

Editor: Albert Y. Zomaya

HANDBOOK OF SENSOR NETWORKS

ALGORITHMS AND ARCHITECTURES

Edited by

Ivan Stojmenović
University of Ottawa

A JOHN WILEY & SONS, INC., PUBLICATION

Library of Congress Cataloging-in-Publication Data:

Handbook of sensor networks : algorithms and architectures / edited by Ivan Stojmenovic.
    p. cm. --- (Wiley series on parallel and distributed computing)
  Includes bibliographical references and index.
  ISBN-13 978-0-471-68472-5 (cloth)
  ISBN-10 0-471-68472-4 (cloth)
  1. Sensor networks.   I. Stojmenovic, Ivan.

  TK7872.D48H358 2005
  681′.2--dc22

                                                                2005005155

Printed in the United States of America

10 9 8 7 6 5 4 3 2 1

To my daughter Milica, son Milos, and wife Natasa, my personal sensor network.
To Val and Emily from Wiley, for their timely and professional cooperation.

■ CONTENTS

■■■■■■ PREFACE

Recent technological advances have enabled the development of low-cost, low-power, and multifunctional sensor devices. These nodes are autonomous devices with integrated sensing, processing, and communication capabilities. A sensor is an electronic device that is capable of detecting environmental conditions such as temperature, sound, chemicals, or the presence of certain objects. Sensors are generally equipped with data processing and communication capabilities. The sensing circuitry measures parameters from the environment surrounding the sensor and transforms them into electric signals. Processing such signals reveals some properties of objects located and/or events happening in the vicinity of the sensor. The sensor sends such sensed data, usually via a radio transmitter, to a command center, either directly or through a data-collection station (a base station or a sink). To conserve the power, reports to the sink are normally sent via other sensors in a multihop fashion. Retransmitting sensors and the base station can perform fusion of the sensed data in order to filter out erroneous data and anomalies, and to draw conclusions from the reported data over a period of time. For example, in a reconnaissance-oriented network, sensor data indicates detection of a target, while fusion of multiple sensor reports can be used for tracking and identifying the detected target.

This handbook is intended for researchers and graduate students in computer science and electrical engineering, and researchers and developers in the telecommunication industry. It provides an opportunity for researchers to explore the currently "hot" field of sensor networks. It is a problem-oriented book, with each chapter discussing computing and communication problems and solutions that arise in rapidly emerging wireless sensor networks. The main purpose of the book is to review various algorithms and protocols that were developed in the area, with the emphasis on the most recent ones.

The handbook is based on a number of stand-alone chapters that together cover the subject matter in a fully comprehensive manner. Edited books are normally collections of chapters freely selected by invited authors. This handbook follows a different approach. First, the sensor network arena was divided into meaningful units, reflecting the state of the art, importance, amount of literature, and, above all, comprehensiveness. Then the most suitable author for each chapter was selected, considering their expertise and presentation skills. The editor also considered the geographical distribution of authors, and representations from industry and top research institutions. Among the authors are researchers from Motorola, Intel, and Fujitsu laboratories, MIT, IIT, Cornell University, University of Illinois, all in the United States, plus researchers from Switzerland, Germany, France, Australia, and Canada.

Sensor networks are currently recognized as one of the priority research areas (for example, a multidisciplinary program on sensors and sensor networks was launched in 2003 at the U.S. National Science Foundation), and research activities recently started booming. A number of ongoing projects are being funded in Europe, Asia, and North America. Before Y2K, research on sensor networks was sporadic, and were treated as a special case of emerging ad hoc networks. Sensor networks were then quickly recognized as an independent topic, their name was added to some event titles, and now events specializing in sensor networks have emerged in the last two years. At least two new journals devoted exclusively to sensor networks appeared in 2005.

As a result of the exponential growth in the number of researchers, publications, conferences, and journals on sensor networks, a number of graduate courses fully or partially concentrating on sensor networks have emerged recently. These courses are mostly based on reading a selected set of recent articles, with the focus on certain topics that reflect the interest of the instructor within the sensor networks domain. It is expected that this book will provide a much needed textbook for such graduate courses. Since the area is gaining popularity, a textbook is needed as a reference source for use by students and researchers. The chapters cover subjects in a comprehensive manner, describing the state of the art and surveying important existing solutions. They provide readable but informative content, with appropriate illustrations, figures, and examples. A number of chapters also provide some problems and exercises for use in graduate courses.

This handbook is intended to cover a wide range of recognized problems in sensor networks, striking a balance between theoretical and practical coverage. The theoretical contributions are limited to the scenarios and solutions that are believed to have practical relevance. The handbook content addresses the dynamic nature of ad hoc and sensor networks. Due to frequent node addition and deletion from networks (changes between active and inactive periods, done to conserve energy, are one of the contributors to this dynamic) and possible node movement, the algorithms that potentially can be used in real equipment must be *localized* and must have minimal communication overhead. The overhead should take both the construction and its maintenance for the structure used in solutions and ongoing protocols into consideration. We believe that only this approach will eventually lead to the design of protocols for real applications. We now explain our design principles and priorities, used to cover the subject matter in this handbook.

A *scalable* solution is one that performs well in a large network. Sensor networks may have hundreds or thousands of nodes. Priority is given to protocols that perform well for small networks, and perform significantly better for large networks (more precisely, are still working as opposed to crashing when other methods are applied). In order to achieve scalability, new design paradigms must be applied. The main paradigm shift is to apply localized schemes, in contrast with most existing protocols, which require global information. In a *localized* algorithm, each node makes protocol decisions solely based on the knowledge about its local neighbors. In addition, the goal is to provide protocols that will minimize the number of messages between nodes, because bandwidth and power are limited. Protocols should use a small constant number of messages, often even none beyond preprocessing "hello" messages. Localized message-limited protocols provide scalable solutions. Typical local information to be considered is one-hop or two-hop neighborhood information (information about direct neighbors and possibly the neighbors of neighbors). Nonlocalized distributed algorithms, on the other hand, typically require global network knowledge, including information about the existence of every edge in the graph. The maintenance of global network information, in the presence of mobility or changes between sleep and active periods, imposes huge communication overhead, which is not affordable for bandwidth and power-limited nodes. In addition to being localized, protocols are also required to be *simple, easy to understand and implement*, and to have *good average-case performance*. Efficient solutions often require position information. It has been widely recognized that sensor networks can function properly only if reasonably accurate position information is provided to the nodes.

BRIEF OUTLINE CONTENT

This handbook consists of 15 chapters. It begins with an introductory chapter that describes various scenarios where sensor networks may be applied, and various application-layer tools for enabling such applications. Applications include habitat monitoring, biomedical sensor engineering, monitoring environments, water and waste management, and military applications. The second chapter is on physical layer and signal processing in sensor networks.

In sensor networks with tiny devices, which are usually designed to run on batteries, the replacement of depleted batteries is not practical. The goal of the third chapter is to explore methods of scavenging ambient power for use by low-power wireless electronic devices in an effort to make the wireless nodes and resulting wireless sensor networks indefinitely self-sustaining.

Chapter 4 describes a vision to build ultra-low-power wireless sensor systems and a self-contained, millimeter-scale sensing and communication platform for a massively distributed sensor network. This vision is based on realistic assumptions about sensors, such as limited ability to provide accurate position information (therefore proposing the concept of cluster position information rather than individual position information), and lack of individual sensor identities (the property commonly recognized but often implicitly assumed in protocols).

The power, computation, and communication limitations of sensor networks make the design and utilization of security and fault-tolerance schemes particularly challenging. Chapter 5 is intended as a starting point for studying sensor network security. It focuses on recent advances in broadcast authentication and key management in sensor networks, which are foundational cryptographic services for sensor network security. It describes random key predistribution techniques proposed for establishing pairwise keys between resource-constrained sensor nodes. Attacks against location discovery and some additional security problems in sensor networks are also discussed.

Chapter 6 reviews research on operating systems and middleware issues in the emerging area of embedded, networked sensors. Chapter 7 addresses the issue of calibration and time synchronization in sensor networks and related problems, such as temporal message ordering. Chapter 8 reviews various medium-access schemes for sensor networks, and the power efficiency aspects of these schemes.

In the position-determination problem, each sensor should be designed to decide about its geographic position based on several reference nodes in the network, in case it has no direct position service such as global positioning system (GPS) attached. The position needs to be determined in cooperation with other sensors, based on hop counts to reference nodes or other information. Chapter 9 reviews triangulation, multilateration, diffusion, and other types of solutions for this problem.

The problem of deciding the best transmission radius of each sensor, and the links that are desirable to have, is a challenging one. For instance, it is known that the probability that a random-unit graph is connected has a sharp transition from 0 to 1, meaning that it is difficult to decide the best uniform transmission radius for network connectivity and congestion avoidance. On the other hand, efficient localized methods exist where each node is designed to decide its own transmission radius and links. Chapter 10 reviews topology construction and maintenance schemes under various sensor architectures.

In a broadcasting (also known as data dissemination) task, a message is sent from one node, which could be a monitoring center, to all the nodes in the network. The activity scheduling problem is one of deciding which sensors should be active and which should go to sleep mode, so that the sensor network's life is prolonged. The best known solutions to these two problems are based on the concept of localized connected dominating sets. Sensors that are randomly placed in an area should be designed to decide which of them should be active and monitor an area, and which of them may sleep and become active at a later time. The connectivity is important so that the measured data can be reported to the monitoring center. Sensors may also be placed deterministically in an area to optimize coverage and reduce their power consumption. Chapter 11 reviews solutions to these three related problems in sensor networks.

Position information enables development of localized routing methods (greedy routing decisions are made at each node, based solely on knowledge of positions of neighbors and destination, with considerable savings in communication overhead and with guaranteed delivery, provided location update schemes are efficient for a given movement pattern. Power consumption can be taken into account in the routing process. Chapter 12 surveys existing position based and power aware routing schemes. It also reviews physical layer aspects of position based routing.

Chapter 13 covers the emerging topic of data-driven routing, for example, directed diffusion. It also covers the emerging topics of constructing and maintaining reporting trees, dynamic evolution of the monitoring region for moving targets, various training options, and receiving reports from a particular area of interest, that is, geocasting.

In order to monitor a region for traffic traversal, sensors can be deployed to perform collaborative target detection. Such a sensor network achieves a certain level of detection performance with an associated cost of deployment. Chapter 14 reviews solutions for the various path-exposure protocols and sensor deployment for increased reliability of measurements. In the object-location problem, sensors collaborate to detect the position of a mobile object. The goal is to derive the location accurately, with a minimum number of sensors involved in the process. This chapter also discusses sensor networks for target classification and tracking, with respect to location-aware data routing to conserve system resources, such as energy and bandwidth. Distributed classification algorithms exploit signals from multiple nodes in several modalities and rely on prior statistical information about target classes.

Data gathering in sensor networks differs from the general ad hoc network's data communication protocols. Sensors in general monitor or measure the same event or data and report it to the monitoring center. Their data may be combined while being routed (data fusion), to save energy and increase reliability of reports. Chapter 15 reviews protocols for data gathering and fusion in sensor networks. This chapter also discusses the challenging problem of transport-layer protocols in sensor networks. Due to severe power and computational limitations, providing quality of service, delay, or jitter guarantees, in routing and data dissemination tasks by sensors is a difficult problem. This chapter also reviews efficient sensor database querying, for example, TinyDB. The sensor system should provide scalable, fault-tolerant, flexible data access and intelligent data reduction, as its design involves a confluence of novel research in database query processing, networking, algorithms, and distributed systems.

ACKNOWLEDGMENTS

The editor is grateful to all the authors for their contribution to the quality of this handbook. The assistance of reviewers for all chapters is also greatly appreciated. The University of Ottawa (with the help of the National Science and Engineering Research Council (NSERC) provided an ideal working environment for the preparation of this handbook. This environment included computer facilities for efficient Internet search, communication by electronic mail, and writing my own contributions.

The editor is thankful to Dr. Albert Zomaya, editor of the Parallel and Distributed Computing book series at Wiley, for his support and encouragement in publishing this handbook at Wiley. Special thanks go to Richard Han and Krishna Sivalingam; this book benefited greatly from their comments and suggestions. Val Moliere (Editor, Wiley-Interscience), Emily Simmons, (Editorial Assistant), and Kirsten Rohstedt (Editorial Program Coordinator) deserve special mention for their timely and professional cooperation, and for their decisive support of this project.

Finally, I thank my children Milos and Milica and my wife Natasa for their encouragement, making this effort worthwhile, and for their patience during the numerous hours at home that I spent in front of the computer.

I hope that the readers will find this handbook informative and worth reading. Comments received by readers will be greatly appreciated.

IVAN STOJMENOVIĆ

School of Information
Technology and Engineering,
University of Ottawa, Ottawa,
Ontario, Canada
Ivan@site.uottawa.ca
www.site.uottawa.ca/~ivan

December 2004

■■■■■ CONTRIBUTORS

Jonathan Bachrach, Artificial Intelligence Laboratory, Massachusetts Institute of Technology, Cambridge, MA 02139, jrb@ai.mit.edu

Shah Bhatti, University of Colorado, Department of Computer Science, Engineering Center, ECOT 717, Campus Box 430 UCB, Boulder, CO 80309-0430

Philipp Blum, Computer Engineering and Networks Laboratory, Department of Information Technology and Electrical Engineering, Swiss Federal Institute of Technology (ETH) Zürich, CH-8092 Zurich, Switzerland

Edgar H. Callaway, Jr., Distinguished Member of the Technical Staff, Florida Communication Research Lab, Motorola Labs, Plantation, FL 33322, ed.callaway@motorola.com

James Carlson, University of Colorado, Department of Computer Science, Engineering Center, ECOT 717, Campus Box 430 UCB, Boulder, CO 80309-0430

Wei-Peng Chen, IP Networking Research, Fujitsu Laboratories of America, Inc., 1240 East Arques Avenue, Sunnyvale, CA 94085, wei-peng.chen@us.fujitsu.com

Hui Dai, University of Colorado, Department of Computer Science, Engineering Center, ECOT 717, Campus Box 430 UCB, Boulder, CO 80309-0430

Jing Deng, University of Colorado, Department of Computer Science, Engineering Center, ECOT 717, Campus Box 430 UCB, Boulder, CO 80309-0430

Luc Frechette, Universite de Sherbrooke, Faculty of Engineering, Department of Mechanical Engineering, 2500 boul. Universite, Sherbrooke, Quebec J1H 2R1 Canada, Luc.Frechette@Usherbrooke.ca

Hannes Frey, University of Trier, System Software and Distributed Systems, Behringstrasse 1, D-54286 Trier, Germany, frey@syssoft.uni-trier.de

Richard Han, University of Colorado, Department of Computer Science, Engineering Center, ECOT 717, Campus Box 430 UCB, Boulder, CO 80309-0430, rhan@cs.colorado.edu

Jennifer Hou, Department of Computer Science, University of Illinois, 3112 Seibel Center, 201 N. Goodwin Avenue, Urbana, IL 61801-2302, jhou@cs.uiuc.edu

An-swol Hu, School of Electrical and Computer Engineering, 326 Rhodes Hall, Cornell University, Ithaca, NY 14853-6701

Ning Li, Department of Computer Science, University of Illinois, 3112 Seibel Center, 201 N. Goodwin Avenue, Urbana, IL 61801-2302

XiangYang Li, Department of Computer Science, Illinois Institute of Technology, Chicago, IL, 60616, xli@cs.iit.edu

Donggang Liu, Department of Computer Science, North Carolina State University, Raleigh, NC 27695-8207

Fernando Martincic, Department of Computer Science, Wayne State University, 5143 Cass Avenue, 431 State Hall, Detroit, MI 48202

Lennart Meier, Computer Engineering and Networks Laboratory, Department of Information Technology and Electrical Engineering, Swiss Federal Institute of Technology (ETH) Zürich, CH-8092 Zurich, Switzerland

Kousha Moaveni-Nejad, Department of Computer Science, Illinois Institute of Technology, Chicago, IL, 60616

Peng Ning, Department of Computer Science, Room 250 Venture III (inside Suite 243) North Carolina State University, Raleigh, NC 27695-8207, pning@ncsu.edu

Stephan Olariu, Department of Computer Science, Old Dominion University, Norfolk, VA 23529-0162, olariu@cs.odu.edu

Kay Römer, Institute for Pervasive Computing, Department of Computer Science, Swiss Federal Institute of Technology (ETH) Zürich, CH-8092 Zurich, Switzerland, roemer@inf.ethz.ch

Jeff Rose, University of Colorado, Department of Computer Science, Engineering Center, ECOT 717, Campus Box 430 UCB, Boulder, CO 80309-0430

Shad Roundy, LV Sensors, Inc., Emeryville, CA, sroundy@lvsensors.com

Loren Schwiebert, Department of Computer Science, Wayne State University, 5143 Cass Avenue, 431 State Hall, Detroit, MI 48202, loren@cs.wayne.edu

Sergio Servetto, School of Electrical and Computer Engineering, 326 Rhodes Hall, Cornell University, Ithaca, NY 14853-6701, servetto@ece.cornell.edu

Anmol Sheth, University of Colorado, Department of Computer Science, Engineering Center, ECOT 717, Campus Box 430 UCB, Boulder, CO 80309-0430

Brian Shucker, University of Colorado, Department of Computer Science, Engineering Center, ECOT 717, Campus Box 430 UCB, Boulder, CO 80309-0430

David Simplot-Ryl, IRCICA/LIFL, Univeriste Lille 1, CNRS UMR 8022, INRIA Futurs, POPS research group, Bât. M3, Cité Scientifique, 59655 Villeneuve d'Ascq Cedex, France, simplot@lifl.fr

Ivan Stojmenović, SITE, University of Ottawa, 800 King Edwards, Ottawa, Ontario K1 N 6N5, Canada, ivan@site.uottawa.ca

Christopher Taylor, Artificial Intelligence Laboratory, Massachusetts Institute of Technology, Cambridge, MA 02139

Ashraf Wadaa, Intel Corporation, Hillsboro, OR

Jie Wu, Department of Computer Science and Engineering, Florida Atlantic University, 777 Glades Road, Boca Raton, FL 33431-6498, jie@cse.fau.edu

Qingwen Xu, Department of Computer Science, Old Dominion University, Norfolk, VA 23529-0162

■■■■■■ **CHAPTER 1**

Introduction to Wireless Sensor Networking

FERNANDO MARTINCIC and LOREN SCHWIEBERT

Wayne State University, Detroit, Michigan

This chapter introduces the topic of wireless sensor networks from the applications perspective. A wireless sensor network consists of a possibly large number of wireless devices able to take environmental measurements such as temperature, light, sound, and humidity. These sensor readings are transmitted over a wireless channel to a running application that makes decisions based on these sensor readings. Authors describe some examples of proposed wireless sensor applications, and consider the following two questions to motivate an application-based viewpoint. What aspects of wireless sensors make the implementation of applications more challenging, or at least different? One widely recognized issue is the limited power available to each wireless sensor node, but there are other challenges such as limited storage or processing. What services are required for a wireless sensor network application to achieve its intended purpose? A number of widely applicable services, such as time synchronization and location determination are briefly discussed in this chapter. Other services are needed to support database requirements, such as message routing, topology management, and data aggregation and storage. As most of these topics are covered in separate chapters, this chapter serves to provide a broad framework to enable the reader to see how these different topics tie together into a cohesive set of capabilities for building wireless sensor network applications.

1.1 INTRODUCTION

A wireless sensor network consists of a possibly large number of wireless devices able to take environmental measurements. Typical examples include temperature,

Handbook of Sensor Networks: Algorithms and Architectures, Edited by Ivan Stojmenović
Copyright © 2005 John Wiley & Sons, Inc.

light, sound, and humidity. These sensor readings are transmitted over a wireless channel to a running application that makes decisions based on these sensor readings. Many applications have been proposed for wireless sensor networks, and many of these applications have specific quality of service (QoS) requirements that offer additional challenges to the application designer. In this chapter, we introduce the topic of wireless sensor networks from the perspective of the application.

Along with some examples of proposed wireless sensor applications, we consider two questions to motivate an application-based viewpoint:

1. *What aspects of wireless sensors make the implementation of applications more challenging, or at least different?*

   One widely recognized issue is the limited power available to each wireless sensor node, but other challenges such as limited storage or processing capabilities play a significant role in constraining the application development.

2. *What services are required for a wireless sensor network application to achieve its intended purpose?*

   A number of widely applicable services, such as time synchronization and location determination are briefly discussed. Other services are needed to support database requirements, such as message routing, topology management, and data aggregation and storage.

Because some of these topics are covered in separate chapters, this discussion serves to provide a broad framework to enable the reader to see how these different topics tie together into a cohesive set of capabilities for building wireless sensor network applications.

1.2 DESIGN CHALLENGES

Several design challenges present themselves to designers of wireless sensor network applications. The limited resources available to individual sensor nodes implies designers must develop highly distributed, fault-tolerant, and energy-efficient applications in a small memory-footprint. Consider the latest-generation MICAz [1,2] sensor node shown in Figure 1.1.

MICAz motes are equipped with an Atmel128L [4] processor capable of a maximum throughput of 8 millions of instructions per second (MIPS) when operating at 8 MHz. It also features an IEEE 802.15.4/Zigbee compliant RF transceiver, operating in the 2.4–2.4835-GHz globally compatible industrial scientific medical (ISM) band, a direct spread-spectrum radio resistant to RF interference, and a 250-kbps data transfer rate. The MICAz runs on TinyOS [5] (v1.1.7 or later) and is compatible with existing sensor boards that are easily mounted onto the mote. A partial list of specifications given by the manufacturers of the MICAz mote is presented in Figure 1.2.

Figure 1.1   MICAz sensor mote hardware. (Image courtesy of Crossbow Technology [3].)

For wireless sensor network applications to have reasonable longevity, an aggressive energy-management policy is mandatory. This is currently the greatest design challenge in any wireless sensor network application. Considering that in the MICAz mote the energy cost associated with transmitting a byte over the transceiver is substantially greater than performing local computation, developers must leverage local processing capabilities to minimize battery-draining radio communication. Several key differences between more traditional ad hoc networks and wireless sensor networks exist [6]:

- Individual nodes in a wireless sensor network have limited computational power and storage capacity. They operate on nonrenewable power sources and employ a short-range transceiver to send and receive messages.
- The number of nodes in a wireless sensor network can be several orders of magnitude higher than in an ad hoc network. Thus, algorithm scalability is an important design criterion for sensor network applications.
- Sensor nodes are generally densely deployed in the area of interest. This dense deployment can be leveraged by the application, since nodes in close proximity can collaborate locally prior to relaying information back to the base station.
- Sensor networks are prone to frequent topology changes. This is due to several reasons, such as hardware failure, depleted batteries, intermittent radio interference, environmental factors, or the addition of sensor nodes. As a result, applications require a degree of inherent fault tolerance and the ability to reconfigure themselves as the network topology evolves over time.

Processor	Atmel ATMega128L @ 8 MHz
Program Flash Memory	128 kilobytes
Measurement Serial Flash	512 kilobytes
Configuration electrically erasable programmable read-only memory (EEPROM)	4 kilobytes
Serial Communications	UART
Analog to Digital Converter	10 bit ADC
Other Interfaces	Digital I/O, I2C, SPI
Processor Current Draw	8 mA in active mode
	$< 1\,\mu A$ in sleep mode
Frequency band	2400MHz to 2483.5MHz
Transmit (TX) data rate	250kbps
RF power	-24dBm to 0dBm
Receive Sensitivity	-90dBm (min), -94dBm (typ)
Adjacent channel rejection	47 dB, +5-MHz channel spacing
	38 dB, -5-MHz channel spacing
Outdoor Range	75m to 100m
Indoor Range	20m to 30m
Radio Current Draw	19.7mA in receive mode
	11mA (TX -10dBm)
	14mA, (TX -5dBm)
	17.4mA (TX 0dBm)
	20 μA in idle mode
	(voltage regulator on)
	1 μA in sleep mode
	(voltage regulator off)
Battery	2 AA batteries
User Interface	red, green, and yellow LED
Size	$2.25 \times 1.25 \times 0.25$ in.
	(w/o battery pack)
Weight	0.7 oz (w/o batteries)
Expansion Connector	51 pin

Figure 1.2 MICAz mote specification [1].

- Wireless sensor networks do not employ a point-to-point communication paradigm because they are usually not aware of the entire size of the network and nodes are not uniquely identifiable. Consequently, it is not possible to individually address a specific node. Paradigms, such as directed diffusion [7,8], employ a data-centric view of generated sensor data. They identify information produced by the sensor network as ⟨attribute, value⟩ pairs. Nodes request data by disseminating interests for this named data throughout the network. Data that matches the criterion are relayed back toward the querying node.

Even with the limitations individual sensor nodes possess and the design challenges application developers face, several advantages exist for instrumenting an area with a wireless sensor network [9]:

- Due to the dense deployment of a greater number of nodes, a higher level of fault tolerance is achievable in wireless sensor networks.
- Coverage of a large area is possible through the union of coverage of several small sensors.
- Coverage of a particular area and terrain can be shaped as needed to overcome any potential barriers or holes in the area under observation.
- It is possible to incrementally extend coverage of the observed area and density by deploying additional sensor nodes within the region of interest.
- An improvement in sensing quality is achieved by combining multiple, independent sensor readings. Local collaboration between nearby sensor nodes achieves a higher level of confidence in observed phenomena.
- Since nodes are deployed in close proximity to the sensed event, this overcomes any ambient environmental factors that might otherwise interfere with observation of the desired phenomenon.

1.3 WIRELESS SENSOR NETWORK APPLICATIONS

Several applications have been envisioned for wireless sensor networks [6]. These range in scope from military applications to environment monitoring to biomedical applications. This section discusses proposed and actual applications that have been implemented by various research groups.

1.3.1 Military Applications

Wireless sensor networks can form a critical part of military command, control, communications, computing, intelligence, surveillance, reconnaissance, and targeting (C4ISRT) systems. Examples of military applications include monitoring of friendly and enemy forces; equipment and ammunition monitoring; targeting; and nuclear, biological, and chemical attack detection.

By equipping or embedding equipment and personnel with sensors, their condition can be monitored more closely. Vehicle-, weapon-, and troop-status information can be gathered and relayed back to a command center to determine the best course of action. Information from military units in separate regions can also be aggregated to give a global snapshot of all military assets.

By deploying wireless sensor networks in critical areas, enemy troop and vehicle movements can be tracked in detail. Sensor nodes can be programmed to send notifications whenever movement through a particular region is detected. Unlike other surveillance techniques, wireless sensor networks can be programmed to be completely passive until a particular phenomenon is detected. Detailed and timely intelligence about enemy movements can then be relayed, in a proactive manner, to a remote base station.

In fact, some routing protocols have been specifically designed with military applications in mind [10]. Consider the case where a troop of soldiers needs to move through a battlefield. If the area is populated by a wireless sensor network, the soldiers can request the location of enemy tanks, vehicles, and personnel detected by the sensor network (Fig. 1.3). The sensor nodes that detect the presence of a tank can collaborate to determine its position and direction, and disseminate this information throughout the network. The soldiers can use this information to strategically position themselves to minimize any possible casualties.

In chemical and biological warfare, close proximity to ground zero is needed for timely and accurate detection of the agents involved. Sensor networks deployed in friendly regions can be used as early-warning systems to raise an alert whenever the presence of toxic substances is detected. Deployment in an area attacked by chemical or biological weapons can provide detailed analysis, such as concentration levels of the agents involved, without the risk of human exposure.

1.3.2   Environmental Applications

By embedding a wireless sensor network within a natural environment, collection of long-term data on a previously unattainable scale and resolution becomes possible. Applications are able to obtain localized, detailed measurements that are otherwise more difficult to collect. As a result, several environmental applications have been proposed for wireless sensor networks [6,9]. Some of these include habitat monitoring, animal tracking, forest-fire detection, precision farming, and disaster relief applications.

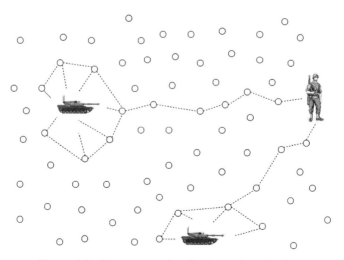

Figure 1.3   Enemy target localization and monitoring.

Habitat monitoring permits researchers to obtain detailed measurements of a particular environment in an unobtrusive manner. For example, applications such as the wireless sensor network deployed on Great Duck Island [11] allow researchers to monitor the nesting burrows of Leach's Storm Petrels without disturbing these seabirds during the breeding season. Deployment of the sensor network occurs prior to the arrival of these offshore birds. Monitoring of the birds can then proceed without direct human contact. Similarly, the PODS project [12,13] at the University of Hawaii uses wireless sensor networks to observe the growth of endangered species of plants. Data collected by the sensor network is used to determine the environmental factors that support the growth of these endangered plants. These two applications are discussed in detail in Sections 1.3.4 and 1.3.5.

Consider a scenario where a fire starts in a forest. A wireless sensor network deployed in the forest could immediately notify authorities before it begins to spread uncontrollably (see Fig. 1.4). Accurate location information [14] about the fire can be quickly deduced. Consequently, this timely detection gives firefighters an unprecedented advantage, since they can arrive at the scene before the fire spreads uncontrollably.

Precision farming [15] is another application area that can benefit from wireless sensor network technology. Precision farming requires analysis of spatial data to determine crop response to varying properties such as soil type [16]. The ability to embed sensor nodes in a field at strategic locations could give farmers detailed soil analysis to help maximize crop yield or possibly alert them when soil and crop conditions attain a predefined threshold. Since wireless sensor networks are designed to run unattended, active physical monitoring is not required.

Figure 1.4 Forest-fire monitoring application.

Disaster relief efforts such as the ALERT flood-detection system [17] make use of remote field sensors to relay information to a central computer system in real time. Typically, an ALERT installation comprises several types of sensors, such as rainfall sensors, water-level sensors, and other weather sensors. Data from each set of sensors are gathered and relayed to a central base station.

1.3.3 Health Applications

Potential health applications abound for wireless sensor networks. Conceivably, hospital patients could be equipped with wireless sensor nodes that monitor the patients' vital signs and track their location. Patients could move about more freely while still being under constant supervision. In case of an accident—say, the patient trips and falls—the sensor could alert hospital workers as to the patient's location and condition. A doctor in close proximity, also equipped with a wireless sensor, could be automatically dispatched to respond to the emergency.

Glucose-level monitoring is a potential application suitable for wireless sensor networks [18]. Individuals with diabetes require constant monitoring of blood sugar levels to lead healthy, productive lives. Embedding a glucose meter within a patient with diabetes could allow the patient to monitor trends in blood-sugar levels and also alert the patient whenever a sharp change in blood-sugar levels is detected. Information could be relayed wirelessly from the monitor to a wristwatch display. It would then be possible to take corrective measures to normalize blood-sugar levels in a timely manner before they get to critical levels. This is of particular importance when the individual is asleep and may not be aware that their blood-sugar levels are abnormal.

The Smart Sensors and Integrated Microsystems (SSIM) project at Wayne State University and the Kresge Eye Institute are working on developing an artificial retina [18]. One of the project goals is to build a chronically implanted artificial retina that allows a visually impaired individual to "see" at an acceptable level. Currently, smart sensor chips equipped with 100 microsensors exist that are used in ex vivo retina testing. The smart sensor comprises an integrated circuit (with transmit and receive capabilities) and an array of sensors. Challenges in this application include establishing a communication link between the retinal implant and an external computer to determine if the image is correctly seen. Regulating the amount of power used by the system to avoid damage to the retina and surrounding tissue is also a primary concern.

1.3.4 Habitat Monitoring on Great Duck Island

Leach's Storm Petrel (Fig. 1.5) is a common elusive seabird in the western North Atlantic. Most of their lives are spent off-shore, only to return to land during the breeding season. During this time, they nest in burrows located in soft, peaty soil, and are active predominantly at night. It is believed Great Duck Island, located 15 km off the coast of Maine, has one of the largest petrel breeding colonies in the eastern United States.

Figure 1.5   Leach's Storm Petrel. (U.S. Geological Survey photo by J. A. Splendelow.)

Petrel activity monitoring is a delicate problem, since disturbance or interference on the part of humans can lead to nest abandonment or increased predation on chicks or eggs.

To circumvent this problem, in the spring of 2002, the Intel Research Laboratory at Berkeley initiated a collaboration with the College of the Atlantic in Bar Harbor and the University of California at Berkeley to deploy a series of wireless sensor networks on the island [11,19,20]. By the summer of 2002, 43 sensor nodes were deployed on the island. The primary purpose of the sensor network was to monitor the microclimates in and around nesting burrows used by the petrels. Thus, researchers could take multiple measurements of biological parameters at frequent intervals, with minimal disturbance to the breeding colony. It was necessary to enter the colony only at the beginning of the study to insert sensor nodes into burrows and other areas of interest. Three major issues explored in this experiment included:

1. Determination of the usage pattern of nesting burrows over the cycle when one or both members of the breeding pair may alternate between incubation and feeding.
2. Determination of changes in the environmental conditions of burrows and surface areas throughout the course of the breeding season.
3. Measuring the differences in the microenvironments with and without large numbers of nesting petrels.

By November 2002, 32 sensor nodes had collected over one million sensor readings. For this particular application, the nodes were equipped with a separate weather board that contained sensors to detect temperature, humidity, barometric pressure, and midrange infrared. Motes periodically sampled and relayed their sensor readings to different base stations located throughout the island. These base stations provided researchers access to real-time environmental data gathered by the sensor nodes via the Internet.

In June 2003, a second-generation network comprising 56 nodes was deployed. This network was further augmented in July 2003 with an additional 49 nodes. Finally, in August 2003, over 60 additional burrow nodes and 25 weather-monitoring nodes were deployed on the island.

1.3.4.1 Hardware The system designers employed Mica motes (Fig. 1.6), which are small devices equipped with a microcontroller, low-power radio, memory, and batteries. The motes are designed with a single-channel 916-MHz radio that provides bidirectional communication at 40 kbps, an Atmel Atmega 103 microcontroller operating at 4 MHz, and 512 kB of nonvolatile storage. Power to the mote is supplied by a pair of AA batteries and a DC boost converter.

To allow sampling of the environment, the Mica mote was equipped with a Mica weather board that contains temperature, photoresistor, barometric pressure, humidity, and passive infrared sensors [11]. To protect the motes from adverse weather conditions, the sensor package was sealed in a 10-micron parylene sealant that protected the electrical contacts from water. The sensors themselves remained exposed so as not to hinder their sensitivity. The coated sensor was then encased in a ventilated acrylic enclosure. The acrylic enclosure was radio and infrared transparent and also elevated the mote off the ground.

Due to the longevity of the proposed application, battery life was budgeted carefully. A conservative estimate of 2200 mAh total capacity was utilized. For illustrative purposes, Table 1.1 lists the costs associated with performing basic Mica mote operations and Table 1.2 lists the costs associated with basic sensor operations [21].

For the habitat monitoring application, an application lifetime of 9 months was desired. Thus, with 2200 mAh of total power available, the sensor motes were

Figure 1.6 Mica sensor node (left) with the Mica Weather Board (right).

TABLE 1.1   Mica Mote Power Requirements for Different Operations

Operation	nAh
30-byte packet transmission	20.000
30-byte packet reception	8.000
1 ms radio listening	1.250
Sensor analog sample	1.080
Sensor digital sample	0.347
Reading sample from ADC	0.011
Flash read data	1.111
4-byte flash write/erase data	83.333

budgeted at 8.148 mAh of power consumption daily. However, sensor motes consume 30 μA in their sleep state [21]. This reduced the daily energy budget to 6.9 mAh available for sensing, communicating, and processing operations. The application was responsible for determining how this energy budget was to be allocated. Without any energy budgeting, a sensor mote operating at a 100% duty cycle can only operate for 7 days [21].

1.3.4.2 Architecture   The wireless sensor network architecture is divided into distinct tiers (Fig. 1.7). The lowest level consists of autonomous motes, equipped with various sensors, that perform basic networking, computing, and sensing tasks. They are organized into a local one-hop network and collectively identified as a *sensor patch*. One of the sensor motes within the sensor patch serves as a gateway between the sensor patch and the base station. It differs from other motes in that it is equipped with a high-gain antenna able to transmit data over a 350-foot link to the base station. The gateway node is also equipped with a solar panel and rechargeable battery in order to be able to operate with a 100% duty cycle. Data relayed to the base station are stored in a database and made available over the Internet.

TABLE 1.2   Individual Sensor Characteristics

Sensor	Accuracy	Changeability	Max Rate (Hz)	Start-Up Time (ms)	Current (mA)
Photoresistor	N/A	10%	2000	10	1.235
I2C temperature	1 K	0.20 K	2	500	0.150
Barometric pressure	1.5 mbar	0.5%	28	35	0.010
Barometric pressure temperature	0.8 K	0.24 K	28	35	0.010
Humidity	2%	3%	500	500–30,000	0.775
Thermopile	3 K	5%	2000	200	0.170
Thermistor	5 K	10%	2000	10	0.126

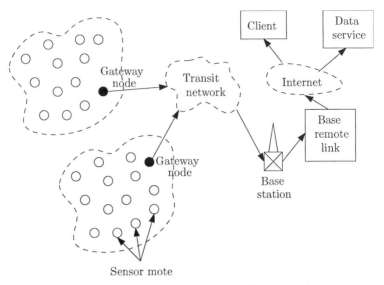

Figure 1.7   System architecture for habitat monitoring.

These collected data are also relayed, via satellite transceiver, to an off-site research facility located in Berkeley, California.

Periodically, motes took readings from each of their sensors. The data were time-stamped and kept in flash memory. Readings were then transmitted in a single 36-byte data packet. After successful transmission, motes entered their lowest power state for the next 70 seconds. The duty cycle was an expected 1.7% for the application. Each sensor mote was powered by two AA batteries with an estimated 2200 mAh capacity.

Several key application requirements identified by the system designers included Internet access, organization of the network as a hierarchy, sensor network longevity, the ability to operate off the grid, remote sensor network management, inconspicuous operation, in situ interaction, sensors and sampling, and data archiving capabilities.

1.3.4.3 Results   Since this is one of the first long-term deployments of the Mica mote platform, it was interesting to see how the wireless sensor network performed. Ironically, although the readings collected by the wireless sensor network proved to be unusable to researchers for making scientific conclusions, the fidelity of the acquired sensor readings gave insight into overall network behavior.

Over 1.1 million readings were collected in a time span of 123 days. During this period, abnormal operation was detected among the sensor node population. Typical problems included nodes generating sensor readings that were outside their predefined range, unreliable and erratic packet delivery, and system node failure.

1.3.5 PODS Project

Rare and endangered species of plants are threatened because they grow in limited select locations. Evidently, these locations have special properties that sustain and support their growth. The PODS project [12,13,22], located at Hawaii Volcanoes National Park, consists of a wireless sensor network deployed to perform long-term studies of these rare and endangered species of plants and their environment.

In Hawaii, the weather gradients are very sharp. In fact, regions of the island exist where rain forests and deserts are located less than 10 miles apart. Thus, it is not surprising that endangered species of plants are restricted to very small areas. Unfortunately, weather stations located throughout the island provide insufficient information for the areas where these endangered plants exist. Consequently, deploying a very dense wireless sensor network in the area of interest allows fine-grained temperature, humidity, rainfall, wind, and solar radiation information to be obtained by researchers.

In this particular wireless sensor network application, two types of data are collected: *weather data*, which are collected every 10 minutes, and *high-resolution images*, which are collected every hour. The data repository is a central server located on a different island than where observations are made. Weather measurements are maintained in a database and the high-resolution images are stored as individual files.

Exception reporting is the type of monitoring of interest to the biological problem studied on the island. Baseline information is developed that describes the expected environmental conditions on the island. This baseline information is reported, including periods during which the environment properly reflects it. The other information gathered are the time periods and degree of variance from the baseline model. These are the periods of most interest, because those intervals are when significant changes to the organisms under observation are likely to occur. Data summarization techniques are employed for these categorized data.

The high-resolution images collected every hour have a resolution of 1600×1200 pixels and serve several important interpretive functions. Images permit casual observations during periods where environmental conditions are reported as normal. During exceptional periods, when the environmental conditions deviate from the norm, images provide an important visual check on the conditions and permit a quick analysis of how the various types of vegetation under observation are responding. Most images are taken close to the endangered plant species. This permits observations of flowering, fruit set, fruit disappearance, leaf flushes, leaf loss, and other significant events. Since the images are stored as individual files, it is a simple matter to review them to confirm observations or review periods that were not being monitored. The data measurements collected are generally unfeasible to obtain via conventional monitoring techniques.

The type of deployable equipment allowed for research in the national park is limited. As a minimum criterion, the equipment cannot pose a threat to any species. Furthermore, it must not interfere or be a distraction to visitors. This is of particular concern since some areas of the island are visited by a large number of tourists.

In some parts of the island, little can be done to hide the instrumentation. Therefore, rocks were chosen as containers that camouflage and house the computer, sensing instruments, and batteries. The availability of small trees along other parts of the island expands the options for concealing sensor nodes. In some cases, short hollow structures, designed to look like branches, were also used to house the sensing equipment.

Upon initial deployment, the wireless sensor network engages in a neighborhood discovery process. This gives each node information about which sensor nodes it can communicate with directly. Next, the sensor network executes a routing protocol so that senders are able to send messages to their desired destination. For this particular application, requirements determine the functionality expected of the underlying routing protocol. Since nodes both send and receive messages, the protocol must provide nodes with routing information so that nodes can send messages specifically to other nodes. Adaptability to changing network topologies is required, as sensor nodes may be added, moved, or become depleted. Finally, the routing protocol needs to be designed such that network connectivity is maintained even when nodes are powered down to conserve battery life. As a result, *Geometric Routing Protocol* [13] and *Multi-path On-Demand Routing Protocol* [13] were developed for this particular application.

1.4 SERVICES

Most large-scale wireless sensor network applications share common characteristics. Services such as time synchronization, location discovery, data aggregation, data storage, topology management, and message routing are employed by these applications. Each is briefly described in this section.

1.4.1 Time Synchronization

Time synchronization is an essential service in wireless sensor networks [23]. In order to properly coordinate their operations to achieve complex sensing tasks, sensor nodes must be synchronized. A globally synchronized clock allows sensor nodes to correctly time-stamp detected events. The proper chronology, duration, and time span between these events can then be determined. Incorrect time stamps, due to factors such as hardware clock drift, can cause the reported events relayed back to the base station to be assembled in incorrect chronological order.

Time synchronization is crucial for efficient maintenance of low-duty power cycles. Sensor nodes can conserve battery life by powering down. When properly synchronized, nodes are able to turn themselves on simultaneously. When powered up, sensor nodes can relay messages to the base station and subsequently power down again to conserve energy. Unsynchronized nodes result in increased delays while they wait for neighboring nodes to turn their radios on, and in the worst case, messages transmitted can be lost altogether.

1.4.1.1 *Design Challenges* Several common challenges exist for the design of time synchronization protocols [23]. In order to perform synchronization, nodes exchange messages with each other. However, factors in the network can cause delays in message delivery. Four sources of error in network time synchronization can be identified. The first factor is *send time*, which includes the amount of time required to construct and transmit a message from the sender. The second factor is *access time*, which includes the delay experienced at the MAC layer, such as waiting for the channel to become idle. The third factor is *propagation time*, which includes the amount of time spent relaying the message across the various network interfaces between the sender and the receiver. Finally, the fourth source of delay is *receive time*, which includes the amount of time required by the receiver to accept and decode the message and transfer it to the host.

1.4.1.2 *Design Metrics* A broad set of design metrics for time synchronization protocols exist [23]. Factors such as energy efficiency, scalability, precision, robustness, lifetime, and scope must all be taken into consideration. As with all protocols designed for wireless sensor networks, energy efficiency is a chief concern. Protocols must be scalable, since sensor networks can potentially contain a very large number of sensor nodes. The precision required may vary depending on the type of sensor network application. For example, in some cases, an ordering of detected events may be required so that a chronology of events can be assembled. In other cases, it may be necessary to time-stamp events at finer resolution. For example, real-time applications, such as target tracking, may require tight synchronization between sensor nodes as they follow the object's movements. Finally, since sensor networks are generally left unattended for long periods of time, time synchronization protocols must be fault-tolerant and adaptive to changing network topologies. For example, as new nodes are introduced and other nodes die, sensor nodes must be able to synchronize themselves seamlessly with their neighbors.

1.4.1.3 *Protocols* Much work has gone into solving the problem of time synchronization among sensor nodes. At a rudimentary level, where a simple causality relationship [24,25] between detected events is desired, even traditional approaches employed in other types of distributed systems, such as vector clocks [26,27], are generally not practical for wireless sensor networks.

Vector clocks are not scalable in resource-constrained sensor networks with an unknown or large number of nodes. The additional overhead required to transmit vector time stamps with each message would quickly deplete a node's battery, rendering it useless. Furthermore, vector clocks are abstract in nature and do not indicate the duration of an event in physical time measurements, such as minutes or seconds. Other complex protocols, such as the network time protocol (NTP) [28], are unsuitable for wireless sensor networks because of their computational requirements.

Protocols such as TSync [29] and reference-broadcast system (RBS) [30] exploit the broadcast nature of wireless sensor networks in order to achieve global time synchronization with a high degree of accuracy.

In refs. [31] and [30], Elson et al. propose RBS, a time synchronization technique that uses a *third party* to perform synchronization among nodes. Individual nodes send reference beacons to their neighbors. The beacon's time of arrival is used by receiving nodes as a reference point for comparing local clocks. Since a reference broadcast arrives at all receivers at essentially the same time, propagation error is minimal. In the simplest form of RBS, a node broadcasts a single pulse to two receivers. Upon receiving the reference broadcast, the receivers exchange their receiving times and attempt to estimate their relative phase offsets. Through simulation, it has been shown that 30 reference broadcasts improves the precision from 11 μs to 1.6 μs when synchronizing a pair of nodes.

In ref. [32], the authors propose a networkwide time synchronization protocol called Timing-Sync Protocol for Sensor Networks (TPSN). The protocol has two phases: *level discovery* and *synchronization*. The level-discovery phase is initiated when the sensor network is deployed. A node is elected as the root node (level 0) and initiates the level-discovery phase by transmitting a level-discovery message, which contains the node ID and level of the sender. Upon receiving this message, a node assigns itself a level that is one level higher than the incoming level-discovery message. Subsequent level-discovery messages received are discarded. This broadcast phase continues until all nodes are assigned a level. The synchronization phase of the algorithm involves a two-way message exchange between a pair of nodes. The authors assume that clock drift and propagation delay (in both directions) between a pair of nodes is constant in the period of time between a single message exchange.

A node initiates synchronization by sending a pulse message that includes the node's level and local time. A node that receives the pulse message responds with an acknowledgment that includes the original time stamp received, the relative clock drift between both nodes, and the propagation delay. The node that initiated the pulse calculates the *actual* ensuing clock drift and propagation delay, and synchronizes itself with the receiving node. The synchronization phase is initiated by the root node. Nodes at the level below the root node exchange messages with the root node and adjust their clocks accordingly. Other nodes at lower levels, upon overhearing that nodes at levels above them are performing time synchronization, also initiate time synchronization. The authors report that their time synchronization protocol is precise within 6.5 μs when implemented on Compaq IPAQs running the Linux operating system. On Mica motes, they report their time synchronization protocol achieves an accuracy of 29.13 μs.

In ref. [33] the authors describe two lightweight synchronization algorithms called Tiny-Sync and Mini-Sync. Both techniques employ the conventional two-way messaging scheme to determine the relative clock drift and offset between the clocks of two sensor nodes.

In ref. [34], the authors describe lightweight tree-based synchronization (LTS), which attempts to minimize the underlying complexity of the time synchronization process, rather than attempting to maximize accuracy. Two approaches are presented in LTS. Both of them require sensor nodes to synchronize their clocks to a reference point. The first approach given is a centralized algorithm that uses the edges of the spanning broadcast to perform pairwise synchronization. The root of the spanning tree is responsible for initiating synchronization. Under the assumption that clock drift is bounded and given the required degree of precision, the reference node calculates the time period a synchronization step is valid.

The second approach presented by the authors is completely distributed. Individual sensor nodes request synchronization with other nodes as needed. When a node decides it is necessary to synchronize its clock with another node, it sends a synchronization request to the closest reference node. As a result, all nodes along the path from the reference node and the node requesting synchronization must have their clocks synchronized for the requesting node to synchronize its local clock properly.

1.4.2 Location Discovery

Location discovery involves sensor nodes deriving their positional information, expressed as global coordinates or within an application-defined local coordinate system. The importance of location discovery is widely recognized [35–40]. It serves as a fundamental basis for additional wireless sensor network services where location awareness is required, such as message routing. Furthermore, in applications such as fire detection, it is generally not sufficient to determine *if* a fire is present, but more importantly, *where*. A brief review of three proposed solutions to location discovery are presented.

1.4.2.1 *Multilateration by Distance Measurements*   Meguerdichian et al. [35] describe a localized algorithm that uses multilateration for solving the problem of location discovery. A node determines its location based on its distance from neighboring nodes that serve as beacons. Beacons are nodes that are location-aware and broadcast their location information periodically. They acquire their location from multilateration procedures or other sources such as GPS. Distances between neighboring nodes are estimated using received signal strength indication (RSSI) or ultrasound techniques. Thus, a node requires only local neighbor information to determine its position.

1.4.2.2 *Ad Hoc Positioning System*   Niculescu and Nath [38] propose their ad hoc positioning system (APS), whereby nodes determine their location in reference to *landmarks* that are location aware. Landmarks can be other sensor nodes, base stations, or beacons that have positional information. Unlike GPS, where direct line of sight is required with a series of satellites in order to triangulate a location, landmark information is propagated through the wireless sensor network in a multihop fashion.

When an arbitrary node in the wireless sensor network has distance estimates to three or more landmarks, it computes its own position in the plane. The node utilizes the centroid of the landmarks as its location estimate. Nodes in direct communication with a landmark infer their distance from it based on the received signal strength of the landmark.

Through message propagation, nodes two hops away from a landmark estimate their distance based on the distance estimates of nodes located next to the landmark. The propagation schemes proposed by the authors eventually flood the entire network until all nodes are able to determine their coordinates.

1.4.2.3   APS using Angle of Arrival   In ref. [39], Niculescu and Nath present two algorithms, *DV-Bearing* and *DV-Radial*, that allow sensor nodes to get a bearing and a radial in relation to a landmark using angle of arrival (AoA) to derive position information. The term "bearing" refers to an angle measurement with respect to another object. A "radial" refers to a reverse bearing which is simply the angle at which an object is seen from another location. The term "heading" refers to the sensor node's bearing with respect to true north and represents its absolute orientation.

AoA sensing requires sensor nodes to be equipped with an antenna array or several ultrasound receivers. This equipment is currently available in small package formats for wireless sensor network nodes such as the one developed for the Cricket Compass Project [41,42]. The theory of operation is based on time difference of arrival (TDoA) and phase difference of arrival. If a node sends an RF signal and an ultrasound signal at about the same time, the receiving node can infer the distance between the sender and itself by measuring the time difference between the arrival of the RF signal and the ultrasound signal. To derive the angle of arrival of the signal, the receiving sensor node uses two ultrasound receivers placed at a known distance from each other.

1.4.3   Data Aggregation

Data aggregation and query dissemination are important issues in wireless sensor networks [43]. Sensor nodes are typically energy constrained. Therefore, it is desirable to minimize the number of messages relayed, because radio transmissions can quickly consume battery power. A naive approach to reporting sensed phenomena is one where all (raw) sensor readings are relayed to a base station for off-line analysis and processing. However, since sensor nodes within the same vicinity often detect the same, common phenomena, it is likely some redundancy in sensor readings will occur [44]. Local collaboration allows nearby sensor nodes to filter and process sensor readings before transmitting them to a base station. Consequently, this process can reduce the number of messages relayed to the base station.

Figure 1.8 represents an animal-tracking application where several sensor nodes are randomly deployed in a forest. When an animal, represented by the solid square, passes through the area being monitored, individual sensor nodes detect the presence of the animal and relay their findings, in a multihop fashion, to the base station located some distance away. In sufficiently dense sensor networks, overlapping areas of coverage are possible. Thus, the animal may be detected by several sensors.

In the scenario presented in Figure 1.8, nodes A, B, C, D, and E sense the presence of a nearby animal. Nodes $B-E$ each send a message to node A with their observed sensor data. Node A forwards the received messages, along with its own set of sensor readings, to the next node along the path to the base station. Thus, node A sends a total

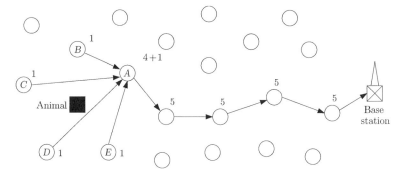

Figure 1.8 Event detection and reporting without data aggregation.

of 5 messages, which are all subsequently relayed from node to node, until they reach the base station. In total, 29 messages are transmitted throughout the network.

A reduction in communication and energy costs is possible if collected sensor data is aggregated prior to relaying. Figure 1.9 is similar to Figure 1.8, except that node A collects sensor readings from nodes $B-E$ and itself, applies an aggregation function ϕ, and then relays the aggregated data. Results are compressed into a single message, which is subsequently transmitted, in a multihop fashion, for further analysis by the base station.

Various types of data aggregation are possible, depending on the level of refinement desired. In-network processing can be designed to perform one or more of the following operations:

- *Aggregate the data into a single binary value.* A Boolean (i.e., true or false) value would be sufficient to indicate if an animal was detected or not.
- *Aggregate the data readings into an area.* Coordinates of a bounding box can be given that defines the area where the sensor readings are observed. Nodes,

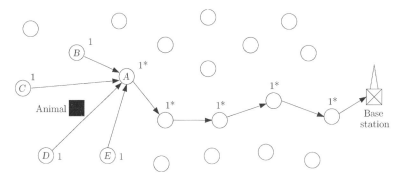

Figure 1.9 Event detection and reporting with data aggregation.

upon receiving this area information, dynamically adjust the size of the bounding box to accommodate their sensor readings before retransmitting.

- *Aggregate the collected data by applying an application-specific aggregation or filtering function.* As an example, the average, maximum, minimum, or sum of sensor values could be calculated en route prior to forwarding any received information.

Energy conservation, as a result of data aggregation, is of particular concern for sensor nodes close to the base station. Without any form of data aggregation, a greater number of messages are transmitted. As a result, their batteries are depleted quickly. Eventually, when nodes that communicate directly with the base station die, the sensor network is rendered unusable, regardless of the remaining power of other nodes (see Fig. 1.10), since no messages can reach the base station.

Data aggregation seeks to combine data arriving from different sources en route. In [44], the authors study the energy savings and latency trade-offs caused by data aggregation and how factors such as source (i.e., event) and sink (i.e., base station) placements and network density affect this trade-off. A complexity analysis of optimal data aggregation in sensor networks is also performed, and although it is shown that optimal data aggregation is NP-hard, polynomial-time solutions exist for certain cases.

The work presented in [45] continuously computes aggregates of wireless sensor network monitoring functions. Aggregates computed include sums, averages, and counts. Network properties considered include loss rates, energy levels, and packet counts. A novel tree construction algorithm is proposed to enable energy-efficient computation of some classes of aggregates, and it is demonstrated, through actual implementation and experiments, that wireless communication artifacts and packet loss significantly impact the computation of these aggregate properties. During experiments conducted on a test bed of 26 sensor nodes, packet loss for each link was measured every minute for two hours under various topology settings.

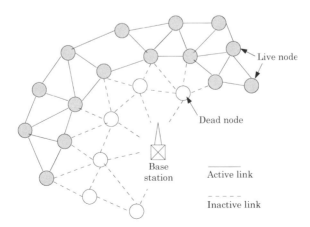

Figure 1.10   Event detection and reporting with data aggregation.

Although the majority of links were good, 10% of the nodes exhibited a packet rate loss greater than 50%. As a result, the value of the COUNT aggregate, which reports the total number of active sensor nodes in the network, fluctuated greatly over time.

The infrastructure presented for wireless sensor network monitoring consists of three classes of software [45]:

1. The first component consists of a tool such as *dump* that is used to collect detailed information about the system state. This is used to provide debugging information about the sensor nodes and also report any logged information kept by the nodes over a period of time.

2. The second category of tool is referred to as *scans*. These constitute a global, albeit aggregated view, of the wireless sensor network and report metrics such as overall resource consumption. An example of a scan is an *escan* whereby a special user-gateway node initiates state information collection from the entire system. However, instead of all nodes relaying their power level information, data collected is aggregated en route in order to minimize the amount of information propagated throughout the network.

3. The final category of tool is referred to as *digests*, which are simply aggregates of some network property. Digests span the entire network, but unlike scans, they are computed continuously. Computed information is propagated throughout the network by piggybacking digests onto regular messages transmitted throughout the sensor network. Clearly, the energy savings achieved is offset by the increased latency.

The second contribution entails the design of protocols to enable computation of network digests. Values such as node energy level, degree of connectivity, and volume of traffic, are considered. Decomposable functions (i.e., functions that can be expressed in terms of another function) such as *sum*, *min*, *max*, *average*, and *count* are applied to these analyzed values. Digest computation is accomplished using *digest diffusion*, which implicitly builds a broadcast tree where computed partial results of decomposable functions are propagated toward the root.

For example, assume a connected homogeneous wireless sensor network exists where all nodes are equipped with thermal sensors that record the ambient temperature. Initially, every node assumes it has observed the highest temperature reading and exchanges this information with its immediate neighbors. A node, upon receiving a temperature measurement from a neighboring node, adjusts the source of the highest reading, if necessary, and propagates this information throughout the broadcast tree. Eventually, all nodes converge on the same maximal temperature reading.

Broadcast tree maintenance is required as nodes fail over the lifetime of the wireless sensor network. Thus, a node periodically broadcasts messages to maintain the digest. Nodes use a time-out value to determine if a neighboring node is no longer transmitting messages to it. Thus, a node may switch to a different parent node when it is no longer receiving messages from its existing parent node.

1.4.4 Data Storage

Data storage presents a unique challenge to developers. Event information collected by individual nodes must be stored at some location, either in situ or externally. In some cases, where an off-line storage area is not available, data must be stored within the wireless sensor network. Ratnasamy et al. [46,47] describe three data-storage paradigms employable in wireless sensor networks:

1. *External Storage.* In this model, when a node detects an event, the corresponding data are relayed to some external storage located outside the network, such as a base station. The advantage of this approach is that queries posed to the network incur no energy expenditure since all data are already stored off-line.

2. *Local Storage.* In this model, when a node detects an event, event information is stored locally at the node. The advantage of this approach is that no initial communication costs are incurred. Queries posed to the wireless sensor network are flooded to all nodes. The nodes with the desired information relay their data back to the base station for further processing.

3. *Data-Centric Storage.* In this model, event information is routed to a predefined location, specified by a geographic hash function (GHT), within the wireless sensor network. Queries are directed to the node that contains the relevant information, which relays the reply to the base station for further processing.

For wireless sensor network applications that are envisioned to be long-lived, even optimized communication schedules can deplete a node's battery within a relatively short period of time (i.e., a couple of months). Consider an environmental application, such as microclimate monitoring, where individual sensor nodes periodically sample their local environment to measure temperature, light, precipitation, pressure, and humidity levels. Over time, the amount of data generated by the sensor network can be substantial. This is particularly true if individual sensor nodes take samples at short regular intervals, such as every 30 minutes.

Ganesan et al. [48] look to provide a distributed, progressively degrading storage model. This is achieved by constructing local, multiresolution summaries of observed sensor data stored hierarchically throughout the wireless sensor network. Queries on summary information are performed in a drill-down fashion: coarse, highly compressed data are stored in nodes at the highest levels in the hierarchy. As more detailed information is required, nodes at lower levels in the hierarchy, with more detailed event information, are queried.

Summary information is created by employing a wavelet-based compression technique, which offers the following advantages:

• A compact representation of data is produced that highlights interesting features in the accumulated data, such as long-term trends, edges, and significant anomalies.

- Spatiotemporal queries can be satisfied with little communication overhead by employing drill-down querying. Basic information from the wireless sensor network is gathered from nodes at the highest level in the hierarchy. As more detailed spatiotemporal information is required, nodes further down the hierarchy are queried for the relevant data.
- Aging, and subsequently discarding summaries selectively, gracefully degrades query performance over time. Since wireless sensor networks are typically resource-constrained, nodes discard older data in favor of newly gathered sensor readings.

1.4.5 Topology Management and Message Routing

Wireless sensor networks can possibly contain hundreds or thousands of nodes. Routing protocols must be designed to achieve an acceptable degree of fault tolerance in the presence of sensor node failures, while minimizing energy consumption. Furthermore, since channel bandwidth is limited, routing protocols should be designed to allow for local collaboration to reduce bandwidth requirements.

Observations made in ref. [49] show that, although intuitively it appears a denser deployment of sensor nodes renders a more effective wireless sensor network, if the topology is not carefully managed, this can lead to a greater number of collisions and potentially congest the network. As a result, there is an increased amount of latency when reporting results and a reduction in the overall energy efficiency of the network. Furthermore, as the number of reported data measurements increases, the accuracy requirements of the application may be surpassed. This increase in the reporting rate by the deployed sensor nodes can actually harm the wireless sensor network performance, rather than prove beneficial.

Message-routing algorithms in ad hoc networks can be separated into two broad categories: greedy algorithms and flooding algorithms [50]. Greedy algorithms apply a greedy path-finding heuristic that may not guarantee a message reaches its intended receiver. One example of greedy routing, proposed by Finn in 1987, is forwarding to a neighbor that is closest to the destination. Additional steps are required to ensure the message is received by its intended recipient. Flooding algorithms employ a controlled packet duplication mechanism to ensure every node receives at least one copy of the message. For these algorithms to terminate, nodes in the sensor network must remember which messages have been previously received.

In ref. [50], the authors present two distributed routing protocols, *face routing* and *greedy-face-greedy* (GFG). Both algorithms guarantee packet delivery as long as the wireless sensor network remains connected and static while the message is relayed from sender to receiver. The medium access is ideal since it guarantees message transmission between two neighbors in a finite time. The communication graph is the unit graph where two nodes can communicate if and only if the distance between them is at most R, where R is the transmission radius of all nodes.

Both algorithms require messages to carry some overhead information. However, sensor nodes themselves do not need to maintain additional routing information.

The algorithms first construct a connected planar subgraph, called a Gabriel graph, of the underlying wireless sensor network in a distributed fashion. Edge e is in the Gabriel graph if and only if the circle with edge e as the diameter contains no other nodes inside it. The Gabriel graph partitions the graph into faces that are bound by polygons and make up the edges of the graph.

In the face-routing algorithm [50], the boundary of the face is traversed in a counterclockwise fashion until an edge is found that intersects with the line that connects the source and destination. The algorithm then continues to scan the next adjoining face in a similar manner. The entire process iterates until the destination is reached.

In the GFG algorithm [50], greedy routing (i.e., forwarding to the neighbor node closest to the destination) is applied as long as the node currently holding the packet has a neighbor closer to the destination node than itself. When current node A does not have such a neighbor, face routing is applied until a node B, closer to the destination node than node A, is encountered. Node B then reverts back to greedy forwarding. This reversal of modes can be repeated until the packet is delivered to its intended destination. Greedy perimeter stateless routing (GPSR) [51] is a routing protocol similar to GFG [50] that incorporates medium-access-layer and mobility considerations.

Greedy routing algorithms have been found to work well in wireless sensor networks due to their efficiency and scalability [52]. Greedy forwarding techniques offer several advantages over naive routing techniques (i.e., flooding):

- Nodes need to maintain only local topology information. This makes the protocol highly scalable, since routing information to all destinations is not maintained locally. Such a routing table would quickly grow in size, consuming the node's limited memory.
- The protocol is adaptable to frequent topology changes, since the routing path can be dynamically adjusted based on the current one-hop neighborhood of a node.
- Since only local information is used, nodes need not be aware of the topology of the entire wireless sensor network.

Network self-organization can be extended further than simple topology management. Assigning *roles* to sensor nodes based on their physical connectivity and sensing capabilities is proposed in [53]. Metrics, such as sensing proximity value, cumulative sensing degree, and other intermediate sensing parameters, allow the wireless sensor network to be partitioned into distinct *sensing zones*. Sensing zones are a collection of sensor nodes with a common sensing objective and a specific sensing quality of service (sQoS). Coordinators are elected to act as leaders within a sensing zone and are responsible for coordinating sensing-zone members and performing network reorganization maintenance. This approach is an improvement over other types of topology management schemes, such as hierarchical topologies, since they may be too rigid for a particular wireless sensor network application.

1.5 WIRELESS SENSOR AND ACTOR NETWORKS

Wireless sensor and actor networks (WSANs) [54] can be considered as an extension of traditional wireless sensor networks. They consist of two major components: *sensor nodes* and *actor nodes*. Sensor nodes are low-cost, low-power devices with limited sensing, computational, and communication capabilities. Actor nodes are resource-rich nodes equipped with more powerful processors, longer-range radio transceivers, and longer-lasting, or possibly renewable, power sources. They may also be able to navigate throughout the area covered by the sensor nodes. The number of sensor nodes generally outnumbers the number of actor nodes by a sizable quantity.

There are several defining characteristics of WSANs. These include:

- *Real-Time Requirements.* Depending on the application, it may be necessary for nodes within the sensor network to respond quickly to detected events. For example, in an environmental monitoring application, if a fire is detected, some sort of corrective action should be initiated as quickly as possible. The data collected by the wireless sensor and actor network must be timely and current when the corrective action is taken.

- *Coordination.* In a wireless sensor network, the process of data collection is coordinated by a central entity, such as a base station. In a wireless sensor and actor network, sensor–sensor coordination, actor–sensor coordination, and actor–actor coordination are required. Sensor nodes report detected events to actor nodes, which in turn, take some appropriate action. This may include coordinating response activities with other actor nodes, providing additional instructions to nearby sensor nodes, or processing sensed event information to relay back to a central base station.

The roles of sensor nodes and actor nodes are to collect data from the environment and react appropriately to sensed events. The *sensor–actor field* defines the area where sensor nodes and actor nodes are distributed. A central base station, sometimes referred to as a *sink*, monitors and coordinates overall network activity.

When a sensor node observes a particular phenomenon, it transmits its findings to a nearby actor node. The actor node processes all incoming data and initiates an appropriate response or processes and relays the information to the sink. The sink can then further process the received information and subsequently issue additional commands to the actor nodes to gather more information, or react to, the detected event.

1.5.1 Architecture

There are two possible types of architectures possible in WSANs:

1. *Semiautomated Architecture.* This architecture bears similarities to the architecture in most wireless sensor networks. A central base station is used

to coordinate the efforts of the actor node and sensor nodes. Queries are issued to the network and results are relayed to the base station for further processing.

2. *Automated Architecture.* This architecture does not require a central base station to coordinate efforts. Actors are programmed to work autonomously and respond to detected events appropriately. This architecture has a few advantages over the semiautomated architecture: it exhibits a lower latency, since sensed information is only relayed to actor nodes; and it has a longer overall network lifetime, since event information is only relayed to the actor node within one hop of the sensor nodes that detected the phenomenon.

Aside from communication between actor nodes and sensor nodes, communication between actor nodes must be coordinated as well in order to achieve the application objectives. Actor nodes, being resource-rich nodes with high transmission power, can transmit information over long distances, unlike sensor nodes. Furthermore, since the number of actor nodes in a wireless sensor and actor network is typically small, communication among actor nodes is analogous to an ad hoc sensor network.

The most crucial aspect of sensor–actor communication is low communication delay due to the proximity between sensor nodes and actor nodes. Other issues to consider include:

- *What are the communication requirements between actor nodes and sensor nodes?* These requirements include factors such as ensuring communication between actor nodes and sensor nodes consume minimal energy, the latency in reporting sensed event information to the actor node(s), and ensuring a proper ordering of event information.

- *Which sensors transmit to which actors?* If an event is detected by multiple sensor nodes, the sensor node may decided to relay information to a single actor node, or perhaps, to a series of actor nodes. Both approaches have their advantages and disadvantages. For example, information sent to a single actor node consumes less overall energy since fewer messages are relayed throughout the wireless sensor and actor network. However, relaying sensed event information to multiple actor nodes provides an increased level of redundancy. This may be a necessity if the network is deployed in a hostile environment where nodes are prone to failure.

- *What is the arrival time of messages?* Consider a hypothetical security application whereby actor nodes are deployed to monitor and patrol an art gallery. If an intruder is detected, one objective of the actor nodes may be to surround and immobilize the intruder. This requires that actor nodes receive notification from sensor nodes that detect the intruder in a timely (i.e., relatively simultaneously) fashion in order to coordinate their movements.

As a consequence, the set of communication protocols for wireless sensor and actor networks should provide real-time services within a specified upper bound for delay, relay messages in an energy-efficient manner among sensor nodes and actor nodes, ensure the proper ordering of events, provide synchronization between sensor nodes reporting an activity to multiple actor nodes, and allow messages to be routed to arbitrary actor nodes.

Depending on the quality of service requirements of the wireless sensor and actor network application, coverage of a sensed event is partitioned into four cases:

1. A minimal set of actor nodes cover the event region
2. A minimum set of sensor nodes cover the event region
3. A minimum set of actor nodes and sensor nodes cover the event region
4. The entire set of actor nodes and sensor nodes in the event region monitor the phenomenon

The first three cases are aimed at reducing the level of redundancy, while the last case aims to provide maximal coverage of a detected event. There are trade-offs with both approaches. The amount of energy consumed in the network is reduced in the first three cases, at the expense of more intense coverage. The last case affords maximal coverage of the detected phenomenon, but at the expense of higher energy consumption.

Aside from communicating with sensor nodes, actor nodes can communicate directly with each other. Communication between actors can occur under various circumstances. For example, an actor node that receives information from a nearby sensor node requires the assistance of additional actor nodes in order to complete its task. Similarly, if multiple actors receive the same event information, the actor nodes can communicate with each other to coordinate their efforts.

1.5.2 Protocol Stack

As of the time of this writing, a de facto protocol stack for wireless sensor networks or wireless sensor and actor networks did not exist [54]. Unfortunately, there is no general consensus within the wireless sensor network research community about the layer structure in wireless sensor networks. It is argued that strict layering guarantees controlled interaction among layers, whereas a cross-layer design can produce spaghetti-like code that is difficult to maintain because modifications must be propagated across all protocols [55]. Furthermore, cross-layer designs can produce unintended interactions among protocols that result in performance degradation.

Other researchers are in favor of adopting a cross-layer design to overcome potential performance problems. The authors in [55] introduce a layered architecture where protocols in different layers cooperate by sharing network-status information while still maintaining separation between various layers. Despite the potential ptifalls, several motivations for employing a cross-layering approach exist [56]:

- *Optimization can be achieved in several layers.* The optimization goals at a particular layer can be designed to work with the optimization goals of other layers above and below.

- *Optimization in one level can require cooperation from other levels to show its effects.* Consider the case where the underlying routing protocol is designed to select the shortest route possible. Although this optimization results in smaller hop distances requiring less energy to transmit message packets, the larger number of messages transmitted can result in a greater amount of contention. If the medium-access control (MAC) layer is not optimized accordingly, the routing protocol may suffer as a consequence.
- *There are possible conflicts between optimization goals in distinct layers.* Some optimization solutions at distinct layers are orthogonal in design. For example, at the network layer, it may be desirable to reduce the amount of overhead maintained at individual nodes. However, this may result in a lower quality of service at the transport layer since less information is broadcast with individual packets. Similarly, employing data-compression techniques may interfere with latency requirements imposed by the application, as the nodes must wait to accumulate and aggregate received information.
- *Some scenarios do not require support from all layers.* Consider a multihop local positioning system (LPS) based on hop-by-hop distance measurements to estimate the relative distance between an arbitrary node and an anchor node. The network layer and transport layer, used to handle the end-to-end data transmissions, are not required in this application. Consequently, these layers can be omitted.

The authors in [54] suggest the protocol stack for sensor nodes and actor nodes consist of three planes:

1. *Communication Plane.* This plane enables the exchange of information between the various nodes within the wireless sensor and actor network. It receives commands from the coordination plane and provides the appropriate link relations between various nodes. The functionality of the communication plane is contained within the constituent transport layer, routing layer, and MAC layer.

   (a) *Transport Layer.* Aside from providing the traditional reliability requirements, the transport-layer protocol is responsible for providing the real-time requirements of the WSAN. For example, if the transport protocol utilized in sensor–actor communication detects a low level of reliability, the transport protocol employed in communication can notify other actors of this situation.

   (b) *Routing Layer.* Sensor nodes that detect an event have to select which actor node(s) will receive the gathered sensor information. This poses a challenge due to the existence of several actor nodes in the network. Once a decision is made, the data are relayed to the appropriate node. The routing protocol is responsible for determining the path messages will take, performing any in-network data aggregation to reduce the number of messages relayed throughout the network, and supporting any real-time communication requirements imposed.

(c) *MAC Layer*. To effectively transmit event information from a large number of sensor nodes to actor nodes, a MAC protocol is essential. In some applications, actor nodes may be mobile. Consequently, actor nodes may leave the transmission area of some sensor nodes. One of the functions of the MAC layer is to ensure connectivity between sensor nodes and actor nodes. Contention-based protocols are generally not suitable for real-time communication between sensor nodes and actor nodes due to the latency imposed by handshaking. Exploiting the periodic nature of sensor network traffic allows for the development of collision-free real-time scheduling algorithms. These are more suitable for wireless sensor and actor networks, since they can reduce the overall delay and provide real-time guarantees.

2. *Coordination Plane*. Data received along the communication plane is forwarded to the coordination plane, which processes the received information and decides on an appropriate action. This enables nodes to collaborate and achieve a higher-level objective. Issues such as sensor–sensor coordination are addressed. These include decisions as to which sensor nodes will relay information to the corresponding actor nodes, how routing of messages in a multihop fashion is handled, how in-network data aggregation is performed, and actor node selection.

3. *Management Plane*. This plane is responsible for monitoring and controlling node functions. This includes functions such as node power management features, node mobility management, and node fault management.

1.6 SENSOR QUERYING AND DATABASE SYSTEMS

Users of wireless sensor network applications are typically interested in continuous streams of information [17] that represent the evolving status of the area under observation as time progresses [57]. Query processing systems such as TinyDB [58], Directed Diffusion [7,8], and Cougar [59] provide users of wireless sensor network applications with a high-level interface for performing queries. This relieves the user from writing complex code to gather information from the sensor network.

Part of the ongoing research into sensor database systems includes distributed query processing [60] and storage mechanisms [48] in sensor networks. The need for scalable self-organized data retrieval and in-network processing is clear. A unified query processing/networking system involves an additional challenge to designers of wireless sensor networks. Different applications have varying requirements in terms of information transfer rates, latency, coverage, and storage. The trade-off between optimizing the network topology and performing efficient query processing is an issue that needs to be resolved.

In TinyDB [58], users specify a set of declarative queries that define the information to be gathered from the wireless sensor network. Queries indicate the type of readings to be obtained, including the subset of nodes the user is interested in, and any simple transformations to be performed over the collected data. They are specified using a language like a structured query language (SQL). A sample query could be expressed as follows:

```
SELECT AVG(temp)
FROM sensors
WHERE location in (0,0,100,100) AND light > 1000 lux
SAMPLE_PERIOD 10 seconds
```

TinyDB queries are generally specified on a PC and then distributed throughout the sensor network by a query executor. The query is disseminated and results are returned in an energy-efficient manner using a variety of in-network processing techniques and cross-layer optimizations. For example, in the preceding sample query, the query executor is responsible for determining which predicate to evaluate first in the sensor network: the *temp* predicate or *light* predicate.

Queries in TinyDB are disseminated through the entire network and collected via a routing tree. The root node of the routing tree is end point of the query, which is generally where the user that issued the query is located. Nodes within the routing tree maintain a parent–child relationship in order to properly propagate results to the root. Research into query processing techniques include the design of an acquisitional query processor for data collection in wireless sensor networks. Information such as where, when, and how often data are physically collected and delivered, can be leveraged to significantly reduce the overall power consumption in the sensor network [61].

Directed diffusion [7,8] employs a different approach to query processing. Rather than utilizing a specific query language, an application specifies a *named interest*, which is used to query the sensor network. Interests contain the query particulars, expressed through a sequence of attribute/value pairs. For example, an interest expressed as:

```
location = [(100,100), (10,200)]
temperature = [10,20]
```

would report the temperature readings from all nodes located within the specified location whose temperature is within the specified limits. The interest is initiated by a *sink* node and flooded throughout the sensor network. A node that lacks data matching an interest forwards it to its neighbor node. The decision as to which node to forward the interest to is based on the contents of the interest. The notion that cues can be embedded within the query itself is one of the core principles behind *data-centric routing*. As the interest is propagated, nodes build routing tables that are used to return matching data to the sink.

1.7 SENSOR NETWORK RELIABILITY

Several applications of wireless sensor networks exist where reliability of data delivery is critical. For example, consider a security application where sensors are required to detect and identify the presence of intruders. Given the critical nature of the application, when an intruder is detected, messages must reach the base station in a timely and reliable manner. Three unique issues must be addressed when discussing data-delivery reliability in wireless sensor networks [62]:

1. *Environmental Considerations.* Wireless sensor networks can be deployed in harsh environments. However, the limited lifetime of individual sensor nodes, low bandwidth, and the size of the sensor network must be considered.

2. *Message Considerations.* Messages relayed throughout a wireless sensor network are generally small compared to ad hoc networks. For example, a simple query that requests information from a specific region of interest might be flooded throughout the sensor network. The reduced message size affects the type of loss–recovery scheme employed in the wireless sensor network.

3. *Reliability Considerations.* Traditional notions of reliability are concerned with reception of 100% of all messages transmitted. However, in a wireless sensor network, reliability may be expressed in terms of data gathered from a particular subregion within the network, or as the fidelity of partial, aggregated results.

1.7.1 PicoRadio Network

The authors in [63] present experimental measurements of radio energy consumption and packet reliability for their prototype PicoRadio network that is composed of PicoNodes [64]. Energy consumption is categorized by the energy consumed when the radio is in different states (i.e., idle, transmitting, or receiving). Packet delivery reliability is measured from a network and link perspective.

1.7.1.1 Hardware The prototype PicoNode consists of a StrongARM SA-1100 microprocessor, a Xilinx C4929XKA field-programmable gate array (FPGA), an Ericsson PBA-313-01/2 Bluetooth radio, 4 MB of DRAM, 4 MB of flash memory, and one of two possible custom sensor boards. The first board is configured with sensors that obtain light, sound, temperature, and humidity measurements. The second possible sensor board is configured with an accelerometer and magnetometer.

1.7.1.2 Protocol Stack The protocol stack utilized by each PicoNode in the sensor network test bed includes [63]:

1. *Physical Layer.* Each PicoNode employs a 100-mW Bluetooth radio that supports 79 channels in the 2.4-GHz ISM frequency band with a maximum data

rate of 1 Mbps. The radios employ Gaussian frequency shift keying modulation with 1 MHz channel spacing.

2. *Data Link Layer.* The data-link layer consists of three major components: the transmit controller and data path (TCD), the receive controller and data path (RCD), and the medium-access control (MAC). The TCD and RCD are responsible for packet buffering, serialization, deserialization, cyclic redundancy checking, and line balancing.

   The MAC uses carrier sense multiple access (CSMA) with preamble sampling (PS) for infrequent message broadcasts. For unicast traffic, a variant of spatial time-division multiple-access (S-TDMA), referred to as *on-demand S-TDMA*, is employed. Packet headers and payloads use an 8-bit cyclic redundancy check (CRC) and a data acknowledgment retransmission scheme with time-outs to help ensure packet reliability.

3. *Network Layer.* The network layer consists of four major components: energy-aware routing (EAR) protocol, location service, neighbour list service (NLS), and queuing service.

   (a) *EAR* is a destination-initiated reactive routing protocol designed to increase the survivability of the sensor network. Routing paths are chosen in a probabilistic fashion where the probability of selecting a route is inversely proportional to the average energy cost of that particular route. This achieves an even energy depletion of the sensor network.

   (b) The *location service* is called *hop-terrain* and makes use of a combination of RSSI and hop counts from reference nodes in order to triangulate a location.

   (c) The NLS maintains a table that maps neighbor-node MAC IDs to network addresses. Each entry in the table contains a link cost metric and a status indicator. The cost metric indicates the average energy required to perform a unicast transmission along a particular link.

   (d) Finally, the *queuing service* manages the timing of events during node initialization, neighbor discovery, location discovery, and MAC ID assignment.

4. *Application Layer.* The application layer consists of a standard sensor board, an optional sensor board, and the required application drivers that provide the interface between adjacent layers. The initial target application for the Pico-Radio project was indoor building monitoring. The test bed comprises of three different types of nodes. The first type is sensor nodes that obtain measurements. The second are controller nodes that issue queries to the network. Finally, anchor nodes provide a location reference by periodically broadcasting their locations to other nodes in the sensor network.

1.7.1.3 Packet Reliability Empirical data about energy consumption and packet reliability of the PicoRadio network was gathered. Three configurations with varying parameters were executed and the results were collected. The first

TABLE 1.3   PicoNode Experiment Configurations

System	Description
Baseline	CSMA and on-demand S-TDMA with $T_f = T$, $S = 20$ ms, and $N_s = 9$
Case 1	CSMA-PS with $T_p = 512$ μs and $T_s = 5$ μs On-demand S-TDMA with $T_f = 256$ ms, $S = 20$ ms, and $N_s = 9$
Case 2	CSMA-PS with $T_p = 512$ μs and $T_s = 5$ μs On-demand S-TDMA with $T_f = 90$ ms, $S = 10$ ms, and $N_s = 9$

configuration is a *baseline* configuration. The subsequent two configurations, denoted *case 1* and *case 2*, have varying parameters. The configurations are summarized in Table 1.3.

The sensor network consisted of 25 PicoNodes placed in an approximately rectangular grid. Spacing between nodes varied from 3 to 7 m, with all nodes placed at roughly the same elevation. At the beginning of each experiment, a controller node broadcasted a query requesting all sensor nodes to relay 200 temperature measurements at intervals of 5 s. T denotes the time period between samples, and N denotes the total number of samples to be taken. These parameters are specified in the query disseminated to each node.

The baseline configuration utilizes CSMA without any preamble sampling. The radio is constantly on, even when the node is not transmitting or the channel is idle. The size of the frame is denoted by T_f, the number of slots within the frame is denoted by N_s, and the size of the slot spacing is denoted by S.

Nodes transmit their data packets during their designated time slots, and data packets acquired from neighboring nodes are forwarded during the designated frame using CSMA. Both case 1 and case 2 utilize CSMA with preamble sampling. Nodes wake up every T_p seconds to sense the channel. If no preamble is detected within the time period denoted by T_s, the node goes back to sleep.

1.7.1.4   *Results*   For the baseline configuration, the end-to-end packet loss ratio (PLR) of individual sensor nodes varied from 0 to 0.2, with an overall average PLR of 0.04 for the entire sensor network. The nodes with the best reliability were those placed closest to the controller. Nodes located farthest from the controller and along the edges of the sensor network exhibited the most packet loss. The hop count for messages to reach the destination varied from a minimum of 1 hop to a maximum of 8 hops.

For case 1, the variation in the PLR was lower, but the overall PLR for the network remained the same. This is because for a given slot spacing, the preamble sampling had a negligible impact on end-to-end packet reliability. In case 2, the PLR ranged from 0 to 0.88, with an overall network average of 0.36. The higher PLR was caused by more packet collisions due to the smaller frame size and slot spacing.

1.8 SENSOR OPERATING SYSTEMS

TinyOS is an open-source operating system designed for wireless embedded sensor networks [5,65]. It features a component-based architecture that enables implementation of sensor network applications. TinyOS features a component library that includes network protocols, distributed services, sensor drivers, and data-acquisition tools. TinyOS features an event-driven execution model and enables fine-grained power management. It has been ported to several platforms with support for various sensor boards.

Currently, over 500 research groups and companies use TinyOS and the sensor motes developed by Crossbow [66]. A partial list of research projects [67] currently under way is presented in Table 1.4. A partial list of companies [67] that use TinyOS in commercial developments is provided in Table 1.5.

TABLE 1.4 TinyOS Research Projects

Project	Description
Calamari [68]	Localization solutions for sensor networks
CotsBots [69]	Inexpensive and modular mobile robots built using off-the-shelf components to investigate distributed sensing and cooperation algorithms in large (>50) robot networks
Firebug [70]	Berkeley civil engineering project for the design and construction of a wildfire instrumentation system using networked sensors
galsC [71]	Language and compiler designed for use with the TinyGALS [72] programming model
Great Duck Island [19]	Remote habitat monitoring of Leach's Storm Petrel
Mate [73]	Application-specific virtual machines for TinyOS networks
PicoRadio [74]	Development of mesoscale low-cost transceivers for ubiquitous wireless data acquisition that minimizes power/energy dissipation
Sensing Structural Integrity [75]	Reporting the location and kinematics of damage during and after an earthquake
Telegraph [76]	Study of various technologies for adaptive data flow such as streaming data from sensors, logs, and peer-to-peer systems
TinyDB [77]	Query processing system for extracting information from a network of TinyOS sensors
TinyGALS [72]	Globally asynchronous and locally synchronous model for programming event-driven embedded systems
XYZ On A Chip [78]	Research focused on airflow measurement technology and the use of sensor networks for controlling indoor temperature

TABLE 1.5 TinyOS Commercial Research Projects

Project	Description
Digital Sun's S. Sense [79]	Soil-moisture sensor system for sprinkler systems to keep grass green while conserving water
Dust Networks [80]	Manufacturers of resilient, self-healing wireless mesh networks optimized for low data-rate applications
Crossbow [66]	Manufacturer of wireless sensor networks and wireless data loggers that use TinyOS
Ember [81]	Developer of wireless semiconductor systems that consist of chips embedded with networking software and low-frequency radio transmitter technology that support wireless mesh monitoring and low-power autohealing management networks
Sensicast [82]	Provider of end-to-end intelligent wireless sensor network solutions to original equipment manufacturers (OEMs) and system integrators
Sensit [83]	Developers of the most highly used wind-eroding mass sensor worldwide

1.9 SUMMARY

This chapter outlined some envisioned, as well as implemented, wireless sensor network applications. A brief overview of the various types of services required by wireless sensor network applications was also presented. Although advances in technology have increased the processing, storage, and communication capabilities of sensor nodes, the main obstacle yet to be overcome is the limited power available to sensor nodes. As battery technology and energy-harvesting techniques improve, wireless sensor network applications will continue to flourish.

As wireless sensor network applications become increasingly more powerful and proliferate, additional services that support their increased functionality will also be required. Several research groups have begun to develop middleware to provide needed services to support wireless sensor networks. Ideally, deployed wireless sensor networks should configure, adjust, and heal themselves automatically with minimal user intervention. Information sharing among independent sensor networks, deployed within the same region, even though they are distinct, is another desirable quality. However, before these scenarios become a reality, much research remains to be done.

ACKNOWLEDGMENTS

This material is based on work supported by the National Science Foundation under Grant ANI-0086020.

REFERENCES

1. Crossbow Technology MPR2400 MICAz, from http://www.xbow.com/products/product_pdf_files/wireless_pdf/6020-0060-01_a_micaz.pdf/, December 2004.

2. See at http://www.xbow.com/products/productsdetails.aspx?sid=101.

3. Crossbow Technology's MicaZ sensor mote, from http://gyro.xbow.com/other/micaz_new.jpg, December 2004.

4. See at http://www.atmel.com.

5. J. Hill, R. Szewczyk, A. Woo, S. Hollar, D. Culler, and K. Pister. System architecture directions for network sensors. In *Proceedings of the 9th International Conference on Architectural Support for Programming Languages and Operating Systems (ASPLOS-IX)*, pages 93–104, Cambridge, Massachusetts, November 2000.

6. I. F. Akyildiz, W. Su, Y. Sankarasubramaniam, and E. Cayirci. Wireless sensor networks: A survey. *Computer Networks*, March 2002.

7. C. Intanagonwiwat, R. Govindan, and D. Estrin. Directed diffusion: A scalable and robust communication paradigm for sensor networks. In *Proceedings of the 6th Annual International Conference on Mobile Computing and Networking*, pages 56–67, ACM Press, 2000.

8. C. Intanagonwiwat, R. Govindan, D. Estrin, J. Heidemann, and F. Silva. Directed diffusion for wireless sensor networking. *IEEE/ACM Transactions on Networking*, 11(1):2–16, 2003.

9. J. Agre and L. Clare. An integrated architecture for cooperative sensing networks. *IEEE Computer*, pages 106–108, May 2000.

10. F. Ye, H. Luo, J. Cheng, S. Lu, and L. Zhang. A two-tier data dissemination model for large-scale wireless sensor networks. In *Proceedings of the 8th Annual International Conference on Mobile Computing and Networking*, pages 148–159, ACM Press, 2002.

11. A. Mainwaring, D. Culler, J. Polastre, R. Szewczyk, and J. Anderson. Wireless sensor networks for habitat monitoring. In *Proceedings of the 1st ACM International Workshop on Wireless Sensor Networks and Applications*, pages 88–97, ACM Press, 2002.

12. E. Biagioni. PODS: Interpreting spatial and temporal environmental information. In *Usability Evaluation and Interface Design: Cognitive Engineering, Intelligent Agents, and Virtual Reality*, Volume I of the *Proceedings of HCI International 2001, the 9th International Conference on Human-Computer Interaction*, pages 317–321, New Orleans, Louisiana, August 2001.

13. E. Biagioni and K. Bridges. The application of remote sensor technology to assist the recovery of rare and endangered species special issue on distributed sensor networks. *International Journal of High Performance Computing Applications*, 16(3), August 2002.

14. D. Niculescu and B. Nath. Ad hoc positioning system (APS), In *Proceedings of GLOBECOM'01 (IEEE)*, pages 2926–2931, San Antonio, Texas, November 2001.

15. K. A. Sudduth. Engineering technologies for precision farming. Presented at the International Seminar on Agricultural Mechanization Technology for Precision Farming, Suwon, Korea, May 1999.

16. C. R. Locke, G. J. Carbone, A. M. Filippi, E. J. Sadler, B. K. Gerwig, and D. E. Evans. Using remote sensing and modeling to measure crop biophysical variability. In *Proceedings of the 5th International Precision Agriculture Conference*, Minneapolis, Minnesota, July 2000.

17. P. Bonnet, J. Gehrke, and P. Seshadri. Querying the physical world. *IEEE Personal Communications*, **7**:10–15, October 2000.

18. L. Schwiebert, S. Gupta, and J. Weinmann. Research challenges in wireless networks of biomedical sensors. In *Proceedings of the 7th Annual International Conference on Mobile Computing and Networking*, pages 151–165, ACM Press, 2001.

19. Habitat monitoring on great duck island, from http://www.greatduckisland.net/, November 2004.

20. R. Szewczyk, J. Polastre, A. Mainwaring, and D. Culler. Lessons from a sensor network expedition. In *Proceedings of the 1st European Workshop on Wireless Sensor Networks (EWSN '04)*, January 2004.

21. J. Polastre. Design and Implementation of Wireless Sensor Networks for Habitat Monitoring. Master's thesis, University of California, Berkeley, May 2003.

22. See at http://www.botany.hawaii.edu/pods/.

23. F. Sivrikaya and B. Yener. Time synchronization in sensor networks: a survey. *IEEE Network*, **18**(4):45–50, July/August 2004.

24. L. Lamport. Time, clocks, and the ordering of events in a distributed system. *Communications of the ACM*, July 1978.

25. K. M. Chandy and L. Lamport. Distributed snapshots: Determining global states of distributed systems. *ACM Transactions on Computer Systems*, February 1985.

26. C. J. Fidge. Partial orders for parallel debugging. In *ACM SIGPLAN/SIGOPS Workshop on Parallel 4 Distributed Debugging*, 1985.

27. F. Mattern. Virtual time and global states of distributed systems. In *International Workshop on Parallel and Distributed Algorithms*, 1989.

28. D. L. Mills. Internet time synchronization: The network time protocol. In *Global States and Time in Distributed Systems*, Zhonghua Yang and T. Anthony Marsland (eds.), pages 91–102, IEEE Computer Society Press, 1994.

29. H. Dai and R. Han. Tsync: A lightweight bidirectional time synchronization service for wireless sensor networks. *Mobile Computing and Communications Review*, **8**(1):125–139, 2004.

30. J. Elson, L. Girod, and D. Estrin. Fine-grained network time synchronization using reference broadcasts. In *Proceedings of 5th Symposium on Operating Systems Design and Implementation (OSDI)*, pages 147–163, December 2002.

31. J. Elson and D. Estrin. Time synchronization for wireless sensor networks. In *Proceedings of the 2001 International Parallel and Distributed Processing Symposium (IPDPS), Workshop on Parallel and Distributed Computing Issues in Wireless and Mobile Computing*, April 2001.

32. S. Ganeriwal, R. Kumar, and M. B. Srivastava. Timing-sync protocol for sensor networks. In *Proceedings of the 1st International Conference on Embedded Networked Sensor Systems (SenSys)*, pages 138–149, ACM Press, 2003.

33. M. L. Sichitiu and C. Veerarittiphan. Simple, accurate time synchronization for wireless sensor networks. In *Proceedings of the IEEE Wireless Communications and Networking Conference (WCNC 2003)*, Volume 2, pages 1266–1273, New Orleans, Louisiana, March 2003.

34. J. van Greunen and J. Rabaey. Lightweight time synchronization for sensor networks. In *Proceedings of the 2nd ACM International Conference on Wireless Sensor Networks and Applications*, pages 11–19, ACM Press, 2003.

35. S. Meguerdichian, S. Slijepcevic, V. Karayan, and M. Potkonjak. Localized algorithms in wireless ad-hoc networks: Location discovery and sensor exposure. In *Proceedings of the 2nd ACM International Symposium on Mobile Ad Hoc Networking and Computing*, pages 106–116, ACM Press, 2001.

36. A. Savvides, C. Han, and M. B. Strivastava. Dynamic fine-grained localization in ad-hoc networks of sensors. In *Proceedings of the 7th Annual International Conference on Mobile Computing and Networking*, pages 166–179, ACM Press, 2001.

37. A. Savvides, H. Park, and M. B. Srivastava. The bits and flops of the n-hop multilateration primitive for node localization problems. In *Proceedings of the 1st ACM International Workshop on Wireless Sensor Networks and Applications*, pages 112–121, ACM Press, 2002.

38. D. Niculescu and B. Nath. Ad hoc positioning system (APS). In *Proceedings of GLOBECOM'01 (IEEE)*, pages 2926–2931, San Antonio, Texas, November 2001.

39. D. Niculescu and B. Nath. Ad hoc positioning system (APS) using AOA. In *Proceedings of IEEE INFOCOM 2003—The Conference on Computer Communications*, **22**(1): 1734–1743, March 2003.

40. D. Niculescu and B. Nath. Localized positioning in ad hoc networks. In *Proceedings of the 1st IEEE International Workshop on Sensor Network Protocols and Applications*, Anchorage, Alaska, April 2003.

41. N. B. Priyantha, A. Miu, H. Balakrishnan, and S. Teller. The cricket compass for context-aware mobile applications. In *Proceedings of the 7th Annual International Conference on Mobile Computing and Networking*, pages 1–14, ACM Press, 2001.

42. Nissanka B. Priyantha, Anit Chakraborty, and Hari Balakrishnan. The cricket location-support system. In *Proceedings of the 6th Annual International Conference on Mobile Computing and Networking*, pages 32–43, ACM Press, 2000.

43. J. Heidemann, F. Silva, C. Intanagonwiwat, R. Govindan, D. Estrin, and D. Ganesan. Building efficient wireless sensor networks with low-level naming. In *Proceedings of the 18th ACM Symposium on Operating Systems Principles*, pages 146–159, ACM Press, 2001.

44. B. Krishnamachari, D. Estrin, and S. Wicker. Impact of data aggregation in wireless sensor networks. In *International Workshop of Distributed Event Based Systems (DEBS)*, July 2002.

45. J. Zhao, R. Govindan, and D. Estrin. Computing aggregates for monitoring wireless sensor networks. In *Proceedings of the 1st IEEE International Workshop on Sensor Network Protocols and Applications*, May 2003.

46. S. Ratnasamy, B. Karp, L. Yin, F. Yu, D. Estrin, R. Govindan, and S. Shenker. Ght: A Geographic hash table for data-centric storage. In *Proceedings of the 1st ACM International Workshop on Wireless Sensor Networks and Applications*, pages 78–87, ACM Press, 2002.

47. S. Ratnasamy, B. Karp, S. Shenker, D. Estrin, R. Govindan, L. Yin, and F. Yu. Data-centric storage in sensornets with ght, a geographic hash table. *Mobile Networks and Applications*, **8**(4):427–442, 2003.

48. D. Ganesan, B. Greenstein, D. Perelyubskiy, D. Estrin, and J. Heidemann. An evaluation of multi-resolution storage for sensor networks. In *Proceedings of the 1st International Conference on Embedded Networked Sensor Systems*, pages 89–102, ACM Press, 2003.

49. S. Tilak, N. B. Abu-Ghazaleh, and W. Heinzelman. Infrastructure tradeoffs for sensor networks. In *Proceedings of the 1st ACM International Workshop on Wireless Sensor Networks and Applications*, pages 49–58, ACM Press, 2002.

50. P. Bose, P. Morin, I. Stojmenovic, and J. Urrutia. Routing with guaranteed delivery in ad hoc wireless networks. *Wireless Networking*, **7**(6):609–616, 2001.

51. B. Karp and H. T. Kung. Gpsr: Greedy perimeter stateless routing for wireless networks. In *Proceedings of the 6th Annual International Conference on Mobile Computing and Networking*, pages 243–254, ACM Press, 2000.

52. G. Xing, C. Lu, R. Pless, and Q. Huang. On greedy geographic routing algorithms in sensing-covered networks. In *Proceedings of the 5th ACM International Symposium on Mobile Ad Hoc Networking and Computing*, pages 31–42, ACM Press, 2004.

53. M. Kochhal, L. Schwiebert, and S. Gupta. Role-based hierarchical self organization for wireless ad hoc sensor networks. In *Proceedings of the 2nd ACM International Conference on Wireless Sensor Networks and Applications*, pages 98–107, ACM Press, 2003.

54. I. F. Akyildiz and I. H. Kasimoglu. Wireless sensor and actor networks: Research challenges. *Ad Hoc Networks*, **2**(4):351–367, October 2004.

55. M. Conti, G. Maselli, G. Turi, and S. Giordano. Cross-layering in mobile ad hoc network design. *Computer (IEEE)*, **37**(2):48–51, February 2004.

56. Y. Zhang and L. Cheng. Cross-layer optimization for sensor networks. *New York Metro Area Networking Workshop 2003*, New York, New York, September 2003.

57. A. Woo, S. Madden, and R. Govindan. Networking support for query processing in sensor networks. *Communications of the ACM*, **47**(6):47–52, 2004.

58. S. Madden, W. Hong, J. Hellerstein, and M. Franklin. Tinydb: A declarative database for sensor networks, from http://telegraph.cs.berkeley.edu/tinydb.

59. Y. Yao and J. Gehrke. The cougar approach to in-network query processing in sensor networks. *ACM SIGMOD Record*, **31**(3):9–18, 2002.

60. X. Li, Y. J. Kim, R. Govindan, and W. Hong. Multi-dimensional range queries in sensor networks. In *Proceedings of the 1st International Conference on Embedded Networked Sensor Systems*, pages 63–75, ACM Press, 2003.

61. S. Madden, M. J. Franklin, J. M. Hellerstein, and W. Hong. The design of an acquisitional query processor for sensor networks. In *Proceedings of the 2003 ACM SIGMOD International Conference on Management of Data*, pages 491–502, ACM Press, 2003.

62. S. J. Park, R. Vedantham, R. Sivakumar, and I. F. Akyildiz. A scalable approach for reliable downstream data delivery in wireless sensor networks. In *Proceedings of the 5th ACM International Symposium on Mobile Ad Hoc Networking and Computing*, pages 78–89, ACM Press, 2004.

63. J. M. Reason and J. M. Rabaey. A study of energy consumption and reliability in a multi-hop sensor network. *Mobile Computing and Communications Review*, **8**(1): 84–97, 2004.

64. J. M. Rabaey, M. J. Ammer, J. L. da Silva, D. Patel, and S. Roundy. Picoradio supports ad hoc ultra-low power wireless networking. *Computer (IEEE)*, **33**(7):42–48, July 2000.

65. TinyOS Community Forum, from http://www.tinyos.net/, November 2004.

66. Crossbow Technology Inc., from http://www.xbow.com, November 2004.

67. TinyOS Community Forum related work, from http://www.tinyos.net/related.html, November 2004.

68. Calamari: a sensor field localization system, from http://www.cs.berkeley.edu/kamin/calamari/, November 2004.

69. CotsBots, from http://www-bsac.eecs.berkeley.edu/projects/cotsbots/, November 2004.

70. FireBug, from http://firebug.sourceforge.net/, November 2004.

71. galsC: A language for event-driven embedded systems, from http://galsc.sourceforge.net/, November 2004.

72. TinyGALS: A programming model for event driven embedded systems, from http://ptolemy.eecs.berkeley.edu/papers/03/tinygals/, November 2004.

73. Mate, from http://www.cs.berkeley.edu/pal/mate-web/, November 2004.

74. PicoRadio, from http://bwrc.eecs.berkeley.edu/research/pico radio/, November 2004.

75. S. D. Glaser, from http://www.ce.berkeley.edu/glaser/curee.pdf, November 2004.

76. The Telegraph Project at UC Berkeley, from http://telegraph.cs.berkeley.edu/, November 2004.

77. TinyDB: A declarative database for sensor networks, from http://telegraph.cs.berkeley.edu/tinydb/, November 2004.

78. XYZ on a chip: Integrated wireless sensor networks for the control of the indoor environment in buildings, from http://www.cbe.berkeley.edu/research/briefs-wirelessxyz.htm, November 2004.

79. Digital Sun, from http://www.digitalsun.com/, November 2004.

80. Dust networks, from http://www.dust-inc.com/products/main.shtml, November 2004.

81. Ember, from http://www.ember.com/index.html, November 2004.

82. Sensicast, from http://www.sensicast.com/, November 2004.

83. Sensit Company, from http://www.sensit.com/, November 2004.

CHAPTER 2

Distributed Signal Processing Algorithms for the Physical Layer of Large-Scale Sensor Networks

AN-SWOL HU and SERGIO D. SERVETTO

Cornell University, Ithaca, New York

The ability to move the sensed data out of the network (the reachback *communication) is one of the basic communication primitives that must be supported by every sensor network. This is most commonly achieved by routing of information through the network to some central data collection point that will act on the sensed information. Authors propose an approach to this problem by developing a method to allow all nodes in the network to cooperatively generate a strong information bearing signal to communicate with a distant data collection point. Such a solution would be extremely robust to the failure of nodes in the network and would allow for the deployment of a homogeneous network of extremely small, low-power nodes for a variety of applications. This approach to the sensor reachback problem is called* cooperative reachback. *In this chapter two aspects of the cooperative reachback problem are considered: time synchronization and reachback modulation schemes. Time synchronization facilitates the design of cooperative reachback modulation schemes. A system model that will apply to both the time synchronization problem and reachback communication is described. Authors then study the properties of waveforms generated by asymptotically dense networks and then move into the development of a time synchronization mechanism for dense networks using these waveform properties. In the asymptotic regime, this time synchronization framework can keep the synchronization mean squared error from increasing with distance from the ideal time source. Next, authors study the performance of this asymptotically optimal scheme for networks of finite size. They then develop a modulation scheme for reachback communication again using these waveform properties and study its performance using simulations.*

Handbook of Sensor Networks: Algorithms and Architectures, Edited by Ivan Stojmenović
Copyright © 2005 John Wiley & Sons, Inc.

2.1   INTRODUCTION

2.1.1   Reachback Communication and Cooperative Reachback

In the deployment of every sensor network, the ability to move the sensed data out of the network is absolutely essential. We call this communication requirement *reachback communication*. This is most commonly achieved through the routing of information through the network to some central data collection point that will act on the sensed information. The scenario implies that the data collection point is within the transmission range of one of the sensor nodes. However, this assumption may not always be desired or practical. Potential applications of sensor networks range from target tracking and classification [1,2] to habitat monitoring [3,4]. With such a varying range of potential applications, the requirement that the data collection node be within the transmission range of a sensor node is quite restricting.

Consider the dense aerial deployment of 100,000 cubic millimeter nodes in the Amazon rain forest for habitat and environmental monitoring. Due to the low-power nature of these sensor nodes, the communication range of each node may only be on the order of meters. No one single node can communicate with a data collection point that may be located on a low-flying aircraft, and thus the conventional manner of routing information to the data collection point for reachback communication will fail. In such a scenario, it is possible to deploy a tiered network where a few more powerful nodes are included for reachback communication purposes. However, such a solution means that the functionality of the network is entirely dependent on the performance of these few communication nodes.

Our approach to this problem is to develop a method to allow all nodes in the network to cooperatively generate a strong information-bearing signal to communicate with a distant data collection point. Such a solution would be extremely robust to the failure of nodes in the network and would allow for the deployment of a homogeneous network of extremely small, low-power nodes for a variety of applications. We call this approach to the sensor reachback problem *cooperative reachback*. The avenue we explore in the development of algorithms for reachback is motivated by recent work on the *sensor broadcast* problem [5a, 5b], where nodes are able to agree on a common stream of bits to transmit.

2.1.2   Large-Scale Dense Sensor Networks

We consider the problem of cooperative reachback in the context of extremely dense large-scale networks. In fact, the approach we take to studying these networks is to assume that the number of nodes in a finite-area network grows unbounded. That is, we consider an asymptotically dense network as a close approximation for realistic large-scale dense networks.

Whereas infinitely large networks consisting of nodes with zero mass are clearly not realizable by physical devices, there is a trend toward miniaturization of these devices. For example, in recent work, a hardware simulation and deployment platform for wireless sensor networks capable of simulating networks with on the order of 100,000 nodes was developed [6]. In addition, for many years now the Smart Dust project has been seeking to build cubic-millimeter motes for a wide range of applications [7]. Furthermore, there is a trend toward the miniaturization of power sources [8]. With large numbers and small nodes, we face a situation involving networks operating at high densities. This implies the need for cooperative reachback capabilities and asymptotic behaviors provide a method to develop these abilities. Techniques developed in the asymptotic regime will have favourable scaling laws, and thus perform well in practical situations with a large, but still finite, numbers of nodes.

2.1.3 Time Synchronization

In this chapter we consider two aspects of the cooperative reachback problem: time synchronization and reachback modulation schemes. We study time synchronization because it allows us to characterize a synchronized network. Once we have such an understanding, we can apply this understanding to the design of cooperative reachback modulation schemes. In fact, it would be ideal to design a mechanism that can achieve cooperative reachback communication and network time synchronization *simultaneously*.

One reason the time synchronization for large-scale dense networks is such a difficult problem is because of scalability issues. As the area of the network and the number of nodes increase, multihop communication will most likely be required for communication between nodes. This also means that timing information will need to be distributed throughout the network in multiple hops, resulting in an accumulation of timing error. For example, with reference-broadcast synchronization (RBS) [9], which performs well for multi-hop synchronization, the average path error still grows as $\sqrt{n}$, where n is the number of hops. This problem of timing error accumulation over multiple hops presents a significant problem for large-scale wireless sensor networks since these networks have a large number of nodes spread out over a wide area. However, the source of this problem may also provide us with a solution. We ask the question: Given an extremely dense network of nodes spread out over a finite area, can we use the large number of nodes to improve synchronization performance?

2.1.4 Related Work

Due to the high level of interest in sensor networks, much recent work has been done in the area of distributed signal processing for sensor and ad hoc networks. This work covers topics ranging from cooperative routing, reachback communication, to time synchronization. First of all, there has been much progress made in taking a cross-layer approach to developing more efficient cooperative and distributed multicasting [10–12] and routing [13,14] techniques. The authors take a new perspective on the problems of routing and multicasting by jointly considering the network layer and the physical layer.

In the area of reachback communication, there has been three significant directions of particular interest. One area is the use of cooperative diversity where nodes achieve uplink transmit diversity by relaying each other's messages. This idea was introduced by Sendonaris et al. in refs. [15] and [16] and extended by Laneman et al. in refs. [17] and [18]. A variation on the concept of cooperative diversity, called *coded cooperation*, was considered in refs. [19] and [20]. The second direction of research has been the use of radar concepts for the upload of information from sensor networks. This idea has been studied in refs. [21] and [22], where synthetic aperture radar (SAR) techniques were employed.

The third direction of research regarding reachback communication has been the study of cooperative reachback. This area of research employs the idea of having nodes in the network cooperatively generate a signal that can more reliably transmit information to a far receiver. In ref. [23], the authors consider the problem of distributed beamforming while accounting for phase errors arising from errors in node placement. It is shown that the expected received signal power grows linearly with the number of nodes in the network, as does the variance of the received power. They conclude that there are large potential gains from distributed beamforming as long as the node placement errors are small compared to the carrier wavelength. In ref. [24], the problem of coherent cooperative transmission from multiple antennas is considered. The authors present a system architecture for such a distributed transmission array and analyze its performance. Another cooperative transmission scheme of interest is presented in ref. [25]. The proposed "opportunistic large arrays" consider the situation where there is one source of information and the remaining nodes act as repeaters. The accumulation of energy as the repeater nodes relay the signal sent by the leader node acts as a physical layer flooding algorithm and a method for reachback communication.

In the area of time synchronization for sensor networks, a great deal of work has been done [9,26–31]. In this work however, we seek to address not only the time synchronization of large-scale sensor networks but also the issue of cooperative reachback. In fact, we propose a method that can maintain time synchronization and reachback communication simultaneously for asymptotically dense networks. A recent piece of work by Hong and Scaglione [32] also deals specifically with the time synchronization problem for large-scale sensor networks and addresses the reachback communication issue. In ref. [32], the authors model the sensor nodes as pulse-coupled oscillators and apply the results of Mirollo and Strogatz [33], who show that a network of pulse-coupled oscillators will converge toward synchrony under the assumptions of no delays, a noise-free environment, identical oscillators, and all-to-all coupling. In ref. [32], the authors extend the theoretical results of Mirollo and Strogatz for better implementation in a wireless sensor network and analyze the system through simulations. In this chapter we use a different system model and analytically prove synchronization before evaluating the results through simulations.

2.1.5   Chapter Organization

The chapter is organized as follows. In Section 2.2 we set up the system model that will apply to both the time synchronization problem and reachback communication.

We study the properties of waveforms generated by asymptotically dense networks in Section 2.3, and then move into the development of a time synchronization mechanism for dense networks in Section 2.4 using these waveform properties. In the asymptotic regime, this time synchronization framework can keep the synchronization-mean-squared error from increasing with distance from the ideal time source. In Section 2.5 we study the performance of this asymptotically optimal scheme for networks of finite size. We then develop a modulation scheme for reachback communication again using these waveform properties in Section 2.6, and study its performance using simulations in Section 2.7. Concluding remarks are presented in Section 2.8.

2.2 SYSTEM MODEL

As will become apparent later in the chapter, our methods for time synchronization and cooperative reachback modulation are intimately related. In fact, the modulation scheme can best be understood as an extension to the synchronization method. As a result, the system model is set up mostly in the context of the time synchronization problem.

2.2.1 Clock Model

We consider a sensor network with N nodes. The clock of one particular node in the network will serve as the ideal time, and to this clock we wish to synchronize all other nodes. This node can be *any* arbitrary node in the network and is not special in any way. The system is defined relative to the clock of this arbitrary node. The synchronization methods presented here synchronize the clocks of all nodes in a network to the clock of one particular node. This is done to make the synchronization scheme self-contained when the only clocks that the network has access to are the clocks of its nodes. If we want the network synchronized to "real time," then the node initiating synchronization would need to have access to it. According to the recommendations of Elson and Römer [26], we allow the local clock of each node to be free-running. We never adjust the local clock frequency or offset, but instead we seek to construct an "operational" clock on top of the free-running local clock. The operational clock of each node will be synchronized to the ideal clock, and it will be defined in terms of that node's local clock.

We will call the node with the ideal clock node 1, and without loss of generality we assume it lies in the center of the network. The clock of node 1, c_1, will be defined as $c_{1,t} = t$, and we also define the counter $c_1(t) = \lfloor t \rfloor$ where $t \in [0, \infty)$. Note that $c_{1,t}$ is continuous while $c_1(t)$ takes only integer values. At any time t_o, $c_1(t_o)$ is the number of ticks the counter of node 1 has made. From the expression for $c_1(t)$, we can easily see that the counter of node 1 ticks on integer values of t. We define the counter $c_1(t)$ to simplify the description of the synchronization procedure, since all synchronization pulses are sent at integer values of t.

Taking c_1 to be the ideal clock, we now define the clock of any other arbitrary node i as c_i. We define c_i as

$$c_{i,t} = \alpha_i(t - \bar{\Delta}_i) + \Psi_i(t) \tag{2.1}$$

where

$\bar{\Delta}_i$ is an unknown constant modeling the fact that it is not known when c_i is started relative to c_1.

$\alpha_i > 0$ is a constant and for each i, $\alpha_i \in [\alpha_{lowbound}, \alpha_{upbound}]$ where $\alpha_{upbound}, \alpha_{lowbound} > 0$ are finite. This bound on α_i means that the frequency offsets between any two nodes cannot be unbounded. We assume that a known function $f_\alpha(s)$ with $s \in [\alpha_{lowbound}, \alpha_{upbound}]$ gives the percentage of nodes with any given α value. Thus, the fraction of nodes with α values in the range s_0 to s_1 can be found by integrating $f_\alpha(s)$ from s_0 to s_1. We also assume that $|f_\alpha(s)| < G_\alpha$, for some constant G_α. We keep this function constant as we increase the number of nodes in the network.

$\Psi_i(t)$ is a zero mean Gaussian process with samples $\Psi_{t_j} \sim \mathcal{N}(0, \sigma^2)$, for $j \in N$, independent and identically distributed. We assume $\sigma^2 < \infty$ and note that σ^2 is defined in terms of the clock of node i. We assume that $\Psi_i(t)$ is Gaussian since the root-mean-square (RMS) jitter is characterized by the Gaussian distribution [34].

Thus, this model assumes that there is a bounded constant frequency offset between the oscillators of any two nodes as well as some random frequency jitter.

The reasoning behind the clock model in equation (2.1) comes from the following oscillator model for the instantaneous frequency $f(t)$,

$$f(t) = f_0 + \Delta f + f_r(t) \tag{2.2}$$

where f_0 is the nominal frequency in hertz, Δf models the frequency accuracy in hertz, and $f_r(t)$ models the short-term stability of the oscillator in hertz. Note that we ignore frequency drift, because we assume that it is negligible for short periods of time, say, on the order of 100 s. From approximate long-term stability numbers plotted in ref. [35], we find that the frequency offset over 100 s is on the order 1×10^{-11}, which is two orders of magnitude less than the short-term frequency stability standard deviation (1×10^{-9}) and over five orders of magnitude less than the frequency accuracy of the SPK-SPG series of oscillators manufactured by SPK Electronics Company (www.spkecl.com). Note that the frequency stability and offset values are given by the formula $f_{offset} = (f_{measured} - f_0)/f_0$. In our clock model, we assume that the oscillator of c_1 is running at f_0 even though f_0 may be varying with time. The oscillators of c_i are then defined relative to that of c_1 by equation (2.2). We assume that $c_{i,t}$ increments an integer value each time the oscillator of node i completes a complete cycle. Thus, we have that

$\alpha_i = 1 + (\Delta f/f_0)$ and $\sigma^2 = (\sigma_{f_r}/f_0)^2$, where σ_{f_r} is the standard deviation of $f_r(t)$ given in hertz.

As mentioned, the clocks c_1 and c_i, for all i will be free-running clocks that will have a synchronized "operational" counter built on top of them. This operational counter is set up in the following manner. We first assume that node 1 at time t_e decides it needs to synchronize the remaining $N - 1$ nodes. Recall that node 1 is any arbitrary node. Any one random node can detect an event and decide to synchronize the network. In this case, that node will effectively be node 1. Node 1 will increment its operational counter to a value of 1 at the next integer time t. That is to say, the operational counter of node 1, denoted by $s_1(t)$, will be $s_1(t) = \lfloor t - n_o \rfloor$, where $n_o = \lfloor t_e \rfloor$. Our goal, ideally, will then be to construct an identical operational counter $s_i(t) = \lfloor t - n_o \rfloor$ at node i. We want the operational counter at the ith node to increment at integer values of t and hold a value equal to $s_1(t)$.

2.2.2 Observation Model

Synchronization will be achieved by the transmission and observation of pulses. We first make the following assumptions about pulse transmission and reception:

- *No Propagation Delay.* We assume no delay between the time a pulse is transmitted and the time it is seen by other nodes. Under certain conditions this may be reasonable, since the propagation time of radio waves traveling at the speed of light over small transmission distances is negligible. However, in general time delays need to be explicitly considered. We leave the rigorous analysis of time delay for future work.

- *No Transmission Delay or Time-Stamping Error.* We assume that a pulse is transmitted at exactly the time the node intends to transmit it. We make this assumption since there will be no delay in message construction or access time [9], since our nodes broadcast the same simple pulse without worrying about collisions. Also, when a node receives a pulse it can determine its clock reading without delay, since any time-stamping error is small and can be absorbed into the random jitter.

Because pulses are exchanged among many different nodes, to clearly describe transmission and reception times in relation to different clocks, we define the following notation (illustrated in Fig. 2.1):

- $t_{j,i}^{c_k}$ is the time, with respect to clock c_k, that the ith node sees its jth pulse.
- $s_{n,i}^{c_k}$ is the time of the nth transition of the operational counter $s_i(t)$ with respect to c_k.
- Let us also say that, in general, any value or variable X^{c_j} means that we are considering the value of X in terms of the timescale of c_j.

To use pulse transmission and reception times to do accurate synchronization, we need to model the relationship between transmissions and receptions. We only

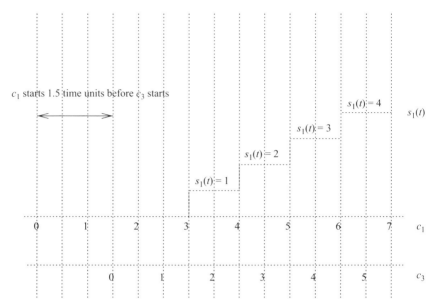

Figure 2.1   This figure illustrates c_1 and c_3 as well as the operational counter of node 1, $s_1(t)$. We assume node 3 is in the broadcast domain of node 1. In this illustration we assume $\Psi_3(t) = 0$ (no random clock jitter), $\alpha_3 = 1$, and $\bar{\Delta}_3 = 1.5$. If we assume a pulse is transmitted by node 1 each time $s_1(t)$ increments, the second pulse will be transmitted at $s_{2,1}^{c_1} = 4$. Since node 3 can hear the pulses of node 1, this pulse will be the second pulse heard by node 3. This occurs at $t_{2,3}^{c_3} = 2.5 = s_{2,1}^{c_3}$.

describe this relationship for a node i within the broadcast domain of node 1 since, as we later show, this is the only important case.

We recall that by definition, $s_{n',1}^{c_1}$ will be an integer and at this time a pulse will be transmitted. Because node i is in the broadcast domain of node 1, we can describe the pulse receive time at node i, with respect to the clock of node i, in terms of the pulse transmission time (or equivalently, the time at which the operational clock of node 1 increments) as the state equations

$$s_{n'+1,1}^{c_1} = s_{n',1}^{c_1} + 1$$
$$t_{n,i}^{c_i} = \alpha_i(s_{n',1}^{c_1} - \bar{\Delta}_i) + \Psi_i(s_{n',1}^{c_1}) \tag{2.3}$$

The first equation of equation (2.3) simply says that if the n'th pulse of node 1 is transmitted at integer $s_{n',1}^{c_1}$ in the time scale of node 1, then the $(n' + 1)$ pulse will be sent at $s_{n',1}^{c_1} + 1$, in the time scale of node 1. The second equation of equation (2.3) makes use of the clock model of node i(2.1) to tell us the time at clock c_i of a pulse transmission by node 1 at $s_{n',1}^{c_1}$, where $s_{n',1}^{c_1}$ is in the timescale of c_1. This second equation effectively converts the time of a pulse transmission from the timescale of c_1 to that of c_i. Under the assumption that node i is in the broadcast domain of node 1, $n' = n$. However, this does not hold in general because in the multihop case the nth pulse observed by the jth node does not necessarily correspond to the nth pulse transmitted by node 1. So, in general, if we assume $n' - n = k$, where $k \in N$, then the expression is saying that the pulse seen by node j at $t_{n,j}^{C_j}$ is occurring at $s_{n+k,1}^{C_j}$.

2.2.3 Propagation Model

To model signal amplitude loss, we assume a general model $K(d)$, where $0 \leq K(d) \leq 1$ for all d. Here $K(d)$ is a fraction of the transmitted magnitude seen at distance d from the transmitter. For example, if the receiver node j is at distance d from node i, and node i transmits a signal of magnitude A, then node j will hear a signal of magnitude $AK(d)$. We derive $K(d)$ from a power path-loss model since any path-loss model captures the *average* received power at a given distance from the transmitter. This average received power is perfect for modeling received signal magnitudes in our problem setup, since we are considering asymptotically dense networks. Due to the large number of nodes at any given distance d from the receiver, using the average received magnitude at distance d as the contribution from each node at that distance will give a good modeling of the amplitude of the aggregate waveform. An example of $K(d)$ is given in Section 2.5.

This $K(d)$ is good for modeling aggregate signal-propagation distances, but to model the magnitude of the aggregate signal at a given node j we choose to use a random variable $K_{j,i}$ with the following properties:

- For a given j, $K_{j,i}$ are independent identically distributed (iid) for all i.
- $K_{j,i}$ is independent from $\Psi_{l,t}$ for all j, i, l, and t.
- $0 \leq K_{j,i} \leq 1$, $0 < E(K_{j,i}) \leq 1$, and $\text{Var}(K_{j,i}) \leq 1$.

Note that the requirements on the random variable $K_{j,i}$ places restrictions on the model $K(d)$. Any function $K(d)$ that yields a $K_{j,i}$ with the preceding requirements can be used to model path loss.

To understand how $K_{j,i}$ and $K(d)$ are related and where the properties of $K_{j,i}$ come from, we give an intuitive explanation of the meaning of $K_{j,i}$: $\Pr(K_{j,i} \in (k, k + \Delta))$ is the fraction of nodes at distances d from node j such that $K(d) \in (k, k + \Delta)$, where Δ is a small constant. This means that, roughly speaking, for any given scaling factor $K_{j,i} = k$, $f_{K_j}(k)\Delta$ is the fraction of received signals with magnitude scaled by approximately k. Thus, if we scale the transmit magnitude A from every node i by an independent $K_{j,i}$, then as the number of nodes, N, gets large, node j will see $Nf_{K_j}(k)\Delta$ signals of approximate magnitude Ak for all possible scaling factors $K_{j,i} = k$. This is because taking a large number of independent samples from a distribution results in a good approximation of the distribution. Thus, for large N, this intuition tells us that by scaling the magnitude of the signal transmitted from every node i by an independent sample of the random variable $K_{j,i}$ gives an aggregate signal at node j that is the same magnitude as if we generated the signal using $K(d)$ directly.

For cooperative reachback, we assume that all nodes are equidistant from the far receiver. This approximation holds for receivers that are far from the network. As a result, we assume that (1) the propagation delay is the same for each node, and (2) the path loss is the same for each node. Thus, for cooperative reachback we do not need to use $K_{j,i}$ and can work directly with $K(d)$.

2.2.4   Synchronization and Communication Pulses

Each node will periodically transmit a scaled version of the pulse $p(t)$ to achieve and maintain synchronization. We call the interval of time during which a synchronization pulse is transmitted a *synchronization phase*. Pulses are only transmitted during the synchronization phases, and at other times the nodes can be dedicated to other tasks. Thus, the smaller the synchronization phase, the better. The actual process of synchronization is described in Section 2.4. We assume that $p(t)$ takes on the shape

$$p(t) = \begin{cases} 1 & -\tau_{nz} < t < 0 \\ 0 & t = 0, \quad t \le -\tau_{nz}, \quad t \ge \tau_{nz} \\ -1 & 0 < t < \tau_{nz} \end{cases} \tag{2.4}$$

for some $\tau_{nz} > 0$, and τ_{nz} is expressed in terms of c_1.

The term τ_{nz} should be chosen large compared to $\max_i \sigma_i^{2,c_1}$, where σ_i^{2,c_1} is the value of σ^2 translated from the timescale of c_i to c_1, that is, $\sigma_i^{2,c_1} \ll \tau_{nz}$. This way, over each synchronization phase, with high probability a zero-crossing will occur. For each node, the duration in terms of c_1 of a synchronization phase will be $2\tau_{nz}$. Note that we assume τ_{nz} is a value that is constant in any consistent timescale. This means that even though nodes have different clocks, identical pulses are transmitted by all nodes. We assume that $p(t)$ is generated by a circuit in each node that emits identical pulses. Each node knows only when to initiate the pulse so that it is sent at the time the node intends it to be sent. We define a pulse to be transmitted at time t if the pulse makes a zero-crossing at time t. Similarly, we define the *pulse receive time* for a node as the time when the observed waveform first makes a zero-crossing. A *zero-crossing* is defined for signals that have a positive amplitude and then transition to a negative amplitude. It is the time that the signal first reaches zero. Note that in this work we study the problem in baseband and do not consider the effects of the carrier. For the exchange of synchronization pulses, we assume that nodes can transmit pulses and receive signals at the same time. This simplifying assumption is not required for the ideas presented here to hold.

For cooperative reachback communication, we modify the type of pulses being transmitted by each node. The modification will fundamentally preserve the zero-crossing property of $p(t)$, and it will be discussed in detail in Section 2.6.

2.2.5   Signal-Reception Model

The aggregate waveform seen by node j at any time t is

$$A_{j,\eta_k}^{c_1}(t) = \sum_{i=1}^{\eta_k} \frac{\kappa A_{\max} K_{j,i}}{\eta_k} p(t - \tau_o - T_i) \tag{2.5}$$

where $A^{c_1}_{j,\eta_k}(t)$ is the waveform seen at node j written in the timescale of c_1 and η_k is the number of nodes contributing to the signal, as it may be possible that only a subset of nodes is transmitting (η_k goes to infinity as N goes to infinity). The term κ is a scaling factor to ensure proper reception of the aggregate waveform by all nodes in the network, and T_i is the random timing jitter suffered by the ith node. We will see that T_i is Gaussian since $\Psi_i(t)$ is a Gaussian process, and $T_i \sim \mathcal{N}(0,\bar{\sigma}_i^2)$ will have $\bar{\sigma}_i^2 < B$ for all i and a constant $B > 0$. If, on the other hand, we assume that node j is receiver Rx a distance d_o from the network, then the aggregate waveform will be modeled as

$$A^{c_1}_{Rx,N}(t) = \sum_{i=1}^{N} \frac{\kappa A_{\max} K(d_o)}{N} p(t - \tau_o - T_i) \qquad (2.6)$$

Note that each node can be told the values of N and η_k before deployment. Assuming the system designer knows the area over which the network will be deployed, the values of η_k can be approximated during the design of the network. Note that an approximation of η_k will not affect any of the analytical results; it is only important that η_k is the correct order of magnitude.

To model the quality of the reception of $A^{c_1}_{j,\eta_k}(t)$ by node j, we model the reception of a signal by defining a threshold γ. The γ is the minimum received maximum signal magnitude required for nodes to perfectly resolve the pulse arrival time. If the maximum received signal magnitude is less than γ, then the node does not make any observations and ignores the received signal waveform. We assume that $\gamma < A_{\max}$, where $A_{\max}$ is the maximum transmit magnitude of a node.

2.2.6 Synchronization Pulse Trains

In equation (2.5) and in the preceding discussions, we have focused on characterizing the aggregate waveform for any one synchronization phase. That is, equation (2.5) is the waveform seen by any node j for the synchronization phase centered around node 1's transmission at $t = \tau_0$, where τ_0 is a positive integer. We can, however, describe a synchronization pulse train in the following form:

$$\bar{A}^{c_1}_{j,\eta_k}(t) = \sum_{q=1}^{\infty} \sum_{i=1}^{\eta_{k,q}} \frac{\kappa A_{\max} K_{j,i}}{\eta_{k,q}} p(t - \tau_q - T_{i,q}) \qquad (2.7)$$

where $\eta_{k,q}$ is the number of contributing nodes at the qth synchronization phase, τ_q is the integer value of t at the qth synchronization phase, and $T_{i,q}$ is the jitter suffered by the ith node in the qth synchronization phase. We seek to create this pulse train with equispaced zero-crossings and use each zero-crossing as a synchronization event.

2.3   STRUCTURAL PROPERTIES OF LARGE COLLECTIONS OF RANDOMLY SHIFTED PULSES

2.3.1   The Structure of $A_\infty(t)$

The aggregate waveform seen at each node j in the network and at the receiver Rx described in Section 2.2.5 both have the form

$$A_N(t) = \frac{1}{N} \sum_{i=1}^{N} CK_i \, p(t - \tau_0 - T_i) \tag{2.8}$$

where C is a constant. As we let the number of nodes grow unbounded, $N \to \infty$, the properties of this limit waveform can be characterized by Theorem 2.1.

Theorem 2.1   Let $p(t)$ be as defined in equation (2.4), $T_i \sim \mathcal{N}(0, \bar{\sigma}^2/\alpha_i^2)$ with $\bar{\sigma}^2/\alpha_i^2 < B < \infty$ for all i, and K_i satisfying the conditions in Section 2.2.3. Then, $\lim_{N \to \infty} A_N(t) = A_\infty(t)$ has the properties

- $A_\infty(t)$ is odd about τ_0, i.e. $A_\infty(\tau_0 - \varepsilon) = -A_\infty(\tau_0 + \varepsilon)$, for some $\varepsilon > 0$.
- $A_\infty(t)$ is continuous
- $A_\infty(\tau_0) = 0$
- $A_\infty(t) > 0$ for $t \in (\tau_0 - \tau, \tau_0)$,   and $A_\infty(t) < 0$ for $t \in (\tau_0, \tau_0 + \tau)$ for some $\tau < \tau_{nz}$

The properties outlined in Theorem 2.1 will be key to the synchronization mechanism and cooperative reachback modulation scheme that we propose. One important issue to note is that Theorem 2.1 holds for any odd-shaped pulse (i.e., $p(-t) = -p(t)$) with compact support. Thus, the generation of rectangular pulses is not required. This fact can be seen from the proof of Theorem 2.1, which is left for ref. [36]. However, we develop and motivate a few important related lemmas in the next section.

2.3.2   The Polarity and Continuity of $A_\infty(t)$

At time $t = \tau_1 \neq \tau_0$, we have that

$$A_N(\tau_1) = \sum_{i=1}^{N} \frac{CK_i}{N} p(\tau_1 - \tau_0 - T_i)$$

$$= \sum_{i=1}^{N} \frac{1}{N} \bar{M}_i(\tau_1)$$

where $\bar{M}_i(\tau_1) \overset{\triangle}{=} CK_i p(\tau_1 - \tau_0 - T_i)$. We have the mean of $\bar{M}_i(\tau_1)$ being

$$E(\bar{M}_i(\tau_1)) = CE(K_i) \int p(\tau_1 - \tau_0 - \psi) f_{T_i}(\psi) \, d\psi \tag{2.9}$$

where $f_{T_i}(\psi)$ is the Gaussian probability density function (pdf)

$$f_{T_i}(\psi) = \frac{1}{(\bar{\sigma}/\alpha_i)\sqrt{2\pi}} \exp\left\{-\frac{(\psi - \tau_0)^2}{2(\bar{\sigma}^2/\alpha_i^2)}\right\}$$

It is clear that the $\bar{M}_i(\tau_1)$'s, for different i's, do not have the same mean and do not have the same variance since the two quantities depend on the α_i value. For generality of notation with $f_\alpha(s)$ from Section 2.2.1, we write the Gaussian distribution for T as

$$f_T(\psi, s) = \frac{1}{(\bar{\sigma}/s)\sqrt{2\pi}} \exp\left\{-\frac{(\psi - \tau_0)^2}{2(\bar{\sigma}^2/s^2)}\right\},$$

and define the notation $\bar{M}_i(\tau_1, s) \triangleq \bar{M}_i(\tau_1)$. We use the results of Lemma 2.1 and a corresponding lemma for $\tau_1 > \tau_0$ to prove the polarity result for $A_\infty(t)$ in ref. [36].

Lemma 2.1 *Given the sequence of independent random variables $\bar{M}_i(\tau_1)$ with $\tau_1 < \tau_0$, $E(\bar{M}_i(\tau_1)) = \mu_i$, and $\text{Var}(\bar{M}_i(\tau_1)) = \sigma_i^2$. Then, for all i,*

$$\gamma_2 > \mu_i > \gamma_1 > 0 \tag{2.10}$$

$$\sigma_i^2 < \gamma_3 < \infty \tag{2.11}$$

for some constants γ_1, γ_2, and γ_3, and

$$\lim_{N\to\infty} \frac{1}{N} \sum_{i=1}^{N} \bar{M}_i(\tau_1) = \eta(\tau_1) > 0$$

almost surely, where

$$\eta(\tau_1) = CE(K_i) \int_{\alpha_{lowbound}}^{\alpha_{upbound}} \int_{\infty}^{\infty} p(\tau_1 - \tau_0 - \psi) f_T(\psi, s) \, d\psi f_\alpha(s) \, ds$$

$$= \int_{\alpha_{lowbound}}^{\alpha_{upbound}} E(\bar{M}_i(\tau_1, s)) f_\alpha(s) \, ds$$

The results of Lemma 2.1 and the corresponding lemma for $\tau_1 > \tau_0$ are intuitive, since given that $p(t)$ is odd, it makes sense for $A_\infty(t)$ to have properties similar to an odd waveform. The proofs are left for the reader and can be found in ref. [36].

Knowing only the polarity of $A_\infty(t)$ is not entirely satisfying, since we would also expect that the limiting waveform be continuous. This, in fact is true, and we see it in the following lemma. Once again, the proof can be found in ref. [36].

Lemma 2.2

$$A_\infty(t) = \lim_{N\to\infty} \frac{1}{N} \sum_{i=1}^{N} CK_i p(t - \tau_0 - T_i) = \lim_{N\to\infty} \frac{1}{N} \sum_{i=1}^{N} \bar{M}_i(t) = \eta(t)$$

is a continuous function of t.

2.4   PHYSICAL-LAYER TIME SYNCHRONIZATION

2.4.1   A Synchronization Protocol

We consider a network of N nodes, uniformly distributed over the $[0, 1] \times [0, 1]$ plane. We describe the mechanism for synchronizing this network to the clock of node 1, which is assumed to be at the center of the network. In Section 2.4.3 we explain why this mechanism is asymptotically optimal and good for the synchronization of dense networks.

Synchronization will be achieved in the following manner. Node 1 will start transmitting pulses and continue to transmit pulses every time the counter $s_1(t)$ increments. After the initial m pulses, the set of nodes in the broadcast domain of 1, not including node 1, will make an optimal estimate of the location of the $(m + 1)$th pulse and transmit at that time. We will call the set of nodes in the broadcast domain of node 1 R_2. The nodes in R_2 will then use their most recent m observations to optimally estimate the time of pulse $m + 2$. The R_2 nodes will continue in this manner. The nodes that can hear the aggregate transmissions from R_2 and node 1, the R_3 nodes, will begin their own predictions and transmissions after observing m pulses. This propagation will then continue until all nodes in the network hear signals. Figure 2.2 illustrates this propagation.

Node 1 will initially transmit with magnitude $A_{\max}$. Once the R_2 nodes begin transmitting, node 1 will scale its transmissions along with the other nodes. The R_2 nodes and node 1 will each transmit with magnitude $(A_{\max}\kappa/\eta_1)$, where η_1 is the number of nodes in $\cup_{i=1}^{2} R_i$, where R_1 is node 1, and κ is a constant that ensures that all nodes in the network will be synchronized after a finite number of hops out from node 1. This trend will continue so that the nodes in $\cup_{i=1}^{k} R_i$ will transmit with magnitude $(A_{\max}\kappa)/\eta_{k-1}$, where η_{k-1} is the number of nodes transmitting. Once all nodes in the network are transmitting, the nodes will be transmitting with magnitude $(A_{\max}\kappa)/N$. Note that each node's knowledge of η_k will be gained from information that is exchanged. The information that needs to be distributed is detailed in ref. [37].

The preceding mechanism is designed for asymptotically dense networks. In applying it to finite-sized ($N < \infty$) networks, we introduce a small amount of feedback into the system to prevent small errors from accumulating. Node 1 is the only node in the network that can observe the aggregate waveform and have access to the ideal clock. We define a tolerance factor, ϱ, such that if node 1's observed zero-crossing is more than ϱ from the ideal zero-crossing, then it informs all nodes in the network to adjust their estimate. Tolerance factor ϱ is defined as

$$\varrho = \frac{\text{Maximum allowed distance between ideal and observed zero-crossing}}{\text{Time between synchronization pulses}}$$

where all times are defined in terms of c_1. It is clear that ϱ is defined in the design of the system so each node knows its value.

When node 1 notices that ϱ has been exceeded, it sends a 1-bit feedback to all nodes. That bit will tell nodes whether the observed zero-crossing occurred before

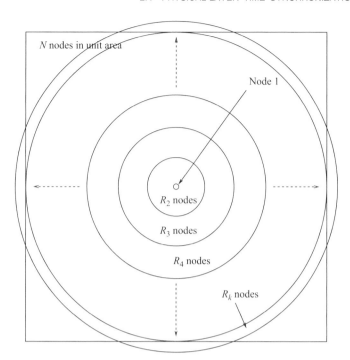

Figure 2.2 This figure illustrates the propagation of the synchronization pulses starting from node 1 at the center of a unit area square with N nodes uniformly distributed over the area. The R_2 nodes hear the pulses from node 1, and the R_3 nodes hear the aggregate signal from node 1 and the nodes in R_2. This propagation continues beyond the R_k nodes until all nodes in the finite area can hear synchronization pulses.

or after the ideal zero-crossing location. If the observed zero-crossing occurred before the ideal, then each node will delay all m of its observations by an adjustment factor. If the observed zero-crossing occurred after the ideal, each node will shift all of its observations back in time by an adjustment factor. This means that if, for example, the observed zero-crossing occurred early, then by having each node delay its set of m observations, the next estimate made by each node will occur later. Since all nodes are making a later estimate, the next aggregate waveform zero-crossing should be delayed as well, bringing it closer to the ideal zero-crossing time. For each node i the adjustment factor is calculated as

Node i adjustment factor $=$

$\varrho \times$ |Difference between most recent two observation times|

Note that these calculations are all done by node i in terms of its own timescale.

It is important to stress two things. First, this added feedback does not in any way affect the asymptotic optimality of the synchronization mechanism. In an asymptotically dense network, the extra feedback and correction mechanism will simply not be needed. It is added only to make the asymptotically optimal synchronization mechanism robust for networks of finite size. Second, the reliance on node 1 to initiate the feedback does not make the synchronization mechanism less robust because node 1 is arbitrary. If this node 1 fails, then the synchronization mechanism can simply be reinitiated using another node as node 1. Thus, the failure of node 1 does not prevent the network from synchronizing.

2.4.2 Optimality Conditions

The reason we propose the synchronization mechanism outlined in Section 2.4.1 is that it does well in an asymptotically dense network. In fact, it is optimal in a manner described in this section. The problem of synchronization is the challenge of having the ith node accurately and precisely predict when node 1 increments its operational counter. In our setup, the reception of a pulse by node i tells it of such an event. Recalling that $\Psi_i(s_{n',1}^{c_i}) \sim \mathcal{N}(0, \sigma^2)$, from equation (2.3) we see that the pulse receive time at node i, $t_{n,i}^{c_i}$, is a Gaussian random variable whose mean is parameterized by the unknown vector $\vartheta = [\alpha_i, s_{n',1}^{c_1}, \bar{\Delta}_i]$. Thus, to achieve synchronization node i will try to estimate the random variable $t_{n,i}^{c_i}$ using a series of m pulse receive times as observations (recall that m is known). Note the observations are also random variables with distributions parameterized by ϑ. We define optimal synchronization as node i making an estimate of $t_{n,i}^{c_i}$, denoted $\hat{t}_{n,i}^{c_i}(t_{n-1,i}^{c_i}, t_{n-2,i}^{c_i}, \ldots, t_{n-m,i}^{c_i})$, which is a function of past observations $t_{n-1,i}^{c_i}, t_{n-2,i}^{c_i}, \ldots, t_{n-m,i}^{c_i}$, that meets the following optimality criteria:

$$E_\vartheta(\hat{t}_{n,i}^{c_i}(t_{n-1,i}^{c_i}, t_{n-2,i}^{c_i}, \ldots, t_{n-m,i}^{c_i})) = E_\vartheta(t_{n,i}^{c_i}) \qquad (2.12)$$

$$\operatorname{argmin}_{\hat{t}_{n,i}^{c_i}} E_\vartheta(\|\hat{t}_{n,i}^{c_i}(t_{n-1,i}^{c_i}, t_{n-2,i}^{c_i}, \ldots, t_{n-m,i}^{c_i}) - t_{n,i}^{c_i}\|^2) \qquad (2.13)$$

for all ϑ. The subscript ϑ means that the expectation is taken over the distributions involved given any possible ϑ. The first optimality condition comes from the fact that given a finite m, it is reasonable to want the expected value of the estimate to be the expected value of the random variable being estimated for all ϑ. As in the justification for unbiased estimators, this condition eliminates unreasonable estimators so that the chosen estimator will perform well, on average, for all values of ϑ. The second condition is the result of seeking to minimize the mean-squared error between the estimate and the random variable being estimated for all ϑ.

However, for optimal synchronization it seems reasonable to require that the mean-squared error of a particular node placed in the broadcast domain of node 1 be same as when that node was placed far away from node 1. Thus, we go beyond the preceding conditions and define an *optimally synchronized network* as one where *all* nodes in the network can achieve the optimality conditions of equations (2.12) and (2.13) *and* the mean-squared error achieved in equation (2.13) for each node is the smallest possible mean-squared error achievable for that node over the area of the network. Thus, the optimality condition for an optimally synchronized network is

$$E_\vartheta(\|\hat{t}_{n',j}^{c_j}(t_{n-1,j}^{c_j}, t_{n-2,j}^{c_j}, \ldots, t_{n-m,j}^{c_j}) - t_{n',j}^{c_j}\|^2)$$
$$= \min_A E_\vartheta(\|\hat{t}_{n',j}^{c_j}(t_{n-1,j}^{c_j}, t_{n-2,j}^{c_j}, \ldots, t_{n-m,j}^{c_j}) - t_{n',j}^{c_j}\|^2) \quad \text{for all } j, \vartheta \qquad (2.14)$$

This means that in order for the network with node j to be optimally synchronized at a given time when $s_1(t)$ increments node j must have its minimum possible mean-squared error over the area of the entire network (A). Thus, we see that an optimally synchronized network is defined only for a given synchronization mechanism. Once a mechanism is determined, the mechanism will achieve optimal synchronization if each node in the network is synchronized with the smallest possible minimum mean-squared error the mechanism can achieve for that node placed anywhere in the network.

2.4.3 One-Hop Synchronization and Multihop Synchronization

Optimal one-hop synchronization can be achieved by designing an optimal estimator (optimal in the sense of equations (2.12) and (2.13)) for estimating the next pulse arrival given m arrivals. This is because the one-hop nodes are within the broadcast domain of node 1 and are able to observe the exact time node 1 increments its operational counter and thus make the best estimate.

We show in ref. [38] that for any m consecutive observations, we have the linear model

$$\mathbf{Y} = \mathbf{H}\theta + \mathbf{W} \qquad (2.15)$$

where

$$\mathbf{Y} = [y_1 \quad y_2 \quad \cdots \quad y_m]^T = [t_{n-m,i}^{c_i} \quad t_{n-m+1,i}^{c_i} \quad \cdots \quad t_{n-1,i}^{c_i}]^T$$

$$\theta = \begin{bmatrix} \theta_1 \\ \theta_2 \end{bmatrix} = \begin{bmatrix} \alpha_i(s_{n'',1}^{c_1} - \bar{\Delta}_i) \\ \alpha_i \end{bmatrix}$$

$$\mathbf{H} = \begin{bmatrix} 1 & 1 & 1 & \cdots & 1 \\ 0 & 1 & 2 & \cdots & m-1 \end{bmatrix}^T$$

with $\mathbf{W} = [w_1 \cdots w_m]^T$. $s_{n'',1}^{c_1}$ is some unknown integer. Based on our assumption that $\Psi_i(t)$ is a Gaussian noise process with independent samples, $\mathbf{W} \sim \mathcal{N}(0, \Sigma)$ with $\Sigma = \sigma^2 \mathbf{I}$. What equation (2.15) fundamentally models is the fact that the vector of pulse reception times of node i, given by $[y_1 \; y_2 \; \cdots \; y_m]^T$, will have a mean that grows linearly.

With the observation model equation (2.15), we want to estimate the next pulse arrival time y_{m+1}, which is jointly distributed with $\mathbf{Y}$ as

$$\begin{bmatrix} \mathbf{Y} \\ y_{m+1} \end{bmatrix} \sim \mathcal{N}\left(\begin{bmatrix} \mathbf{M} \\ \theta_1 + m\theta_2 \end{bmatrix}, \begin{bmatrix} \Sigma & 0 \\ 0 & \sigma^2 \end{bmatrix} \right)$$

The optimal estimator for y_{m+1} will be a uniformly minimum variance unbiased (UMVU) estimator of $\theta_1 + m\theta_2$. As shown in ref. [38], this can be found as $\hat{\phi}_{ML} = \hat{\theta}_{1ML} + m\hat{\theta}_{2ML}$, where

$$\hat{\theta}_{ML} = (\mathbf{H}^T\mathbf{\Sigma}^{-1}\mathbf{H})^{-1}\mathbf{H}^T\mathbf{\Sigma}^{-1}\mathbf{Y} = (\mathbf{H}^T\mathbf{H})^{-1}\mathbf{H}^T Y$$

We also show that $\hat{\phi}_{ML}$ is Gaussian with $E_\theta(\hat{\phi}_{ML}) = \theta_1 + m\theta_2$ and

$$\text{Var}_\theta(\hat{\phi}_{ML}) = \mathbf{C}\sigma^2(\mathbf{H}^T\mathbf{H})^{-1}\mathbf{C}^T = \frac{2\sigma^2(2m+1)}{m(m-1)}$$

Please also note that the preceding optimal estimation is carried out by node i according to c_i. Intuitively, it is most important to realize that the mean of $\hat{\phi}_{ML}$ translated to the timescale of c_1 is an integer value. This means that $E(\hat{\phi}_{ML})$ is when $s_1(t)$ increments.

To see how $\hat{\phi}_{ML}$ is related to T_i from Theorem 2.1, start by seeing that

$$\hat{\phi}_{ML} \sim \mathcal{N}\left(\alpha_i(s_{n'',1}^{c_1} - \bar{\Delta}_i) + m\alpha_i, \frac{2\sigma^2(2m+1)}{m(m-1)}\right)$$

Using equation (2.1), we can translate $\hat{\phi}_{ML}$ into the time scale of c_1 as

$$\hat{\phi}_{ML}^{c_1} = \frac{(\hat{\phi}_{ML} - \Psi_i)}{\alpha_i} + \bar{\Delta}_i$$

This means that

$$\hat{\phi}_{ML}^{c_1} \sim \mathcal{N}\left(s_{n'',1}^{c_1} + m, \frac{\sigma^2}{\alpha_i^2}\left(1 + \frac{2(2m+1)}{m(m-1)}\right)\right)$$

Since $s_{n'',1}^{c_1} + m$ is the ideal crossing time in the timescale of c_1, it is τ_0. Thus,

$$\hat{\phi}_{ML}^{c_1} = \tau_0 + T_i$$

Therefore, we see that

$$\text{Var}(T_i) = \frac{\sigma^2}{\alpha_i^2}\left(1 + \frac{2(2m+1)}{m(m-1)}\right) = \frac{\bar{\sigma}^2}{\alpha_i^2} \tag{2.16}$$

where $\bar{\sigma}^2$ from Theorem 2.1 is

$$\bar{\sigma}^2 = \sigma^2\left(1 + \frac{2(2m+1)}{m(m-1)}\right)$$

For multihop synchronization, we first note that an optimally synchronized network would be possible if every node in the network, no matter its distance from node 1, could somehow hear the synchronization pulses emitted by node 1. This means that our goal for multihop synchronization would be to somehow allow the nodes outside the broadcast domain of node 1 to observe node 1's synchronization pulses. We find that this is possible, as $N \to \infty$, by considering the zero-crossing of the aggregate waveform generated by all nodes in the network.

Recall from Section 2.2.5 that the aggregate waveform observed by any node j is

$$A_{j,\eta_k}^{c_1}(t) = \sum_{i=1}^{\eta_k} \frac{\kappa A_{\max} K_{j,i}}{\eta_k} p(t - \tau_o - T_i)$$

where η_k was defined in Section 2.4.1. Note that the variance of T_i in the timescale of c_1 is in the form required by Theorem 2.1. Also, the variance is upper bounded by some constant, since α_i is lower bounded by $\alpha_{lowbound}$. Thus, the properties of $A_{j,\infty}^{c_1}(t)$ are characterized by Theorem 2.1. Figure 2.3 illustrates the properties.

The result of Theorem 2.1 has significant implications for synchronization. First note that since $N \to \infty$, $A_{j,\eta_k}^{c_1}(\tau_o) \to 0$, node j sees a zero crossing that occurs at an integer value in the time scale of c_1 (as $N \to \infty$, we have $\eta_k \to \infty$). Now following our synchronization mechanism outlined in Section 2.4.1, we know that when the R_2 nodes start transmitting synchronization pulses, the pulse transmission time $(\tau_o + T_i)$ for any node i in R_2 will satisfy the requirements of Theorem 2.1. This is because node i is in the broadcast domain of node 1, and from earlier in this section we know that its optimal estimate of the next pulse arrival time is a finite-mean Gaussian random variable. Furthermore, the mean is the exact time node 1 increments its operational counter. Thus, we can apply Theorem 2.1 to the transmissions of the R_2 nodes and any node l in $R_2 \cup R_3$ will see a received signal $A_{l,\eta_2}^{c_1}(\tau_1) = 0$ for $N \to \infty$, where τ_1 is the time when node 1 next increments $s_1(t)$. Since a node i in R_3 can effectively see the exact time the pulse from node 1 makes a zero-crossing, its

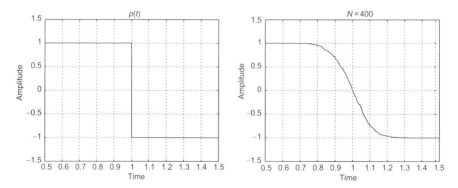

Figure 2.3 The pulse $p(t)$ is shown on the left figure, with $\tau_0 = 1$ and $A_{\max} = 1$. On the right we have a realization of $A_N(t)$ ($N = 400$), and we assume that $K_{j,i} = 1$ (no path loss) and $T_i \sim \mathcal{N}(0, 0.01)$ for all i. As expected from Theorem 2.1, we notice that the zero-crossing of the simulated waveform is almost exactly at $t = 1$.

estimate $\hat{t}_{n,i}^{c_i}$ will have a minimum mean-squared error equal to if it were in the broadcast domain of node 1. This is the minimum mean-squared error for all nodes in the network that is achievable by the synchronization mechanism. This is why the only optimal estimator needed was the estimator outlined earlier in this section for nodes in the broadcast domain of node 1. Now because the estimate made by node i in R_3 is the same as if it would have made in the broadcast domain of node 1, the pulse transmission time for node i again satisfies the requirements of Theorem 2.1. Hence, again Theorem 2.1 can be applied. This cycle will then repeat until all nodes are synchronized, and then the cycle will continue to keep the nodes synchronized. It is important to note that with this dense network, the network is *optimally synchronized* at each step of the synchronization process since every node has access to the transition times of node 1.

2.5   TIME SYNCHRONIZATION SIMULATION

2.5.1   Simulator Implementation

As mentioned, we study the synchronization problem in the asymptotically dense regime, since it closely approximates the behavior of networks with large, but still finite, densities. As a result, an obvious question is how well the limiting regime actually approximates finite-density networks. In an effort to answer this question, we implement a simulator to study the performance of our synchronization mechanism on finite-density networks. This is a key step toward building a massively distributed software radio. Note that the simulation results are presented in time units that are unspecified. The reason for this is that any time units can be used and the results will still hold. What is most important to note in the simulations is that even with a finite number of nodes, the synchronization error closely approximates the limiting results presented in Section 2.4.3 thus showing that the limit regime gives us analytical results that closely model finite-sized networks.

The time synchronization simulator is implemented in MATLAB, and the N nodes are uniformly distributed over a circle with area 30. The node parameters are independently and randomly generated using $\bar{\Delta}_i \sim \mathcal{N}(0,0.1)$ and $\alpha_i = |X|$, where $X \sim \mathcal{N}(1,0.01)$. The jitter variance is set to be $\mathrm{Var}(\Psi_i) = 0.01$ for all i. In the generation of the aggregate waveform we use the following parameters:

$$\tau_{n,z} = 0.2 \qquad A_{max} = 1 \qquad \kappa = 8 \qquad K_{j,i} = 1$$

In determining the transmission range of the aggregate waveform, we assume $K(d)$ to be

$$K(d) = \begin{cases} 1 & d < \epsilon \\ \sqrt{\dfrac{\epsilon^\beta}{d^\beta}} & d \geq \epsilon \end{cases} \tag{2.17}$$

Recall that $K(d)$ models the signal amplitude loss, thus it is clear that equation (2.17) is derived from the standard path-loss model where signal power decays as $1/d^\beta$

for $2 \leq \beta \leq 4$. For simulations we use

$$\beta = 2 \qquad \epsilon = 0.1 \qquad \gamma = 0.2$$

Last, for simulations we set the tolerance factor to be $\varrho = 0.05$. The details of the simulator implementation can be found in ref. [37].

2.5.2 Simulation Results

Before presenting the results of the simulations, we first describe how we measure the performance of the synchronization mechanism. Recall that ideally we would want all nodes to transmit a synchronization pulse at the exact same time. This means that in the ideal situation, when we translate each node i's estimate of the next zero-crossing location into the timescale of node 1, it should be the next integer value of t. In reality, this is not the case and we use a measure, which we call the *average squared distance* (*ASD*), to quantify the average distance of the nodes' estimates from the ideal integer time of c_1. The ASD is calculated as follows:

$$ASD = \frac{1}{\bar{N}} \sum_{i=1}^{\bar{N}} (\hat{t}_i^{c_1} - t_0)^2$$

where $\bar{N} \leq N$ is the number of nodes currently making estimates, t_0 is the integer value of t where node 1 will next increment its operational counter, and $\hat{t}_i^{c_1}$ is the ith estimating node's estimate of t_0 in the timescale of c_1.

The first simulation result that we present in Figure 2.4 serves as motivation for the modified synchronization mechanism that includes feedback. We see in the first

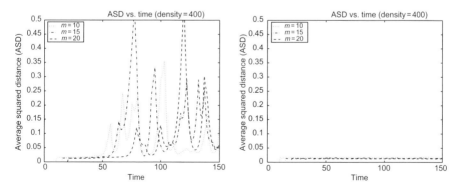

Figure 2.4 Left: A plot of ASD versus time for the synchronization mechanism without feedback. The results were averaged over 10 runs. We see that synchronization is held for a period of time, but not indefinitely. Right: A plot of ASD versus time for the synchronization mechanism with feedback. We note that ASD is bounded and synchronization can be maintain indefinitely.

panel of Figure 2.4 that for $m = 10, 15, 20$, synchronization is maintained over a period of time. In fact, for $m = 20$, synchronization is maintained for over 70 time units. On average, as seen in the first panel of Figure 2.4, the larger the m value, the longer synchronization can be maintained. However, in all cases synchronization is eventually lost. This is due to the fact that small errors in the aggregate waveform zero-crossing location accumulate. For example, if an observed zero-crossing arrives late, then the next aggregate waveform zero-crossing may arrive late as well, since all nodes are making an estimate using the delayed zero-crossing. Thus, these errors accumulate and eventually the aggregate waveform zero-crossing might be delayed so much that the nodes can no longer observe the zero-crossing. We also note that the length of time synchronization can be maintained may vary a great deal and is difficult to predict from run to run. As a result, by introducing feedback we can correct this drifting zero-crossing. An illustration of ASD versus time for the mechanism with feedback is presented in the right-hand panel of Figure 2.4. There we run the simulation once and notice that in all cases the ASD is bounded and synchronization is maintained indefinitely. Figure 2.5 is a close-up of the right-hand panel of Figure 2.4, showing the "sawtooth" waveform for $m = 10$ and $m = 15$. Each "tooth" coincides with one time that the feedback triggered by node 1 adjusted each node's observations. In fact,

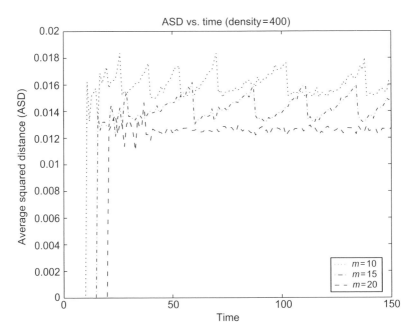

Figure 2.5 A plot of ASD versus time for the synchronization mechanism with feedback. We see a "sawtooth" waveform for $m = 10, 15$, and each "tooth" occurs at a time where a correction was made.

the simulator tells us that for $m = 10$, there were six corrections, $m = 15$ had four, and $m = 20$ so far did not require any corrections to the node observations.

Another key property of the synchronization mechanism with feedback is that it performs well for a wide range of network sizes. In Figure 2.6 we plot the ASD versus time for network sizes varying from $N = 300$ (density $= 10$, area $= 30$) to $N = 18,300$ (density $= 610$, area $= 30$). In steady state all nodes are transmitting, and we notice that the ASD curve for the 300-node network is at most 0.0005 greater than the ASD curve for the 18,300-node network. This means that on average the ASD varies only by at most 0.0005 for network sizes in this range, and thus the mechanism is well suited for network sizes as small as a few hundred nodes. Of course, as expected, the mechanism must make more active corrections based on feedback from the network. In fact, we find that the average number of corrections made for the 150 time units of the simulation was 18.2 for $N = 300$ and 2.9 for $N = 18,300$. As a result, even though the mechanism performs well for networks of only a few hundred nodes, it does require more active adjustments on the part of the mechanism. Such a result is in line with our comment at the end of Section 2.4.1 since in the limit as $N \rightarrow \infty$ the feedback and correction mechanism will not be needed.

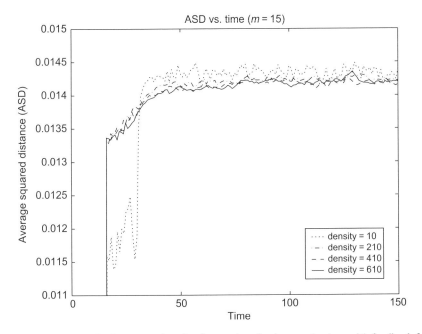

Figure 2.6 A plot of ASD versus time for the synchronization mechanism with feedback for different network sizes. Each plot was averaged over 500 runs. In steady state we see that the mechanism performs well for a wide range of network sizes since the difference in ASD for a network of $N = 300$ nodes and a network of 18,300 nodes is at most 0.0005.

Before concluding this simulation section, we connect the simulation results to the analytical results presented in Section 2.4.3. From equation (2.16), we see that variance of the time estimate in terms of c_1 should be the jitter variance σ^2 multiplied by a function of m and α_i. Since the ASD approximates this synchronization error averaged over all nodes, we would expect the simulation results to closely approximate the analytical value. Using $m = 10$, $\alpha_i = 1$, and using the simulation value $\sigma^2 = \text{Var}(\Psi_i) = 0.01$, equation (2.16) gives us a value of about 0.01467. From the right panel of Figure 2.4 we see that the ASD for $m = 10$ is consistently less than 0.02. Thus, even for a finite number of nodes, the synchronization mechanism gives us a synchronization error that closely approximates the limiting analytical results.

Last, to better understand the simulation results in terms of some realistic numbers, we consider a system that uses a 1-MHz oscillator and sends a synchronization pulse every microsecond. Using an oscillator from the SPK-SPG series of oscillators, we have $f_0 = 1 \times 10^6$ Hz and $\Delta f = \pm 100$ Hz. Using a figure from ref. [35] we take the standard deviation of $f_r(t)$ to be 0.001 Hz. For our clock model this translates into $\alpha_i \in (1 - 100 \times 10^{-6}, 1 + 100 \times 10^{-6})$, $\sigma^2 = 1 \times 10^{-18}$, and $c_i(t)$ is counting in microseconds. This means that our ASD value will be on the order of 1×10^{-18} μs, which translates into a timing jitter standard deviation of about 1 ps.

2.6 DISTRIBUTED FREQUENCY SHIFT KEYING

The time synchronization mechanism described in Section 2.4 forms the core on top of which distributed frequency shift keying (dFSK) is built. In this section, we first show how an aggregate waveform suitable now for both synchronization *and* communication is generated, and then show how bits are modulated onto this new waveform.

2.6.1 Waveform Generation

We observe that synchronization is achieved and maintained based solely on every node i's ability to observe a zero-crossing that occurs at the exact time $s_1(t)$ increments. It is possible to retain this property while generating an aggregate waveform that is suitable for reachback communication.

Consider a network that has already been synchronized and is simply maintaining synchronization. If instead of cooperatively generating a pulse train with zero-crossings at integer values of t, the nodes generate an aggregate waveform similar to that illustrated in Figure 2.7, time synchronization can still be maintained since the waveform in Figure 2.7 has zero-crossings at integer values. Clearly, this waveform can be easily generated, since each node can simply make a step in its transmission waveform at the time it normally would have sent a pulse. Thus, this waveform can be used for time synchronization and is generated simply by modifying the type of waveform transmitted by each node.

From Figure 2.7, we see an interval between two transitions where the signal is effectively flat with a magnitude of $E(K_{i,n})A_{\max}$. Let us call the transitions that

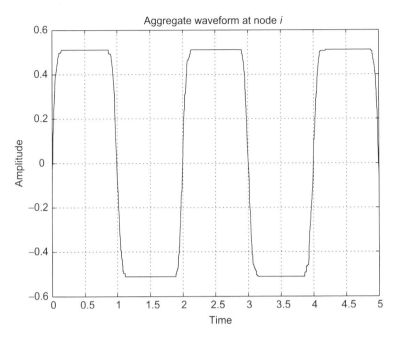

Figure 2.7   This figure illustrates the aggregate waveform with $N = 200$ nodes seen at node i. In this example $A_{\max} = 1$, and we assume that the jitter variance of every node is the same in the timescale of c_1 with standard deviation 0.05. For illustration purposes, we take $K_{i,n}$ to be exponentially distributed with $\lambda = 2$. Notice that even with only $N = 200$ nodes the zero-crossings occur almost exactly in the correct place.

occur at the times where synchronization pulses would have occurred the *primary transitions*. This flat interval between two primary transitions comes from the assumption that the time between two synchronization pulses is long relative to the synchronization pulse duration. From the figure we see that by putting extra transitions between the primary transitions, it is possible to modulate information onto the aggregate waveform. This idea is illustrated in Figure 2.8. For use with the synchronization technique, we would like the waveform to be smooth at symbol boundaries. As a result, we choose to have the waveform always transition from negative to positive at the primary transitions. Because of this requirement, we must have the number of zero-crossings, R, between $t = \tau_0$ and $t = \tau_0 + 1$ take on the form $R = 2q + 1$, where q is a nonnegative integer. Here τ_0 is any integer value of t. Note that the time it takes the aggregate waveform to make a primary transition will limit the maximum value of R. The details of this relationship can be found in ref. [39].

2.6.2   Modulation Scheme

To study exactly how this modulation scheme would work, we focus on one interval between $t = \tau_o$ and $t = \tau_o + 1$. We consider the case where this cooperative

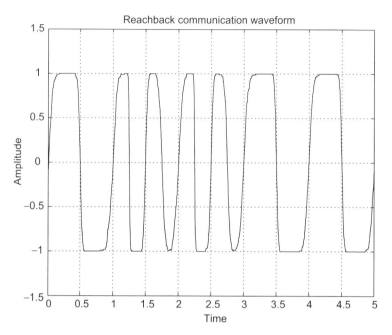

Figure 2.8 This figure illustrates 5 symbol periods, with $R = 1, 3, 3, 1, 1$, respectively. For example, such a waveform could be used to send the bit stream 01100.

reachback communication system will do M-ary signaling. This means that $q \in \{0, 1, \ldots, M - 1\}$ and $R = 2q + 1 \in \{1, 3, \ldots, 2M - 1\}$. Symbol S_{q_o} is a symbol waveform with $2q_o + 1$ zero-crossings between $t = \tau_o$ and $t = \tau_o + 1$. Figure 2.8 illustrates an example of a waveform modulated in this manner.

An important point to note is that each node i looks for a zero-crossing only in a small interval around its estimate of where the primary zero-crossing should be. As a result, when other zero-crossings are placed between $t = \tau_o + k$ and $t = \tau_o + k + 1$, node i still only observes the zero-crossings at $t = \tau_o + k$ and $t = \tau_o + k + 1$. Thus, the same synchronization properties are still maintained using the zero-crossings at $t = \tau_o + k$ for $\tau_o, k \in Z$, while the other zero-crossings are used for communicating with the far receiver.

2.7 dFSK RECEIVER SIMULATION

An interesting question regarding the design of a receiver for the dFSK waveforms described in Section 2.6 is how the number of nodes, N, in the network affects the performance. The dFSK waveforms were generated by infinitely dense networks, and hence the waveforms were limiting waveforms. How would the probability of bit error change if we instead had a finite number of nodes generating a waveform that is only a crude approximation of the limiting waveform? We look for the answer by designing an optimal receiver for the limiting waveform and then using this receiver to decode signals generated by networks of finite size.

We consider binary signaling using two waveforms $s_0(t)$ and $s_1(t)$. As outlined in Section 2.6.2, $s_0(t)$ has one extra zero-crossing and $s_1(t)$ has three. We will design an optimal receiver for detecting $s(t) \in \{s_0(t), s_1(t)\}$ when the received signal is

$$r(t) = s(t) + N(t)$$

where $N(t)$ is zero-mean white Gaussian noise with power spectral density $N_o/2$. We assume that $s_0(t)$ and $s_1(t)$ are sent with equal probability.

Two sets of waveforms $s_0(t)$ and $s_1(t)$ will be used for the simulations. The first set of limiting waveforms shown in Figure 2.9 (a and b) are generated using

$$N = 4000 \qquad K_i = 1 \qquad A_{\max} = 1 \qquad T_i \sim \mathcal{N}(0, 0.0025)$$

for all i, while the second set in Figure 2.9 (c and d) is generated using

$$N = 4000 \qquad K_i = 1 \qquad A_{\max} = 1 \qquad T_i \sim \mathcal{N}(0, 0.0225)$$

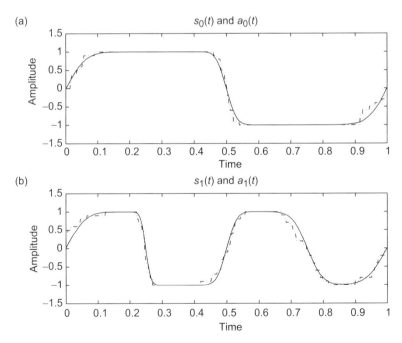

Figure 2.9 (a) The limit waveform $s_0(t)$ and its approximation $a_0(t)$ generated by $N = 20$ nodes using $T_i \sim \mathcal{N}(0, 0.0025)$. (b) The limit waveform $s_1(t)$ and its approximation $a_1(t)$ generated by $N = 20$ nodes using $T_i \sim \mathcal{N}(0, 0.0025)$. (c) The limit waveform $s_0(t)$ and its approximation $a_0(t)$ generated by $N = 20$ nodes using $T_i \sim \mathcal{N}(0, 0.0225)$. (d) The limit waveform $s_1(t)$ and its approximation $a_1(t)$ generated by $N = 20$ nodes using $T_i \sim \mathcal{N}(0, 0.0225)$.

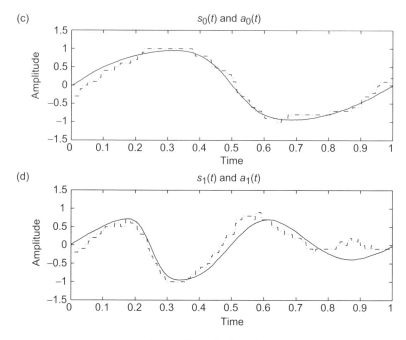

Figure 2.9   *Continued.*

The difference between these two sets of waveforms is the variance of the jitter T_i. We would like to determine if jitter variance has an impact on the performance of the reciever. Note that for simulation purposes we assume that the limit waveforms are generated by $N = 4000$ nodes, since these waveforms are indeed close to the limit waveform under the simulation parameters. In Figure 2.9 we also show the waveforms $a_0(t)$ and $a_1(t)$, which are approximations of $s_0(t)$ and $s_1(t)$, respectively, generated by much smaller networks. We will analyze the performance of the optimal detector when the signals generated by the network are not $s_i(t)$, but instead are $a_i(t)$, for $i = 0, 1$.

We first follow standard communication theory to design the optimal detector for two known signals $s_0(t)$ and $s_1(t)$ corrupted by additive white Gaussian noise. A detailed reference to this approach can be found in ref. [40]. We find that the optimal detector is illustrated in Figure 2.10, where that orthonormal basis functions are $\phi_0(t)$ and $\phi_1(t)$, and $\bar{s}_0$ and $\bar{s}_1$ are the signal space representations for $s_0(t)$ and $s_1(t)$, respectively.

The probability of bit error can be calculated as

$$P_e = Q\left(\frac{\Delta}{2\sqrt{N_0/2}}\right) = Q\left(\sqrt{\frac{\Delta^2}{2N_0}}\right)$$

where $\Delta = \|\bar{s}_0 - \bar{s}_1\|$ and Q is the standard Q-function. Recall that this is assuming $s(t) \in \{s_0(t), s_1(t)\}$, so the actual limit waveform is sent and that each waveform is

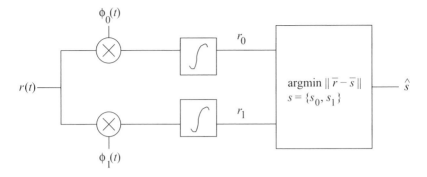

Figure 2.10   A diagram for the optimal detector, where $r(t)$ is received and projected onto the two orthonormal basis functions to get the signal space representation $\bar{r} = [r_0 \quad r_1]$. We then decode to 0 or 1, depending if $\bar{r}$ is closer to $\bar{s}_0$ or $\bar{s}_1$, respectively.

sent with probability 0.5. We plot this probability of error as a function of E_b/N_0 in Figure 2.11 for the two different sets of $s_0(t)$ and $s_1(t)$ where

$$E_b = \frac{\int s_0^2(t)\,dt + \int s_1^2(t)\,dt}{2}$$

is the average energy per bit. That is, we plot

$$P_e = Q\left(\sqrt{K \frac{E_b}{N_0}}\right) \tag{2.18}$$

where $K = \Delta^2/(2E_b)$ by varying N_0.

   The remaining curves in Figure 2.11 are simulations of the probability of error when we send $a(t) \in \{a_0(t), a_1(t)\}$ instead of $s(t) \in \{s_0(t), s_1(t)\}$. We generate $a(t)$ with networks of varying size, N. We use the same detector illustrated in Figure 2.10, but this time $r(t) = a(t) + N(t)$, where $a(t) \in \{a_0(t), a_1(t)\}$. Recall that $a_i(t)$ is an approximation of $s_i(t)$, where $i = 0, 1$, that is generated using finite N. In the simulations of Figure 2.11 we use $N = 10, 40, 70, 100$. In the left panel we use $T_i \sim \mathcal{N}(0, 0.0025)$, and it is clear that using smaller N does not significantly impact the bit error rate of the detector. In fact, the P_e versus E_b/N_0 plots for $N = 10$, 40, 70, and 100 are almost completely on top of each other and they coincide with the theoretical bit-error-rate plot generated from equation (2.18). What this means is that for the coherent detection of dFSK signals with small jitter variance, N does not have a significant impact on P_e and signals generated from very small networks can be detected using the detector shown in Figure 2.10. However, in the right figure of Figure 2.11 we use $T_i \sim \mathcal{N}(0, 0.0225)$ and notice that for $N = 10$ the P_e is higher than the theoretical rate. We see that as N increases, the P_e approaches the theoretical rate and at $N = 100$ the two are very close. Thus, for larger jitter variance, the value of N has a more significant impact on the probability of bit error. This is expected since with larger jitter variance, $a_i(t)$ will be more corrupted for a given

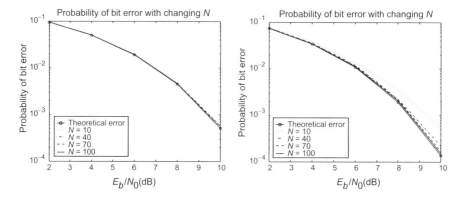

Figure 2.11 We illustrate the theoretical bit-error-rate curve generated by equation (2.18). We also plot the P_e versus E_b/N_0 curves for networks of size $N = 10, 40, 70, 100$. Left: We note that with $T_i \sim \mathcal{N}(0, 0.0025)$ all four curves nearly fall right on top of each other, showing that N does not significantly affect P_e. These curves were generated from 500,000 runs. Right: Here we plot using $T_i \sim \mathcal{N}(0, 0.0225)$, and we notice that for small N, the probability of error is noticeably worse than the theoretical error curve. As we increase N, the P_e curves approach the theoretical values. These curves were generated from 800,000 runs.

N. This can be seen in Figure 2.9. Note, however, that increased jitter variance does not always negatively impact the system. From Figure 2.11 we also notice that for $T_i \sim \mathcal{N}(0, 0.0225)$ the theoretical P_e curve actually gives a lower probability of bit error for a given signal-to-noise ratio E_b/N_0.

2.8 CONCLUSION

In this chapter we have presented the framework for a cooperative reachback system. A synchronization mechanism designed specifically for large-scale dense sensor networks was presented first. The mechanism generates an aggregate waveform with equispaced zero-crossings that can be observed by all nodes in the network, and the zero-crossings are used as synchronization events. The choice of using zero-crossings for synchronization was motivated by the fact that low-power zero-crossing detector circuits can be constructed. The performance of this method of synchronization scales well with the number of nodes in the network. A key feature is that it eliminates the error accumulation that occurs with most traditional synchronization methods that route timing information throughout the network. Simulation results show that through a minor modification to the synchronization mechanism developed for asymptotically dense networks, the synchronization ideas can be effectively applied to networks of finite size.

Using the waveform properties that were studied, we were able to modify the pulse shapes emitted by the nodes in the network to create a waveform suitable for reachback communication. The waveform encodes data in the frequency of the zero-crossings, and hence we call the method distributed frequency shift keying. The important feature of dFSK is that it perfectly complements the time synchronization method, since the network is able to do time synchronization and reachback communication simultaneously without extraneous computation. We also see that a receiver designed to detect the limiting dFSK waveforms can be effectively employed to detect waveforms generated by networks of finite size.

Note that the theory presented for the generation of waveforms for dFSK can be applied directly to time synchronization as well. That is, it is possible to start the synchronization processes by sending a continuous waveform instead of a series of pulses. This allows for passband time synchronization with the center frequency limited only by hardware performance and the magnitude of the timing jitter.

Future extensions to this work involve the consideration of nonzero signal-propagation times. There is a need to consider time delays in sending signals between nodes in the network and the different propagation delays in sending information to the far receiver. Once we have explicitly considered the issue of propagation delay, then these concepts can be implemented in our acoustic sensor network test bed. We will seek to validate the simulation results for the synchronization of a finite sized network.

ACKNOWLEDGMENTS

This work was supported by the National Science Foundation, under awards CCR-0238271 (CAREER), CCR-0330059, and ANR-0325556.

REFERENCES

1. X. Wang and H. Qi. Acoustic target classification using distributed sensor arrays. In *Proceedings of the International Conference on Acoustic, Speech, and Signal Processing (ICASSP)*, Orlando, Florida, 2002.

2. Y. Tian and H. Qi. Target detection and classification using seismic signal processing in unattended ground sensor systems. In *Proceedings of the International Conference on Acoustic, Speech, and Signal Processing (ICASSP)*, Orlando, Florida, 2002.

3. H. Wang, D. Estrin, and L. Girod. Preprocessing in a tiered sensor network for habitat monitoring. *EURASIP Journal on Applied Signal Processing*, 4:392–401, 2003.

4. A. Cerpa, J. Elson, D. Estrin, L. Girod, M. Hamilton, and J. Zhao. Habitat monitoring: Application driver for wireless communications technology. In *Proceedings of the 1st ACM SIGCOMM Workshop on Data Communications in Latin America and the Caribbean*, San Jose, Costa Rica, 2001.

5a. S. D. Servetto. Sensing Lena—Massively distributed compression of sensor images. In *Proceedings of IEEE International Conference on Image Processing (ICIP)*, Barcelona, Spain, September 2003.

5b. S. D. Servetto. Distributed signal proccessing algorithms for the sensor broadcast problem. In *Proceedings of the 37th Annual Conference on Information Sciences and Systems (CISS)*, Baltimore, MD, March 2003.

6. C. Kelly IV, V. Ekanayake, and R. Manohar. SNAP: A sensor network asynchronous processor. In *Proceedings of the 9th International Symposium on Asynchronous Circuits and Systems*, Vancouver, BC, 2003.

7. B. Warneke, M. Last, B. Liebowitz, and K. S. J. Pister. Smart dust: Communicating with a cubic-millimeter computer. *Computer (IEEE)*, **34**(1):44–51, 2001.

8. H. Li, A. Lal, J. Blanchard, and D. Henderson. Self-reciprocating radioisotope-powered cantilever. *Journal of Applied Physics*, **92**(2):1122–1127, 2002.

9. J. Elson, L. Girod, and D. Estrin. Fine-grained network time syncrhonization using reference broadcasts. In *Proceedings of the 5th Symposium Operating Systems Design and Implementation (OSDI)*, Boston, Massachusetts, 2002.

10. J. E. Wieselthier, G. D. Nguyen, and A. Ephremides. Algorithms for energy-efficient multicasting in static ad hoc wireless networks. *Mobile Networks and Applications*, **6**:251–263, 2001.

11. J. E. Wieselthier, G. D. Nguyen, and A. Ephremides. Energy-efficient broadcast and multicast trees in wireless networks. *Mobile Networks and Applications*, **7**:481–492, 2002.

12. I. Maric and R. D. Yates. Cooperative multihop broadcast for wireless networks. *IEEE Journal on Selected Areas in Communications*, **22**(7):1080–1088.

13. A. E. Khandani, J. Abounadi, E. Modiano, and L. Zheng. Cooperative routing in wireless networks. In *Proceedings of the Allerton Conference on Communications, Control and Computing*, 2003.

14. A. Srinivas and E. Modiano. Finding minimum energy disjoint paths in wireless ad-hoc networks. *ACM Wireless Networks*, forthcoming.

15. A. Sendonaris, E. Erkip, and B. Aazhang. User cooperation diversity—part I: System description. *IEEE Transactions on Communications*, **51**(11):1927–1938, 2003.

16. A. Sendonaris, E. Erkip, and B. Aazhang. User cooperation diversity—Part II: Implementation aspects and performance analysis. *IEEE Transactions on Communications*, **51**(11):1939–1948, 2003.

17. J. N. Laneman, G. W. Wornell, and D. Tse. An efficient protocol for realizing cooperative diversity in wireless networks. In *Proceedings of the IEEE International Symposium on Information Theory (ISIT)*, Washington, D.C., 2001.

18. J. N. Laneman, D. Tse, and G. W. Wornell. Cooperative diversity in wireless networks: Low-complexity protocols and outage behavior. *IEEE Transactions on Information Theory*, forthcoming.

19. M. Janani, A. Hedayat, T. Hunter, and A. Nosratinia. Coded cooperation in wireless communications: Space-time transmission and iterative decoding. *IEEE Transactions on Signal Processing*, **52**(2):362–371, 2004.

20. T. Hunter and A. Nosratinia. Coded cooperation under slow fading, fast fading, and power control. In *Proceedings of the 36th Asilomar Conference on Signals, Systems and Computers*, 2002.

21. B. Ananthasubramaniam and U. Madhow. Virtual radar imaging for sensor networks. In *Proceedings of the International Conference on Information Processing in Sensor Networks (IPSN)*, Berkeley, California, 2004.

22. L. R. Varshney and S. D. Servetto. A Distributed Transmitter for the Sensor Reachback Problem Based on Radar Signals. Paper presented at the NSF-RPI Workshop on Pervasive Computing and Networking, Troy, New York, April 2004.

23. G. Barriac, R. Mudumbai, and U. Madhow. Distributed beamforming for information transfer in sensor networks. In *Proceedings of the 3rd International Conference on Information Processing for Sensor Networks (IPSN)*, pages 81–88, Berkeley, California, 2004.

24. Y.-S. Tu and G. J. Pottie. Coherent cooperative transmission from multiple adjacent antennas to a distant stationary antenna through AWGN channels. In *Proceedings of the IEEE 55th Vehicular Technology Conference*, Birmingham, Alabama, 2002.

25. A. Scaglione and Y. Hong. Opportunistic large arrays: Cooperative transmission in wireless multihop ad hoc networks to reach far distances. *IEEE Transactions on Signal Processing*, **51**(8):2082–2092, 2003.

26. J. Elson and K. Römer. Wireless sensor networks: A new regime for time synchronization. In *Proceedings of the 1st Workshop on Hot Topics in Networks (HotNets-I)*, Princeton, New Jersey, 2002.

27. S. Ganeriwal, R. Kumar, S. Adlakha, and M. B. Srivastava. *Network-Wide Time Synchronization in Sensor Networks*. Technical Report NESL 01-01-2003, University of California, Los Angeles, 2003.

28. K. Römer. Time synchronization in ad hoc networks. In *Proceedings of the 2nd ACM International Symposium on Mobile Ad Hoc Networking and Computing*, 2001.

29. R. Karp, J. Elson, D. Estrin, and S. Shenker. *Optimal and Global Time Synchronization in Sensornets*. CENS Technical Report 0012, Center for Embedded Network Sensing, April 2003.

30. J. V. Greunen and J. Rabaey. Lightweight time synchronization for sensor networks. In *Proceedings of the 2nd ACM International Workshop on Sensor Networks and Applications (WSNA)*, pages 11–19, San Diego, California, 2003.

31. S. Ganeriwal, R. Kumar, and M. B. Srivastava. Timing-sync protocol for sensor networks. In *Proceedings of the ACM SenSys'03*, Los Angeles, California, 2003.

32. Y. Hong and A. Scaglione. A scalable synchronization protocol for large scale sensor networks and its applications. *IEEE Journal on Selected Areas in Communications*, forthcoming.

33. R. E. Mirollo and S. H. Strogatz. Synchronization of pulse-coupled biological oscillators. *SIAM Journal on Applied Mathematics*, **50**(6):1645–1662, 1990.

34. N. Roberts. Phase noise and jitter: A primer for digital designers, from http://www.eedesign.com/showArticle.jhtml?articleID=16501598, 2003.

35. *Fundamentals of Quartz Oscillators*, Application Note 200-2, Electronic Counters Series, Hewlett Packaging, 1997.

36. A. Hu and S. D. Servetto. On the Scalability of Cooperative Time Synchronization in Pulse-Connected Networks. Submitted for publication.

37. A. Hu and S. D. Servetto. Algorithmic aspects of the time synchronization problem in large-scale sensor networks. *Mobile Networks and Applications*, forthcoming.

38. A. Hu and S. D. Servetto. Asymptotically optimal time synchronization in dense sensor networks. In *Proceedings of the 2nd ACM International Workshop on Sensor Networks and Applications (WSNA)*, San Diego, California, 2003.

39. A. Hu and S. D. Servetto. dFSK: Distributed frequency shift keying modulation in dense sensor networks. In *Proceedings of the IEEE International Conference on Communication (ICC)*, Paris, France, June 2004.

40. S. G. Wilson. *Digital Modulation and Coding*. Prentice-Hall, 1996.

41. S. D. Servetto. Lattice quantization with side information: Codes, asymptotics, and applications in sensor networks. *IEEE Transactions on Information Theory*, forthcoming. (Available from http://cn.ece.cornell.edu/.)

CHAPTER 3

Energy Scavenging and Nontraditional Power Sources for Wireless Sensor Networks

SHAD ROUNDY

LV Sensors, Inc., Emeryville, California

LUC FRECHETTE

Universite de Sherbrooke, Sherbrooke, Canada

Wireless sensor networks are poised to become a very significant enabling technology in many sectors. While there has been a significant research effort in this area for a number of years, only more recently have companies begun to offer standard wireless sensor platforms and customized wireless sensor network solutions. Almost all of the available platforms are designed to run on batteries that have a very limited lifetime. However, longer lifetimes are necessary if wireless sensor networks are to become a ubiquitous part of our environment. While progress can be made by reducing the power consumption, eventually alternative power sources will need to be employed. This chapter reviews many potential power sources for wireless sensor nodes. Traditional power sources, such as batteries, are reviewed along with emerging technologies and currently untapped sources. Potential power sources are classified as energy reservoirs, power-distribution methods, or power-scavenging methods, which enable wireless nodes to be completely self-sustaining. Several power sources capable of providing power on the order of 100 $\mu W/cm^3$ for very long lifetimes are feasible. It is the authors' opinion that no single power source will suffice for all applications, and that the choice of a power source needs to be considered on an application-by-application basis.

Handbook of Sensor Networks: Algorithms and Architectures, Edited by Ivan Stojmenović
Copyright © 2005 John Wiley & Sons, Inc.

3.1 INTRODUCTION

The last several years have witnessed a large research effort based around the vision of ubiquitous networks of wireless sensor and communication nodes [1–3]. As the size and cost of such wireless sensor nodes continue to decrease, the likelihood of their use becoming widespread in buildings, industrial environments, automobiles, aircraft, and so forth, increases. However, as their size and cost decrease, and as their prevalence increases, effective power supplies become a larger problem.

The scaling down in size and cost of complementary metal-oxide semiconductor (CMOS) electronics has far outpaced the scaling of energy density in batteries, which are by far the most prevalent power sources currently used. Therefore, the power supply is quickly becoming the largest and most expensive component of the emerging wireless sensor nodes being proposed and designed. The cost of batteries is compounded by the fact that batteries must be either replaced or recharged on a regular basis. This regular maintenance could easily become the single greatest cost of installing a wireless sensor network for many applications. If wireless sensor networks are to truly become ubiquitous, replacing batteries in every device every year or two is simply cost prohibitive.

The purpose of this chapter, then, is to review existing and potential power sources for wireless sensor networks. Current state-of-the-art, ongoing research, and theoretical limits for many potential power sources are discussed. One can classify possible methods of providing power for wireless nodes into three groups: store energy on the node (i.e., a battery), distribute power to the node (i.e., a wire), and scavenge available ambient power at the node (i.e., a solar cell). Power sources that fall into each of these three categories are reviewed. Of course, combinations of the three methods are also possible. In fact, even in an energy-scavenging method some onboard energy storage must be available.

A direct comparison of vastly different types of power source technologies is difficult. For example, comparing the efficiency of a solar cell to that of a battery is not very useful. However, in an effort to provide a general understanding of a wide variety of power sources, the following metrics will be used for comparison: power density, energy density (where applicable), and average power density over a year of use. Additional considerations are the complexity of the power electronics needed and whether secondary energy storage is needed.

3.2 POWER CONSUMPTION

Before considering power sources, it is useful to consider the power demand of a typical wireless sensor node. Assuming that the radio transmitter operates at approximately 0 dBm (which would roughly correspond to an average distance of 10 m between nodes), the peak power consumption of the radio transmitter will be around 2–3 mW, depending upon its efficiency. Using low-power techniques [4], the receiver should not consume more than 1 mW. Including the dissipation of the sensors and peripheral circuitry, a maximum peak power of 5 mW is quite reasonable. Given a maximum data rate for the radio of 100 kbit/s, and an average traffic load per node of 1 kbit/s (these numbers are based on real radio prototypes and a realistic smart home scenario), every node operates at a duty cycle of approximately 1%. During the remaining 99%, the only activities taking place in a node are a number of background tasks: low-speed timers, channel monitoring, and node synchronization. The latter actually is the dominant power-consuming source of the node if not handled appropriately. Using advanced "wake-up radio techniques" or semi-asynchronous beaconing techniques; the average "standby" power of the node can be limited to 50 μW or lower. Combining peak and standby power dissipation leads to an average power dissipation of approximately 100 μW.

Several small low-power wireless platforms are currently available commercially. Companies providing wireless sensor platforms include Dust Networks [5], Crossbow [6], Xsilogy [7], Ember [8], and Millenial Net [9]. The power needed to operate these platforms depends on how and where they are used. Based on the authors' investigations, they generally require an average power consumption of about one order of magnitude higher than the 100 μW proposed earlier (generally one to several mW). However, research projects have demonstrated that a wide range of applications is possible within a power budget of approximately 100 μW. For the purposes of this discussion, it will be assumed that a rough standard of acceptability for a power source is its ability to provide 100 μW within the size constraints of the application for a lifetime prescribed by the application. For purposes of comparison, it is necessary to normalize the power potential of the different technologies that will be discussed. While each application will have different constraints on the overall size of the wireless sensor, a standard volume of 1 cm^3 has been chosen as a reasonable size constraint for many applications. Therefore, power (or energy) per cm^3 will be used as a primary metric of comparison.

3.3 ENERGY RESERVOIRS

Energy storage, in the form of electrochemical energy stored in a battery, is the predominant means of providing power to wireless devices today. Batteries are probably the easiest power solution for wireless electronics because of their versatility. However, several other forms of energy storage may be useful for wireless sensor nodes. Regardless of the form of the energy storage, the lifetime of the node will be determined by the fixed amount of energy stored on the device. While it is cost effective in some applications to repeatedly change or recharge batteries, if wireless sensor nodes are to become a ubiquitous part of the environment, it will no longer be cost effective. The primary metric of interest for all forms of energy storage will be usable energy per unit volume (J/cm^3) and the closely related average power per unit volume per unit time (μW/cm^3/year) of operation. An additional issue is that the instantaneous power that an energy reservoir can supply is usually dependent on its size. Therefore, in some cases, such as microbatteries, the maximum instantaneous power density (μW/cm^3) is also an issue for energy reservoirs.

TABLE 3.1 Energy Density of Three Primary Battery Chemistries

Chemistry	Zinc–Air	Lithium	Alkaline
Energy (J/cm^3)	3780	2880	1200

3.3.1 Macroscale Batteries

Electrochemical batteries have been the dominant form of power storage and delivery for electronic devices for decades, thus their consideration for use in wireless sensor networks is natural. Primary batteries are perhaps the most versatile of all small power sources. Table 3.1 shows the energy density for a few common primary battery chemistries. Figure 3.1 shows the average power available from these battery chemistries versus lifetime. Figure 3.1 includes leakage and shelf-life effects. Note that while zinc–air batteries have the highest energy density, their lifetime is very short.

Because batteries have a fairly stable voltage, electronic devices can often be run directly from the battery without any intervening power electronics. While this may not be the most robust method of powering the electronics, it is often used and is advantageous in that it avoids the extra power consumed by power electronics.

Macroscale secondary (rechargeable) batteries are commonly used in consumer electronic products such as cell phones, personal digital assistants (PDAs), and notebook computers. Table 3.2 gives the energy density of a few common rechargeable battery chemistries.

It should be remembered that rechargeable batteries are a *secondary* power source. Therefore, in the context of wireless sensor networks, another primary power source must be used to charge them. In most cases it would be cost prohibitive to manually recharge each device. More likely, an energy-scavenging source on the node itself, such as a solar cell, would be used to recharge the battery. One item to

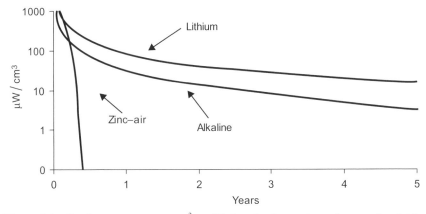

Figure 3.1 Continuous power per cm^3 vs. lifetime for three primary battery chemistries.

TABLE 3.2   Energy Density of Three Secondary Battery Chemistries

Chemistry	Lithium	NiMHd	NiCd
Energy (J/cm^3)	1080	860	650

consider when using rechargeable batteries is that electronics to control the charging profile must often be used. These electronics add to the overall power dissipation of the device. However, like primary batteries, the output voltages are stable and power electronics between the battery and the load electronics can often be avoided.

3.3.2   Microscale Batteries

Microscale batteries could be grouped into two categories. The first includes extremely thin (and often flexible) batteries that can be used in very small assembled devices. The second includes thin-film batteries that are fabricated on a wafer substrate (often silicon) and have the potential of being monolithically integrated with electronics. Examples of the first kind include the products of Paper Power [10] and Cymbet [11]. Research on thin-film batteries of the second kind are included in the following discussion.

One of the main stumbling blocks to reducing the size of microbatteries is power output due to surface area limitations of microscale devices. This is true for either type of microbatteries, as classified here. However, it is even more acute for on-chip microbatteries. For low-power sensor nodes, maximum power output is generally not an issue. Therefore, the capacity, or energy density, of the battery dominates its consideration for use. However, the maximum current output of a battery depends on the surface area of the electrodes. Because microbatteries are so small, the electrodes have a small surface area, and their maximum current output is also very small. This problem can be alleviated to a certain degree by placing a large capacitor in parallel with the battery capable of providing short bursts of current. However, the capacitor itself consumes additional volume, and therefore may not be desirable in many applications.

The challenge of maintaining (or increasing) performance of on-chip microbatteries while decreasing size is being addressed on multiple fronts. Bates et al. at Oak Ridge National Laboratory have created a process by which a primary thin-film lithium battery can be deposited onto a chip [12]. The thickness of the entire battery is on the order of tens of μm, but the areas studied are in the cm^2 range. This battery is in the form of a traditional Volta pile, with alternating layers of lithium manganese oxide (or lithium cobalt oxide), lithium phosphate oxynitride, and lithium metal. Maximum potential is rated at 4.2 V with continuous/max current output on the order of 1 mA/cm^2 and 5 mA/cm^2 for the $LiCoO_2$–Li-based cell.

Work is being done on thick-film batteries with a smaller surface area by Harb et al. [13], who have developed microbatteries of Ni/Zn with an aqueous NaOH electrolyte. Thick films are on the order of 0.1 mm, but overall thicknesses are minimized by use of three-dimensional (3D) structures. While each cell is only rated at 1.5 V, geometries have been duty-cycle optimized to give acceptable power outputs at small overall theoretical volumes (4 mm by 1.5 mm by 0.2 mm), with good durability demonstrated by the electrochemical components of the battery. The main challenges lie in maintaining a microfabricated structure that can contain an aqueous electrolyte.

Radical 3D structures are also being investigated to maximize power output. Hart et al. [14] have theorized a 3D battery made of series alternating cathode and anode rods suspended in a solid electrolyte matrix. Theoretical power outputs for a 3D microbattery are shown to be many times larger than a two-dimensional (2D) battery of equal size (with far lower ohmic ionic transport distances, thus lower ohmic losses).

For example, a 1-cm^2 thin film with each electrode having a thickness of 22 μm and a 5-μm electrolyte would have a maximum current density on the order of 5 mA. If the battery is restructured to have the same total volume, with square packing electrode rods (as Hart et al. have proposed) with a 5-μm radius and a 5-μm surface-to-surface distance, geometry dictates that the energy capacity is reduced to 39% of the thin-film capacity (due to a higher volume percentage of electrode material for the standard thin-film battery). However, while the energy density is lower for the 3D battery, the power density is higher due to a higher surface area. In fact, the 3D battery would have a total electrode area of 3.5 cm^2, an increase of 350%. The increase in surface area alone improves the current density to 17.5 mA. Moreover, the ionic transport scale in the 2D structure is about 350% longer than the 3D case because the electrodes for the 3D case are much thinner. Therefore, decreased ohmic losses could further improve the maximum throughput to 20 mA at 4.2 V. However, the inherent nonuniformities in current distribution in 3D batteries (exacerbated by the particular complexity of this cell) may lead to difficulties with regard to device reliability on primary battery systems and cycle life in secondary battery systems.

3.3.3 Ultracapacitors

Ultracapacitors represent a compromise of sorts between rechargeable batteries and standard capacitors. Capacitors can provide significantly higher power densities than batteries; however, their energy density is lower by about two orders of magnitude. Ultracapacitors (also called supercapacitors or electrochemical capacitors) achieve significantly higher energy density than standard capacitors, but retain many of the favorable characteristics of capacitors, such as long life, high current density, and short charging time.

Rather than just storing charge across a dielectric material, as capacitors do, ultracapacitors store ionic charge in an electric double layer to increase their effective capacitance. The energy density of commercially available ultracapacitors is about one order of magnitude higher than standard capacitors and about one order of magnitude lower than rechargeable batteries (or about 50 to 100 J/cm^3). Because of their increased lifetimes, short charging times, and high power densities, ultracapacitors could be very attractive as a secondary power source in place of rechargeable batteries in some wireless sensor node applications. Corporations working on such ultracapacitors include NEC [15] and Maxwell [16].

3.3.4 Microfuel Cells

Hydrocarbon-based fuels have very high energy densities compared to batteries. For example, methanol has an energy density of 17.6 kJ/cm^3, which is about six times that of a lithium battery. Therefore, fuel cells are potentially very attractive for wireless sensor nodes. Like batteries, fuel cells produce electrical power from a chemical reaction. A standard fuel cell uses hydrogen atoms as fuel. A catalyst promotes the separation of the electron in the hydrogen atom from the proton. The proton diffuses through an electrolyte (often a solid membrane), while the electron is available for use by an external circuit. The protons and electrons recombine with oxygen atoms on the other side (the oxidant side) of the electrolyte to produce water molecules. This process is illustrated in Figure 3.2. While pure hydrogen can be used as a fuel, other hydrocarbon fuels are often used. For example, in direct methanol fuel cells (DFMC) the anode catalyst draws the hydrogen atoms out from the methanol.

Most single fuel cells tend to output open-circuit voltages around $1.0–1.5$ V. Of course, like batteries, the cells can be placed in series for higher voltages. The voltage is quite stable over the operating lifetime of the cell, but it does fall off with increasing current draw. Figure 3.3 shows the voltage versus current load for a typical fuel cell. Notice that as the current density increases, the dominant loss mechanism also changes. Because the voltage drops with current, it is likely that some additional power electronics will be necessary if replacing a battery with a fuel cell.

Large-scale fuel cells have been used as power supplies for decades. For example, the Apollo spacecraft used alkaline fuel cells for electricity. More recently,

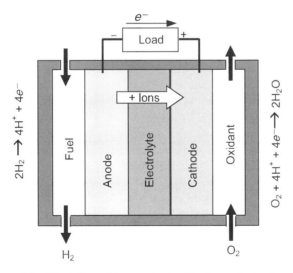

Figure 3.2 Illustration of how a standard hydrogen fuel cell works.

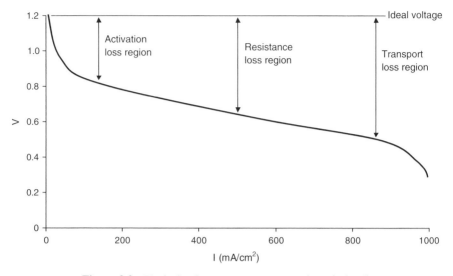

Figure 3.3 Typical voltage vs. current curve for a fuel cell.

fuel cells have been developed as alternative power supplies for automobiles. Cells using a variety of fuels and electrolytes have been successfully used at the macroscale. Recently, fuel cells have gained favor as a replacement for consumer batteries [17]. Small, but still macroscale, fuel cells are likely to soon appear in the market as battery rechargers and battery replacements [18].

The research trend is toward microfuel cells that could possibly be closely integrated with wireless sensor nodes. Like microbatteries, a primary metric of comparison in microfuel cells is power density in addition to energy density. As with microbatteries, the maximum continuous current output is dependent on the electrode surface area. Microfabricated fuel cells offer an advantage in surface-to-volume ratio, thereby giving them a higher power density. Likewise microfabricated features can potentially improve gas diffusion and lower the internal resistance [19], both of which improve efficiency.

Fuel cells tend to operate better at higher temperatures, which are more difficult to maintain for microfuel cells. Efficiencies of large-scale fuel cells have reached approximately 45% electrical conversion efficiency and nearly 90% if cogeneration is employed [20]. Efficiencies for microscale fuel cells will certainly be lower. The maximum obtainable efficiency for a microfuel cell is still uncertain. Demonstrated efficiencies are generally below 1% [21].

Many research groups are working on microfabricated partial systems that typically include an electrolyte membrane, electrodes, and channels for fuel and oxidant flow. Recent examples include the hydrogen-based fuel cells developed by Hahn et al. [22] and Lee et al. [23]. Both systems implement microfabricated electrodes and channels for fuel and oxidant flow. The system by Hahn et al. produces power on the order of $100 \, \text{mW/cm}^2$ from a device $0.54 \, \text{cm}^2$ in size. The system by Lee et al. produces $40 \text{ mW}/\text{cm}^2$. It should be noted that the fundamental characteristic here is power per unit area rather than power per unit volume, because the devices are fundamentally planar. Complete fuel storage systems are not part of their studies, and therefore an energy or power per unit volume metric is not appropriate. Fuel-conversion efficiencies are not reported.

Hydrogen storage on a small scale is a difficult problem that has not yet been solved. It is primarily for this reason that methanol-based microfuel cells are also being investigated by numerous groups. Holloday et al. [21] have demonstrated a research methanol fuel processor with a total size on the order of several mm^3. This fuel processor has been combined with a thin fuel cell, 2 cm^2 in area, to produce roughly 25 mA at 1-V with 0.5% overall efficiency. They are targeting a 5% efficient cell. Additionally, Mench et al. [24] have proposed a complete 3D methanol fuel cell with a volume of 1 cm^3. The system would contain all necessary elements except a methanol reservoir. The projected power output is 1 W/cm^3 at a projected efficiency of 30%; however, to the authors' knowledge, this has not been demonstrated. It should be noted that this is a stacked fuel cell and that if fuel volume were included, the power density would be lower.

Given the energy density of fuels such as methanol, fuel cells need to reach efficiencies of at least 20% in order to be more attractive than primary batteries. Nevertheless, at the microscale, where battery efficiencies are also lower, a lower efficiency fuel cell could still be attractive. Finally, providing for sufficient fuel and oxidant flows is a very difficult task in microfuel cell development. The ability to microfabricate electrodes and electrolytes does not guarantee the ability to realize a microfuel cell. The problem of microfabricating the fuel reservoir and all of the plumbing is arguably a more difficult task than the microfabrication of electrodes. To the authors' knowledge, a self-contained on-chip fuel cell has yet to be demonstrated.

3.3.5 Micro-Heat Engines

At large scales, fossil fuels are the dominant source of energy used for electric-power generation, mostly due to the low cost per joule, high energy density, abundant availability, storability, and ease of transport. Power plants typically convert the chemical energy of the fuel into thermal energy through combustion, then convert thermal to mechanical power by driving a heat engine that implements a thermodynamic cycle (such as gas turbines or internal combustion engines). The engine then entrains a magnetic generator to produce the electrical power. To date, the complexity and multitude of components involved in such a process have hindered the miniaturization of heat engines and power generation approaches based on combustion of hydrocarbon fuels. As the scale of a mechanical system is reduced, the tolerances must reduce accordingly, and the assembly process becomes increasingly challenging. This results in increasing costs per unit power and/or deteriorated performance.

The extension of silicon microfabrication technology from microelectronics to microelectromechanical systems (MEMS) is changing this paradigm. Complex microsystems that integrate mechanical, chemical, thermal, fluidic, and electro-magnetic functions on-chip can be batch fabricated with micron-scale precision using photolithography, etching, and other microfabrication techniques. In the mid-1990s, Epstein et al. proposed that microengines, that is, dime-size heat engines, for portable power generation and propulsion could be fabricated using MEMS technology [25]. The initial concept consisted of using silicon deep reactive ion etching, fusion wafer bonding, and thin-film processes to microfabricate and integrate high-speed turbomachinery, with bearings, a generator, and a combustor within a cubic-centimeter volume. An application-ready power supply would also require auxiliary components, such as a fuel tank, engine and fuel controller, electrical power conditioning with short-term storage, thermal management and packaging. Expected performance is 10–20 W of electrical power output at thermal efficiencies on the order of 5–20%. Figure 3.4 shows a microturbine test device used for turbomachinery and air bearing development.

Multiple research groups across the globe have also undertaken the development of various micro-heat engine–based power-generation approaches. Ongoing micro-engine projects include microgas turbine engines [25,26], Rankine steam turbines [27], rotary Wankel internal combustion engines [28], free and spring-loaded piston internal combustion engines [29,30], and thermal-expansion–actuated piezo-electric power generators [31,32], to name a few. In addition, various static

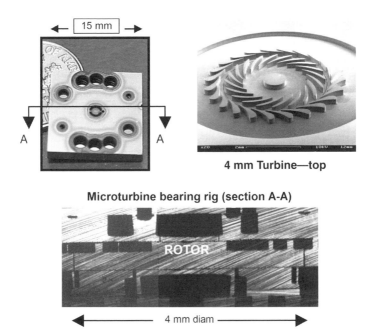

4 mm Turbine—top

Microturbine bearing rig (section A-A)

Figure 3.4 Microturbine development device, which consists of a 4-mm-diameter single-crystal silicon rotor enclosed in a stack of five bonded wafers used for microair bearing development.

approaches to convert heat into electricity are in development for small scales, including thermoelectric [33,34], thermionic [35], and thermophotovoltaic [36] components coupled with a heat source.

Most of these and similar efforts are at initial stages of development, and performance has not been demonstrated. However, predictions range from 0.1 to 10 W of electrical power output, with typical masses $\sim 1-5$ g and volumes ~ 1 cm^3. Microengines are not expected to grow smaller in size due to manufacturing and efficiency constraints. At small scales, viscous drag on moving parts and heat transfer to the ambient air and between components increase, which adversely impacts efficiency.

The main system level parameter that emerges for wireless sensor applications is the energy-conversion efficiency, h (ratio of output electrical power to what is available from the fuel). For a duration, t, and average power level, P, the mass of fuel required is simply the product of duration and average power level, divided by the fuel heating value, h_{fuel}, and efficiency: $m_{fuel} = (t^*P)/(h_{fuel}{}^*\eta)$. Typical values of expected fuel requirements are presented in Table 3.3 for a 10-year mission consuming an average power of 1 mW (efficiency of 10% is assumed). The fuel requirement tends to dominate the envelope of the complete system, given the small engine size and mass. If refueling is possible during the mission, then the overall size of the power supply is dramatically reduced, and tends toward the size of the engine and auxiliary components for short autonomous periods.

Alternatively, if high-quality (temperature) heat is available from the surroundings, the engine could scavenge it instead of burning fuel. Examples of such sources include waste heat from large engines and solar irradiation. Lower efficiencies are, however, expected if the heat-source temperatures are lower than those created by combustion products (1000–1500 K). This situation is considered further in the Section 3.5 on power scavenging.

Given the relatively large power level, a single microengine would only need to operate at low duty cycles (less than 1% of the time) to periodically recharge a battery. The total operating time is therefore on the order of hundreds of hours, which alleviates lifetime issues for the engine. It should also be noted that the inefficiency of a heat engine will result in heat discharge to its surroundings. For example, an engine with 1-W output power operating at 10% efficiency is consuming 10 W from the fuel and discharging 9 W of heat during periods of operation. Specific applications must allow release of this heat. Combining micro-heat engines with

TABLE 3.3 Fuel for 10-Year Mission at 1-mW Average Power Provided by a 10% Efficient Micro-Heat Engine

Fuel	Net Specific Energy ($h_{fuel}{}^*\eta$) (Wh/kg)	Fuel Mass (g)	Fuel Volume (cm^3)
Gasoline	1324	66	94
Butane	1270	69	99
Hydrogen	3337	26	972

thermoelectrics that convert some of this waste heat would lead to greater overall efficiency, but with a cost and size penalty for adding such components.

Overall, the greatest benefits of micro-heat engines are their high power density (0.1–2 W/g, without fuel) and their use of fuels allowing high-density energy storage for compact, long-duration power supplies. For long missions, the power density is not as important as efficiency. Microengines will therefore require many years of development before reaching the expected efficiencies and being applicable for real-life applications.

3.3.6   Radioactive Power Sources

Some radioactive materials contain extremely high-energy densities. As with hydrocarbon fuels, energy derived from radioactive materials has been used on a much larger scale for decades. However, it has not been exploited on a small scale, as would be necessary to power wireless sensor networks. The use of radioactive materials can pose a serious health hazard, and is a highly political and controversial topic. It should, therefore, be noted that the goal here is neither to promote nor to discourage investigation into radioactive power sources, but to present their potential, and the research being done in the area.

The most probable method of generating power from radioactive materials on a small scale is to make use of their natural decay. The total energy emitted by radioactive decay of a material can be expressed as in equation (3.1).

$$E_t = A_c E_e T \qquad (3.1)$$

where E_t is the total emitted energy, A_c is the activity, E_e is the average energy of emitted particles, and T is the time period over which power is collected. Table 3.4 lists several potential radioisotopes, their half-lives, specific activities, and energy densities based on radioactive decay. It should be noted that materials with lower activities and longer half-lives will produce lower power levels for more time than materials with comparatively short half-lives and high specific activities. The half-life of the material has been used as the time over which power would be collected. Only alpha and beta emitters have been included, because

TABLE 3.4   Comparison of Radioisotopes

Material	Half-Life (years)	Activity Volume Density (Ci/cm^3)	Energy Density (J/cm^3)
^{238}U	4.5×10^9	6.34×10^{-6}	2.23×10^{10}
^{63}Ni	100.2	506	1.6×10^8
^{32}Si	172.1	151	3.3×10^8
^{90}Sr	28.8	350	3.7×10^8
^{32}P	0.04	5.2×10^5	2.7×10^9

of the heavy shielding needed for gamma emitters. Finally, uranium-238 is included for purposes of comparison only.

While the energy density numbers reported for radioactive materials are extremely attractive, it must be remembered that in most cases the energy is being emitted over a very long period of time. Second, efficient methods of converting this power to electricity at small scales do not exist. Therefore, efficiencies would likely be extremely low.

Li and Lal [37] have used the ^{63}Ni isotope to actuate a conductive cantilever. As the beta particles (electrons) emitted from the ^{63}Ni isotope collect on the conductive cantilever, there is an electrostatic attraction. At some point, the cantilever contacts the radioisotope and discharges, causing the cantilever to oscillate. Up to this point, the research has only demonstrated the actuation of a cantilever, and not electric power generation. However, electric power could be generated from an oscillating cantilever. The reported power output, defined as the change over time in the combined mechanical and electrostatic energy stored in the cantilever, is 0.4 pW from a 4 mm × 4 mm thin film of ^{63}Ni. This power level is equivalent to 0.52 μW/cm^3. However, it should be noted that using 1 cm^3 of ^{63}Ni is impractical. The reported efficiency of the device is 4×10^{-6}.

3.4 POWER DISTRIBUTION

In addition to storing power on a wireless node, in certain circumstances power can be distributed to the node from a nearby energy-rich source. It is difficult to characterize the effectiveness of power-distribution methods by the same metrics (power or energy density), because in most cases the power received at the node is more a function of how much power is emitted rather than the size of the power receiver at the node. Nevertheless, an effort is made to characterize the effectiveness of a few power distribution methods as they apply to wireless sensor networks.

3.4.1 Electromagnetic (RF) Power Distribution

The most common method (other than wires) of distributing power to embedded electronics is through the use of RF radiation. Many passive electronic devices, such as electronic ID tags and smart cards, are powered by a nearby energy-rich source that transmits RF energy to the passive device. The device then uses that energy to run its electronics [38,39]. This solution works well, as evidenced by the wide variety of applications where it is used, if there is a high-power scanner or other source in very close proximity to the wireless device. It is, however, less effective in dense ad hoc networks where a large area must be flooded with RF radiation to power many wireless sensor nodes.

Using a very simple model and neglecting any reflections or interference, the power received by a wireless node can be expressed by equation (3.2) [40]:

$$P_r = \frac{P_0 \lambda^2}{4 \pi R^2} \tag{3.2}$$

where P_0 is the transmitted power, λ is the wavelength of the signal, and R is the distance between transmitter and receiver. Assume that the maximum distance between the power transmitter and any sensor node is 5 m, and that the power is being transmitted to the nodes in the 2.4–2.485-GHz frequency band, which is the unlicensed industrial, scientific, and medical band in the United States. Federal regulations limit ceiling-mounted transmitters in this band to 1 W or lower. Given a 1-W transmitter, and a 5-m maximum distance, the power received at the node would be 50 μW, which is probably on the borderline of being useful for wireless sensor networks. However, in reality the power transmitted will fall off at a rate faster than $1/R^2$ in an indoor environment. A more likely figure is $1/R^4$. While the 1-W limit on a transmitter is by no means general for indoor use, it is usually the case that some sort of safety limitation would need to be exceeded in order to flood a room or other area with enough RF radiation to power a dense network of wireless devices.

3.4.2 Wires, Acoustic Emitters, Light, and More

Other means of transmitting power to wireless sensor nodes might include wires, acoustic emitters, and light or lasers. However, none of these methods seem appropriate for wireless sensor networks. Running wires to a wireless communications device defeats the purpose of wireless communications. Energy in the form of acoustic waves has a far lower power density than is sometimes assumed. A sound wave of 100 dB in sound level only has a power level of 0.96 μW/cm^2. One could also imagine using a laser or other focused light source to direct power to each of the nodes in the sensor network. However, to do this in a controlled way, distributing light energy directly to each node, rather than just flooding the space with light, would likely be too complex and not cost effective. If an entire space is flooded with light, then this source of power becomes attractive. However, this situation has been classified as "power scavenging" and will be discussed in the following section.

3.5 POWER SCAVENGING

Unlike power sources that are fundamentally energy reservoirs, power-scavenging (also referred to as energy scavenging or energy harvesting) sources are usually characterized by their power density rather than energy density. Energy reservoirs have a characteristic energy density, and how much average *power* they can provide is then dependent on the lifetime over which they are operating. In contrast, power-scavenging sources have a characteristic power density, and the total amount of *energy* they provide depends on how long the source is in operation. Therefore, the primary metric for comparison of scavenged sources is power density, not energy density.

3.5.1 Photovoltaics (Solar Cells)

At midday on a sunny day, the incident light on the Earth's surface has a power density of roughly 100 mW/cm^2. Single-crystal silicon solar cells exhibit efficiencies of

TABLE 3.5   Power from a Cadmium Telluride Solar Cell at Various Distances from a 60 W Incandescent Bulb and Under Standard Office Lighting Conditions

Distance	8 in.	12 in.	18 in.	Office Light
Power (μW/cm^2)	503	236	111	7.2

15%–20% [41] under high light conditions, as one would find outdoors. Common indoor lighting conditions have far lower power density than outdoor conditions. Common office lighting provides about 100 μW/cm^2 at the surface of a desk. Single-crystal silicon solar cells are better suited to high light conditions and the spectrum of light available outdoors [41]. Thin-film amorphous silicon or cadmium telluride cells offer better efficiency indoors because their spectral response more closely matches that of artificial indoor light. Still, these thin-film cells only offer about 10% efficiency. Therefore, the power available from photovoltaics ranges from about 15 mW/cm^2 outdoors to about 10 μW/cm^2 indoors. Table 3.5 shows the measured power outputs from a cadmium telluride solar cell (Panasonic BP-243318) at varying distances from a 60-W incandescent bulb.

A single solar cell has an open circuit voltage of about 0.6 V. Individual cells are easily placed in series, especially in the case of thin-film cells, to get almost any desired voltage needed. A current vs. voltage (I–V) curve for a typical five cell array (wired in series) is shown below in Figure 3.5. Unlike the voltage, current densities are directly dependent on the light intensity.

Solar cells provide a fairly stable DC voltage through much of their operating space. Therefore, they can be used to directly power electronics in cases where the current load is such that it allows the cell to operate on high voltage side of the "knee" in the I–V curve and where the electronics can tolerate some deviation in source voltage. More commonly, solar cells are used to charge a secondary battery. Solar cells can be connected directly to rechargeable batteries through a simple

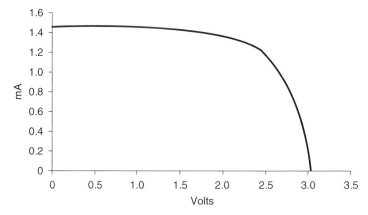

Figure 3.5   Typical I–V curve from a cadmium telluride solar array (Panasonic BP-243318).

series diode to prevent the battery from discharging through the solar cell. This extremely simple circuit does not ensure that the solar cell will be operating at its optimal point (which is at the knee in the $I-V$ curve), and so power production will be lower than the maximum possible. Second, rechargeable batteries will have a longer lifetime if a more controlled charging profile is employed. However, controlling the charging profile and the operating point of the solar cell both require more electronics, which use power themselves. An analysis needs to be done for each individual application to determine what level of power electronics would provide the highest net level of power to the load electronics. Longevity of the battery is another issue to be considered in this analysis.

3.5.2   Temperature Gradients

Naturally occurring temperature variations can also provide a means by which energy can be scavenged from the environment. The maximum efficiency of power conversion from a temperature difference is equal to the Carnot efficiency, which is given as equation (3.3):

$$\eta = \frac{T_{high} - T_{low}}{T_{high}} \tag{3.3}$$

Assuming a room temperature of 20°C, the efficiency is 1.6% from a source 5°C above room temperature and 3.3% for a source 10°C above room temperature.

A reasonable estimate of the maximum amount of power available can be made, assuming heat conduction through silicon material. Convection and radiation would be quite small compared to conduction at small scales and low temperature differentials. The amount of heat flow (power) is given by equation (3.4):

$$q' = k\frac{\Delta T}{L} \tag{3.4}$$

where k is the thermal conductivity of the material and L is the length of the material through which the heat is flowing. The conductivity of silicon is approximately 140 W/mK. Assuming a 5°C temperature differential and a length of 1 cm, the heat flow is 7 W/cm². If Carnot efficiency could be obtained, the resulting power output would be 117 mW/cm³. While this is an excellent result compared with other power sources, one must realize demonstrated efficiencies are well below the Carnot efficiency. This is particularly true of microscale devices.

A number of researchers have developed systems to convert power from temperature differentials to electricity. The most common method is through thermoelectric generators that exploit the Seebeck effect to generate power. For example, Stordeur and Stark [42] have demonstrated a microthermoelectric generator capable of generating 15 μW/cm² from a 10°C temperature differential. Furthermore, they report a technology limit of about 30 μW/cm² for the technology used. The first wristwatches powered by body heat have been manufactured by Seiko and Citizen.

The energy consumption of the Seiko watch is specified at 1 μW, with a driving voltage of 1.5 V [43]. Citizen [44] employs a thermoelectric generator with a contact area of 0.7 cm × 0.7 cm and a height of 1.5 mm that generates a voltage of 0.5 V and has a power output of 13.8 μW under load at a temperature difference of 1 K. Additionally, Applied Digital Solutions has developed a thermoelectric generator soon to be marketed as a commercial product. The generator is reported to be able to produce 40 μW of power from a 5°C temperature differential using a device 0.5 cm^2 in area and a few millimeters thick [45]. The output voltage of the device is approximately 1 V. Finally, the thermal-expansion–actuated piezoelectric generator referred to earlier [31] has also been proposed as a method to convert power from ambient temperature gradients to electricity. While all of these devices exhibit efficiencies well below the theoretical maximum efficiency, power densities in the range of 50 to 100 μW/cm^2 of device area have been demonstrated. This level of power could be enough to power a wireless sensor in an environment where thermal gradients of 1°C to 5°C are common.

3.5.3 Human Power

An average human body burns about 10.5 MJ of energy per day. (This corresponds to an average power dissipation of 121 W.) Starner has proposed tapping into some of this energy to power wearable electronics [46]. For example, wristwatches are powered using both the kinetic energy of a swinging arm and the heat flow away from the surface of the skin [47].

The conclusion of studies undertaken at MIT suggests that the most energy-rich and most easily exploitable source occurs at the foot during heel strike and in the bending of the ball of the foot [48]. This research has led to the development of piezoelectric shoe inserts capable of producing an average of 330 μW/cm^2 while a person is walking. The shoe inserts have been used to power a low-power wireless transceiver mounted to the shoes. While this power source is of great use for a wireless node worn on a person's foot, the problem of how to get the power from the shoe to any other point of interest still remains.

The sources of power mentioned earlier are passive power sources in that the human does not need to do anything other than what he or she would normally do to generate power. There is also a class of power generators that could be classified as active human power in that they require the human to perform an action that they would not normally perform. For example, Freeplay [49] markets a line of products that are powered by a constant-force spring that the user must wind up. While these types of products are extremely useful, they are not very applicable to wireless sensor networks, because it would be impractical and not cost efficient to individually wind up every node.

3.5.4 Wind/Airflow

Wind power has been used on a large scale as a power source for centuries. Large windmills are still common today. The potential power from moving air is quite easily calculated, as shown in equation (3.5):

$$P = \frac{1}{2}\rho A v^3$$

(3.5)

where P is the power, ρ is the density of air, A is the cross sectional area, and v is the air velocity. At standard atmospheric conditions, the density of air is approximately 1.22 kg/m^3. Figure 3.6 shows the power per square centimeter versus air velocity.

Large-scale windmills operate at maximum efficiencies of about 40%. The theoretical maximum efficiency is 59%. Efficiency is dependent on wind velocity for a given design, and average operating efficiencies are usually about 20%. Windmills are generally designed such that maximum efficiency occurs at wind velocities around 8 m/s (or about 18 mph). At low air velocity, efficiency can be significantly lower than 20%. Figure 3.6 also shows power output assuming 20% and 5% efficiency in conversion.

The authors are aware of only one project to generate power from airflow at small scales for use by wireless sensors. Federspiel and Chen [50] used a small (about 10 cm in diameter) airflow turbine to generate power intended for use by a Mica Mote [6]. The reported power output and efficiency values are shown in Table 3.6. The reported values in Table 3.6, taken together with the calculation in Figure 3.6 indicate that power densities from air velocity are quite promising. As there are many possible applications in which a fairly constant airflow of a few meters per second exists, it seems that research leading to the development of devices to convert airflow to electrical power at small scales is warranted.

3.5.5 Pressure Variations

Variations in pressure can also be used to generate power. For example, one could imagine a closed volume of gas that undergoes pressure variation as the daily

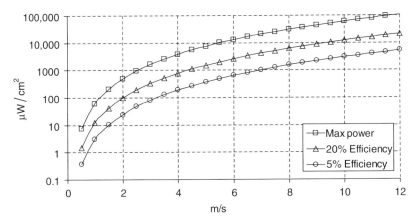

Figure 3.6 Maximum power density from airflow. Power densities assuming 20% and 5% conversion efficiencies are also shown.

TABLE 3.6 Air Speed, Power, and Efficiency Values Reported by Federspiel and Chen [50]

Air Velocity (m/s)	Power (μW/cm^2)	Efficiency (%)
2.5	100	5
4	215	5.5
5	350	11

temperature changes. Likewise, atmospheric pressure varies throughout the day. The change in energy for a fixed volume of ideal gas due to a change in pressure is simply given by

$$\Delta E = \Delta P V \qquad (3.6)$$

where ΔE is the change in energy, ΔP is the change in pressure, and V is the volume. A quick survey of atmospheric conditions around the world reveals that an average atmospheric pressure change over 24 hours is about 0.2 in. Hg or 677 Pa, which corresponds to an energy change of 677 μJ/cm^3. If the pressure cycles through 0.2 in. Hg once per day, for a frequency of 1.16×10^{-5}, the power density would then be 7.8 nW/cm^3.

An average temperature variation over a 24-h period would be about 10°C. The change in pressure to a fixed volume of ideal gas from a 10°C change in temperature is given by

$$\Delta P = \frac{mR\Delta T}{V} \qquad (3.7)$$

where m is mass of the gas, R is gas constant, and ΔT is the change in temperature. If 1 cm^3 of helium gas were used, a 10°C temperature variation would result in a pressure change of 1.4 MPa. The corresponding change in energy would be 1.4 J per day, which corresponds to 17 μW/cm^3. While this is a simplistic analysis and assumes 100% conversion efficiency to electricity, it does give an idea of what might be theoretically expected from naturally occurring pressure variations.

To the authors' knowledge, there is no research underway to exploit naturally occurring pressure variations to generate electricity. Some clocks, such as the "Atmos clock," are powered by an enclosed volume of fluid that undergoes a phase change under normal daily temperature variations. The volume and pressure change corresponding to the phase change of the fluid mechanically actuates the clock. However, this is on a large scale, and no effort is made to convert the power to electricity.

3.5.6 Time-Varying Structural Strain

Power can be scavenged from a surface undergoing a time-varying strain in a number of ways. The most straightforward method is to attach a smart material element to the surface that is undergoing strain. A number of smart materials exist, including piezoelectric, magnetostrictive, and electroactive materials. However, although there are many possible solutions, piezoelectric materials are by far the most common. Because they have been widely used for a long time, available piezoelectric materials exhibit very good electromechanical coupling. Furthermore, piezoelectric materials can directly produce voltages on the right order of magnitude for circuit applications. They are, therefore, the natural choice to pursue for strain-induced energy-scavenging solutions.

It is useful to develop a simple analytical model for strain-based power generation in order to generate rough estimates of power generation and gain design insight. If we assume, as is common, that the maximum power output available from a piezoelectric element occurs when operating at one half its open-circuit voltage and one half its closed-circuit current, we can develop a very simple expression for maximum output power.

The established constitutive equations for a linear piezoelectric material in reduced-matrix form are

$$\{S\} = [s^E]\{T\} + [d]^t\{E\} \tag{3.8}$$

$$\{D\} = [d]\{T\} + [\varepsilon^T]\{E\} \tag{3.9}$$

where $\{S\}$ is the six-dimensional (6D) strain vector, $\{T\}$ is the vector of stresses, $\{D\}$ is the 3D electric displacement vector, $\{E\}$ is the electric-field vector, $[s^E]$ is the six-by-six compliance matrix evaluated at constant electric field, $[d]$ is the three-by-six matrix of piezoelectric strain coefficients, and $[\varepsilon^T]$ is the three-by-three dielectric-constant matrix evaluated at constant stress. In many applications, the dominant stress state is one dimensional (1D), allowing equations 3.8 and 3.9 to be expressed as simple scalar equations. This simplification will be used in the following discussion.

The open-circuit voltage (V_{oc}) resulting from a strain (S) is given by equation (3.10):

$$V_{oc} = \frac{-dt_p Y}{(1 + k^2)\varepsilon} S \tag{3.10}$$

where t_p is the thickness of the piezoelectric element, Y is the elastic constant of the material (Young's modulus, $Y = 1/s$), and k is the piezoelectric coupling coefficient ($k = d^2 Y/\varepsilon$).

The short-circuit current developed is given by equation (3.11):

$$I_{sc} = fdAYS \tag{3.11}$$

where f is the frequency of the periodic strain, and A is the area of the piezoelectric patch.

Using the assumption made earlier, the maximum output power (P) is then given by equation (3.12):

$$P = \left(\frac{1}{2}V_{oc}\right)\left(\frac{1}{2}I_{sc}\right) = \frac{fk^2 vY}{4(1+k^2)}S^2 \tag{3.12}$$

where v is the volume of the piezoelectric element ($v = t_p A$). Note that the actual power output will simply be the value calculated by equation (3.12) multiplied by the efficiency of the power conditioning for the load.

As shown by equation (3.12), the power output is directly proportional to the volume of piezoelectric material used. Not surprisingly, the power output is very sensitive to both the coupling coefficient and the magnitude of the strain.

Figure 3.7 contains constant power-density contours calculated using the basic model shown in equation (3.12). As such, they represent the maximum power output possible from a surface undergoing a periodic strain at a given magnitude and frequency. The calculations shown in Figure 3.7 assume a reasonable coupling coefficient (k_{31}) of 0.3, a piezoelectric patch of 1 mm in thickness, and a Young's modulus of 52 GPa (PZT-5H). As can be seen from the graph, power outputs on the order of 100 μW are possible given high enough strains and frequencies.

Strain-based power generation has made its way into both the research and commercial sectors. Elvin et al. [51] have reported producing a self-powered wireless strain sensor; however, the power production from the sensor is well below the maximum values shown in Figure 3.7. MicroStrain Inc. is also marketing a wireless sensor that incorporates strain-based power harvesting [52].

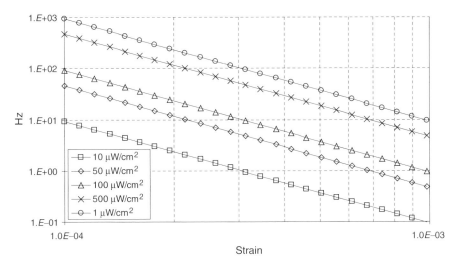

Figure 3.7 Power profiles vs. strain and frequency for a piezoelectric strain-based generator.

3.5.7   Vibrations

Low-level mechanical vibrations are present in many environments. Examples include automobiles, manufacturing and assembly equipment, aircraft, trains, heating, ventilation, and air-conditioning (HVAC) ducts, exterior windows, and small household appliances. Table 3.7 shows results of measurements on several different vibration sources performed by the authors. It will be noticed that the primary frequency of all sources is between 60 and 200 Hz. Acceleration amplitudes range from about 1 to 10 m/s^2.

A simple general model for power conversion from vibrations has been presented by Williams and Yates [53]. Note that this model is limited to generators containing a sprung proof mass and assumes that the force exerted on that mass by the electromechanical coupling is proportional to the velocity of the proof mass. The second assumption is quite valid for electromagnetic generators, but less so for piezoelectric and electrostatic generators. Other researchers are attempting to remedy these shortcomings developing technology-independent models that are not subject to these assumptions [54,55]. Nevertheless, the model proposed by Williams and Yates has been widely accepted and used in the research community. Furthermore, the model can be used to quickly obtain rough estimates for potential power output based on a given vibration source. Therefore, despite the fact that the model may lack accuracy for some types of generators, it will be used here as a basis to estimate the potential power output from common vibration sources. The final equation for power output from this model is shown here as equation (3.13):

$$
P = \frac{m\zeta_e \left(\dfrac{\omega}{\omega_n}\right)^3 A^2}{\omega \left[\left(2(\zeta_e + \zeta_m)\dfrac{\omega}{\omega_n}\right)^2 + \left(1 - \left(\dfrac{\omega}{\omega_n}\right)^2\right)^2 \right]}
\tag{3.13}
$$

TABLE 3.7   Summary of Several Vibration Sources

Vibration Source	Peak Acceleration (m/s^2)	Frequency (Hz)
Base of 3-axis machine tool	10	70
Kitchen blender casing	6.4	121
Clothes dryer	3.5	121
Door frame just as door closes	3	125
Small microwave oven	2.25	121
HVAC vents in office building	0.2–1.5	60
Wooden deck with foot traffic	1.3	385
Breadmaker	1.03	121
External windows next to a busy street	0.7	100
Notebook computer while CD is being read	0.6	75
Washing machine	0.5	109
Second story floor of a wood-frame office building	0.2	100
Refrigerator	0.1	240

where P is the power output, m is the oscillating proof mass, A is the acceleration magnitude of the input vibrations, ω is the frequency of the driving vibrations, ω_n is the resonance frequency of the generator, ζ_m is the mechanical damping ratio, and ζ_e is an electrically induced damping ratio. The primary idea behind this model is that the energy removed from the oscillating proof mass by the electro-mechanical coupling behaves as a linear viscous damper with damping ratio ζ_e. The mechanical damping ratio (ζ_m) represents the viscous loss in the system. The power output of the system as calculated by equation (3.13) is highly dependent on the resonance frequency of the system. Figure 3.8 shows the power output for a converter with a proof mass of 8.5 g (0.5 cm^3 of tungsten alloy), damping ratios of 0.015 (for both ζ_e and ζ_m), and resonance frequency of 100 Hz. An input vibration source of 2.25 m/s^2 in magnitude was used.

If it is assumed that the resonance frequency of the generator is either designed or tuned to match the dominant frequency of the input vibrations, equation (3.13) reduces to the expression in equation (3.14):

$$P = \frac{m\zeta_e A^2}{4\omega(\zeta_e + \zeta_m)^2} \tag{3.14}$$

Three interesting relationships are evident from this model.

1. Power output is proportional to the oscillating mass of the system.
2. Power output is proportional to A^2/ω.
3. Power is maximized for $\zeta_e = \zeta_m$.

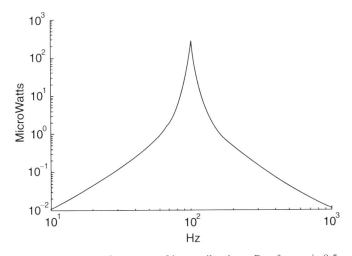

Figure 3.8 Power output vs. frequency of input vibrations. Proof mass is 8.5 g, damping ratios are 0.015, and magnitude of input vibrations is 2.25 m/s^2.

Point two indicates that the generator should be designed to resonate at the lowest frequency peak in the vibrations spectrum provided that higher frequency peaks do not have a higher acceleration magnitude. Many spectra measured by Roundy et al. [56] verify that generally the lowest frequency peak has the highest acceleration magnitude. The equivalent electrical damping ratio (ζ_e) is dependent on both design parameters and the specific load circuit applied. Point 3 indicates that there is an optimal level of electrically induced damping, and that this optimal level is equal to the amount of mechanical damping or pure loss. Therefore, to maximize the power output, the designer should try to minimize the mechanical damping (ζ_m) and design the overall system such that the level of damping seen by the generator as a result of the electromechanical coupling matches the level of damping attributable to pure loss. This principle will have slightly different implications for different types of generators (e.g., piezoelectric, electromagnetic, electrostatic).

Figures 3.9 and 3.10 provide a range of power densities that can be expected from vibrations similar to those listed previously in Table 3.7. The data shown in the figures are based on calculations from the model of Williams and Yates, and do not consider the technology that is used to convert the mechanical kinetic energy to electrical energy. As with the calculations in Figure 3.9, a proof mass of 8.5 g and damping ratios of 0.015 were used.

Several researchers have developed devices to scavenge power from vibrations. Devices include electromagnetic [57–59], electrostatic [60–62], and piezoelectric [63–65] methods to convert mechanical motion into electricity. Furthermore, there are now a number of companies beginning to offer generators based on scavenging power from vibrations [66–68]. Each approach offers benefits and drawbacks. In general, electrostatic converters suffer from the necessity to reliably maintain very small air gaps and generally exhibit lower power density, electromagnetic converters often suffer from low output voltages, and piezoelectric converters

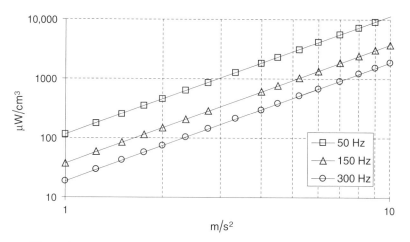

Figure 3.9 Power density vs. vibration amplitude for three frequencies.

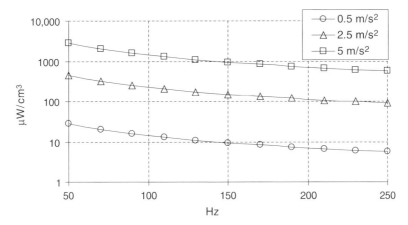

Figure 3.10   Power density vs. frequency of vibration input for three amplitudes.

usually rely on relatively brittle ceramics. Based on theory, simulations, and experiments, the opinion expressed by the authors is that piezoelectric generators offer the best long-term solution for most applications. An example of a wireless transceiver and piezoelectric vibration–based generator that powers the transceiver is shown in Figure 3.11 [69]. The generator has a size of 1 cm^3 and produces 200 μW from input vibrations of 2.25 m/s^2 at 120 Hz. The transmit power of the wireless transceiver is 0 dBm.

Although quite a number of vibration-based generators have been demonstrated in the literature, a number of research issues remain to be explored further. Under many circumstances, the driving frequency will be known before the device is designed and fabricated, and the appropriate resonance characteristics can thus be "built in." In other situations, however, this frequency will not be known a priori, or it may change over time. It is also relevant to consider the mass fabrication of such devices for use by other investigators. It would clearly be advantageous to create a single design that operates effectively over a range of vibration frequencies. Thus methods to improve the bandwidth of a generator without sacrificing the peak

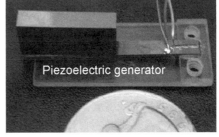

Figure 3.11   Piezoelectric generator, power circuit, and radio powered from vibrations of 2.25 m/s^2 at 120 Hz.

power output or methods to actively tune the resonance frequency of the generator are important research areas. With few exceptions, researchers have focused on a very limited set of potential geometries for analysis. A broader study of potential design geometries for each of the three fundamental types of generators would be beneficial. Finally, the transmission of power from the generator to the sensor node is of critical importance. While a few researchers have studied this problem [53,56,60], further optimization of power circuits is possible and will yield improved power output.

3.6 SUMMARY

An effort has been made to give an overview of the many potential alternative power sources for wireless sensor networks. Traditional power sources, such as batteries, have been considered along with potential sources on which little or no work has been done. Because some sources are fundamentally characterized by energy density (such as batteries), while others are characterized by power density (such as solar cells), a direct comparison with a single metric is difficult. Adding to this difficulty is the fact that some power sources do not make much use of the third dimension (such as solar cells), so their fundamental metric is power per square centimeter rather than power per cubic centimeter. Nevertheless, in an effort to compare all possible sources, a summary table is shown below as Table 3.8. Note that power density is listed as $\mu W/cm^3$; however, it is understood that in certain instances the number reported really represents $\mu W/cm^2$. Such values are marked with a superscript a. Note also that with only two exceptions, values listed are numbers that have been demonstrated or are based on experiments rather than theoretical optimal values. The authors were not able to find demonstrated or experimental values for strain-induced generators or air pressure–induced generators. Therefore theoretical values have been used, and these values are italicized. In many other cases the theoretical best values are explained in the chapter text.

Almost all wireless sensor nodes currently available are powered by batteries. This situation presents a substantial roadblock to the widespread deployment of wireless sensor networks, because the replacement of batteries is cost prohibitive. Furthermore, a battery that is large enough to last the lifetime of the device would dominate the overall system size and cost, and thus is not very attractive. It is therefore essential that alternative power sources be considered and developed.

This chapter has attempted to characterize a wide variety of such sources. It is the authors' opinion that no single alternative power source will solve the problem for all, or even a large majority of cases. However, many attractive and creative solutions do exist that can be considered on an application-by-application basis.

ACKNOWLEDGMENTS

The authors gratefully acknowledge the helpful input of Dan Steingart, Prof. Paul Wright, and Prof. Jan Rabaey. Dan was particularly helpful in doing background research for the

TABLE 3.8   Comparison of Various Potential Power Sources for Wireless Sensor Networks

Power Source	P/cm^3 ($\mu W/cm^3$)	E/cm^3 (J/cm^3)	$P/cm^3/yr$ ($\mu W/cm^3/Y$)	Secondary Storage Needed	Voltage Regulation	Comm. Available
Primary battery	—	2880	90	No	No	Yes
Secondary battery	—	1080	34	—	No	Yes
Microfuel cell	—	3500	110	Maybe	Maybe	No
Ultracapacitor	—	50–100	1.6–3.2	No	Yes	Yes
Heat engine	—	3346	106	Yes	Yes	No
Radioactive (^{63}Ni)	0.52	1640	0.52	Yes	Yes	No
Solar (outside)	$15{,}000^a$	—	—	Usually	Maybe	Yes
Solar (inside)	10^a	—	—	Usually	Maybe	Yes
Temperature	$40^{a,b}$	—	—	Usually	Maybe	Limited
Human Power	330	—	—	Yes	Yes	No
Air flow	350^c	—	—	Yes	Yes	No
Pressure Variation	17^d	—	—	Yes	Yes	No
Vibrations	200	—	—	Yes	Yes	Limited
Strain induced	200	—	—	Yes	Yes	Limited

Note: Values shown are actual demonstrated numbers except in two cases, which have been italicized.
[a]Denotes sources whose fundamental metric is power per *square* centimeter rather than per *cubic* centimeter.
[b]Demonstrated from a 5°C temperature differential.
[c]Based on reported values at an air velocity of 5 m/s and 11% conversion efficiency.
[d]Based on a 1-cm^3 closed volume of helium undergoing a 10°C temperature change once per day.

microbatteries section. Professors Wright and Rabaey were actively involved with and have greatly supported the vibration-based energy-scavenging work for a number of years.

REFERENCES

1. J. Rabaey, J. Ammer, T. Karalar, S. Li, B. Otis, M. Sheets, and T. Tuan. Picoradios for wireless sensor networks: The next challenge in ultra-low-power design. In *Proceedings of the International Solid-State Circuits Conference*, San Francisco, California, February 2002.

2. B. Warneke, B. Atwood, and K. S. J. Pister. Smart dust mote forerunners. In *Proceedings of the 14th Annual International Conference on Microelectromechanical Systems (MEMS 2001)*, pages 357–360, Interlaken, Switzerland, January 2001.

3. J. Hill and D. Culler. Mica: A wireless platform for deeply embedded networks, *IEEE Micro*, **22**(6):12–24, 2002.

4. B. Otis and J. Rabaey. A 300 μW 1.9 GHz oscillator utilizing micro-machined resonators. In *IEEE Proceedings of the 28th European Solid State Circuits Conference*, vol. 28, September 2002.

5. See at http://www.dust-inc.com, 2004.

6. See at http://www.xbow.com, 2004.

7. See at http://www.xsilogy.com, 2004.

8. See at http://www.ember.com, 2004.

9. See at http://www.millenial.net, 2004.

10. See at http://www.paperpower.com, 2004.

11. See at http://cymbet.com, 2004.

12. J. Bates, N. Dudney, B. Neudecker, A. Ueda, and C. D. Evans. Thin-film lithium and lithium-ion batteries. *Solid State Ionics*, **135**:33–45, 2000.

13. J. N. Harb, R. M. LaFollete, R. H. Selfridge, and L. L. Howell. Mircobatteries for self-sustained hybrid micropower supplies. *Journal of Power Sources*, **104**:46–51, 2002.

14. R. W. Hart, H. S. White, B. Dunn, and D. R. Rolison. 3-D microbatteries. *Electrochemistry Communications*, **5**:120–123, 2003.

15. See at http://www.nec-tokin.com/english/product/product_list.html, 2004.

16. See at http://www.maxwell.com/ultracapacitors/, 2004.

17. A. Heinzel, C. Hebling, M. Muller, M. Zedda, and C. Muller. Fuel cells for low power applications. *Journal of Power Sources*, **105**:250–255, 2002.

18. See at http://www.toshiba.co.jp/about/press/2003_03/pr0501.htm, 2003.

19. S. Kang, S.-J. J. Lee, and F. B. Prinz. Size does matter: The pros and cons of miniaturization. *ABB Review*, **2**:54–62, 2001.

20. K. Kordesh and G. Simader. *Fuel Cells and Their Applications*. VCH Publishers, 2001.

21. J. D. Holloday, E. E. Jones, M. Phelps, and J. Hu. Microfuel processor for use in a miniature power supply. *Journal of Power Sources*, **108**:21–27, 2002.

22. See at http://www.pb.izm.fhg.de/hdi/040_groups/group4/fuelcell_micro.html.

23. S. J. Lee, A. Chang-Chien, S. W. Cha, R. O'Hayre, Y. I. Park, Y. Saito, and F. B. Prinz. Design and fabrication of a micro fuel cell array with "flip-flop" interconnection. *Journal of Power Sources*, **112**:410–418, 2002.

24. M. M. Mench, Z. H. Wang, K. Bhatia, and C. Y. Wang. Design of a micro direct methanol fuel cell (μDMFC). In *Proceedings of the ASME International Mechanical Engineering Congress and Exposition (IMECE)*, New York, November 2001.

25. A. H. Epstein et al. Micro-Heat Engine, Gas Turbine, and Rocket Engines—The MIT Microengine Project. Paper AIAA 97-1773, presented at the 28th AIAA Fluid Dynamics Conference, Snowmass Village, Colorado, June 1997.

26. K. Isomura, M. Murayama, H. Yamaguchi, N. Ijichi, H. Asakura, N. Saji, O. Shiga, K. Takahashi, S. Tanaka, T. Genda, and M. Esashi. Development of Microturbocharger and Microcombustor for a Three-Dimensional Gas Turbine at Microscale. Paper GT-2002-30580, presented at the ASME-IGTI 2002 TURBO EXPO, Amsterdam, Netherlands, June 2002.

27. C. Lee, S. Arslan, Y.-C. Liu, and L. G. Fréchette. Design of a microfabricated Rankine cycle steam turbine for power generation. In *Proceedings of the ASME International*

Mechanical Engineering Congress and Exhibition (IMECE), Washington, D.C., November 2003.

28. K. Fu, A. J. Knobloch, F. C. Martinez, D. C. Walther, C. Fernandez-Pello, A. P. Pisano, and D. Liepmann. Design and fabrication of a silicon-based MEMS rotary engine. In *Proceedings of the ASME International Mechanical Engineering Congress and Exhibition (IMECE)*, New York, November 2001.

29. L. M. Matta, M. Nan, S. P. Davis, D. V. McAllister, B. T. Zinn, and M. G. Allen. Miniature Excess Enthalpy Combustor for Microscale Power Generation. AIAA Paper 2001-0978, presented at the 39th Aerospace Sciences Meeting and Exhibit, Reno, Nevada, January 2001.

30. T. Toriyama, K. Hashimoto, and S. Sugiyama. Design of a resonant micro recipro-cating engine for power generation. In *Proceedings of Transducers'03, the 12th International Conference on Solid-State Sensors and Actuators*, Boston, Massachusetts, June 2003.

31. S. Whalen, M. Thompson, D. Bahr, C. Richards, and R. Richards. Design, fabrication and testing of the P3 micro heat engine. *Sensors and Actuators*, **104**(3):200–208, 2003.

32. D. Santavicca, K. Sharp, J. Hemmer, B. Mayrides, D. Taylor, and J. Weiss. A solid piston micro-engine for portable power generation. In *Proceedings of the ASME International Mechanical Engineering Congress and Exhibition (IMECE)*, Washington, D.C., November 2003.

33. S. B. Schaevitz, A. J. Franz, K. F. Jensen, and M. A. Schmidt. A combustion-based MEMS thermoelectric power generator. In *Proceedings of Transducers'01 the 11th International Conference on Solid-State Sensors and Actuators*, pages 30–33, Munich, Germany, June 2001.

34. L. Sitzki, K. Borer, S. Wussow, E. Schuster, P. D. Ronney, and A. Cohen. Combustion in Microscale Heat Recirculating Burners. Paper No. 2001-1087, presented at the 39th AIAA Aerospace Sciences Meeting, Reno, Nevada, January 2001.

35. C. Zhang, K. Najafi, L. P. Bernal, and P. D. Washabaugh. Micro combustion-thermionic power generation: Feasibility, design and initial results. In *Proceedings of Trans-ducers'03, the 12th International Conference on Solid-State Sensors and Actuators*, Boston, Massachusetts, June 2003.

36. O. M. Nielsen, L. R. Arana, C. D. Baertsch, K. F. Jensen, and M. A. Schmidt. A thermo-photovoltaic micro-generator for portable power applications. In *Proceedings of Trans-ducers'03, the 12th International Conference on Solid-State Sensors and Actuators*, Boston, Massachusetts, June 2003.

37. H. Li and M. Lal. Self-reciprocating radio-isotope powered cantilever. *Journal of Applied Physics*, **92**(2):1122–1127, 2002.

38. D. Friedman, H. Heinrich, and D.-W. Duan. A low-power CMOS integrated circuit for field-powered radio frequency identification. In *Proceedings of the 1997 IEEE Solid-State Circuits Conference*, pages 294–295, 474, 1997.

39. See at http://www.hitachi.co.jp/Prod/mu-chip/, 2003.

40. A. A. Smith. *Radio Frequency Principles and Applications: The Generation, Propa-gation, and Reception of Signals and Noise*. IEEE Press, 1998.

41. J. F. Randall. On Ambient Energy Sources for Powering Indoor Electronic Devices. Ph.D. thesis, Ecole Polytechnique Federale de Lausanne, Switzerland, May 2003.

42. M. Stordeur and I. Stark. Low power thermoelectric generator—self-sufficient energy supply for micro systems. In *Proceedings of the 16th International Conference on Thermoelectrics*, pages 575–577, 1997.

43. M. Kishi, H. Nemoto, T. Hamao, M. Yamamoto, S. Sudou, M. Mandai, and S. Yamamoto. Microthermoelectric modules and their application to wristwatches as an energy source. In *Proceedings of the 18th International Conference on Thermoelectrics (ICT)*, pages 301–307, Baltimore, Maryland, August 1999.

44. Citizen Eco-Drive Thermo. Citizen press release, Basel, March 2003.

45. D. Pescovitz. The power of small tech. *Smalltimes*, **2**(1):21–31, 51–54, 2002.

46. T. Starner. Human-powered wearable computing. *IBM Systems Journal*, **35**(3):618–629, 1996.

47. See at http://www.seikowatches.com, 2004.

48. N. S. Shenck and J. A. Paradiso. Energy scavenging with shoe-mounted piezoelectrics. *IEEE Micro*, **21**:30–41, 2001.

49. See at http://www.freeplay.net, 2004.

50. C. C. Federspiel and J. Chen. Air-powered sensor. In *Proceedings of IEEE Sensors 2003*, Toronto, October 2003.

51. N. G. Elvin, A. A. Elvin, and M. Spector. A self-powered mechanical strain energy sensor. *Smart Materials and Structures*, **10**:293–299, 2001.

52. See at http://www.microstrain.com/white_strain_energy_harvesting.htm, 2004.

53. C. B. Williams and R. B. Yates. Analysis of a micro-electric generator for microsystems. In *Proceedings of Transducers 95/Eurosensors IX*, pages 369–372, 1995.

54. P. D. Mitcheson, T. C. Green, E. M. Yeatman, and A. S. Holmes. Architectures for vibration-driven micropower generators. *Journal of Microelectromechanical Systems*, **13**(3):1–12, 2004.

55. S. Roundy. On the effectiveness of vibration based energy harvesting. *Journal of Intelligent Material Systems*, forthcoming.

56. S. Roundy, P. K. Wright, and J. Rabaey. A study of low level vibrations as a power source for wireless sensor nodes. *Computer Communications*, **26**(11):1131–1144, 2003.

57. R. Amirtharajah and A. P. Chandrakasan. Self-powered signal processing using vibration-based power generation. *Journal of Solid-State Circuits*, **33**(5):687–695, 1998.

58. M. El-hami, P. Glynne-Jones, N. W. White, M. Hill, S. Beeby, E. James, A. D. Brown, and J. N. Ross. Design and fabrication of a new vibration-based electromechanical power generator. *Sensors and Actuators A (physical)*, **92**:335–342, 2001.

59. N. N. H. Ching, H. Y. Wong, W. J. Li, P. H. W. Leong, and Z. Wen. A laser-micromachined multi-modal resonating power transducer for wireless sensing systems. *Sensors and Actuators A (physical)*, **97–98**:685–690, 2002.

60. S. Meninger, J. O. Mur-Miranda, R. Amirtharajah, A. P. Chandrakasan, and J. H. Lang. Vibration-to-electric energy conversion. *IEEE Transactions on Very Large Scale Integration (VLSI) Systems*, **9**:64–76, 2001.

61. M. Miyazaki, H. Tanaka, G. Ono, T. Nagano, N. Ohkubo, T. Kawahara, and K. Yano. Electric-energy generation using variable-capacitive resonator for power-free LSI: Efficiency analysis and fundamental experiment. In *Proceedings of the 2003 International Symposium on Low Power Electronics and Design (ISLPED 2003)*, pages 193–198, Seoul, Korea, August 2003.

62. P. D. Mitcheson, P. Miao, B. H. Stark, E. M. Yeatman, A. S. Holmes, and T. C. Green. MEMS electrostatic micropower generator for low frequency operation. *Sensors and Actuators A (physical)*, forthcoming.

63. P. Glynne-Jones, S. P. Beeby, E. P. James, and N. M. White. The modelling of a piezoelectric vibration powered generator for microsystems. In *Proceedings of Transducers '01, the 11th International Conference on Solid-State Sensors and Actuators*, Munich, Germany, June 2001.

64. G. K. Ottman, H. F. Hofmann, and G. A. Lesieutre. Optimized piezoelectric energy harvesting circuit using step-down converter in discontinuous conduction mode. *IEEE Transactions on Power Electronics*, **18**(2):696–703, 2003.

65. S. Roundy and P. K. Wright. A piezoelectric vibration based generator for wireless electronics. *Smart Materials and Structures*, **13**:1131–1142, 2004.

66. See at http://www.ferrosi.com, 2004.

67. See at http://www.continuumcontrol.com, 2004.

68. See at http://www.intellisense.com, 2004.

69. S. Roundy, B. Otis, Y.-H. Chee, J. Rabaey, and P. K. Wright. A 1.9 GHz transmit beacon using environmentally scavenged energy. In *Proceedings of the 2003 International Symposium on Low Power Electronics and Design (ISPLED 2003)*, Seoul, Korea, August 2003.

■■■■■ CHAPTER 4

A Virtual Infrastructure for Wireless Sensor Networks

STEPHAN OLARIU and QINGWEN XU

Old Dominion University, Norfolk, Virginia

ASHRAF WADAA

Intel Corporation, Hillsboro, Oregon

IVAN STOJMENOVIĆ

University of Ottawa, Ontario, Canada

Overlaying a virtual infrastructure over a physical network is a time-honored strategy for conquering scale. There are, essentially, two approaches for building such an infrastructure. The first is to design the virtual infrastructure in support of a specific protocol, routing, for example. However, more often than not, the resulting infrastructure is not useful for other purposes. The alternative approach is to design the general-purpose virtual infrastructure with no particular protocol in mind. The challenge, of course, is to design the virtual infrastructure in such a way that it can be leveraged by a multitude *of different protocols.*

The main goal of this chapter is to propose a lightweight and robust virtual infrastructure for a network, consisting of tiny energy-constrained commodity sensors massively deployed in an area of interest. In addition, we present evidence that the proposed virtual infrastructure can be leveraged by a number of protocols ranging from routing to data aggregation.

4.1 INTRODUCTION

Recent advances in nanotechnology have made it possible to develop a large variety of microelectromechanical systems (MEMS), miniaturized low-power devices that integrate sensing, special-purpose computing, and wireless communications

Handbook of Sensor Networks: Algorithms and Architectures, Edited by Ivan Stojmenović
Copyright © 2005 John Wiley & Sons, Inc.

capabilities [1–5]. It is expected that these small devices, referred to as *sensors*, will be mass-produced, making their production cost-negligible. Individual sensors have a small, nonrenewable energy supply and, once deployed, must work unattended. For most applications, we envision a massive deployment of sensors, perhaps in the thousands or even tens of thousands [6–9].

Aggregating sensors into sophisticated computational and communication infrastructures, called *wireless sensor networks*, will have a significant impact on a wide array of applications, ranging from military, to scientific, to industrial, to health care, to domestic, establishing ubiquitous wireless sensor networks that will pervade society, redefining the way in which we live and work [10–13]. The novelty of wireless sensor networks and their tremendous potential for relevance to a multitude of application domains has triggered a flurry of activity in both academia and industry. We refer the reader to refs. [7,14–19] for a summary of recent applications of wireless sensor networks.

The fundamental goal of a sensor network is to produce, over an extended period of time, globally meaningful information from raw local data obtained by individual sensors. Importantly, this goal must be achieved in the context of prolonging as much as possible the useful lifetime of the network and ensuring that the network remains highly available and continues to provide accurate information in the face of security attacks and hardware failure. The sheer number of sensors in a sensor network combined with the unique characteristics of their operating environment (anonymity of individual sensors, limited energy budget, and a possibly hostile environment), pose unique challenges to the designers of protocols. For one thing, the limited energy budget at the individual sensor level mandates the design of ultralightweight data gathering, aggregation, and communication protocols. An important guideline in this direction is to perform as much local data processing at the sensor level as possible, avoiding the transmission of raw data through the sensor network.

Recent advances in hardware technology are making it plain that the biggest challenge facing the wireless sensor network community is the development of ultralightweight communication protocols ranging from training, to self-organization, to network maintenance and governance, to security, to data collection and aggravation, to routing [12,20,21].

4.1.1 The Name of the Game: Conquering Scale

Overlaying a virtual infrastructure over a physical network is a time-honored strategy for conquering scale. There are, essentially, two approaches to this exercise. The first is to design the virtual infrastructure in support of a specific protocol. However, more often than not, the resulting infrastructure is not useful for other purposes. The alternate approach is to design a general-purpose virtual infrastructure with no particular protocol in mind. The challenge, of course, is to design the virtual infrastructure in such a way that it can be leveraged by a *multitude* of different protocols [22].

To the best of our knowledge, research studies addressing wireless sensor networks have thus far taken only the first approach. To wit, in ref. [15] a set of paths is dynamically established as a result of the controlled diffusion of a query from a source node into the network. Relevant data are routed back to the source node, and possibly aggregated, along these paths. The paths can be viewed as a form of data-dissemination and aggregation infrastructure. However, this infrastructure serves the sole purpose of routing and data aggregation, and it is not clear how it can be leveraged for other purposes. A similar example is offered by ref. [23], where sensors use a discovery procedure to dynamically establish secure communications links to their neighbors; collectively, these links can be viewed as a secure communications infrastructure. As before, it is not clear that the resulting infrastructure can be leveraged for other purposes.

We view the principal contribution of this chapter at the conceptual level. Indeed, we introduce a simple and natural general-purpose virtual infrastructure for wireless sensor networks, consisting of a massive deployment of anonymous sensors. The proposed infrastructure consists of a dynamic coordinate system and a companion clustering scheme. We also show that the task of endowing the wireless sensor network with the virtual infrastructure—a task that we shall refer to as *training*—can be performed by a protocol that is at the same time lightweight and secure. In addition, we show that a number of fundamental tasks, including routing and data aggregation, can be performed efficiently once the virtual infrastructure is in place.

The remainder of this chapter is organized as follows: Section 4.2 discusses the sensor model used throughout the work. Section 4.3 discusses wireless sensor networks, as a conglomerate of individual sensors that have to self-organize and self-govern. In particular, we discuss interfacing wireless sensor networks with the outside world, as well as a brief preview of the training process. Next, Section 4.4 offers a brief overview of location awareness in wireless sensor networks. We also provide a lightweight protocol allowing the sensors to acquire fine-grain location information. Section 4.5 presents an overview of the general-purpose virtual infrastructure for wireless sensor networks. Specifically, Subsection 4.5.1 discusses the details of our dynamic coordinate system, the key component of our general-purpose virtual infrastructure; and Subsection 4.5.2 discusses the clustering scheme induced by the dynamic coordinate system. Section 4.6 is the backbone of the entire chapter, presenting the theoretical underpinnings of the training process. Section 4.8 proposes routing and data-aggregation algorithms in a trained wireless sensor network. Section 4.9 takes a close look at the problem of energy expenditure related to routing data in a wireless sensor network. Finally, Section 4.10 offers concluding remarks and maps out areas for future investigations.

4.2 THE SENSOR MODEL

We assume a sensor to be a device that possesses three basic capabilities: sensory, computation, and wireless communication. The sensory capability is necessary to acquire data from the environment; the computational capability is necessary for aggregating data, processing control information, and managing both sensory and communication activity. Sensor clocks drift at a bounded rate allowing only short-lived and group-based synchronization, where a group is loosely defined as the collection of sensors that *collaborate* to achieve a given task. The details of a light-weight synchronization protocol for wireless sensor networks will be the subject of another chapter in this book.

We assume that individual sensors operate subject to the following fundamental constraints:

- Sensors are *anonymous*—they do not have fabrication-time identities.
- Sensors are tiny, commodity devices that are mass-produced in an environment where testing is a luxury.
- Each sensor has a nonrenewable energy budget; when the on-board energy supply is exhausted, the sensor becomes nonoperational.
- In order to save energy, each sensor is in *sleep* mode most of the time, waking up at random points in time for short intervals under the control of an internal timer.
- Each sensor has a modest transmission range, perhaps a few meters. This implies that outbound messages sent by a sensor can reach only the sensors in its proximity, typically a small fraction of the sensors deployed.
- Once deployed, the sensors must work *unattended*, it is either infeasible or impractical to devote attention to individual sensors.

At any point in time, a sensor, will be engaged in performing one of a finite set of possible operations, or will be asleep. Example operations are sensing (data acquisition), routing (data communication; sending or receiving), and computing (e.g., data aggregation). We assume each operation performed by a sensor consumes a known fixed amount of energy and that a sleeping sensor performs no operation and consumes essentially no energy.

It is worth mentioning that while the energy budget can supply short-term applications, sensors dedicated to work over years may need to scavenge energy from the ambient environment. Indeed, it was shown recently that energy scavenging from vibration, kinetics, magnetic fields, seismic tremors, pressure, and so on, will become reality in the near future [24,25].

4.2.1 Genetic Material

We assume that just prior to deployment (perhaps onboard the aircraft that drops them in the terrain) the sensors are injected with the following *genetic material*:

- A standard public-domain pseudorandom number generator
- A set of *secret* seeds to be used as parameters for the random number generator
- A perfect hash function ϕ
- An initial time, at which point all the clocks are synchronous; later, synchronization is lost due to clock drift

The way in which this genetic material is used by individual sensors will be discussed in detail later in the chapter. For a more detailed discussion and applications to securing sensor networks we refer the interested reader to refs. [26] and [27].

4.3 STRUCTURE AND ORGANIZATION OF A WIRELESS SENSOR NETWORK

We envision a massive deployment of sensors, perhaps in the thousands or even tens of thousands. The sensors are aggregated into sophisticated computational and communication infrastructures, called wireless sensor networks, whose goal is to produce globally meaningful information from data collected by individual sensors. However, the massive deployment of sensors, combined with anonymity of individual sensors, limited energy budget and, in many applications, a hostile environment, pose daunting challenges to the design of protocols for wireless sensor networks. For one thing, the limited energy budget at the individual sensor level mandates the design of ultralightweight communication protocols. Likewise, issues concerning how the data collected by individual sensors could be queried and accessed, and how concurrent sensing tasks could be executed internally, are of particular significance. An important guideline in this direction is to perform as much local data processing as possible at the sensor level, avoiding the transmission of raw data through the network. Indeed, it is known that it costs 3 J of energy to transmit 1 kb of data a distance of 100 m. Using the same amount of energy, a general-purpose processor with the modest specification of 100 million instructions/watt executes 300 million instructions [20,21].

As a consequence, the wireless sensor network must be multihop, and only a limited number of the sensors count the sink among their one-hop neighbors. For reasons of scalability, it is assumed that no sensor knows the topology of the network.

4.3.1 Interfacing Wireless Sensor Networks

We assume that the wireless sensor network is connected to the outside world (e.g., point of command and control, the Internet, etc.) through a *sink*. The sink may or may not be collocated with the sensors in the deployment area. In case of a noncollocated sink, the interface with the outside world may be achieved by a vehicle driving by the area of deployment, or a helicopter, aircraft, or low earth orbit (LEO) satellite overflying the sensor network, and collecting information from a select group of reporting nodes. In such scenarios communication between individual sensors is by radio, while the reporting nodes are communicating with the noncollocated sink by radio, infrared, or laser [8,9]. One can easily contemplate a collection of mobile sinks for fault tolerance.

When the sink is *collocated* with the wireless sensor network, it can also be in charge of performing any necessary training and maintenance operations. Throughout this chapter we shall assume that the sink is collocated with the sensors, and we shall refer to it occasionally as *training agent* (TA, for short), especially in contexts where the sink engages in training operations. Moreover, we shall assume that the sink is centrally placed in the deployment area. This is for convenience only; it will be clear that the virtual infrastructure induced by the sink is topologically invariant to translating the sink out of its central position. A corollary of this is that our approach works equally well with eccentric sinks as well as with moving ones. We shall not elaborate this point further in this chapter.

4.3.2 Synchronization

The problem of synchronizing sensors has deep implications on the types of applications for which wireless sensor networks are a suitable platform. Not surprisingly, the synchronization problem has received a good deal of well-deserved attention in the recent literature [28,29]. To the best of our knowledge, all the synchronization strategies used are *active* in the sense that time awareness is propagated from sensor to sensor in the network. Our strategy is *passive* in the sense that the sensors synchronize to a master clock running at the sink. In addition to being simpler, our method promises to be far more accurate as we avoid the snowballing effect of errors inherent to active propagation.

Using the genetic material, each sensor can generate (pointers into) three sequences of random numbers as follows:

1. A sequence $t_1, t_2, \ldots, t_i, \ldots$ of time-epoch lengths
2. A sequence $n_1, n_2, \ldots, n_i, \ldots$ of frequency sets drawn from a huge universe, for example, the industrial, scientific, medical (ISM) band
3. For every i ($i \geq 1$), a permutation $f_1^i, f_2^i, \ldots$ of frequencies from n_i

The interpretation of these sequences is: time is ruled into epochs: during the ith time epoch, of length t_i, frequency set n_i is used, subject to the hopping sequence $f_1^i, f_2^i, \ldots$. Thus, as long as a sensor is synchronous to the TA, it knows the current time epoch, the offset into the epoch, the frequencies, and the hopping pattern for that epoch.

Suppose that the TA dwells τ microseconds on each frequency in the hopping sequence. For every i ($i \geq 1$), we let l_i stand for t_i/τ (assumed to be an integer); thus, epoch t_i involves a hopping sequence of length l_i. Think of epoch t_i as being partitioned into l_i slot, each slot using its own frequency selected by the hopping pattern from the set n_i. We refer the reader to Figure 4.1 where some of these ideas are illustrated. For example, time epoch t_{i-1} uses a set of frequencies $n_{i-1} = \{1, 3, 4, 5, 12, 13, 14, 15, 16\}$. Similarly, t_i uses the set of frequencies $n_i = \{2, 3, 6, 7, 10, 11, 12, 14\}$, while epoch t_{i+1} uses $n_{i+1} = \{4, 5, 8, 9, 13, 16\}$. The figure also illustrates the specific frequencies used in each slot.

It is clear that determining the epoch and the offset of the TA in the epoch is sufficient for synchronization. Our synchronization protocol is predicated on the assumption that sensor clock drift is bounded. Specifically, assume that whenever a sensor wakes up and its *local* clock shows epoch t_i, the master clock at the TA

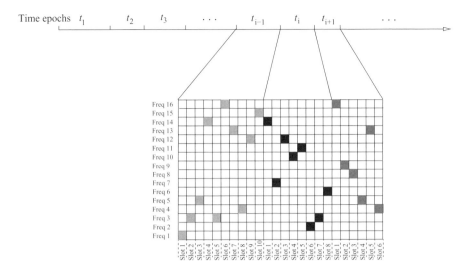

Figure 4.1 Sensor synchronization.

is in one of the time epochs t_{i-1}, t_i, or t_{i+1}. Using its genetic information, the sensor knows the last frequencies λ_{i-1}, λ_i, and λ_{i+1} on which the TA will dwell in the time epochs t_{i-1}, t_i, and t_{i+1}, respectively. Its strategy, therefore, is to tune in, cyclically, to these frequencies, spending $\tau/3$ time units on each of them. It is clear that eventually the sensor meets the TA on one of these frequencies. Assume, without loss of generality, that the sensor meets the TA on frequency λ in some (unknown) slot s of one of the epochs t_{i-1}, t_i, or t_{i+1}. To verify the synchronization, the sensor will attempt to meet the TA in slots $s + 1$, $s + 2$, and $s + 3$ at the start of the next epoch. If a match is found, the sensor declares itself synchronized. Otherwise, the sensor will repeat the process just delineated.

It is important to understand that the synchronization protocol outlined is probabilistic: even if a sensor declares itself synchronized, there is a slight chance that it is not. However, a missynchronization will be discovered quickly and the sensor will reattempt to synchronize.

4.4 LOCATION AWARENESS IN WIRELESS SENSOR NETWORKS

Consider a circular deployment area along with a centrally placed TA equipped with a long-range radio and a steady energy supply, that can communicate with the sensors in the deployment area. Recall that, as noted before, the role of the TA is played by the collocated sink.

It was recognized that some applications require that the collected sensory data be supplemented with location information, encouraging the development of communication protocols that are location-aware and perhaps location-dependent [7,30–33]. The practical deployment of many wireless sensor networks results in sensors initially *unaware* of their location: they must acquire this information post-deployment. Further, due to limitations in form factor, cost per unit and energy budget, individual sensors are not expected to be global positioning system (GPS)-enabled. Moreover, in many probable application environments, including those inside buildings, hangars, or warehouses, satellite access is drastically limited.

The *location awareness* problem, then, is for individual sensors to acquire location information either in absolute form (e.g., geographic coordinates) or relative to a reference point. The *localization* problem is for individual sensors to determine, as closely as possible, their geographic coordinates in the area of deployment. Prominent solutions to the localization problem are based on multilateration or multiangulation [30–36]. Most of these solutions assume the existence of several *anchors* that are aware of their location (perhaps by endowing them with a GPS-like device). Sensors receiving location messages from at least three sources can approximate their own locations. For a good survey of localization protocols for wireless sensor networks, we refer the reader to ref. [37].

For the sake of completeness, we now outline a very simple localization protocol for wireless sensor networks that does not rely on multiple anchors.

4.4.1 A Simple Localization Protocol for Wireless Sensor Networks

The task of *localization* is performed immediately after deployment. If the sensors are considered stationary, localization is a one-time operation.[1] Unlike the vast majority of existing protocols that rely heavily on multilateration or multiangulation and on the existence of a minimum of three anchors with known geographic position, our protocol only requires one anchor—the TA—whose role can be played by a collocated sink. The key idea of our protocol is to allow each sensor to determine its position in a polar coordinate system centered at the TA. In particular, each sensor determines its *polar angle* with respect to a standard polar axis as well as a *polar distance* to the TA.

Referring to Figure 4.2, assume without loss of generality that the TA is centrally located.[2] The TA knows its own geographic coordinates, is not energy constrained and it has (highly) directional transmission capabilities.

For some predetermined time, the TA transmits a rotating beacon, as illustrated in Figure 4.2. The rotation is uniform with a period of T time units, known to all the sensors in the deployment area. Every time the beacon coincides with the polar axis the TA transmits a synchronization signal on a channel λ, known to the sensors.

In outline, the protocol is as follows. A generic sensor a wakes up according to its internal clock. It listens to channel λ for T time units. Let t_0 be the moment at which

---

[1] In fact, even if the sensors are stationary, they may move from their original deployment position due to such factors as wind, rain, and small ground tremors.

[2] The reader should have no difficulty confirming that this is assumed for convenience and the eccentric TA case is perfectly similar.

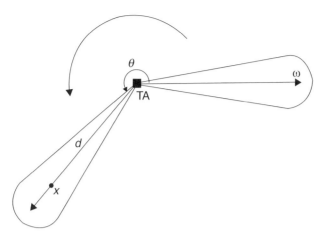

Figure 4.2   The localization protocol.

it hears the synchronization beacon. At that point it switches to channel μ, on which the rotating beacon is transmitted. Assume that the rotating beacon is received by sensor a at time t_1. The polar angle θ corresponding to a is

$$\theta = \frac{2\pi(t_1 - t_0)}{T} \qquad (4.1)$$

Similarly, the polar distance ρ can be determined by using the well-known formula

$$\rho = \left(\frac{P_T}{cP_R}\right)^{1/\alpha} \qquad (4.2)$$

where

P_T and P_R represent, respectively, the transmitted and received energy levels

c and α are constants that depend on the atmospheric conditions at the moment when the localization takes place. These values may be passed on by the TA, along with P_T.

It is worth noting that a sensor may perform several determinations of θ and ρ and use averages to improve the accuracy of the localization. Indeed, once t_1 is known, the sensor can go to sleep until time $t_1 + T$, at which it knows that it needs to wake up to receive the beacon again.

In some other applications, exact geographic location is not necessary: all that individual sensors need is *coarse-grain* location awareness. There is an obvious trade-off: coarse-grain location awareness is lightweight, but the resulting accuracy is only a rough approximation of the exact geographic coordinates. In this chapter we show that sensors acquire coarse-grain location awareness by the *training* protocol that imposes a coordinate system onto the network. An interesting by-product of our training protocol is that it provides a partitioning into clusters and a structured topology with natural communication paths. The resulting topology will make it simple to avoid collisions between transmissions of nodes in different clusters, between different paths and also between nodes on the same path. This is in contrast with the majority of papers that assume routing along spanning trees with frequent collisions.

4.5 THE VIRTUAL INFRASTRUCTURE

The main goal of this section is to present a broad overview of the main components of the proposed general-purpose virtual infrastructure for wireless sensor networks.

4.5.1 A Dynamic Coordinate System

To help with organizing the virtual infrastructure we assume a centrally placed TA, equipped with a long-range radio and a steady energy supply, that can communicate with both the sink and the sensors in the deployment area.

Referring to Figure 4.3(a) the coordinate system divides the wireless sensor network area into equiangular wedges. In turn, these wedges are divided into *sectors* by means of concentric circles or coronas centered at the TA (sink). As will be discussed in Subsection 4.5.2, the sensors in a given sector map to a cluster, the

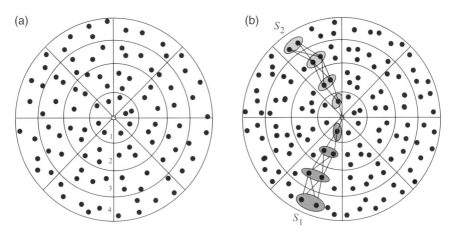

Figure 4.3 Different perspectives of the dynamic coordinate system: (a) the dynamic system, and (b) routing in a wireless sensor network.

mapping between clusters and sectors being one-to-one. The task of training a wireless sensor network involves establishing:

Coronas. The deployment area is covered by k coronas determined by k concentric circles of radii $0 < r_1 < r_2 < \cdots < r_k \le t_x$ centered at the sink.

Wedges. The deployment area is ruled into a number of angular wedges centered at the sink.

As illustrated in Figure 4.3(a), at the end of the training period each sensor has acquired two coordinates: the identity of the corona in which it lies, as well as the identity of the wedge to which it belongs. It is important to note that the locus of all the sensors that have the same coordinates determines a cluster.

4.5.2 The Cluster Structure

Clustering was proposed in large-scale networks as a means of achieving scalability through a hierarchical approach. For example, at the medium access layer, clustering helps increase system capacity by promoting the spatial reuse of the wireless channel; at the network layer, clustering helps reducing the size of routing tables and striking a balance between reactive and proactive routing. It is intuitively clear that wireless sensor networks benefit a great deal from clustering; indeed, separating concerns about intercluster management and the intracluster management can substantially decrease and load balance the management overhead. Given the importance of clustering, a number a clustering protocols for wireless sensor networks have been proposed in the recent literature [38–40]. However, virtually all clustering protocols for wireless sensor networks assume tacitly or explicitly that individual sensors have identities.

The dynamic coordinate system suggests a simple and robust *clustering scheme*: a cluster is the locus of all sensors having the same coordinates. It is important to note that clustering is obtained for free once the coordinate system is established. Also, our clustering scheme does not assume synchronization and accommodates sensor anonymity: sensors need not know the identity of the other sensors in their cluster. For an illustration, refer again to Figure 4.3(a). Each sector in the dynamic coordinate system represents a cluster; indeed, as is easily visible, the sensors in a sector share the same coordinates: the same corona number and the same wedge number.

Recently Olariu et al. [27] showed that one can augment the virtual infrastructure with a task-based management system where clusters are tasks with sensing, routing, or collective data storage.

4.6 THE LIGHTWEIGHT TRAINING PROTOCOL

The model for a wireless sensor network that we adopt assumes that after deployment the sensors must be trained before they can be operational. Recall that sensors do not have identities and are initially unaware of their location. It follows that untrained nodes are not addressable and cannot be targeted to do work in the network. The main goal of this section is to present, in full detail, our lightweight, highly scalable training protocol for wireless sensor networks. The key advantage of this protocol is that each sensor participating in the training incurs an energy cost that is logarithmic in the number of clusters and wedges defined by the protocol. Being energy-efficient, this training can be repeated on a scheduled or ad hoc basis, providing robustness and dynamic reorganization.

After deployment the individual sensors sleep until wakened by their individual timers. Thus, each sensor sleeps for a random period of time, wakes up briefly, and if it hears no messages of interest, selects a random number x and returns to sleep x time units. Clocks are not synchronized, but over any time interval $[t, t + \Delta t]$ a percentage directly proportional to Δt of the nodes are expected to wake up briefly. During this time interval the sink continuously repeats a call to training, specifying the current time and a rendezvous time. Thus, in a probabilistic sense a certain percentage of the sensor population will be selected for training. The time interval Δt can be adjusted to control the percentage of sensors that is selected. Using the synchronization protocol described in Subsection 4.3.2 the selected sensors reset their clocks and set their timer appropriately before returning to sleep.

4.6.1   The Corona Training Protocol

The main goal of this subsection is to present the details of the corona training protocol. The wedge training protocol being quite straightforward will not be discussed further in this chapter.

Let k be an integer[3] known to the sensors and let the k coronas be determined by concentric circles of radii $0 < r_1 < r_2 < \cdots < r_k \le t_x$ centered at the sink.

The idea of the corona training protocol is for each individual sensor to learn the identity of the corona to which it belongs. For this purpose, each sensor learns a string of $\log k$ bits, from which the corona number can be determined easily. To see how this is done, it is useful to assume time ruled into slots $s_1, s_2, \ldots, s_{k-1}$ and that the sensors synchronize to the master clock running at the sink, as discussed in Subsection 4.3.2.

In time slot s_1 all the sensors are awake and the sink uses a transmission range of $r_{k/2}$. As a net effect, in the first slot the sensors in the first $k/2$ coronas will receive the message above a certain threshold, while the others will not. Accordingly, the sensors that receive the signal set $b_1 = 0$, the others set $b_1 = 1$.

Consider a k-leaf binary tree T and refer to Figure 4.4. In the figure the leaves are represented by *boxes* numbered left to right from 1 to k. It is very important to note that the intention here is for the k boxes to represent, in left-to-right order, the k coronas. The training protocols is for individual sensors to determine the "box" (i.e., the corona) to which they belong.

---

[3]For simplicity, we shall assume that k is a power of 2.

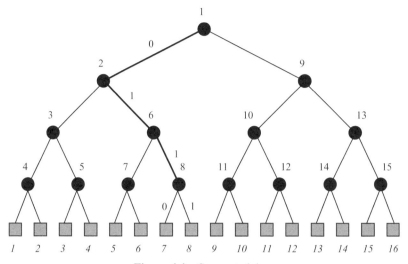

Figure 4.4 Corona training.

The edges of T are labeled by 0s and 1s in such a way that an edge leading to a left subtree is labeled by a 0 and an edge leading to a right subtree is labeled by a 1. Let l ($1 \le l \le k$) be an arbitrary leaf, and let $b_1, b_2, \ldots, b_{\log k}$ be the edge labels of the unique path leading from the root to l. It is both well known and easy to prove by a standard inductive argument that

$$l = 1 + \sum_{j=1}^{\log k} b_j \frac{k}{2^j} \tag{4.3}$$

As an illustration, applying equation (4.3) to leaf 7, we have $7 = 1 + 0 * 2^3 + 1 * 2^2 + 1 * 2^1 + 0 * 2^0$.

Referring again to Figure 4.4, let the *interior* nodes of T be numbered in *preorder* from 1 to $k - 1$, and let T' be the tree consisting of the interior nodes only.[4] Let u be an arbitrary node in T', and let $b_1, b_2, \ldots, b_{i-1}$ be the edge labels on the unique path from the root to u. We take note of the following technical result.

Lemma 4.1: Let $p(u)$ be the preorder number of u in T'. Then, we have

$$p(u) = 1 + \sum_{j=1}^{i-1} c_j$$

---

[4]In other words, T' is the tree obtained from T by ignoring the last level (i.e., the "boxes").

where

$$
c_j = \begin{cases} 1 & \text{if } b_j = 0 \\ \dfrac{k}{2^j} & \text{if } b_j = 1 \end{cases}
$$

Proof: The proof is by induction on the depth i of node u in T'. To settle the basis, note that for $i = 1$, u must be the root and $p(u) = 1$, as expected.

For the inductive step, assume the statement true for all nodes in T' of depth less that u. Indeed, let v be the parent of u and consider the unique path of length $i - 1$ joining the root to u. Clearly, nodes u and v share $b_1, b_2, \ldots, b_{i-2}$ and, thus, $c_1, c_2, \ldots, c_{i-2}$. By the inductive hypothesis,

$$
p(v) = 1 + \sum_{j=1}^{i-2} c_j \tag{4.4}
$$

On the other hand, since v is the parent of u, we can write

$$
p(u) = p(v) + \begin{cases} 1 & \text{if } u \text{ is the left child of } v \\ \dfrac{k}{2^{i-1}} & \text{otherwise} \end{cases} \tag{4.5}
$$

Notice that if u is the left child of v we have $b_{i-1} = 0$ and $c_{i-1} = 1$; otherwise, $b_{i-1} = 1$ and $c_{i-1} = k/2^{i-1}$. This observation, along with equations (4.4) and (4.5) combined, allows us to write

$$
p(u) = 1 + \sum_{j=1}^{i-2} c_j + c_{i-1} = 1 + \sum_{j=1}^{i-1} c_j
$$

completing the proof of the lemma. ∎

Let u be an arbitrary node of T' and let $n(u)$ denote its inorder number in T'. Let m be the left-to-right rank among the leaves of T of the rightmost leaf of the left subtree of T rooted at u.

Lemma 4.2: $n(u) = m$.

Proof: We proceed by induction on the inorder number of a node in T'. Indeed, if $n(u) = 1$, then u must be the leftmost leaf in T' and, thus, its left subtree in T consists of the leftmost leaf of T', settling the base case.

Assume that the statement is true for all nodes of T' with inorder number smaller than that of u. we shall distinguish between the following two cases:

Case 1: v is an ancestor of u in T'. Let $T'(v)$ be the subtree of T' rooted at v. In this case, u must be the leftmost leaf in the right subtree of $T'(v)$. Let q be the left-to-right rank among the leaves of T of the rightmost leaf of the left subtree of $T'(v)$. By the inductive hypothesis, $n(v) = q$. Since u is a leaf in T', it has exactly two children in T, namely, the leaves of ranks $q + 1$ and $q + 2$. Thus, in this case, $n(u) = n(v) + 1 = q + 1$, as claimed.

Case 2: u is an ancestor of v in T'. Let $T'(u)$ be the subtree of T' rooted at u. In this case, v must be the rightmost leaf in the left subtree of $T'(u)$. Assume that $n(v) = r$. Observe that v has exactly two leaf children T. By the induction hypothesis, these children have ranks r and $r + 1$. Thus, in this case, $n(u) = n(v) + 1 = r + 1$, as claimed.

This completes the proof of the lemma. ∎

To illustrate Lemma 4.2, refer again to Figure 4.4 and let u be the internal node labeled "6." Recall that the tree T' consists of the tree T with the level removed. It is easy to verify that "6" is, in fact, the inorder number of u in T'. By Lemma 4.2 this coincides with the label of the box that is the leftmost leaf in the right subtree of $T'(v)$ rooted at u.

With these technicalities out of the way, we now return to the corona training protocol. In our setting, the preorder and inorder numbers of internal nodes in T correspond, respectively, to time slots in the training protocol and to the transmission ranges used by the sink. More precisely, consider an arbitrary integer i, $(2 \leq i \leq \log k - 1)$, and assume that at the end of time slot s a sensor has learned the leftmost $i - 1$ bits $b_1, b_2, \ldots, b_{i-1}$. The following important result is implied by Lemma 4.1 and Lemma 4.2.

Theorem 4.1: Having learned bits $b_1, b_2, \ldots, b_{i-1}$, a sensor must wake up in time slot $z = 1 + \sum_{j=1}^{i-1} c_j$ to learn bit b_i. Moreover in time slot z the sink uses a transmission range of $r_{inorder(z)}$.

To illustrate Theorem 4.1, refer again to Figure 4.4 where the internal nodes are labeled by their preorder numbers. Consider the node labeled 2. It is easy to verify that its inorder number is 4. Thus, all the nodes in the subtree rooted at 2 will be awake in slot 2 and the sink will transmit with a transmission range of r_4. Consequently, the sensors at a distance from the sink not exceeding r_4 will receive the signal, while the others will not.

It is also worth noting that only the sensors that need to be awake in a given time slot will stay awake; the others will sleep, minimizing the energy expenditure. Yet another interesting feature of the training protocol we just described is that individual sensors sleep for as many contiguous slots as possible before waking up, thus avoiding repeated wake–sleep transitions that are likely to waste energy.

At the same time, in case the corona training process has to be aborted before it is complete, Theorem 4.1 guarantees that if the training process restarts at some later point, every sensor knows the exact time slots when it has to wake up in order to learn its missing bits.

Making the training protocol secure is especially important, since training is a prerequisite for subsequent network operations. Recently, Jones et al. [26] and Wadaa et al. [41,42] have shown that the training protocol described earlier can be made secure.

4.7 TASK-BASED DATA PROCESSING AND COMMUNICATION

The goal of this section is to describe a task-based data-processing and communication system for wireless sensor networks that exploits the virtual infrastructure introduced in this chapter. For this purpose, we shall adopt the view that the wireless sensor network performs tasks mandated by a remote end user. The end user issues queries expressed in terms of high-level abstractions, to be answered by the network. The middleware, running at the sink, provides the interface between the application layer (where the end user resides) and the wireless sensor network. Specifically, the sink parses the queries from the application layer, considers the current capabilities of the network including the remaining energy budget and negotiates a contract with the application layer before committing the network [42]. After a contract has been agreed upon, the middleware translates the corresponding query into low-level tasks, assigned to individual clusters. The clusters must then perform these tasks and send the aggregated data back to the sink for consolidation. The consolidated information is then passed on to the application layer.

4.7.1 Associating Sensors with Tasks

For our purposes a task is a tuple $T(A, S, E)$, where

- A describes the action to be performed (i.e., detecting physical intrusion into the deployment area).
- S specifies the identity of the cluster tasked with data collection (sensing).
- E specifies the minimum energy level required of sensors participating in the task.

The suitably aggregated data collected by the sensors is to be routed to the sink before being uploaded to the end user. In addition to the sensors in cluster S, a number of sensors are selected to act as *routers*, relaying the data collected to the sink. Collectively, these sensors are the *workforce* $W(T)$ associated with T.

The process by which $W(T)$ is selected follows. During a time interval of length Δ the sink issues a *call for work* containing the parameters of T. The sensors in the same wedge as S and with corona numbers smaller than that of S that happen to be awake during the interval Δ and that satisfy the conditions specified (membership in S and energy level) stay awake and constitute $W(T)$. It is intuitively clear that by knowing the number of sensors, the density of deployment and the expected value of sleep periods, one can fine-tune Δ in such a way that $W(T)$ is commensurate with the desired grade of service. It is extremely important to note that, as discussed in Subsection 4.3.2, a by-product of the call for work is that all the sensors in $W(T)$ are synchronized for the duration of the task.

For an illustration of the concepts discussed in this subsection, we refer to Figure 4.3(b). In the figure two tasks are in progress. One of these tasks has mandated sensors in cluster S_1 to collect data in support of a query. The sensors associated with this task as routers are those in the outlined sets in the same wedge as S_1. Since the width of each corona does not exceed the maximum transmission range t_x, communication between sensors in adjacent coronas is assumed. Also note that the sensors that constitute the workforce of this transaction are synchronized. As for the transmission of data, all the sensors in the same sector transmit at the same time. As will be discussed in detail in Subsection 4.8.2, one of the benefits of our scheme is that data aggregation can be accomplished in a straightforward manner.

The figure features a second task that involves data collection in a cluster S_2 along with its workforce. As will be discussed in the next subsection, there is no collision between the two tasks, as they use a different set of frequencies.

4.7.2 Task-Based Synchronization

The generic synchronization protocol discussed earlier in this chapter can be used as a building block for a more sophisticated task-based synchronization protocol. The motivation is to support multitasking. Indeed, it is often desirable for the sensors in a cluster to perform several tasks in parallel.[5] However, any attempt at synchronization using the generic synchronization protocol will result in all the concurrent tasks using *exactly* the same frequency set and the same hopping sequence, creating frequent collisions and the need for subsequent retransmission.

Suppose that we wish to synchronize the workforce $W(T)$ of a task T that uses some color class c and that the generic synchronization protocol would show that the actual time epoch is t_i. The idea is to use the perfect hash function ϕ to compute a *virtual* time epoch t_j with $j = \phi(i, k(c), T)$ to be used by $W(T)$. Therefore, the sensors in $W(T)$ will act as if the real time were t_j, using the frequency set n_j and the frequency hopping sequence $f_1^j, f_2^j, \ldots$. Thus, different concurrent tasks will employ different frequency sets and hopping sequences minimizing the occurrence of collisions.

4.8 ROUTING AND DATA AGGREGATION

The main goal of this section is to show that once a wireless sensor network has been trained, both routing and data aggregation become easy and straightforward.

---

[5]However, the sets of sensors allocated to these tasks *must* be disjoint.

4.8.1 Routing

The routing problem in sensor networks differs rather substantially from routing in other types of wireless networks. For one thing, individual sensors are anonymous, lacking identities; thus, standard addressing methods do not work directly. For another reason, the stringent energy limitations present in the sensor network render the vast majority of conventional routing protocols impractical.

Given the importance of routing, it is not surprising to see that a number of routing protocols specifically designed for wireless sensor networks were proposed in the literature [15,43–46]. For example, in ref. [15] Intanagonwiwat et al. describe *directed diffusion* and a companion routing protocol based on interest tables at the expense of maintaining a cache of information indexed by interest area at each node. Shah and Rabaey [46] responds to client requests by selecting paths that maximize the longevity of the network rather than minimize total energy consumed by a path with path options established by local flooding. Other routing protocols include rumor routing [43], and multipath routing [44], among others. As we are about to demonstrate, our training protocol provides a novel solution to the routing problem by yielding energy-efficient paths-based routing.

Recall that sensor networks are multihop. Thus, in order for the sensing information to be conveyed to the sink, routing is necessary. Our cluster structure allows a very simple routing process, as described in the following paragraphs. The idea is that the information is routed within its own wedge along a virtual path joining the outermost sector to the sink, as illustrated in Figure 4.3(b). The collection of all the virtual paths (one per wedge) defines a tree. In this tree, each internal node, except for the root, has exactly one child, eliminating medium access control (MAC)–level contention in sending sensor information to the sink.

Recently, a number of MAC-layer protocols for wireless sensor networks have been proposed in the literature [47–49]. In fact, in our routing scheme by appropriately staggering transmissions in neighboring wedges, collision and, therefore, the need for retransmissions is completely eliminated. Thus, our training protocol implies an efficient MAC protocol as well.

4.8.2 Data Aggregation

Once sensory data is collected by a multitude of sensors, the next important task is to consolidate the data in order to minimize the amount of traffic to the sink. We place the presentation in the context of our work model. To be more specific, we assume that the cluster identified by (i,j)—that is, the set of sensors located in sector $A_{i,j}$— are tasked to perform a certain task T. A number of sensors in sectors $A_{i,1}, A_{i,2}, \ldots, A_{i-1,j}$ are selected to act as *routers* of the data collected by the sensors in $A_{i,j}$ to the sink. Collectively, these sensors are the *support* sensors of task T.

It is, perhaps, of interest to describe the process by which the sensors associated with T are selected. To begin, during a time interval of length Δ the sink will issue a *call for work* specifying the identity j of the wedge in which the task is to be performed, as well as the identity i of the corona in which data are to be collected.

The sensors in wedge j that happen to wake up during the interval Δ and that have an appropriate energy level stay awake and will participate in the task either as data collectors or as routers, depending on their respective position within the wedge. It is intuitively clear that by knowing the number of sensors, the density of deployment and the expected value of sleep periods, one can fine-tune Δ in such a way that a suitable number of routers will be awake in wedge j in support of $\mathcal{T}$. Likewise, we can select the set $\mathcal{D}$ of data collecting sensors in $A_{i,j}$. Let $\mathcal{S}$ denote the set of support sensors for $\mathcal{T}$. It is appropriate to recall that a by-product of the call for work is that all the sensors in $\mathcal{S}$ are synchronized. In order to make the task secure the sensors in $\mathcal{S}$ will share a secret key that allows them access to a set of time epochs, a set of frequencies to be used in each time epoch, and a hopping sequence to be used within each epoch. For details, we refer the reader to Section 4.2.

Assume that the results of the data collection specific to task $\mathcal{T}$ can be partitioned into 2^m, $(m \geq 0)$, disjoint groups. Thus, each sensor performing data collection will encode its data in a string of m bits.

Since, typically, $\mathcal{D}$ contains a large number of sensors, it is important to *fuse* individual results into a final result that will be sent to the sink. We now outline a possible solution to the data-aggregation problem. Using the algorithms of Nakano and Olariu [50,51] which do not require sensors to have identities, the sensors in $\mathcal{D}$ acquire *temporary* identities ranging from 1 to $|\mathcal{D}|$. Using their newly acquired identities, individual data values are being transmitted to the sensor whose identity is 1, which will perform data aggregation and will send the final result to the sink. The advantage of this data-aggregation scheme is that there is no data loss and all the collected values will be correctly fused. There are, however, many disadvantages. For one thing, the initialization algorithm of [50] requires every sensor in $\mathcal{D}$ to expend an amount of energy proportional with $\log|\mathcal{D}|$. For another, the final result of the data collection is concentrated in a single sensor (i.e., the sensor with temporary identity 1), which is a single point of failure.

We now propose a much simpler data-aggregation scheme that involves some data loss, but that is fault tolerant and does not require the sensors in $\mathcal{D}$ to have unique identities. The idea is that the sensors in $\mathcal{D}$ transmit the data collected bit by bit, starting, say, left to right, as follows: a value of 0 is not transmitted, while a 1 will be transmitted. The sensors in $A_{i-1,j}$ that have been elected as routers in support of task $\mathcal{T}$ pick up the values transmitted. The following disambiguation scheme is used:

- No bit is received—in this case, a 0 is recorded.
- A bit of 1 is received—in this case, a 1 is recorded.
- A collision is recorded—in this case a 1 is recorded.

It is clear that as a result of this disambiguation scheme, every sensor in $A_{i-1,j}$ that is in support of $\mathcal{T}$ stores the logical OR of the values stored by sensors in $\mathcal{D}$. Note also that while there was loss of information in the process of fusing data, no further loss can occur in traversing the path from $A_{i-1,j}$ to the sink: this is because all routers in $A_{i-1,j}$ transmit the same bit string.

4.8.3   An Example

For an example of data aggregation consider a wireless sensor network that is tasked to monitor and report the temperature in cluster $A_{i,j}$. Referring to Table 4.1, for the application at hand temperatures below 111°F are considered to be noncritical, and if such a temperature is reported, no specific action is to be taken. By contrast, temperatures above 111°F are considered to be critical, and they trigger a further monitoring action. The encoding featured in Table 4.1 is specifically designed to reflects the relative importance of various temperature ranges. For example, the temperature ranges in the noncritical zone are twice as large as those in the critical zone. Also, notice that the leftmost bit differentiates critical from noncritical temperatures. Thus, if the sink receives a reported temperature whose leftmost bit is a 1, then further action is initiated; if, on the other hand, the leftmost bit is 0, then no special action is necessary.

Let us see how our data aggregation works in this context. Referring to Figure 4.5, assume that a group of three sensors in $A_{i,j}$ have collected data and are about to transmit them to the sensors in $A_{i-1,j}$. The values collected are encoded, respectively, as 0110, 0101, and 0110. Thus, none of the values indicates a critical situation. After transmission and disambiguation, the sensors in $A_{i-1,j}$ will store 0111, which is the logical OR of the values transmitted. Notice that although the data-aggregation process involves loss of information, we do not loose critical information. This is because the logical OR of noncritical temperatures must remain noncritical. Conversely, if the logical OR indicates a critical temperature, one of the fused temperatures must have been critical, and thus action must be initiated. It is also interest-

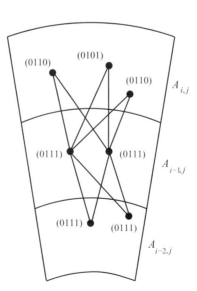

Figure 4.5   Data aggregation.

TABLE 4.1 Temperature Ranges and Their Encoding

Temperature	51–60	61–70	71–80	81–90	91–100	101–110	111–115	116–120	121–125	126–130	131–135	136–140	141–145	146–150
Code	0010	0011	0100	0101	0110	0111	1000	1001	1010	1011	1100	1101	1110	1111

ing to note that when the sensors in $A_{i-1,j}$ transmit to those in $A_{i-2,j}$, no further loss of information occurs.

4.8.4 Lossless Aggregation

It is worth noting that there is an interesting interplay between the amount of loss in data aggregation and the amount of energy expended to effect it. As we are about to show, if we are willing to expend slightly more energy, lossless data aggregation can be achieved.

The corresponding trade-off is interesting in its own right, being characteristic of choices that present themselves in the design of protocols for wireless sensor networks. For illustration purposes, assume that it is necessary to determine the maximum of the bit codes stored by the sensors in $A_{i,j}$ and refer to Figure 4.6.

To solve this problem, all the sensors in $A_{i,j}$ that have collected relevant information engage in the following protocol, which is guaranteed to aggregate the values into the maximum. Assume that each sensor stores a d-bit code for the range.

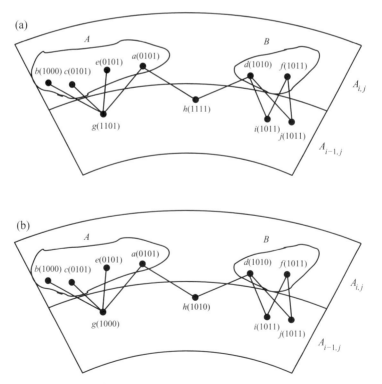

Figure 4.6 Lossless data aggregation.

Protocol (Correct_Maximum): For every position p starting with the most significant bit to the least:

1. Sensors in $A_{i,j}$ that have a 0 in position p listen for two time slots; if in any of these slots a 1 or a collision message is received, they terminate their participation in the protocol.
2. Sensors that have a 1 in position p transmit in the first time slot and sleep in the second.
3. Sensors in $A_{i-1,j}$ do the following:
   3.1. Any sensor that has received a 1 or a collision in the first time slot, echoes a 1 in the second.
   3.2. Any sensor that has not received a transmission in the first slot sleeps in the second slot.

To see why the two time slots for transmitting a single bit are necessary consider the situation depicted in Figure 4.6(a) and the following simple "algorithm":

Protocol (Incorrect_Maximum): For every position p starting with the most significant bit to the least:

1. Sensors in $A_{i,j}$ that have a 0 in position p listen; if a 1 or a collision message is received, they terminate their participation in the protocol.
2. Sensors that have a 1 in position p transmit.

Figure 4.6(a) depicts the case where, due to energy depletion the sensors that participate in the protocol are sparsely deployed. Implicit in the protocol Incorrect_Maximum is that every sensor can hear the transmission of every other sensor. In particular, notice that in group A sensor a does not hear the transmission of sensor b and continues transmitting even though it should not. Indeed, for this reason, the value received by sensor g in $A_{i-1,j}$ is not the correct maximum of values stored by the sensors in group A. A similar situation occurs when sensor h in $A_{i-1,j}$ heard the transmission of sensors a in group A and d in group B. Clearly h stores a value that corresponds to no maximum.

Notice how protocol Correct_Maximum is sidestepping this difficulty. The transmission of a single bit is separated into two time slots: first, all the sensors in $A_{i,j}$ transmit their corresponding bit. In the second slot, the sensors in $A_{i-1,j}$ echo back the values received. Since the sensor in $A_{i,j}$ that store a 0 listen for two time slots, they will realize that some sensor in $A_{i,j}$ has a 1 in that bit position and, consequently, they should drop out. The result is illustrated in Figure 4.6(b).

4.9 EVALUATING ROUTING-RELATED ENERGY EXPENDITURE

The main goal of this section is to explore the problem of energy expenditure related to routing data in a wireless sensor network. Indeed, we adopt a task-based model

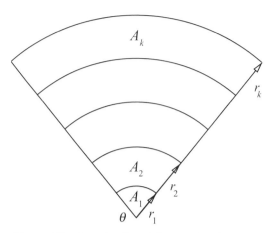

Figure 4.7 A wedge W and the associated sectors.

[27,41,42] whereby the sensor network is subjected to a set T of tasks. Each task involves the nodes in a sector (i.e., a cluster) and involves performing local sensing by the sensors, data aggregation, and sending the resulting information to the sink. Recall that, as discussed in Section 4.8, one of the key benefits of our training is that transmitting the result of the task from a sector to the sink amounts to routing the information along a path lying within the same wedge (see also Fig. 4.3(b)). Thus, we associate each task with such a path. We will now analyze the energy expended by sensors to fulfill their path-related duties.

Throughout the remainder of this chapter we assume a sensor network deployed in a circular area and a collocated sink placed at its center. Consider a wedge W subtended by an angle of θ and refer to Figure 4.7. The wedge W is partitioned into k sectors $A_1, A_2, \ldots, A_k$ by its intersection with k concentric circles, centered at the sink, and of monotonically *increasing* radii $r_1 < r_2 < \cdots < r_k$. It is important to note that r_k, the deployment radius, is a system parameter, and thus a constant for a particular sensor network.

For convenience of notation we write $r_0 = 0$ and interpret A_0 as the sink itself. Let t_x denote the *maximum* transmission range of a sensor.[6]

Let n denote the total number of sensors deployed in wedge W. We assume a uniform deployment with density ρ. In particular, with A standing for the area of wedge W, we can write

$$n = \rho A = \frac{\rho \theta}{2} r_k^2 \qquad (4.6)$$

Let $n_1, n_2, n_3, \ldots, n_k$ stand for the number of nodes deployed in the sectors $A_1, A_2, A_3, \ldots, A_k$, respectively. Since the deployment is uniform, it is easy to

---

[6]Of course, t_x is a system parameter that depends on the particular type of sensors deployed.

confirm that for every i $(1 \leq i \leq k)$,

$$n_i = \rho A_1 = \frac{\rho \theta}{2}(r_i^2 - r_{i-1}^2). \tag{4.7}$$

Let N denote the number of sector-to-sink paths (henceforth, simply denoted by *paths*) that the wedge W sees during the lifetime of the sensor network. By our previous discussion there is a one-to-one map between paths and tasks. Thus, N equals the total number T of tasks that the wedge can handle during the lifetime of the network.

We make the following assumptions motivated by the uniformity of the deployment:

- Each sensor in W is equally likely to be the source of a path to the sink
- For $2 \leq i \leq k$, each sensor in sector A_{i-1} is equally likely to serve as the next hop for a path that involves a node in A_i.

By virtue of the first assumption, the expected number of paths originating at a node in W is

$$\frac{N}{n} \tag{4.8}$$

Consider sector A_1. Since the N paths have the sink as their destination, the nodes in sector A_1 must collectively participate in all the N paths. Since A_1 contains n_1 nodes, the expected number of transmissions per node is N/n_1. Assuming a power-degradation factor of α, $2 \leq \alpha \leq 6$, the energy expended by a node in A_1 per path served is $r_1^\alpha + c$ for some *nonnegative* constant c. Thus, the total energy E_1 consumed by a node in A_1 to fulfill its routing duties is

$$E_1 = \frac{N}{n_1}\left[r_1^\alpha + c\right]$$

which, by equation (4.7), can be written as

$$E_1 = \frac{N}{n_1}\left[r_1^\alpha + c\right] = \frac{2N}{\rho \theta r_1^2}\left[r_1^\alpha + c\right] = \frac{2N}{\rho \theta}\left[r_1^{\alpha-2} + \frac{c}{r_1^2}\right] \tag{4.9}$$

It is very important to note that equation (4.9) allows us to determine the optimal value r_1^{opt} of r_1 that minimizes the value of E_1. For later reference, we note that this value is

$$r_1^{opt} = \begin{cases} t_x & \text{if } \alpha = 2 \\ \min\left\{\left(\frac{2c}{\alpha - 2}\right)^{1/\alpha}, t_x\right\} & \text{if } 2 < \alpha \leq 6 \end{cases} \tag{4.10}$$

Let $\overline{T}$ denote the total number of tasks performed by the entire wireless sensor network (not just wedge W) during its lifetime, and let $\overline{N}$ be the corresponding number of node-to-sink paths. Assuming that the $\overline{T}$ tasks are uniformly distributed throughout the sensor network, we can write

$$\frac{\overline{N}}{2\pi} = \frac{N}{\theta} \tag{4.11}$$

By equations (4.9) and (4.11) combined, the total energy needed by a node in A_1 to handle its routing duties is

$$E_1 = \frac{2N}{\rho\theta}\left[r_1^{\alpha-2} + \frac{c}{r_1^2}\right] = \frac{\overline{N}}{\rho\pi}\left[r_1^{\alpha-2} + \frac{c}{r_1^2}\right] \tag{4.12}$$

Let E denote the total energy budget of a sensor. Since the sensors in A_1 must have sufficient energy to handle their routing duties, by using equation (4.12) we can write

$$\frac{\overline{N}}{\rho\pi}\left[r_1^{\alpha-2} + \frac{c}{r_1^2}\right] < E$$

Recalling that in our working model there is a one-to-one correspondence between tasks and sector-to-sink paths, this inequality can be written in its equivalent form

$$\frac{\overline{T}}{\rho\pi}\left[r_1^{\alpha-2} + \frac{c}{r_1^2}\right] < E \tag{4.13}$$

4.9.1   Reasoning About the System Parameters

Inequality equation (4.13) can be interpreted in several ways, each expressing a different view of the limiting factors inherent to the sensors deployed. The goal of this subsection is to look at some of possible interpretations of inequality (4.13).

1. *Network Longevity*: We interpret $\overline{T}$, the number of transactions that the system can sustain during its lifetime as the *network longevity*. Thus, inequality (4.13) allows us to write

$$\overline{T} < \frac{\rho\pi E r_1^2}{r_1^\alpha + c} \tag{4.14}$$

which tells us that the longevity of the system is upper bounded by the ratio (4.14). More specifically, the longevity is directly proportional to the deployment density and to the reciprocal of $r_1^\alpha + c$. Consequently, if we wish to design a wireless sensor network that must sustain a given number $\overline{T}$ of transactions, we must select the deployment density as well as the radius of the first corona accordingly. We also need to chose sensors packing an amount of energy compatible with ratio (4.14).

2. *Maximum Transmission Range Close to the Sink*: First, assuming a *known* deployment density[7] ρ, inequality (4.13) shows that for a given energy budget E, in order to guarantee a desired network longevity of $\overline{T}$ tasks, the (maximum) transmission radius of sensors deployed in close proximity to the sink must satisfy

$$r_1^{\alpha-2} + \frac{c}{r_1^2} < \frac{\pi\rho E}{\overline{T}} \qquad (4.15)$$

with the additional constraint that $r_1 \le t_x$ where, recall, t_x stands for the *maximum* transmission range of a sensor.

3. *Deployment Density*: Likewise, for a selected radius r_1 ($t_x \ge r_1$), and for a given energy budget E, in order to guarantee a network *longevity* of $\overline{T}$ tasks, the deployment density ρ must satisfy the inequality

$$\rho > \frac{\overline{T}[r_1^\alpha + c]}{E\pi r_1^2} \qquad (4.16)$$

This latter inequality can also be used (perhaps in conjunction with (14) to plan future re-deployments as the existing sensors exhaust their energy budget.

4.9.2 Energy Expenditure

In this subsection we turn to the task of evaluating the energy expenditure per node in an arbitrary sector A_i with $i \ge 1$. Since the case $i = 1$ was handled in the previous section, we now assume $i \ge 2$.

Observe that nodes in a generic sector A_i ($2 \le i \le k$) are called on to serve two kinds of paths:

1. Paths originating in a sector A_j with $i < j \le k$
2. Paths originating at a node in A_i

It is easy to confirm that the number of paths involving nodes in A_i includes all paths except those originating in one of the sectors $A_1, A_2, \ldots, A_{i-1}$. Therefore, the total number of paths that the nodes in A_i must handle is

$$N - \frac{N}{n}(n_1 + n_2 + \cdots + n_{i-1})$$

---

[7]It is important to note that given the deployment area, the density can be engineered beforehand by simply deploying a suitable number of sensors uniformly at random.

By equations (4.6) and (4.7) combined with elementary manipulations, this expression can be written as

$$N\left[1 - \frac{\sum_{i=1}^{k}(r_i^2 - r_{i-1}^2)}{r_k^2}\right] = N\left[1 - \frac{r_{i-1}^2}{r_k^2}\right] \tag{4.17}$$

Recall that sector A_i contains n_i nodes. This implies that each node in A_i must participate in

$$\frac{N}{n_i}\left[1 - \frac{r_{i-1}^2}{r_k^2}\right]$$

paths. Using equation (4.7), the number of paths handled by each node in A_i can be written as

$$\frac{2N}{\rho\theta}\left[1 - \frac{r_{i-1}^2}{r_k^2}\right]\frac{1}{r_i^2 - r_{i-1}^2} \tag{4.18}$$

Observe that the width of sector A_i is $r_i - r_{i-1}$. It follows that the transmission range needed to send information between A_i and A_{i-1} is $r_i - r_{i-1}$. We shall adopt a most general power-degradation model according to which the energy expended by a node in A_i to send information to sensors in A_{i-1} is

$$(r_i - r_{i-1})^\alpha + c$$

where c is a nonnegative constant.

Let the total amount of energy expended by a node in A_i be E_i. By equations (4.11) and (4.18), we have

$$E_i = \frac{N}{\pi\rho}\left[1 - \frac{r_{i-1}^2}{r_k^2}\right]\frac{1}{r_i^2 - r_{i-1}^2}[(r_i - r_{i-1})^\alpha + c]$$

Simple manipulations show that

$$E_i = \frac{N}{\pi\rho}\left[1 - \frac{r_{i-1}^2}{r_k^2}\right]\left[\frac{(r_i - r_{i-1})^{\alpha-1}}{r_i + r_{i-1}} + \frac{c}{r_i^2 - r_{i-1}^2}\right] \tag{4.19}$$

For later reference we will find it convenient to write

$$E_i = E_i' + E_i''$$

where

$$E_i' = \frac{\overline{N}}{\pi\rho}\left[1 - \frac{r_{i-1}^2}{r_k^2}\right]\frac{(r_i - r_{i-1})^{\alpha-1}}{r_i + r_{i-1}} \qquad (4.20)$$

and

$$E_i'' = \frac{\overline{N}}{\pi\rho}\left[1 - \frac{r_{i-1}^2}{r_k^2}\right]\frac{c}{r_i^2 - r_{i-1}^2} \qquad (4.21)$$

We also assume that for all i, $1 \leq i \leq k$, every sensor in sector A_i should be within transmission range from *some* sensor in sector A_{i-1}. In particular, every sensor in sector A_1 must be within transmission range from the sink.[8]

4.9.3 Optimizing the Size of Coronas

The main goal of this section is to show how to select the radii $r_1, r_2, \ldots, r_k$ in such a way that *total* energy spent per sector-to-sink routing path is minimized. For this purpose, let ε_i denote the total amount of energy expended by the nodes along a generic path transferring data from sector A_i to the sink. Write $r_0 = 0$ and assume that A_0 is the sink node itself; since in transmitting from A_j to A_{j-1} ($2 \leq j \leq i$), the amount of energy spent is $(r_j - r_{j-1})^\alpha + c$, it follows that

$$\varepsilon_i = \sum_{j=1}^{i}\left[(r_j - r_{j-1})^\alpha + c\right] \qquad (4.22)$$

Recall the Lagrange identity [ref. 52, p. 64]:

$$\sum_{1 \leq p < q \leq i}(a_p b_q - a_q b_p)^2 = \left(\sum_{p=1}^{i} a_p^2\right)\left(\sum_{p=1}^{i} b_p^2\right) - \left(\sum_{p=1}^{i} a_p b_p\right)^2$$

For every j ($1 \leq j \leq i$), write $a_j = (r_j - r_{j-1})^{\alpha/2}$ and $b_j = 1$. Noticing that

- $\sum_{p=1}^{i} a_p^2 = \varepsilon_i - ic$
- $\sum_{p=1}^{i} b_p^2 = i$

and substituting in Langrage's identity, we obtain

$$\sum_{1 \leq p < q \leq i}(a_p - a_q)^2 = i(\varepsilon_i - ic) - \left(\sum_{p=1}^{i} a_p\right)^2$$

---

[8]For convenience of notation we write $r_0 = 0$ and interpret A_0 as the sink itself.

Thus, we can write

$$i(\varepsilon_i - ic) = \sum_{p=1}^{i} (a_p)^2 + \sum_{1 \le p < q \le i} (a_p - a_q)^2 \qquad (4.23)$$

Clearly, the left-hand side of equation (4.23) is minimized whenever

$$\sum_{1 \le p < q \le i} (a_p - a_q)^2 = 0$$

which occurs if and only if

$$a_1 = a_2 = a_3 = \cdots = a_i$$

Now, recalling that the optimal value of r_1 from equation (4.10) is

$$r_1^{opt} = \begin{cases} t_x & \text{if } \alpha = 2 \\ \min\left\{ \left(\dfrac{2c}{\alpha - 2} \right)^{1/\alpha}, t_x \right\} & \text{if } 2 < \alpha \le 6 \end{cases}$$

We can set for every

$$\begin{aligned} j(1 \le j \le i), \\ r_j - r_{j-1} = r_1^{opt} \end{aligned} \qquad (4.24)$$

It is easy to see that equation (4.24) implies

$$r_i = i \times r_1^{opt} \qquad (4.25)$$

and so, substituting in equation (4.23), we obtain

$$\varepsilon_i = i \times \min\left\{ \frac{c\alpha}{\alpha - 2}, \ t_x^\alpha + c \right\}$$

To summarize, we state the following result.

Theorem 4.2 In order to minimize the total amount of energy spent on routing along a path originating at a sensor in corona A_i and ending at the sink, all the coronas must have the same width and the optimal amount of energy is i times the energy needed to send the desired information between adjacent coronas.

4.10 CONCLUDING REMARKS AND DIRECTIONS FOR FURTHER WORK

In this chapter we have proposed a general-purpose virtual infrastructure for a massively deployed collection of anonymous sensors. The key component of the virtual infrastructure is a dynamic coordinate system that suggests a simple and robust clustering scheme. We have also shown that *training* the sensors—the process of learning their coordinates—can be performed by a protocol that is lightweight. Being energy efficient, this training can be repeated on either a scheduled or ad hoc basis to provide robustness and dynamic reorganization.

We also showed that in a trained wireless sensor network the tasks of routing and data aggregation can be performed by very simple and energy-efficient protocols.

It is important to point out that Olariu et al. [27] have shown that the virtual infrastructure can be leveraged by a number of applications, including in-network data storage and security-related problems. This is an extremely important problem, as the information provided by the sensor network may be used for decision making in military or civilian environments where human life is at stake.

The genetic material discussed in Subsection 4.2.1 has many other applications. One of then is *generational learning* discussed in [53,54] in the context of modeling wireless sensor networks, and by Jones et al. [55] in the context of biology-inspired protocols for wireless sensor networks.

REFERENCES

1. C. C. Enz, A. El-Hoiydi, J.-D. Decotignie, and V. Peiris. WiseNET: An ultralow power wireless sensor network solution. *Computer (IEEE)*, **37**(8):62–69, 2004.

2. See at http://www.darpa.mil/mto/mems/.

3. See at http://www.stanford.edu/class/ee321/ho/MEMS-14-sensors.pdf.

4. See at http://www.xs4all.nl/g̃answijk/chipdir/m/sensor.htm.

5. V. V. Zhirnov and D. J. C. Herr. New frontiers: Self-assembly and nano-electronics. *Computer (IEEE)*, **34**(1):34–43, 2001.

6. J. Hill, M. Horton, R. Kling, and L. Krishnamurthy. The platforms enabling wireless sensor networks. *Communications of the ACM*, **47**(6):41–46, 2004.

7. F. Akyildiz, W. Su, Y. Sankarasubramanian, and E. Cayirci. Wireless sensor networks: A survey. *Computer Networks*, **38**(4):393–422, 2002; *IEEE Wireless Communications*, **9**(1):40–48, 2002.

8. J. M. Kahn, R. H. Katz, and K. S. J. Pister. Mobile networking for smart dust. In *Proceedings of the 5th Annual ACM/IEEE International Conference on Computing and Networking (MobiCom'99)*, Seattle, Washington, August 1999.

9. B. Warneke, M. Last, B. Leibowitz, and K. Pister. SmartDust: Communicating with a cubic-millimeter computer. *Computer (IEEE)*, **34**(1):44–51, 2001.

10. D. Culler, D. Estrin, and M. Srivastava. Overview of sensor networks. *Computer (IEEE)*, **37**(8):41–49, 2004.

11. D. Culler and W. Hong. Wireless sensor networks. *Communications of the ACM*, **47**(6): 30–33, 2004.

12. National Research Council. *Embedded, Everywhere: A Research Agenda for Systems of Embedded Computers* Committee on Networked Systems of Embedded Computers, for the Computer Science and Telecommunications Board, Division on Engineering and Physical Sciences, Washington, D.C., 2001.

13. P. Saffo. Sensors, the next wave of innovation. *Communications of the ACM*, **40**(2):93–97, 1997.

14. J. Agre and L. Clare. An integrated architecture for cooperative sensing networks. *IEEE Computer*, **33**(5):106–108, 2000.

15. C. Intanagonwiwat, R. Govindan, D. Estrin, J. Heidemann, and F. Silva, Directed diffusion for wireless sensor networking. *IEEE/ACM Transactions on Networking*, **11**(1): February, 2003.

16. K. Martinez, J. K. Hart, and R. Ong. Sensor network applications. *Computer (IEEE)*, **37**(8):50–56, 2004.

17. C.-C. Shen, C. Srisathapornphat, and C. Jaikaeo. Sensor information networking architecture and applications. *IEEE Personal Communications*, pages 52–59, August 2001.

18. R. Szewczyk, E. Osterweil, J. Polatre, M. Hamilton, A. Mainwaring, and D. Estrin. Habitat monitoring with sensor networks. *Communications of the ACM*, **47**(6):34–40,

19. S. Tilak, N. B. Abu-Ghazaleh, and W. Heinzelman. A taxonomy of wireless micro-sensor network models. *Mobile Computing and Communications Review*, **6**(2):28–36,

20. G. J. Pottie and W. J. Kaiser. Wireless integrated sensor networks. *Communications of the ACM*, **43**(5):51–58, 2000.

21. K. Sohrabi, J. Gao, V. Ailawadhi, and G. Pottie. Protocols for self-organization of a wireless sensor network. *IEEE Personal Communications*, pages 16–27, October 2000.

22. L. Wang and S. Olariu. Towards a general-purpose virtual infrastructure for mobile aD-HOC networks. In *Ad Hoc and Sensor Networks*, Y. Xiao and Y. Pan (eds.), Nova Science Publishers, January 2005.

23. K. H. Chan, A. Perrig, and D. Song. Random key pre-distribution schemes for sensor networks. In *Proceedings of the IEEE Symposium on Security and Privacy*, Berkeley, California, May 2003.

24. S. Roundy, P. K. Wright, and J. Rabaey. *Energy Scavenging for Wireless Sensor Networks with Special Focus on Vibrations*. Kluwer Academic Press, 2004.

25. N. S. Shenck and J. A. Paradiso. Energy scavenging with shoe-mounter piezoelectrics. *IEEE Micro*, **21**:30–41, 2001.

26. K. Jones, A. Wadaa, S. Olariu, L. Wilson, and M. Eltoweissy. Towards a new paradigm for securing wireless sensor networks. In *Proceeings of the New Security Paradigms Workshop (NSPW'2003)*, Ascona, Switzerland, August 2003.

27. S. Olariu, A. Wadaa, L. Wilson, and M. Eltoweissy. Wireless sensor networks: Leveraging the virtual infrastructure. *IEEE Network*, **18**(4):51–56, 2004.

28. M. Sichitiu and C. Veerarithiphan. Simple accurate synchronization for wireless sensor networks. In *Proceedings of the IEEE Wireless Communications and Networking Conference (WCNC 2003)*, New Orleans, Louisiana, March 2003.

29. F. Sivrukaya and B. Yener. Time synchronization in sensor networks: A survey. *IEEE Network*, **18**(4):45–50, 2004.

30. N. Bulusu, J. Heidemann, and D. Estrin. GPS-less low cost outdoor localization for very small devices. *IEEE Personal Communications*, **7**(5):28–34, 2000.

31. N. Bulusu, J. Heidemann, and D. Estrin. Scalable coordination for wireless sensor networks: Self-configuration localization systems. In *Proceedings of the 6th International Symposium on Communication Theory and Applications (ISCTA 2001)*, Ambleside, Lake District, UK, July 2001.

32. S. Capkun, M. Hamdi, and J.-P. Hubeaux. GPS-free positioning in mobile ad-hoc networks. *Cluster Computing*, **5**(2):157–167, 2002.

33. L. Girod, V. Bychkovskiy, J. Elson, and D. Estrin. Locating tiny sensors in time and space: A case study. In *Proceedings of the International Conference on Computer Design (ICCD 2002)*, Freiburg, Germany, September 2002.

34. L. Doherty, H. S. J. Pister, and L. E. Ghaoui. Convex position estimation in wireless sensor networks. In *Proceedings of IEEE INFOCOM 2001*, **3**:1655–1663, April 2001.

35. D. Niculescu. Positioning in ad hoc sensor networks. *IEEE Network*, **18**(4):24–29, 2004.

36. C. Savarese, J. Rabaey, and K. Langendoen. Robust positioning algorithms for distributed ad-hoc wireless sensor networks. In *Proceedings of the USENIX Technical Annual Conference*, pages 317–328, Monterey, California, June 2002.

37. K. Langendoen and N. Reijers. Distributed localization algorithm. In *Embedded Systems Handbook*, R. Zurawski (ed.), CRC Press, forthcoming.

38. S. Bandyopadhyay and E. Coyle. An efficient hierarchical clustering algorithm for wireless sensor networks. In *Proceedings of IEEE INFOCOM 2003—The Conference on Computer Communications*, **22**(1):1713–1723, March 2003.

39. D. Coore, R. Nagpal, and R. Weiss. *Paradigms for Structure in an Amorphous Computer*, MIT Artificial Intelligence laboratory Technical Report AI-1616, October 1997.

40. S. Ghiasi, A. Srivastava, X. Yang, and M. Sarrafzadeh. Optimal energy-aware clustering in sensor networks. *Sensors*, **2**:258–269, 2002.

41. A. Wadaa, S. Olariu, L. Wilson, K. Jones, and Q. Xu. On training wireless sensor networks. In *Proceedings of the 3rd International Workshop on Wireless, Mobile and Ad Hoc Networks (WMAN'03)*, Nice, France, April 2003.

42. A. Wadaa, S. Olariu, L. Wilson, K. Jones, and M. Eltoweissy. Training a sensor networks. *Mobile Networks and Applications*, February 2005, forthcoming.

43. D. Braginsky and D. Estrin. Rumor Routing Algorithm for Sensor Networks. Paper submitted to the International Conference on Distributed Computing Systems (ICDCS-22), November 2001.

44. D. Ganesan, R. Govindan, S. Shenker, and D. Estrin. Highly resilient, energy-efficient multipath routing in wireless sensor networks. *ACM Mobile Computing and Communications Review*, **5**(4), 2001.

45. J. Kulik, W. Heinzelman, and H. Balakrishnan. Negotiation-based protocols for disseminating information in wireless sensor networks. *Wireless Networks*, **8**(3), March 2002.

46. R. C. Shah and J. Rabaey. Energy aware routing for low energy ad hoc sensor networks. In *Proceedings of the IEEE Wireless Communications and Networking Conference (WCNC 2002)*, Orlando, Florida, March 2002.

47. E. Shih, S. Cho, N. Ickes, R. Min, A. Sinha, A. Wang, and A. Chandrakasan. A physical layer driven protocol and algorithm design for energy-efficient wireless sensor networks. In *Proceedings of the 7th Annual ACM/IEEE International Conference on Computing and Networking (MobiCom 2001)*, Rome, Italy, July 2001.

48. A. Woo and D. E. Culler. A transmission control scheme for media access in sensor networks. In *Proceedings of the 7th Annual ACM/IEEE International Conference on Computing and Networking (MobiCom 2001)*, Rome, Italy, July 2001.

49. W. Ye, J. Heidemann, and D. Estrin. An energy-efficient MAC protocol for wireless sensor networks. In *Proceedings of the 7th Annual ACM/IEEE International Conference on Computing and Networking INFOCOM 2002*, New York, June, 2002.

50. K. Nakano and S. Olariu. Randomized initialization protocols for radio networks. In *Handbook of Wireless Networks and Mobile Computing*, Stojmenović (ed.), pages 195–218, John Wiley & Sons, 2002.

51. K. Nakano and S. Olariu. Uniform leader election for radio networks. *IEEE Transactions on Parallel and Distributed Systems*, **13**:516–526, 2002.

52. R. G. Graham, D. E. Knuth, and O. Patashnik. *Concrete Mathematics*, Addison-Wesley, 1989.

53. D. Gracanin, M. Eltoweissy, S. Olariu, and A. Wadaa. On Modeling Wireless Sensor Networks. Paper presented at 18th International Parallel and Distributed Processing Symposium (IPDPS '04), Workshop 12: Fourth International Workshop on Algorithms for Wireless, Mobile, Ad Hoc and Sensor Networks (WMAN), Santa Fe, NM, April 2004.

54. D. Gracanin, M. Eltoweissy, S. Olariu, and A. Wadaa. Dependability support in wireless sensor networks. In *Dependable Systems*, H. Diab and A. Y. Zomaya (eds.), John Wiley & Sons, 2005.

55. K. Jones, K. N. Lodding, S. Olariu, A. Wadaa, L. Wilson, and M. Eltoweissy. Biomimetic models for wireless sensor networks. In *Handbook of BioInspired Algorithms*, S. Olariu and A. Y. Zomaya (eds.), CRC Press, 2005.

■■■■■■ CHAPTER 5

Broadcast Authentication and Key Management for Secure Sensor Networks

PENG NING

North Carolina State University, Raleigh, North Carolina

DONGGANG LIU

University of Texas at Arlington, Arlington, Texas

This chapter is intended as a starting point for studying sensor network security. It focuses on recent advances in broadcast authentication and key management in sensor networks, which are foundational cryptographic services for sensor network security. Authors describe μTESLA and multi-level μTESLA protocols developed for scalable broadcast authentication in sensor networks. They then describe random key predistribution techniques proposed for establishing pairwise keys between resource constrained sensor nodes. Attacks against location discovery and some additional security problems in sensor networks are then discussed.

5.1 INTRODUCTION

Recent technological advances have made it possible to develop distributed sensor networks consisting of a large number of low-cost, low-power, and multifunctional sensor nodes that communicate at short distances through wireless links [1]. Such sensor networks may also include a few more powerful nodes called *base stations* to facilitate computation as well as communication with the outside world. Sensor networks are ideal candidates for a wide range of applications. Example applications include monitoring of critical infrastructures such as the power grid, data acquisition in hazardous environments, and military operations. The desirable features of distributed sensor networks have attracted many researchers to develop protocols and algorithms that can fulfill the requirements of these applications.

Handbook of Sensor Networks: Algorithms and Architectures, Edited by Ivan Stojmenović
Copyright © 2005 John Wiley & Sons, Inc.

It is necessary to guarantee the security of sensor networks as well as sensing applications in hostile environments, especially when the failures of these applications (e.g., critical infrastructure protection) may result in catastrophic events with impacts affecting safety, security, the economy, and society at large. In other words, sensor networks and sensing applications should work as expected, and offer gracefully degrading services in hostile environments where there are malicious attackers, even if some nodes fail or are compromised.

However, several unique features of sensor networks make it very challenging to provide security in sensor networks. First, sensor nodes are typically resource constrained due to the need to lower the cost. As a result, it is usually undesirable to use expensive mechanisms such as public key cryptography on such nodes. Second, sensor networks are often deployed in an unattended fashion, possibly exposed to physical attacks. Sensor nodes may be captured, and any secret information on a captured node can potentially be disclosed to attackers. Thus, any security mechanism for sensor networks has to be resilient to compromised nodes. Third, most sensor network applications depend on local communication and computation because of the resource constraints on sensor nodes. However, a determined attacker may attack any node in a sensor network and use the information gathered from the compromised nodes to attack noncompromised ones in a certain region. This further adds to the imbalance between the threat and the defense in sensor networks.

Security in sensor networks has attracted a lot of attention in the past several years. Key management is one of the most fundamental security services for sensor networks. The performance of several traditional key management approaches has been examined on different hardware platforms [2]. To provide practical key management techniques for sensor networks, researchers have developed a number of random pairwise key predistribution techniques [3–9], which can handle the resource constraints on sensor nodes and are resilient to node compromises. Moreover, a protocol suite called LEAP was developed to help establish individual keys between sensor nodes and a base station, pairwise keys between sensor nodes, cluster keys within a local area, and a group key shared by all nodes [10]. Another fundamental security service is broadcast authentication, which verifies the integrity and the source of broadcast messages to multiple receivers. A protocol named μTESLA [11] has been adapted for sensor networks from TESLA, a multicast stream authentication protocol [12,13], and later improved for higher scalability in refs. [14] and [15].

Since the primary goal of sensor networks is to collect data from physical phenomena, it is critical to ensure the authenticity and integrity of the collected data in hostile environments, even if some nodes have been compromised. A number of mechanisms were proposed to improve the security for in-network processing [16]. An interleaved hop-by-hop authentication mechanism [17] and a statistical hop-by-hop authentication mechanism [18] were developed to mitigate malicious data injection in sensor networks. Several techniques were proposed to use redundant information sources to detect malicious aggregators or sensor nodes for secure data aggregation [19–21]. Moreover, researchers have been investigating potential attacks against sensor networks and possible countermeasures, including DOS attacks [22], attacks against routing mechanisms in sensor networks [23], and Sybil attacks [24].

Despite the recent advances in sensor network security, sensor network security is still not reality. Indeed, quite a number of problems remain unsolved or require better solutions. Besides more effective and efficient techniques for some of the preceding problems, such as key management and broadcast authentication, it is necessary to have novel techniques to protect critical services such as clock synchronization and location discovery, mitigate or defeat signal jamming, denial of service, and other attacks, and provide additional capabilities such as intrusion detection.

This chapter is intended as a starting point for studying sensor network security. In the remainder of this chapter, we focus on recent advances in broadcast authentication and key management in sensor networks, which are foundational cryptographic services for sensor network security. In Section 5.2, we introduce μTESLA and multilevel μTESLA protocols developed for scalable broadcast authentication in sensor networks. In Section 5.3, we describe random key predistribution techniques proposed for establishing pairwise keys between resource-constrained sensor nodes. In Section 5.4, we discuss additional security problems in sensor networks. In Section 5.5, we conclude this chapter and identify a few research areas that require further investigation.

5.2 BROADCAST AUTHENTICATION IN SENSOR NETWORKS

Broadcast authentication is an essential service in sensor networks. Because of the large number of sensor nodes and the broadcast nature of wireless communication, it is usually desirable for the base stations to broadcast commands, queries, and data to sensor nodes. The authenticity of such commands and data is critical for the normal operation of sensor networks. If convinced to accept forged or modified commands or data, sensor nodes may perform unnecessary or incorrect operations, and cannot fulfill the intended purposes of the network. Thus, in hostile environments (e.g., battlefield, antiterrorists operations), it is necessary to enable sensor nodes to authenticate broadcast messages received from the base station.

Providing broadcast authentication in sensor networks turns out to be a nontrivial task. On the one hand, public key–based digital signatures (e.g., the Rivest–Shamir–Adleman (RSA) algorithm [25]), which are typically used for broadcast authentication in traditional networks, are too expensive to be used in sensor networks, due to the intensive computation involved in signature verification and the resource constraints on sensor nodes. On the other hand, secret key-based mechanisms (e.g., HMAC [26]) cannot be directly applied to broadcast authentication, since otherwise a compromised receiver can easily forge any message from the sender.

In the following, we describe the μTESLA and multilevel μTESLA protocols, which were proposed for broadcast authentication in sensor networks. μTESLA employs a chain of authentication keys linked to each other by a pseudorandom function [27], which is by definition a one-way function. Each key in the key chain is the image of the next key under the pseudorandom function. μTESLA achieves broadcast authentication through delayed disclosure of authentication keys in the key chain. Multilevel μTESLA is aimed at improving the scalability of μTESLA in large sensor networks. It includes several variations, all of which are extended from the basic μTESLA protocol.

5.2.1 μTESLA

An asymmetric mechanism such as public key cryptography is generally required for broadcast authentication [12]. Otherwise, a malicious receiver can easily forge any packet from the sender, as discussed earlier. μTESLA introduces asymmetry by delaying the disclosure of symmetric keys [11]. A sender broadcasts a message with a message authentication code (MAC) generated with a secret key K, which is disclosed after a certain period of time. When a receiver gets this message, if it can ensure that the packet was sent before the key was disclosed, the receiver buffers this packet and authenticates the packet when it later receives the disclosed key. To continuously authenticate broadcast packets, μTESLA divides the time period for broadcast into multiple intervals, assigning different keys to different time intervals. All packets broadcast in a particular time interval are authenticated with the same key assigned to that time interval. Figure 5.1 illustrates the division of the time line and the assignment of authentication keys.

To authenticate the broadcast messages, a receiver first authenticates the disclosed keys. μTESLA uses a one-way key chain for this purpose. The sender selects a random value K_n as the last key in the key chain and repeatedly performs a pseudorandom function F to compute all the other keys: $K_i = F(K_{i+1}), 0 \leq i \leq n - 1$, where the secret key K_i (except for K_0) is assigned to the ith time interval. Because of the one-way property of the pseudorandom function, given K_j in the key chain, anybody can compute all the previous keys $K_i, 0 \leq i \leq j$, but nobody can compute any of the later ones $K_i, j + 1 \leq i \leq n$. Thus, with the knowledge of the initial key K_0, which is called the *commitment* of the key chain, a receiver can authenticate any key in the key chain by merely performing pseudorandom function operations. When a broadcast message is available in the ith time interval, the sender generates a MAC for this message with a key derived from K_i, broadcasts this message along with its MAC, and discloses the key K_{i-d} for time interval I_{i-d} in the broadcast message (where d is the disclosure lag of the authentication keys).

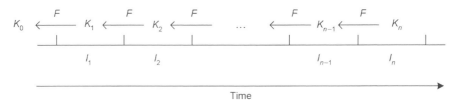

Figure 5.1 μTESLA.

Each key in the key chain will be disclosed after some delay. As a result, the attacker can forge a broadcast packet by using the disclosed key. μTESLA uses a security condition to prevent such situations. When a receiver receives an incoming broadcast packet in time interval I_i, it checks the security condition $\lfloor (T_c + \Delta - T_1)/T_{int} \rfloor < i + d - 1$, where T_c is the local time when the packet is received, T_1 is the start time of the time interval 1, T_{int} is the duration of each time interval, and Δ is the maximum clock difference between the sender and itself. If the security condition is satisfied, that is, the sender has not disclosed the key K_i yet, the receiver accepts this packet. Otherwise, the receiver simply drops it.

μTESLA is an extension to TESLA [12]. The only difference between TESLA and μTESLA is in their key chain commitment distribution schemes. TESLA uses asymmetric cryptography to bootstrap new receivers, which is impractical for current sensor networks due to its high computation and storage overheads. μTESLA depends on symmetric cryptography (with the master key shared between the sender and each receiver) to bootstrap the new receivers individually. TESLA was later extended to include an immediate authentication mechanism [13]. The basic idea is to include an image under a pseudorandom function of a late message content in an earlier message so that once the earlier message is authenticated, the later message content can be authenticated immediately after being received. This extension can also be applied to μTESLA.

The original TESLA protocol uses broadcast to distribute the initial parameters (e.g., the key chain commitment) required for broadcast authentication. The authenticity of these parameters is guaranteed by a digital signature generated by the sender. However, due to the low bandwidth of a sensor network and the low computational resources at each sensor node, μTESLA cannot distribute these initial parameters using public key cryptography. Instead, the base station has to unicast the initial parameters to the sensor nodes individually. This feature severely limits the application of μTESLA in large sensor networks. For example, the implementation of μTESLA in [11] has 10 kbps bandwidth and supports 30-byte messages. To bootstrap 2000 nodes, the base station has to send or receive at least 4000 packets to distribute the initial parameters, which takes at least $(4000 \times 30 \times 8)/10,240 = 93.75$ s, even if the channel utilization is perfect. Such a method certainly cannot scale up to very large sensor networks, which may have thousands of nodes.

5.2.2 Multilevel μTESLA

Multilevel μTESLA was developed to improve the scalability of μTESLA in large sensor networks. The basic idea is to *predetermine* and *broadcast* the parameters such as the key chain commitments instead of unicast-based message transmissions. In the following, we first present the basic approach, and then describe several techniques to deal with potential attacks against multilevel μTESLA.

For simplicity, we use two-level μTESLA to illustrate the idea. This can easily be extended to multilevel μTESLA. Refer to ref. [15] for further details.

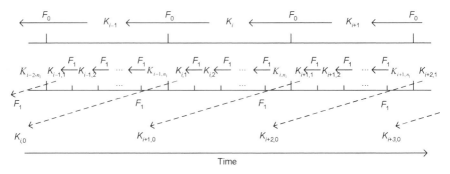

Figure 5.2   Organizing key chains in two levels.

5.2.2.1 Basic Approach

The two-level μTESLA consists of a high-level key chain and multiple low-level key chains. The low-level key chains are intended for authenticating broadcast messages, while the high-level key chain is used to distribute and authenticate commitments of the low-level key chains. The high-level key chain uses a long enough interval to divide the time line so that it can cover the lifetime of a sensor network without having too many keys. The low-level key chains have short enough intervals so that the delay between the receipt of broadcast messages and the verification of the messages is tolerable. Figure 5.2 illustrates two levels of key chains.

The lifetime of a sensor network is divided into n_0 (long) intervals of duration Δ_0, denoted as $I_1, I_2, \ldots,$ and I_{n_0}. The high-level key chain has $n_0 + 1$ elements $K_0, K_1, \ldots, K_{n_0}$, which are generated by randomly picking K_{n_0} and computing $K_i = F_0(K_{i+1})$ for $i = 0, 1, \ldots, n_0 - 1$, where F_0 is a pseudorandom function. The key K_i is associated with each time interval I_i. We denote the starting time of I_i as T_i. Thus, the starting time of the high-level key chain is T_1.

Since the duration of the high-level time intervals is usually very long compared with the network delay and clock discrepancies, we choose to disclose a high-level key K_i used for I_i in the following time interval I_{i+1}. Thus, we use the following security condition to check whether the base station has disclosed the key K_i when a sensor node receives a message authenticated with K_i at time t: $t + \delta_{\max} < T_{i+1}$, where $\delta_{\max}$ is the maximum clock discrepancy between the base station and the sensor node.

Each time interval I_i is further divided into n_1 (short) intervals of duration Δ_1, denoted as $I_{i,1}, I_{i,2}, \ldots, I_{i,n_1}$. If needed, the base station generates a low-level key chain for each time interval I_i by randomly picking K_{i,n_1} and computing $K_{i,j} = F_1(K_{i,j+1})$ for $j = 0, 1, \ldots, n_1 - 1$, where F_1 is a pseudorandom function. The key $K_{i,j}$ is intended for authenticating messages broadcasted during the time interval $I_{i,j}$. The starting time of the key chain $\langle K_{i,0} \rangle$ is predetermined as T_i. The disclosure lag for the low-level key chains can be determined in the same way as μTESLA and TESLA [11,12]. For simplicity, we assume all the low-level key chains use the same disclosure lag d. Further assume that messages broadcasted during $I_{i,j}$ are indexed as (i, j). Thus, the security condition for a message authenticated with $K_{i,j}$

and received at time t is: $i' < (i-1) * n_1 + j + d$, where $i' = \lfloor (t - T_1 + \delta_{max})/ \Delta_1 \rfloor + 1$, and δ_{max} is the maximum clock discrepancy between the base station and the sensor node.

When sensor nodes are initialized, their clocks are synchronized with the base station. In addition, the starting time T_1, the commitment K_0 of the high-level key chain, the duration Δ_0 of each high-level time interval, the duration Δ_1 of each low-level time interval, the disclosure lag d for the low-level key chains, and the maximum clock discrepancy δ_{max} between the base station and the sensor nodes throughout the lifetime of the sensor network are distributed to the sensor nodes.

In order for the sensor nodes to use a low-level key chain $\langle K_{i,0} \rangle$ during the time interval I_i, they must authenticate the commitment $K_{i,0}$ before T_i. To achieve this goal, the base station broadcasts a *commitment distribution message*, denoted as CDM_i, during each time interval I_i. (In this chapter, we use commitment distribution message and its abbreviation CDM interchangeably.) This message consists of the commitment $K_{i+2,0}$ of the low-level key chain $\langle K_{i+2,0} \rangle$ and the key K_{i-1} in the high-level key chain. Specifically, the base station constructs the CDM_i message as follows:

$$\text{CDM}_i = i|K_{i+2,0}|\text{MAC}_{K_i'}(i|K_{i+2,0}|)K_{i-1},$$

where "|" denotes message concatenation, and K_i' is derived from K_i with a pseudo-random function other than F_0 and F_1. Thus, to use a low-level key chain $\langle K_{i,0} \rangle$ during I_i, the base station needs to generate the key chain during I_{i-2} and distribute $K_{i,0}$ in CDM_{i-2}.

Since the high-level authentication key K_i is disclosed in CDM_{i+1} during the time interval I_{i+1}, each sensor node needs to store CDM_i until it receives CDM_{i+1}. Each sensor node also stores a key K_j, which is initially K_0. After receiving K_{i-1} in CDM_i, the sensor node authenticates it by verifying that $F_1^{i-1-j}(K_{i-1}) = K_j$. Then the sensor node replaces the current K_j with K_{i-1}.

Suppose a sensor node has received CDM_{i-2}. Upon receiving CDM_{i-1} during I_{i-1}, the node can authenticate CDM_{i-2} with K_{i-2} disclosed in CDM_{i-1}, and thus verify $K_{i,0}$. As a result, the sensor node can authenticate broadcast messages sent by the base station using the μTESLA key chain $\langle K_{i,0} \rangle$ during the high-level time interval I_i. Intuitively, this approach uses μTESLA in two different levels. The high-level key chain relies on the initialization phase of the sensor nodes to distribute the key chain commitment, and it only has a single key chain throughout the lifetime of the sensor network. The low-level key chains depend on the high-level key chain to distribute and authenticate the commitments.

The approach in the current form does not tolerate message losses very well. There are two types of message losses: the losses of normal messages, and the losses of CDM messages. Both may cause problems. First, the low-level keys are not entirely chained together. Thus, losses of key disclosure messages for later keys in a low-level key chain cannot be recovered even if the sensor node can receive keys in some later low-level key chains. For example, consider the last key K_{i,n_1} that is used to authenticate the packet in the key chain of time interval I_i. If the packets that disclose K_{i,n_1} are lost, the sensor node then has no way to authenticate this packet. As a result, a sensor node may not be able to authenticate a stored message even if it receives some key disclosure messages later. In contrast, with μTESLA a receiver can authenticate a stored message as long as it receives a later key. Second, if CDM_{i-2} does not reach a sensor node, the node will not be able to use the key chain $\langle K_{i,0} \rangle$ for authentication during the entire time interval I_i, which is usually pretty long (to make the high-level key chain short).

To address the first problem, we further connect the low-level key chains to the high-level one. Specifically, instead of choosing each K_{i,n_1} randomly, we derive each K_{i,n_1} from a high-level key K_{i+1} (which is to be used in the next high-level time interval) through another pseudorandom function F_{01}. That is, $K_{i,n_1} = F_{01}(K_{i+1})$. As a result, a sensor node can recover any authentication key $K_{i,j}$ as long as it receives a CDM message that discloses $K_{i'}$ with $i' \geq i + 1$, even if it does not receive any later low-level key $K_{i,j'}$ with $j' \geq j$. Thus, the first problem can be resolved. Figure 5.3 illustrates the key chains in two-level μTESLA.

The second problem does not have an ultimate solution. If the base station cannot reach a sensor node at all during a time interval I_i, CDM_i will not be delivered to the sensor node. However, the impact of temporary communication failures can be reduced by standard fault-tolerant approaches. Multilevel μTESLA has the base station periodically broadcast the CDM message during each time interval. Assuming that the frequency of this broadcast is F, each CDM message is therefore broadcasted $F \times \Delta_0$ times. To simplify the analysis, we assume the probability that a sensor node cannot receive a broadcast of a CDM message is p_f. Thus, the probability that a sensor node cannot receive any copy of the CDM message is reduced to $p_f^{F \times \Delta_0}$.

Note that even if a sensor node cannot receive any CDM message during a time interval I_i, it still has the opportunity to authenticate broadcast messages in time intervals later than I_{i+1}. Not having the CDM message in time interval I_i only prevents a sensor node from authenticating broadcast messages during I_{i+1}. As long as the sensor node gets a CDM message, it can derive all the low-level keys in the previous time intervals.

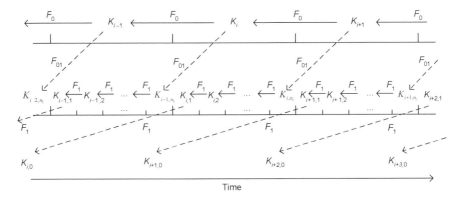

Figure 5.3   Key chains in two-level μTESLA.

The security of multilevel μTESLA follows directly from the security of μTESLA. Note that the high-level key chain is only used to authenticate the commitment of each low-level key chain. As long as the security condition of each μTESLA key chain is satisfied, the two-level μTESLA has the same degree of security as all the μTESLA instances involved in this scheme. Thus, similar to μTESLA and TESLA, a sensor node can detect forged messages by verifying the MAC with the corresponding authentication key once the sensor node receives it. In addition, replay attacks can be easily defeated if a sequence number is included in each message.

5.2.2.2 DOS Attacks Against Multilevel μTESLA

One limitation of multilevel μTESLA is that if a sensor node misses all copies of CDM_i during the time interval I_i, it cannot authenticate any data packets received during I_{i+2} before it receives an authentic $K_j, j > i + 2$. (Note that the sensor node does not have to receive an authentic CDM message. As long as the sensor node can authenticate a high-level key K_j with $j > i + 2$, it can derive the low-level keys through the pseudorandom functions F_0, F_{01}, and F_1.) Since the earliest high-level key K_j that satisfies $j > i + 2$ is K_{i+3}, and K_{i+3} is disclosed during I_{i+4}, the sensor node has to buffer the data packets received during I_{i+2} for at least the duration of one high-level time interval.

This makes the CDM messages attractive targets for attackers. An attacker may disrupt the distribution of CDM messages, and thus prevent the sensor nodes from authenticating broadcast messages during the corresponding high-level time intervals. Although the high-level key chain and the low-level ones are chained together, and such sensor nodes may store the broadcast messages and authenticate them once they receive a later commitment distribution message, the delay between the receipt and the authentication of the messages may introduce a problem: An attacker may send a large number of forged messages to exhaust the sensor nodes' buffer before they can authenticate the buffered messages, and force them to drop some authentic messages.

The simplest way for an attacker to disrupt the CDM messages is to jam the communication channel. We may have to resort to techniques such as frequency hopping if the attacker completely jams the communication channel. The attacker may also jam the communication channel only when the CDM messages are being transmitted. If the attacker can predict the schedule of such messages, it would be much easier for the attacker to disrupt such message transmissions. Thus, the base station needs to send the CDM messages randomly or in a pseudorandom manner that cannot be predicted by an attacker that is unaware of the random transmission. For simplicity, we assume that the base station sends the CDM messages randomly.

An attacker may forge commitment distribution messages to confuse the sensor nodes. If a sensor node does not have a copy of the actual CDM_i, it will not be able to get the correct $K_{i+2,0}$, and cannot use the low-level key chain $\langle K_{i+2,0} \rangle$ during the time interval I_{i+2}.

Consider a CDM message: $CDM_i = i|K_{i+2,0}|MAC_{K_i'}(i|K_{i+2,0})|K_{i-1}$. Once seeing such a message, the attacker learns i and K_{i-1}. Then the attacker can replace the actual $K_{i+2,0}$ or $MAC_{K_i'}(i|K_{i+2,0})$ with arbitrary values $K_{i+2,0}'$ or MAC', and forge another message: $CDM_i' = i|K_{i+2,0}'|MAC'|K_{i-1}$. Assume a sensor node has an authentic copy of CDM_{i-1}. The sensor node can verify K_{i-1} with K_{i-2}, since K_{i-2} is included in CDM_{i-1}. However, the sensor node has no way to verify the authenticity of $K_{i+2,0}'$ or MAC' without the corresponding key, which will be disclosed later. In other words, the sensor node cannot distinguish between the authentic CDM_i messages and those forged by the attacker. If the sensor node does not save an authentic copy of CDM_i during I_i, it will not be able to get an authenticated $K_{i+2,0}$, even if it receives the authentication key K_i in CDM_{i+1} during I_{i+1}. As a result, the sensor node cannot use the key chain $\langle K_{i+2,0} \rangle$ during I_{i+2}.

Multilevel μTESLA uses two techniques to deal with the disk operating system (DOS) attacks. One is a random selection technique to tolerate DOS attacks, and the other uses precomputation to defeat such attacks. These two approaches can be combined to provide additional trade-offs.

5.2.2.3 DOS-Tolerant Multilevel μTESLA
DOS-tolerant multilevel μTESLA involves an initial filtering and a random selection process to improve the reliable broadcast of commitment distribution messages. For the CDM_i messages received during each time interval I_i, each sensor node first tries to discard as many forged messages as possible. There is a simple test for a sensor node to identify some forged CDM_i messages during I_i. The sensor node can verify if $F_0^{i-1-j}(K_{i-1}) = K_j$, where K_{i-1} is the high-level key disclosed in CDM_i and K_j is a previously disclosed high-level key. (Note that such a K_j always exists, since the commitment K_0 of the high-level key chain is distributed during the initialization of the sensor nodes.) Messages that fail this test are certainly forged and should be discarded.

The simple test can filter out some forged messages; however, they do not rule out the forged messages discussed earlier. To further improve the possibility that the sensor node has an authentic CDM_i message, the base station uses a random selection method to store the CDM_i messages that pass the test just cited. The goal is to make the DOS attacks so difficult that the attacker would rather use constant signal jamming instead to attack the sensor network. In other words, we want to prevent the DOS attacks that can be achieved by sending a few packets. Some of the strategies are also applicable to the low-level key chains as well as the (extended) TESLA and μTESLA protocols.

Without loss of generality, we assume that each copy of CDM_i has been weakly authenticated in the time interval I_i by using the preceding test. Assume there are m buffers for CDM packets. During each time interval I_i, a sensor node can save the first m copies of CDM_i. For the kth copy with $k > m$, the sensor node keeps it with probability m/k. If a copy is to be kept, the sensor node randomly selects one of the m buffers and replaces the corresponding copy. It is easy to verify that if a sensor node receives n copies of CDM_i, all copies have the same probability m/n to be kept in one of the buffers.

During the time interval I_{i+1}, a sensor node can verify if it has an authentic copy of CDM_i once it receives and weakly authenticates a copy of CDM_{i+1}. Specifically, the sensor node uses the key K_i disclosed in CDM_{i+1} to verify the MAC of the buffered copies of CDM_i. Once it authenticates a copy, the sensor node can discard all the other buffered copies.

If a sensor node cannot find an authentic copy of CDM_i after the verification just discussed, it can conclude that all buffered copies of CDM_i are forged and discard all of them. The sensor node then needs to repeat the random selection process for the copies of CDM_{i+1}. Thus, a sensor node needs at most $m + 1$ buffers for CDM messages with this strategy: m buffers for copies of CDM_i, and one buffer for the first weakly authenticated copy of CDM_{i+1}.

It is easy to see that each sensor node needs to verify the MACs for at most m times. The number of pseudorandom function operations required to weakly authenticate the CDM messages depends on the total number of (true and forged) CDM messages a sensor node receives. With m buffer random selections, the probability that a sensor node has an authentic copy of CDM_i can be estimated as $P(CDM_i) = 1 - p^m$, where $p = $ (No. forged copies)/(No. total copies).

5.2.2.4 *DOS-Resistant Multilevel* μ*TESLA*   DOS-resistant multilevel μTESLA is intended for base stations with sufficient computational and storage resources. When at least one copy of each CDM message can reach the sensor nodes, DOS-resistant multilevel μTESLA can completely defeat the DOS attacks without buffering and random selection.

The solution can be considered a variation of the immediate authentication extension to TESLA [13]. The idea is to include in CDM_i the image $H(CDM_{i+1})$ for each i, where H is a pseudorandom function. As a result, if a sensor node can authenticate CDM_i, it can get authentic $H(CDM_{i+1})$ and then authenticate CDM_{i+1} when it is received. Specifically, the base station constructs CDM_i for the high-level time interval I_i as follows:

$$CDM_i = i|K_{i+1,0}|H(CDM_{i+1})|MAC_{K_i'}(i|K_{i+1,0}|H(CDM_{i+1}))|K_{i-1},$$

where "|" denotes message concatenation, H is a pseudorandom function other than F_0 and F_1, and K_i' is derived from K_i with a pseudorandom function other than H, F_0, and F_1.

Suppose a sensor node has received CDM_i. Upon receiving CDM_{i+1}, the sensor node can authenticate CDM_i with K_i disclosed in CDM_{i+1}. Then the sensor node can immediately authenticate CDM_{i+1} by verifying that applying H to CDM_{i+1} results in the same $H(CDM_{i+1})$ included in CDM_i. As a result, the sensor node can authenticate a commitment distribution message immediately after receiving it.

Alternatively, if $H(CDM_1)$ is predistributed before deployment, the sensor node can immediately authenticate CDM_1 when receiving it, and then use $H(CDM_2)$ included in CDM_1 to authenticate CDM_2, and so on. One can see that in this case, a sensor node does not use the disclosed high-level keys in CDM messages directly. However, including such keys in CDM messages is still useful. Indeed, when a sensor node fails to receive or keep an authentic CDM message, it can use the random-selection mechanism and the approach described in the previous paragraph to recover from the failure.

The cost, however, is that the base station has to precompute the CDM messages in the reverse order. That is, in order to include $H(CDM_{i+1})$ in CDM_i, the base station has to have CDM_{i+1}, which implies that it also needs CDM_{i+2}, and so on. Therefore, the base station needs to compute both the high-level and the low-level key chains completely to get the commitments of these key chains, and construct all the CDM messages in the reverse order before distributing the first one of them. (Note that the base station only needs to compute the high-level key chain but not all the low-level ones during initialization. The base station can delay the computation of a low-level key chain until it needs to distribute the commitment of that key chain.)

This imposes additional computation during the initialization phase. Assume that all the key chains have 1000 keys. The base station needs to perform about 1,001,000 pseudorandom function operations to generate all the key chain commitments, and 1000 pseudorandom function operations and 1000 MAC operations to generate all the CDM messages. Due to the efficiency of pseudorandom functions, such computation is still practical if the base station is relatively resourceful. For example, using MD5 as the pseudorandom function, a modern personal digital assistant (PDA) can finish the preceding computation in several seconds. Moreover, the base station does not have to save the low-level key chains. Indeed, to reduce the storage overhead, the base station may compute a low-level key chain (again) when the key chain is needed. Thus, the base station only needs to store the high-level key chain and the MACs of all the CDM messages. Further assume both the authentication key and the image of a pseudorandom function are 8 bytes. To continue the earlier example, the base station needs $(8 + 8) \times 1000 = 16,000$ bytes to store the high-level key chain and the MACs.

The immediate authentication of CDM_i depends on the successful receipt of CDM_{i-1}. However, if a sensor node cannot receive an authentic CDM_i due to communication failure or an attacker's active disruption, the sensor node has to fall back to the DOS-tolerant multilevel μTESLA. This implies that the base station still needs to distribute CDM messages multiple times in a random manner.

Now let us assess how difficult it is for a sensor node to recover if it fails to receive an authentic CDM message. We assume an attacker will launch a DOS attack to deter this recovery. To recover from the failure, the sensor node has to buffer an authentic CDM message by the end of a later high-level time interval and then authenticate this message. For example, suppose a sensor node buffers an authentic CDM_{i+j}. If it receives a disclosed key in interval I_{i+j+1}, it can authenticate CDM_{i+j} immediately and gets $H(CDM_{i+j+1})$. The sensor node then recovers from the failure. Thus, if a sensor node fails to receive an authentic CDM_i, the probability that it recovers from this failure within the next l high-level time intervals is $1 - p^{m \times l}$, where

$$p = \frac{\text{No. forged copies of each CDM message}}{\text{No. total copies of each CDM message}}$$

and m is the number of buffers for CDM messages.

The base station needs to broadcast each CDM message multiple times to alleviate communication failures and to help sensor nodes recover from failures under potential DOS attacks. A sensor node in this scheme only needs a large number of CDM buffers temporarily during recovery. Moreover, a sensor node only needs to recover one authentic CDM message in order to go back to normal operations, and the sensor node may recover over several high-level time intervals. Thus, the bandwidth required for CDM messages can be much less than DOS-tolerant multilevel μTESLA.

DOS-resistant multilevel μTESLA introduces additional computation requirement before deployment, though it can defeat the DOS attacks when at least one copy of each CDM message reaches the sensor nodes. Fortunately, such computation is affordable if the base station is relatively resourceful. It is also possible to perform such computation on powerful machines and then download the result to the base station before deployment. In addition, the communication overhead and the storage overhead on sensor nodes in this scheme is potentially much less than that in DOS-tolerant μTESLA. Thus, when the required computational resources are available (on either the base station or some other machines), DOS-resistant multilevel μTESLA is more desirable. Otherwise, DOS-tolerant multilevel μTESLA could be used to mitigate the DOS attacks.

5.3 KEY PREDISTRIBUTION IN SENSOR NETWORKS

As one of the most fundamental security services, pairwise key establishment enables the sensor nodes to communicate securely with each other using cryptographic techniques such as encryption and authentication. However, due to the resource constraints on sensor nodes, it is undesirable for them to use traditional pairwise key establishment techniques such as public key cryptography and key distribution center (KDC).

Instead of the traditional approaches to key establishment, a promising alternative is *key predistribution*, where keying materials are predistributed to sensor nodes before deployment. As two extreme cases, one may set up a global key among the network so that two sensor nodes can establish a key based on this global key, or assign each sensor node a unique random key with each of the other nodes. However, the former is vulnerable to the compromise of a single node, and the latter introduces substantial storage overhead on sensor nodes.

There have been a number of recent advances in key predistribution for sensor networks, starting from the probabilistic key predistribution scheme proposed by Eschenauer and Gligor [3]. In this section, we describe several of these advances.

5.3.1 Random Key Predistribution Based on Key Pools

Eschenauer and Gligor developed the first random key predistribution scheme based on probabilistic key sharing [3]. The main idea is to let each sensor node randomly pick a set of keys from a key pool before the deployment so that any two sensor nodes have a certain probability to share at least one common key. By configuring the size of the key pool and the size of the key ring, which includes the keys a sensor node selects, this scheme provides good security properties with reasonable overhead. For convenience, we call this scheme *basic probabilistic key predistribution*.

The basic probabilistic key predistribution scheme consists of three phases: *key predistribution*, *shared key discovery*, and *path-key establishment*. The key predistribution phase is performed to generate the key pool and distribute keys to sensor nodes. After being deployed, if two sensor nodes need to establish a pairwise key, they first attempt to do so through shared key discovery, during which they discover whether they share a common key. If they do, there is no need to start path-key establishment; otherwise, these two nodes start the path-key establishment phase to establish a pairwise key with the help of other sensor nodes.

Eschenauer and Gligor studied the connectivity between sensor nodes resulting from the basic key predistribution as random graphs [3]. Consider a graph of n sensor nodes where there is an edge between two nodes if and only if they share a common key. Let p be the probability that two sensor nodes have a shared key, and n be the number of sensor nodes in a network. This graph can be modeled as a random graph $G(n, p)$ of n nodes for which the probability that an edge exists between two nodes is p. According to the classic results on random graph theory by Erdos and Rényi, there exists a value of p such that the probability that a random graph is fully connected moves from 0 to 1 in a large random graph. To further consider the fact that a sensor node has limited communication range, the preceding random graph can be modified to have an edge between two nodes only if they are neighbors. This requires a larger p to ensure that the random graph of the sensors is fully connected. The additional analysis by Eschenauer and Gligor also indicates that there are two critical parameters during the phase of key predistribution: the size of the key pool (P), and the size of each key ring (k). The probability of sharing at least one key between two nodes increases when k increases give a fixed P, and decreases when P increases given a fixed k. The adoption of the basic probabilistic key predistribution then involves determining the values of k and P for the network size and memory constraints on sensor nodes.

A limitation of the basic probabilistic key predistribution scheme is that an attacker may learn the pairwise keys shared between noncompromised nodes when the attacker compromises a number of nodes. This is because all the sensor nodes draw their keys from a common key pool, and by learning the keys from compromised nodes, the attacker will be able to get keys used by the noncompromised nodes.

Chan et al. [4] developed a q-composite key predistribution scheme by extending the basic probabilistic key predistribution scheme, aiming at addressing the problem just discussed. The q-composite key predistribution also uses a key pool; however, it also requires two nodes to compute a pairwise key from at least q predistributed keys that they share. As a result, an attacker has to learn at least q keys shared between two noncompromised nodes in order to recover the pairwise key they use. Chan et al.'s analysis indicates that the q-composite scheme offers stronger resilience against random node compromises than the basic probabilistic scheme when the number of compromised nodes is small.

5.3.2 Random Pairwise Keys

The basic probabilistic key predistribution and the q-composite scheme do not allow node-to-node authentication, since all nodes select keys from a common key pool and there is no unique key and identity for node authentication. The random pairwise keys scheme was proposed to address this limitation.

The random pairwise keys scheme randomly picks pairs of sensor nodes and assigns each pair a unique random key. Based on Erdos and Rényi's results about random graphs, we can easily determine a probability p that allows a sensor networks to be fully connected with a probability. In other words, given the storage for m keys and the preceding probability p, there is a high probability for a sensor network to be fully connected as long as the network size n is smaller than m/p.

In addition to allowing node-to-node authentication, the random pairwise keys scheme has several other nice properties. Due to the simplicity of the approach, it is possible to give a thorough security and performance analysis. Moreover, the random pairwise keys scheme is resistant to node compromises. Since different pairs of nodes share different keys, an attacker does not learn the keys shared between noncompromised nodes by learning the keys stored on compromised ones. Revocation in this scheme also becomes very straightforward; a sensor node only needs to delete a key in order to revoke a compromised node.

A limitation of the random pairwise keys scheme is the strict limit on the network size. The maximum supported network size is strictly limited by the available memory for keys on sensor nodes and the probability for two sensor nodes to share a common key.

5.3.3 Polynomial Pool–Based Key Predistribution

Polynomial pool–based key predistribution can be considered to be a combination of the polynomial-based key predistribution [28] with the aforementioned random key predistribution techniques based on key pools. Instead of using a pool of random keys, this approach employs a pool of random polynomials. Due to the nice threshold property provided by the polynomial-based key predistribution, the polynomial pool–based approach offers more resilience against compromised nodes. In the following, before discussing this approach, we first give an overview of polynomial-based key predistribution.

5.3.3.1 *Polynomial-Based Key Predistribution*   Polynomial-based key predistribution [28] was developed for group key predistribution. Since our goal is to establish pairwise keys, for simplicity, we only discuss the special case of pairwise key establishment in the context of sensor networks.

To predistribute pairwise keys, the (key) setup server randomly generates a bivariate t-degree polynomial $f(x, y) = \sum_{i,j=0}^{t} a_{ij} x^i y^j$ over a finite field F_q, where q is a prime number that is large enough to accommodate a cryptographic key, such that it has the property of $f(x, y) = f(y, x)$. (In the following, we assume all the bivariate polynomials have this property without explicitly saying so.) It is assumed that each sensor node has a unique ID. For each node i, the setup server computes a *polynomial share* of $f(x, y)$, that is, $f(i, y)$. This polynomial share is predistributed to node i. Thus, for any two sensor nodes i and j, node i can compute the key $f(i, j)$ by evaluating $f(i, y)$ at point j, and node j can compute the same key $f(j, i) = f(i, j)$ by evaluating $f(j, y)$ at point i. As a result, nodes i and j can establish a common key $f(i, j)$.

In this approach, each sensor node i needs to store a t-degree polynomial $f(i, x)$, which occupies $(t + 1) \log q$ storage space. To establish a pairwise key, both sensor nodes need to evaluate the polynomial at the ID of the other sensor node. There is no communication overhead during the pairwise key establishment process. The security proof in ref. [28] ensures that this scheme is unconditionally secure and t-collusion resistant. That is, a coalition of no more than t compromised sensor nodes knows anything about the pairwise key between any two noncompromised nodes.

The polynomial-based key predistribution scheme just discussed has some limitations. In particular, it can only tolerate the collusion of no more than t compromised nodes, where the value of t is limited by the available memory space and the computation capability on sensor nodes. Indeed, the larger a sensor network is, the more likely it is that an adversary will compromise more than t sensor nodes and then the entire network.

It is theoretically possible to use the general group key distribution protocol in ref. [28] in sensor networks. However, the storage cost for a polynomial share is exponential in terms of the group size, making it prohibitive in sensor networks. In this chapter, we focus on the problem of pairwise key establishment.

5.3.3.2 A Framework for Key Predistribution Based on Random Polynomials

A general framework called *polynomial pool–based key predistribution* was proposed to develop secure and practical key establishment techniques, using a pool of random bivariate polynomials. In the following, we first discuss this general framework, and then present two efficient examples of this framework.

The polynomial pool–based key predistribution is inspired by the studies in refs. [3] and [4]. The basic idea can be considered as the combination of the polynomial-based key predistribution and the key pool idea used in refs. [3] and [4]. However, our framework is more general in that it allows different choices to be instantiated within this framework.

Intuitively, this general framework generates a pool of random bivariate polynomials and assigns shares on a subset of bivariate polynomials in the pool to each sensor node. The polynomial pool has two special cases. When it has only one polynomial, the general framework degenerates into the polynomial-based key predistribution. When all the polynomials are 0-degree ones, the polynomial pool degenerates into the key pool used in refs. [3] and [4].

Similar to the basic probabilistic key predistribution scheme, pairwise key establishment in this framework has three phases: *setup, direct key establishment*, and *path-key establishment*. The setup phase is performed to initialize the nodes by distributing polynomial shares to them. After being deployed, if two sensor nodes need to establish a pairwise key, they first attempt to do so through direct key establishment. If they can successfully establish a common key, there is no need to start path-key establishment; otherwise, these two nodes start path-key establishment, trying to establish a pairwise key with the help of other sensor nodes.

- *Phase 1: Setup.* The setup server randomly generates a set $\mathcal{F}$ of bivariate t-degree polynomials over the finite field F_q. To identify different polynomials, the setup server may assign each polynomial a unique ID. For each sensor node i, the setup server picks a subset of polynomials $\mathcal{F}_i \subseteq \mathcal{F}$, and assigns the shares of these polynomials to node i. The main issue in this phase is the *subset assignment* problem, which specifies how to pick a subset of polynomials from $\mathcal{F}$ for each sensor node.

- *Phase 2: Direct Key Establishment.* A sensor node starts phase 2 if it needs to establish a pairwise key with another node. If both sensor nodes have shares on the same bivariate polynomial, they can establish the pairwise key directly using the polynomial-based key predistribution. The main issue in this phase is the *polynomial share discovery* problem, which specifies how to find a common bivariate polynomial, of which both nodes have polynomial shares. For convenience, we say two sensor nodes have a *secure link* if they can establish a pairwise key through direct key establishment. A pairwise key established in this phase is called a *direct key*.

- *Phase 3: Path-Key Establishment.* If direct key establishment fails, two sensor nodes need to start phase 3 to establish a pairwise key with the help of other sensor nodes. To establish a pairwise key with node j, a sensor node i needs to find a sequence of nodes between itself and node j such that any two adjacent nodes in this sequence can establish a direct key. For the sake of presentation, we call such a sequence of nodes a *key path* (or simply a *path*), since the purpose of such a path is to establish a pairwise key. Then either node i or j initiates a key establishment request with the other node through the intermediate nodes along the path. A pairwise key established in this phase is called an *indirect key*. A subtle issue is that two adjacent nodes in the path may not be able to communicate with each other directly. This framework assumes that they can always discover a route between themselves so that the messages from one node can reach the other. The main issue in this phase is the *path discovery* problem, which specifies how to find a path between two sensor nodes.

In the following, we describe two random key predistribution schemes that fall in the framework of polynomial pool–based key predistribution.

5.3.3.3 Random Subset Assignment Key Predistribution
As an instantiation of the polynomial pool–based key predistribution, the random subset assignment scheme can also be considered as an extension to the basic probabilistic

scheme. Instead of randomly selecting keys from a large key pool and assigning them to sensor nodes, this method randomly chooses polynomials from a random polynomial pool and assigns their polynomial shares to sensor nodes. However, these two schemes also differ from each other. In the basic probabilistic key predistribution scheme, the same key may be shared by multiple sensor nodes. In contrast, in the random subset assignment scheme, there is a unique key for each pair of sensor nodes. If no more than t shares on the same polynomial are disclosed, none of the pairwise keys constructed using this polynomial between two noncompromised sensor nodes is disclosed.

Now let us describe this scheme by instantiating the three components in the general framework.

1. *Subset Assignment.* The setup server randomly generates a set $\mathcal{F}$ of s bivariate t-degree polynomials over the finite field F_q. For each sensor node, the setup server randomly picks a subset of s' polynomials from $\mathcal{F}$ and assigns shares as well as the IDs of these s' polynomials to the sensor node.

2. *Polynomial Share Discovery.* Since the setup server does not predistribute enough information to the sensor nodes for polynomial share discovery, sensor nodes that need to establish a pairwise key have to discover a common polynomial with real-time discovery techniques. To find a common bivariate polynomial, the source node discloses a list of polynomial IDs to the destination node. If the destination node finds that they have shares on the same polynomial, it informs the source node the ID of this polynomial; otherwise, it replies with a message that contains a list of its polynomial IDs, which also indicates that the direct key establishment has failed.

3. *Path Discovery.* If two sensor nodes fail to establish a direct key, they need to start the path-key establishment phase. During this phase, the source node tries to find another node that can help it set up a pairwise key with the destination node. Basically, the source node broadcasts two lists of polynomial IDs. One list includes the polynomial IDs at the source node, and the other includes the polynomial IDs at the destination node. These two lists are available at both the source and the destination nodes after the polynomial share discovery. If one of the nodes that receives this request is able to establish direct keys with both the source and the destination nodes, it replies with a message that contains two encrypted copies of a randomly generated key: one encrypted by the direct key with the source node, and the other encrypted by the direct key with the destination node. Both the source and the destination nodes can then get the new pairwise key from this message. (Note that the intermediate node acts as an ad hoc KDC in this case.) In practice, we can require that a sensor node only contact its neighbors within a certain range.

The random subset assignment scheme has some nice properties. (We refer the reader to ref. [6] for a detailed analysis of these properties.) In particular, when the fraction of randomly compromised secure links is small (e.g., less than 50%), given the same storage constraint, the random subset assignment scheme provides a significantly higher probability of establishing secure communication between noncompromised nodes than the basic probabilistic and the q-composite key predistribution schemes. Figure 5.4(a) and 5.4(b) compare the security performance of the random subset assignment scheme with these previous schemes. These figures clearly show that before the number of compromised sensor nodes reaches a certain point, the random subset assignment scheme performs much better than both of the other schemes. When the number of compromised nodes exceeds a certain point, the other schemes have fewer compromised links or keys than the random subset assignment scheme. Nevertheless, under such circumstances, none of these schemes provides sufficient security due to the large fraction of compromised links (over 60%) or the large fraction of compromised (direct or indirect) keys (over 80%). Thus, the

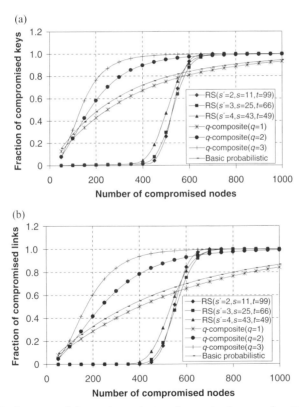

Figure 5.4 Performance of the random subset assignment scheme under attacks. RS refers to the random subset assignment scheme. Assume the network size is $N = 20,000$, that each node has available storage for 200 keys, and that the probability of sharing a direct key between two nodes is $p = .33$. (a) Fraction of compromised links between noncompromised nodes vs. number of compromised nodes. (b) Fraction of compromised keys (direct or indirect) between noncompromised nodes vs. number of compromised nodes.

random subset assignment scheme clearly has advantages over the basic probabilistic and the q-composite schemes.

The random subset assignment scheme also has several advantages over the random pairwise keys scheme [4]. In terms of security performance, the random pairwise keys scheme does not allow reuse of the same key by multiple pairs of sensor nodes [4]. Thus, the compromise of some sensor nodes does not lead to the compromise of direct keys shared between noncompromised nodes. However, the random pairwise keys scheme has a strict upper bound on the network size for a given storage constraint and a desired probability to share common keys between nodes. In contrast, the random subset assignment scheme allows the network to grow by trading off with security. Moreover, in the random subset assignment scheme, sensor nodes can be added dynamically without having to contact the previously deployed sensor nodes. However, in the random pairwise keys scheme, if it is necessary to dynamically deploy sensor nodes, the setup server has to either reserve space for sensor nodes that may never be deployed, which reduces the probability that two deployed nodes share a common key, or inform some previously deployed nodes of additional pairwise keys, which introduces additional communication overhead. Thus, in certain applications, the random subset assignment scheme is a more attractive choice than the random pairwise keys scheme.

5.3.3.4 Grid-Based Key Predistribution

Grid-based key predistribution is another instantiation of the general framework. Suppose a sensor network has at most N sensor nodes. The grid-based scheme constructs an $m \times m$ grid and generates $2m$ bivariate polynomials $\{f_i^c(x, y), f_i^r(x, y)\}_{i=0,\ldots,m-1}$, where $m = \lceil \sqrt{N} \rceil$. As shown in Figure 5.5(a), each row i in the grid is associated with a polynomial $f_i^r(x, y)$, and each column i is associated with a polynomial $f_i^c(x, y)$. The setup server assigns each

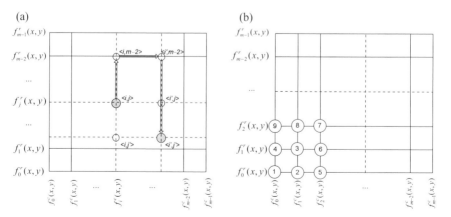

Figure 5.5 Grid-based key predistribution: (a) the grid, and (b) an example order of node assignment.

sensor node in the network to a unique coordinate in this grid. For the node at the coordinate (i,j), the setup server distributes the polynomial shares of $f_i^c(x,y)$ and $f_j^r(x,y)$ to this node. As a result, sensor nodes can perform share discovery and path discovery based on this information.

For convenience, we encode the coordinate of a sensor node into a single-valued node ID. Let $l = \lceil \log_2 m \rceil$. Any valid coordinate can be represented as two l-bit binary strings (one from the column, and the other from the row in the grid). We then denote the ID of a sensor node as the concatenation of these two binary strings. Syntactically, we represent an ID constructed from the coordinate (i,j) as $\langle i,j \rangle$. For the sake of presentation, we also denote ID i as $\langle i_c, i_r \rangle$, where i_c and i_r are the first and last l bits of i, respectively.

The grid-based key predistribution scheme is also generalized as the hypercube-based scheme, which has n dimensions in hypercube instead of two dimensions in the grid. However, we do not discuss it, but refer the readers to ref. [29]. The details of the grid-based key predistribution scheme are presented below.

- *Subset Assignment.* The setup server randomly generates $2m$ t-degree bivariate polynomials $\mathcal{F} = \{f_i^c(x,y), f_i^r(x,y)\}_{i=0,\dots,m-1}$ over a finite field F_q. For each sensor node, the setup server picks an unoccupied coordinate (i,j) in the grid and assigns it to the node. The ID of this sensor node is ID $= \langle i,j \rangle$. The setup server then distributes $\{ID, f_i^c(j,x), f_j^r(i,x)\}$ to this node. To facilitate path discovery and guarantee that there is at least one key path between two nodes if there are no compromised nodes and that any two nodes can communicate with each other, we require that the coordinates assigned to sensor nodes are densely selected within a rectangle area in the grid. Figure 5.5(b) shows a possible order to assign coordinates to sensor nodes. It is easy to see that if there exist nodes at $\langle i,j \rangle$ and $\langle i',j' \rangle$, then there must be a node at either $\langle i,j' \rangle$ or $\langle i',j \rangle$, or both.

- *Polynomial Share Discovery.* To establish a pairwise key with node v, node u checks whether $u_c = v_c$ or $u_r = v_r$. If $u_c = v_c$, both nodes u and v have polynomial shares of $f_{u_c}^c(x,y)$, and they can use the polynomial-based key predistribution scheme to establish a direct key. Similarly, if $u_r = v_r$, they both have polynomial shares of $f_{u_r}^r(x,y)$, and can establish a direct key accordingly. If neither of these conditions is true, nodes u and v go through path discovery to establish an indirect key.

- *Path Discovery.* Nodes u and v need to use the path discovery if $u_c \neq v_c$ and $u_r \neq v_r$. However, we note that either node $\langle u_c, v_r \rangle$ or $\langle v_c, u_r \rangle$ can establish direct keys with both nodes u and v. Indeed, if there is no compromised node, it is guaranteed that there exists at least one node that can be used as an intermediate node between any two sensor nodes due to the node-assignment algorithm. For example, in Figure 5.5(a), both node $\langle i',j \rangle$ and $\langle i,j' \rangle$ can help node $\langle i,j \rangle$ establish a pairwise key with node $\langle i',j' \rangle$. Note that nodes u and v can predetermine the possible intermediate nodes without communicating with others.

Dynamic Key-Path Discovery   Although the path discovery algorithm just described can predetermine the key paths that have one intermediate node, both of the preceding intermediate nodes may have been compromised, or are out of communication range in some situations. However, there are still alternative key paths. In particular, we can reuse the predetermined paths at other nodes to find a secure key path. For example, in Figure 5.5(a), besides node $\langle i',j \rangle$ and $\langle i,j' \rangle$, node $\langle i, m-2 \rangle$ has a predetermined path to node $\langle i',j' \rangle$ through node $\langle i', m-2 \rangle$. Thus, it can help node $\langle i,j \rangle$ set up a common key with node $\langle i',j' \rangle$. Indeed, there are up to $2(m-2)$ such nodes in the grid.

Due to the resource constraints on sensor nodes, we focus on the key paths that can be discovered by reusing the predetermined key paths at other nodes. Specifically, a sensor node S can use the following dynamic path discovery to find a key path to node D, using the predetermined key paths to D at a node with which S can establish a direct key using a noncompromised polynomial. This choice reduces the code size at sensor nodes, since we can reuse the code of computing predetermined key paths between sensor nodes. The algorithm can be performed multiple times to increase the chance of success.

1. The source node S randomly selects a noncompromised node that can establish a direct key with S using a noncompromised polynomial. Node S also generates a random number r, and maintains a counter c with initial value 0. If none of the nodes is selected, the protocol stops; otherwise, it goes to the next step.

   (The countervalue c is used to identify the randomly generated keys, since this algorithm can be performed several times to increase the chance of success, and different keys are used in different rounds for security purposes.)

2. For the selected intermediate node u, S increments the counter c and computes $K_c = F(r, c)$, where F is a pseudorandom function [27]. Then S sends u the IDs of S and D, c, and K_c in a message encrypted and authenticated with the direct key $K_{S,u}$ shared between S and u.

3. If u receives and authenticates such a message, it knows that S wants to establish a pairwise key with D. Node u then tries to establish a key with D using one of its predetermined key paths to D. If this fails, u notifies S that the key path discovery fails; otherwise, u sends D the IDs of S and D, c, and K_c in a message encrypted and authenticated with the key between u and D. To save communication overhead, this message can be piggy-backed in the key establishment message between u and D.

4. When the destination node D receives such a message, it knows that S wants to establish a pairwise key $K_{S,D}$ with it. Then it sets $K_{S,D} = K_c$, and informs S of the countervalue c. Finally, S and D can use $K_{S,D}$ to make their communication secure.

This grid-based scheme has a number of attractive properties. First, it guarantees that any two sensor nodes can establish a pairwise key when there are no compromised sensor nodes, assuming that the sensor nodes can communicate with each other. Second, this scheme is sensitive to node compromises. Even if some nodes are compromised, there is still a high probability that a pairwise key can be reestablished between two noncompromised nodes. Third, a sensor node can directly determine whether it can establish a direct key with another node, and if it can, which polynomial should be used. As a result, there is no communication overhead during polynomial share discovery.

5.3.4 Improving Random Key Predistribution with Expected Deployment Locations

Despite the recent advances, key management in distributed sensor networks is still not an entirely solved problem. This is especially because the performance of these schemes, particularly the probability of establishing a common key between communicating sensors and the ability to tolerate compromised sensors, are highly dependent on the memory available on sensor nodes. Because of the resource constraints on sensor nodes and the need to lower the cost of sensor networks, it is always desirable to reduce the memory required by key management and allocate more resources for the sensor network applications.

In some applications, the sensors may have low mobility, and we may be able to predetermine the location of the sensors to a certain extent. In this case, we can use the sensors' location information to improve the performance of pairwise key predistribution. In this subsection, we describe a simple location-aware deployment model as well as a pairwise key management scheme that can take advantage of the location information. More details of random key predistribution using deployment knowledge can be found in refs. [7] and [8].

5.3.4.1 *A Location-Aware Deployment Model*   We assume that sensors are deployed in a two-dimensional area called the *target field*, and two sensors can communicate with each other if they are within each other's *signal range*. The location of a sensor can be represented by a coordinate in the target field. Each sensor has an *expected location* that can be predicted or predetermined. After the deployment, a sensor is placed at an *actual location* that may be different from its expected location. We call the difference between the expected location and the actual location of a sensor the *deployment error* for the sensor. Thus, this model can be characterized by the following three parameters:

1. *Signal Range d_r.* A sensor can receive messages from another sensor if the former is located within the signal range of the latter. We model the signal range of a sensor as a circle centered at its actual location with the radius d_r. For simplicity, we assume that the radius d_r defining the signal range is a networkwide parameter, and denote the signal range by d_r. We say two sensors are neighbors if they are physically located within each other's signal range.

2. *Expected Location* (L_x, L_y). The expected location (L_x, L_y) of a sensor is a coordinate in the two-dimensional target field; it specifies where the sensor is expected to be deployed. Sometimes, a sensor can be expected to be deployed within an area instead of a particular location. In this case, we assume the sensor is expected to be deployed at any location in that area with equal probability.

3. *Deployment Error* ϵ. We model the deployment error ϵ with a *probability density function*. The sensor expected to be deployed at (L_x, L_y) may appear at a particular area with a certain probability, which is calculated by the integration of probability density function ϵ over this area. In some cases, the sensor may have certain mobility, and appear somewhere near its expected location with a certain probability. The actual location of a sensor at any point in time may also be modeled by the probability density function. We assume the deployment error is also a networkwide parameter.

5.3.4.2   *Closest Pairwise Keys Scheme*

The closest pairwise keys scheme is a variation of the random pairwise keys scheme that takes advantage of the sensors' expected locations. The basic idea is to have each node share pairwise keys with c other nodes whose expected locations are closest to this node's expected location, where c is a system parameter determined by the memory constraint.

Assume there is a setup server responsible for key predistribution, which is also aware of the networkwide signal range and deployment error, and the expected location of each sensor before deployment. Further assume each sensor has a unique, integer-valued ID. We also use a sensor ID to refer to a particular sensor. For convenience, we call a pairwise key shared directly between two neighbor nodes a *direct key*, and a pairwise key established through other intermediate nodes an *indirect key*.

The closest pairwise keys scheme predistributes pairwise keys between pairs of sensors so that two sensors have a predistributed pairwise key if they have a high probability of appearing in each other's signal range. Although reasonable, this idea is difficult to implement, since it is nontrivial to get the probability that two sensors are neighbors. Indeed, this probability depends on the distribution of the deployment error, which is generally not available and may vary in different applications. To simplify the situation, we predistribute pairwise keys between pairs of sensors whose expected locations are close to each other, hoping that the closer the expected locations of two sensors, the more possible that they are physically located in each other's signal range. We will then use a simple deployment error model to analyze the probability that two neighbor sensors share a pairwise key. The basic scheme follows:

- *Key Predistribution.* Based on the expected locations of the sensors, the setup server predistributes pairwise keys on each sensor to facilitate establishing pairwise keys during the normal operation. Specifically, for each sensor u, the setup server first discovers a set S of c other sensors whose expected locations are closest to the expected location of u. For each sensor v in S, the setup server randomly generates a unique pairwise key $K_{u,v}$ if no pairwise key between u and v has been assigned. The setup server then distributes $(v, K_{u,v})$ and $(u, K_{u,v})$ to sensors u and v, respectively.

- *Direct Key Establishment.* After the deployment of the sensor network, if two sensors u and v want to set up a pairwise key to secure the communication between them, they only need to check whether they have a predeployed pairwise key with the other party. This information is obtained from the setup server at the predistribution phase. The algorithm to identify such a common key is trivial, because each pairwise key in a particular sensor was associated with a sensor ID.

- *Indirect Key Establishment.* After deployment, if two neighbor sensors u and v do not share a predistributed pairwise key, they can find an intermediate neighbor sensor that shares pairwise keys with both of them to help establish a *session key*. Basically, either of these two sensors can broadcast a request message with their IDs. Without loss of generality, we assume u sends this request. Suppose sensor i receives this request, and i shares a pairwise key $k_{u,i}$ with u, and a pairwise key $k_{v,i}$ with v. Sensor i then generates a random session key k and sends a message back to u, which contains $E_{k_{u,i}}(k)$ and $E_{k_{v,i}}(k)$. These are the session key k encrypted with $k_{u,i}$ and $k_{v,i}$, respectively. Upon receiving this reply message, sensor u can get the session key by decrypting $E_{k_{u,i}}(k)$, and inform sensor v by forwarding $E_{k_{v,i}}(k)$ to v. (Note that sensor i acts as a KDC in this case.) Sensor u may receive multiple replies; it can choose any one of them.

The analysis in ref. [8] indicates that if the expected locations are known and the deployment of sensors follow a certain distribution centered at the expected locations, the closest pairwise keys scheme can significantly improve the probability that two neighbor nodes share a common key. Since each predistributed pairwise key between two sensor nodes is randomly generated, no matter how many sensors are compromised, the direct keys between noncompromised sensors are still secure. This is a property inherited from the random pairwise keys scheme.

The essential approach of the closest pairwise keys scheme is to use deployment knowledge to improve the probability of sharing keys between neighbor nodes. The same general idea can also be applied to the basic probabilistic key predistribution, the q-composite, the polynomial pool–based key predistribution, and the multispace key predistribution schemes. Additional schemes using deployment knowledge can be found in refs. [7] and [8].

5.3.5 Further Reading on Key Predistribution in Sensor Networks

As mentioned earlier, key management is one of the most fundamental security services in sensor networks, and key predistribution is considered a promising approach for efficient and resilient key establishment in sensor networks. What we have described is by no means the complete picture. For additional information on key predistribution in sensor networks, refer to refs. [3–10] and [30].

5.4   DEMAND FOR MORE SECURITY RESEARCH

Existing sensor network security research has been mostly on foundational cryptographic services, or applying cryptographic techniques to protect sensor network applications. Examples include resilient data aggregation [19–21], secure in-network processing [16], and hop-by-hop authentication for filtering out false data-injection attacks [17,18]. These are certainly critical issues for sensor network security.

However, more issues are required for secure and resilient sensor networks. In particular, we need novel solutions to protect fundamental services such as location discovery and clock synchronization, which are also resilient in the presence of failures and malicious attacks. Moreover, intrusion detection in sensor networks is particularly important due to the fact that unattended sensor nodes can be easily captured and compromised. Since the ultimate goal of sensor networks is to collect data from physical environments, secure data collection and management is another critical issue as well.

To illustrate the need for additional research, in the rest of this section we discuss attacks against location discovery in sensor networks.

5.4.1   Attacks Against Location Discovery

Sensors' locations play a critical role in numerous sensor network applications. For example, target tracking applications require sensors' locations to estimate the moving direction of target objects. As another example, in geographical routing protocols (e.g., FACE [31,32], greedy perimeter stateless routing (GPSR) [32–34]), sensor nodes make routing decisions at least partially based on their own and their neighbors' locations. Indeed, many sensor network applications will not work without sensors' location information.

A number of location discovery protocols [33,35–42] have been proposed for wireless sensor networks in recent years. These protocols share a common feature: They all use some special nodes, called *beacon nodes*, which are assumed to know their own locations (e.g., through global positioning system (GPS) receivers or manual configuration). These protocols work in two stages. In the first stage, nonbeacon nodes receive radio signals called *beacon signals* from the beacon nodes. The packet carried by a beacon signal, which we call a *beacon packet*, usually includes the location of the beacon node. The nonbeacon nodes then estimate certain measurements (e.g., distance between the beacon and the nonbeacon nodes) based on features of the beacon signals (e.g., received signal strength indicator, time difference of arrival). We refer to such a measurement and the location of the corresponding beacon node collectively as a *location reference*. In the second stage, a sensor node determines its own location when it has enough number-of-location references from different beacon nodes. A typical approach is to consider the location references as constraints that a sensor node's location must satisfy, and estimate it by finding a mathematical solution that satisfies these constraints with minimum estimation error. Existing approaches either employ *range-based* methods [35–39], which use the exact measurements obtained in stage one, or *range-free* ones [33,40–42], which only need the existences of beacon signals in stage one.

Despite the recent advances, location discovery for wireless sensor networks in hostile environments, where there may be malicious attacks, has been mostly overlooked. As a matter of fact, all of the existing location discovery protocols become vulnerable in the presence of malicious attacks. As illustrated in Figure 5.6, an attacker may provide an incorrect location reference by pretending to be valid beacon nodes (Figure 5.6(a)), compromising beacon nodes (Figure 5.6(b)), or replaying the beacon packets that were intercepted in different locations (Figure 5.6(c)). In either of these cases, nonbeacon nodes will determine their locations incorrectly.

Without protection, an attacker may easily mislead the location estimation at sensor nodes and subvert the normal operation of sensor networks. The security of location discovery can certainly be enhanced by authentication. Specifically, each beacon packet should be authenticated with a cryptographic key known only to the sender and the intended receivers, and a nonbeacon node accepts a beacon

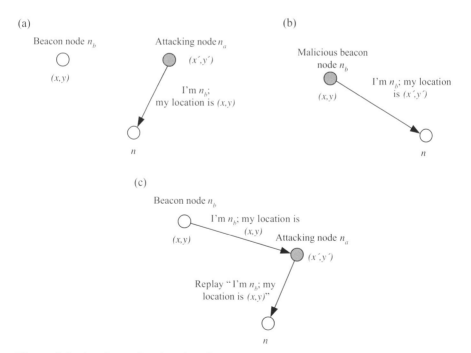

Figure 5.6 Attacks against location discovery services: (a) masquerade beacon nodes, (b) compromised beacon nodes, and (c) replay attack.

signal only when the beacon packet carried by the beacon signal can be authenticated. However, only having authentication does not guarantee the security of location discovery, either. As discussed earlier, an attacker may forge beacon packets with keys learned through compromised nodes, or replay beacon signals intercepted in different locations.

Most of the localization schemes for sensor networks have a certain ability to tolerate measurement errors (e.g., by averaging the effect of problematic location references over all location references). For example, minimum mean-square estimation (MMSE) has been used in most of the range-based and some range-free localization schemes to improve the accuracy of location estimation when a sensor node has redundant location references. However, these methods cannot properly handle malicious location references, which typically include very large errors not seen in natural faults.

To demonstrate the impact of malicious attacks, we performed an experiment through simulation with the MMSE-based location estimation method in [35]. We used nine honest beacon nodes and one malicious beacon node randomly deployed in a 30 m × 30 m field. The node that estimates location is positioned at the center of the field. The malicious beacon node always declares a false location that is x meters away from its real location, where x is a parameter representing the location error created by an attacker. To model the distance measurement error, we assume such an error is uniformly distributed between $-e_{\max}$ and $e_{\max}$. Figure 5.7 shows the location estimation error (i.e., the distance between a sensor's real location and the estimated location) introduced by the malicious beacon node when the location error x created by an attacker increases. We can clearly see that the malicious node affects the estimated location significantly by declaring incorrect locations. Since an attacker can introduce arbitrarily large errors by declaring false locations in beacon packets, the preceding result implies that the attacker can introduce arbitrarily large errors into a nonbeacon node's location estimation.

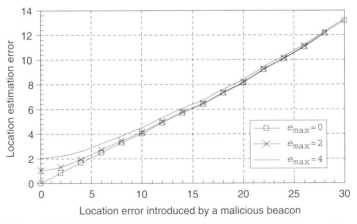

Figure 5.7   Location estimation error introduced by malicious attacks.

Such malicious attacks will generate similar impacts on the other localization schemes. This is because an attacker may introduce arbitrary errors into location estimation process, while all the existing localization techniques assume bounded errors, which are only true in benign environments. As discussed earlier, such attacks cannot be simply prevented by cryptographic techniques due to the threat of compromised nodes and replay attacks. Thus, it is highly desirable to have additional mechanisms to improve the security of location discovery in sensor networks.

5.5 CONCLUSION

Sensor network security is a challenging problem, particularly due to the resource constraints on sensor nodes, the threat of node compromises resulting from unattended deployment, and the imbalance between the threat and the defense in sensor networks. Foundational cryptographic services such as broadcast authentication and key management are definitely a necessary condition to ensure secure and resilient sensor network applications. Other foundational services in sensor network applications also deserve intensive investigation. Examples of such services include secure clock synchronization, secure location discovery, secure aggregation and in-network processing, cluster formation, and cluster head election. Moreover, intrusion detection in sensor networks is highly desirable, particularly due to the fact that unattended sensor nodes may be easily captured and compromised. We expect to see more advances in sensor network security in the next several years.

Research in sensor network security is likely to generate an impact beyond sensor networks themselves. Experiences gained here will offer techniques and insights in handling security problems arising in many other applications involving resource-constrained devices interfacing with a malicious physical world.

REFERENCES

1. I. F. Akyildiz, W. Su, Y. Sankarasubramaniam, and E. Cayirci. Wireless sensor networks: A survey. *Computer Networks*, **38**(4):393–422, 2002.

2. D. W. Carman, P. S. Kruus, and B. J. Matt. *Constrains and Approaches for Distributed Sensor Network Security*. Technical report, NAI Labs, 2000.

3. L. Eschenauer and V. D. Gligor. A key-management scheme for distributed sensor networks. In *Proceedings of the 9th ACM Conference on Computer and Communications Security*, pages 41–47, November 2002.

4. H. Chan, A. Perrig, and D. Song. Random key predistribution schemes for sensor networks. In *IEEE Symposium on Research in Security and Privacy*, pages 197–213, 2003.

5. W. Du, J. Deng, Y. S. Han, and P. Varshney. A pairwise key pre-distribution scheme for wireless sensor networks. In *Proceedings of the 10th ACM Conference on Computer and Communications Security (CCS'03)*, pages 42–51, October 2003.

6. D. Liu and P. Ning. Establishing pairwise keys in distributed sensor networks. In *Proceedings of the 10th ACM Conference on Computer and Communications Security (CCS'03)*, pages 52–61, October 2003.

7. W. Du, J. Deng, Y. S. Han, S. Chen, and P. Varshney. A key management scheme for wireless sensor networks using deployment knowledge. In *Proceedings of IEEE INFOCOM 2004*, Hong Kong, March 2004.

8. D. Liu and P. Ning. Location-based pairwise key establishments for static sensor networks. In *2003 ACM Workshop on Security in Ad Hoc and Sensor Networks (SASN '03)*, page 720082, October 2003.

9. R. D. Pietro, L. V. Mancini, and A. Mei. Random key assignment for secure wireless sensor networks. In *2003 ACM Workshop on Security in Ad Hoc and Sensor Networks (SASN '03)*, October 2003.

10. S. Zhu, S. Setia, and S. Jajodia. LEAP: Efficient security mechanisms for large-scale distributed sensor networks. In *Proceedings of the 10th ACM Conference on Computer and Communications Security (CCS'03)*, pages 62–72, October 2003.

11. A. Perrig, R. Szewczyk, V. Wen, D. Culler, and D. Tygar. SPINS: Security protocols for sensor networks. In *Proceedings of 7th Annual International Conference on Mobile Computing and Networks*, July 2001.

12. A. Perrig, R. Canetti, D. Song, and D. Tygar. Efficient authentication and signing of multicast streams over lossy channels. In *Proceedings of the 2000 IEEE Symposium on Security and Privacy*, May 2000.

13. A. Perrig, R. Canetti, D. Song, and D. Tygar. Efficient and secure source authentication for multicast. In *Proceedings of the Network and Distributed System Security Symposium*, February 2001.

14. D. Liu and P. Ning. Efficient distribution of key chain commitments for broadcast authentication in distributed sensor networks. In *Proceedings of the 10th Annual Network and Distributed System Security Symposium*, pages 263–276, February 2003.

15. D. Liu and P. Ning. Multi-level μTESLA: Broadcast authentication for distributed sensor networks. *ACM Transactions in Embedded Computing Systems*, **3**(4):800–836, 2004.

16. J. Deng, R. Han, and S. Mishra. Security support for in-network processing in wireless sensor networks. In *2003 ACM Workshop on Security in Ad Hoc and Sensor Networks (SASN '03)*, October 2003.

17. S. Zhu, S. Setia, S. Jajodia, and P. Ning. An interleaved hop-by-hop authentication scheme for filtering false data in sensor networks. In *Proceedings of the 2004 IEEE Symposium on Security and Privacy*, May 2004.

18. F. Ye, H. Luo, S. Lu, and L. Zhang. Statistical en-route filtering of injected false data in sensor networks. In *Proceedings of IEEE INFOCOM 2004*, March 2004.

19. L. Hu and D. Evans. Secure aggregation for wireless networks. In *Workshop on Security and Assurance in Ad Hoc Networks*, January 2003.

20. B. Przydatek, D. Song, and A. Perrig. SIA: Secure information aggregation in sensor networks. In *Proceedings of the 1st ACM Conference on Embedded Networked Sensor Systems (SenSys '03)*, November 2003.

21. W. Du, J. Deng, Y. S. Han, and P. K. Varshney. A witness-based approach for data fusion assurance in wireless sensor networks. In *Proceedings of the IEEE Global Communications Conference (GLOBECOM 03)*, December 2003.

22. A. D. Wood and J. A. Stankovic. Denial of service in sensor networks. *IEEE Computer*, **35**(10):54–62, 2002.

23. C. Karlof and D. Wagner. Secure routing in wireless sensor networks: Attacks and countermeasures. In *Proceedings of the 1st IEEE International Workshop on Sensor Network Protocols and Applications*, May 2003.

24. J. Newsome, R. Shi, D. Song, and A. Perrig. The sybil attack in sensor networks: Analysis and defenses. In *Proceedings of the IEEE International Conference on Information Processing in Sensor Networks (IPSN 2004)*, April 2004.

25. R. L. Rivest, A. Shamir, and L. A. Adleman. A method for obtaining digital signatures and public-key cryptosystems. *Communications of the ACM*, **21**(2):120–126, 1978.

26. H. Krawczyk, M. Bellare, and R. Canetti. HMAC: Keyed-Hashing for Message Authentication. Internet RFC 2104, February 1997.

27. O. Goldreich, S. Goldwasser, and S. Micali. How to construct random functions. *Journal of the ACM*, **33**(4):792–807, October 1986.

28. C. Blundo, A. De Santis, Amir Herzberg, S. Kutten, U. Vaccaro, and M. Yung. Perfectly-secure key distribution for dynamic conferences. In *Advances in Cryptology—CRYPTO '92, LNCS 740*, pages 471–486, 1993.

29. D. Liu, P. Ning, and R. Li. Establishing pairwise keys in distributed sensor networks. *ACM Transactions on Information and System Security*, forthcoming.

30. S. A. Camtepe and B. Yener. Combinatorial design of key distribution mechanisms for wireless sensor networks. In *Proceedings of the 9th European Symposium on Research in Computer Security (ESORICS '04)*, 2004.

31. P. Bose, P. Morin, I. Stojmenović, and J. Urrutia. Routing with guaranteed delivery in ad hoc wireless networks. In *Proceedings of the 3rd ACM International Workshop on Discrete Algorithms and Methods for Mobile Computing and Communications*, pages 48–55, 1999.

32. P. Bose, P. Morin, I. Stojmenović, and J. Urrutia. Routing with guaranteed delivery in ad hoc wireless networks. *ACM Wireless Networks*, **7**(6):609–616, 2001.

33. N. Bulusu, J. Heidemann, and D. Estrin. GPS-less low cost outdoor localization for very small devices. *IEEE Personal Communications*, pages 28–34, October 2000.

34. B. Karp and H. T. Kung. GPSR: Greedy perimeter stateless routing for wireless networks. In *Proceedings of ACM MobiCom 2000*, 2000.

35. A. Savvides, C. Han, and M. Srivastava. Dynamic fine-grained localization in ad-hoc networks of sensors. In *Proceedings of ACM MobiCom '01*, pages 166–179, July 2001.

36. A. Savvides, H. Park, and M. Srivastava. The bits and flops of the n-hop multilateration primitive for node localization problems. In *Proceedings of ACM WSNA '02*, September 2002.

37. D. Niculescu and B. Nath. Ad hoc positioning system (APS) using AOA. In *Proceedings of IEEE INFOCOM 2003—The Conference on Computer Communications*, **22**(1):1734–1743, March 2003.

38. A. Nasipuri and K. Li. A directionality based location discovery scheme for wireless sensor networks. In *Proceedings of the ACM WSNA'02*, September 2002.

39. L. Doherty, K. S. Pister, and L. E. Ghaoui. Convex optimization methods for sensor node position estimation. In *Proceedings of IEEE INFOCOM 2001*, pages 1655–1663, Anchorage, Alaska, April 2001.

40. D. Niculescu and B. Nath. DV based positioning in ad hoc networks. *Telecommunication Systems*, **22**:1–4, 267–280, 2003.

41. R. Nagpal, H. Shrobe, and J. Bachrach. Organizing a global coordinate system from local information on an ad hoc sensor network. In *IPSN'03*, 2003.

42. T. He, C. Huang, B. M. Blum, J. A. Stankovic, and T. F. Abdelzaher. Range-free localization schemes in large scale sensor networks. In *Proceedings of ACM MobiCom 2003*, 2003.

■■■■■■■ **CHAPTER 6**

Embedded Operating Systems for Wireless Microsensor Nodes

BRIAN SHUCKER, JEFF ROSE, ANMOL SHETH, JAMES CARLSON,
SHAH BHATTI, HUI DAI, JING DENG, and RICHARD HAN
University of Colorado at Boulder, Boulder, Colorado

Sensor nodes fall somewhere in between the single application devices that need no operating system, and the more capable, general purpose devices with the resources to run a traditional embedded operating system. This is reflected in the design of sensornet operating systems, which provide a limited number of common services for application developers. These common services typically include hardware management of sensors, radios, and I/O buses, and devices such as external flash. Other services needed by applications include task coordination, power management, adapting to resource constraints, and networking. In this chapter, authors examine the principles behind the design of sensornet operating systems, their basic architecture, and features that are unique to the sensor network domain. These principles are illustrated by examining two very different sensor operating systems, TinyOS and MOS (the MANTIS operating system).

6.1 INTRODUCTION

The first question one might ask about operating systems for sensor networks is, "Why do sensor nodes need an operating system at all?" Indeed, there are many embedded devices that do not use an operating system. For devices customized to a particular application, such as digital cameras or microwaves, it is much simpler to run the application code directly on the microcontroller. On the other hand, more general-purpose devices such as personal digital assistants (PDAs) use embedded versions of full-blown operating systems, such as Windows Mobile PocketPC OS

Handbook of Sensor Networks: Algorithms and Architectures, Edited by Ivan Stojmenović
Copyright © 2005 John Wiley & Sons, Inc.

or Embedded Linux, to provide basic services that are common to multiple applications.

Sensor nodes fall somewhere in between the single-application devices that need no operating system, and the more capable, general-purpose devices with the resources to run a traditional embedded operating system. This is reflected in the design of sensornet operating systems, which may not even be considered operating systems in the traditional sense. A sensor operating system provides a limited number of common services for application developers. These common services typically include hardware management of sensors, radios, and I/O buses, and devices such as external flash. Other services needed by applications include task coordination, power management, adaptation to resource constraints, and networking. In this chapter, we examine the principles behind the design of sensornet operating systems, their basic architecture, and features that are unique to the sensor network domain. We will illustrate these principles by examining two very different sensor operating systems, TinyOS [1] and MOS [2], the MANTIS operating system.

Figure 6.1 identifies the design space that sensor operating systems are targeting. Minimum requirements of operating systems in terms volatile RAM for run-time execution and permanent storage for code are shown in Figure 6.1. The design space of sensor operating systems is in the far lower left of the figure, where both TinyOS and MOS are clustered. Sensor operating systems impose the most minimal needs in terms of RAM and flash, and are equivalent in memory requirements to smart card operating systems. The design space of PDA-class operating systems such as PalmOS and PocketPC OS are located about two magnitudes higher in terms of memory requirements. Most embedded operating systems span the gap between sensor-class operating systems and PDA-class operating systems. Real-time embedded operating systems (RTOS) like QNX and VXWorks roughly fall into the gap between the PDA-class operating systemes and sensor-class operating

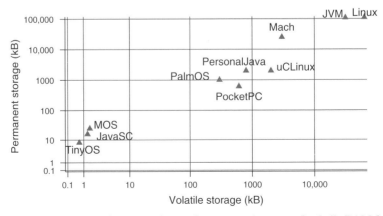

Figure 6.1   Minimum requirements of operating systems in terms of volatile RAM for run-time execution and permanent storage for code.

systemes, requiring approximately 100 kB and 286 kB of ROM, respectively, but are not shown because we were not able to determine their run-time footprints. Also, we expect μCOS and AVRX embedded operating systems to fall into the same range between sensor-class operating systems and PDA-class operating systems. μCLinux requires resources on the order of a PDA-class operating system. Much larger operating systems, such as Linux, are also shown for comparison. Windows-class operating systems fall into the Linux category as well.

Section 6.2 provides an overview of common features of microsensor node hardware. In Section 6.3, we explore the fundamental principles of sensor node operating system design, and Section 6.4 describes a set of features that are unique to sensor node operating systems. We follow with a discussion of two operating systems, TinyOS (Section 6.6) and MANTIS OS (Section 6.7). A detailed comparison of the relative advantages and disadvantages of event-driven and multithreaded sensor operating systems is presented in Section 6.8.

6.2 MICROSENSOR NODE HARDWARE

In developing operating systems for embedded microsensor nodes, the chosen hardware has a direct effect on many aspects of the system design. The hardware of a wireless microsensor node typically consists of five major subsystems, namely, the microcontroller, sensors, radio, power unit, and permanent storage. Here we discuss the primary aspects of the microcontroller and other devices that have the greatest effect on the operating system.

6.2.1 Flash, SRAM, and EEPROM

Most low-power microcontrollers today utilize the Harvard architecture [3], which means the processing unit uses physically separate storage and signaling pathways for the executable instructions and the data [4–7]. Today, this typically results in the usage of flash memory for persistent storage of application code, the text segment, and static RAM (SRAM) for data storage at run time.

In terms of both time and energy, flash memory is inexpensive to read but expensive to write, which makes it well suited for program memory. Also, it may or may not be possible for the microcontroller to write to its own flash memory; reprogramming may require additional hardware. In either case, flash memory is written in blocks (pages), but may be read at random. The number of writes to a particular block is limited, as the blocks wear out over time. However, flash memory is nonvolatile, so it retains data when powered off.

SRAM is random access and is generally fast for reads and writes, but is volatile, consumes more energy than flash, and costs more. SRAM does not wear out in the same fashion as flash memory. For these reasons, SRAM is used in small quantities as data memory. Extremely limited data memory is one of the major constraints in sensor network programming.

Additional nonvolatile storage may take the form of a separate flash bank, or another memory technology such as electrically erasable programmable read-only memory (EEPROM). From an operating system perspective, EEPROM behaves similarly to flash: it is nonvolatile, survives a limited number of writes, and may be slow to write. Both flash and EEPROM come in fairly large units with serial interfaces as well as parallel interfaces; the serial devices may be used in a sensor node similarly to a disk.

6.2.2   Peripheral Interfaces

Modern microcontrollers come with a number of peripheral interfaces, all integrated on the same chip with the CPU and memory. Peripheral devices, such as the radio and sensors, connect to the microcontroller through one or more of these interfaces. Memory-mapped peripherals are rarely used.

Common interfaces include UART (universal asynchronous receiver transmitter), SPI (serial peripheral interface), and I^{2}C (inter-integrated circuit), among others. Each type has unique characteristics, such as maximum speed, number of devices per bus, and signaling protocol. However, the details are largely hidden from the operating system designers, since the interfaces are implemented in hardware.

6.2.3   Radios

Radio-frequency (RF) devices may or may not have a large impact on the performance of a sensor operating system, depending on how much of the RF protocol is handled in hardware. For example, the common Mica2 sensor node uses a CC1000 radio [8] that provides a raw bit interface. Besides requiring somewhat complex software to operate, the CC1000 constantly interrupts the CPU with noise bits whenever it is on. This increases the energy cost of activating the radio, and also reduces available CPU bandwidth. By contrast, the more recent MicaZ node uses a packet-based CC2420 radio, which handles entire packet transfers in hardware. This greatly reduces software complexity and allows the CPU to be in a low-power mode even while the radio is sending or receiving a packet.

6.2.4   Sensors

Sensors are difficult to characterize as a group, because they come in such a wide variety. Sensors may be digital or analog in nature, may or may not require significant attention from the CPU, and may have power requirements ranging from insignificant to dominant. In order to support the multitude of available sensors, an operating system must be designed with flexibility and modularity in mind.

6.2.5   Power Systems

Sensor nodes typically run on batteries, which makes effective power management a key challenge in operating system design, as described in Section 6.3.4. While methods of supplying continuous power in outdoor in situ deployments have been investigated, for example, solar panels and other energy-scavenging techniques, it is still necessary for operating systems to conserve power as much as possible. There are some situations, such as in certain indoor sensor networks, when power can be obtained from wall outlets, though communication continues to be conducted via wireless multihop networking.

6.2.6 Contrast with PC Systems

Sensor nodes that are deployed in situ do not typically support output devices for interaction, such as displays. User interaction is not a paramount requirement, as it is in PC and PDA systems. The nodes are instead left physically unattended for most of their lifetime. Similarly, in situ sensor nodes do not typically provide input devices for user interaction such as keyboards or mice.

6.3 PRINCIPLES OF SENSOR OPERATING SYSTEM DESIGN

Modern sensor operating systems differ from conventional PC operating systems in a number of ways. These differences stem from the unique hardware and energy constraints typically encountered in sensor networking. The following sections describe the general principles behind sensor operating systems.

6.3.1 Managing Hardware

The first task of any operating system is to manage the hardware resources available on the machine. Sensor operating systems are no exception to this rule. The operating system provides abstract services, such as reading sensors, sending and receiving data over the radio, and using timers.

One of the primary hardware constraints on a typical microcontroller is the lack of a memory management unit (MMU). Additionally, most sensor node controllers only have a single operating mode, whereas a typical processor has both user and supervisor modes. This eliminates the distinction between executing kernel code and executing application code, so hardware management may be implemented as a library of function calls. While these function calls can provide a clean, abstract interface to the hardware, they do not provide protection from users who access the hardware directly, either by accident or with malicious intent. This lack of protection has implications for system reliability, debugging, coordination of multiple tasks, and security. Such concerns will be addressed in more detail below.

6.3.2 Task Coordination

Another major problem solved by the operating system is that of coordination of multiple tasks. This primarily consists of two sub-problems: scheduling and synchronization. The operating system must decide when to allocate the CPU to each task, and it must provide mechanisms for the user to attain guarantees about execution order and mutual exclusion when necessary.

Some systems solve both problems by restricting the user to a single task; this is the case when running application code directly on a microcontroller (in a sense, the degenerate case of operating systems). In such a model, there may be multiple logically distinct tasks present, but the coordination of those tasks is handled in an application-specific way by the application programmer, rather than managed in a standard way by the operating system. Some sensor operating systems, such as TinyOS, solve the hardware management problem, but make no attempt to solve the task coordination problem. Others, such as MOS, include task management features.

There are two costs associated with task management: a small amount of CPU bandwidth (consumed by the kernel to make scheduling decisions and context switches) and a significant amount of memory. The memory cost is high because multiple tasks, each with its own static memory requirements and execution stack, may need to coexist in memory at the same time. While it is possible to swap the memory of suspended tasks out to another device, such as flash memory, this greatly increases the context switching time and thus the CPU overhead of the scheduling system.

Task coordination does, however, move significant complexity away from the applications and into the operating system. This may be a drawback if applications are relatively simple, such as a single application thread and network stack. On the other hand, if applications make use of such features, it makes sense to implement them once in the operating system and test them thoroughly, eliminating the need for application programmers to reimplement and debug such complex code multiple times.

6.3.3 Resource Constraints

Sensor nodes must operate under resource constraints that are not major concerns on PCs, or even on smaller embedded devices such as PDAs. For example, the Mica2 node has only 4 kB of data memory (SRAM), 128 kB of program memory (flash), and an 8-bit, 7.3 MHz CPU. Each of these limitations affects the design of sensor operating systems.

It has been argued that one should not design around these resource constraints, because the hardware is subject to Moore's law and will have increased in capability by the time sensor networks are widely adopted. However, current sensor nodes are expensive—on the order of $100 per node. The advances predicted by Moore's law can be applied to lower the cost, rather than increase the capability of a sensor node. An operating system that runs within tight resource constraints *will always run on cheaper hardware*, which can be economically deployed in larger numbers.

6.3.3.1 Data Memory Data memory is an extremely scarce resource, especially in multithreaded systems that must keep multiple execution stacks present at the same time. Besides the small size of physical memory, sensor nodes may also be limited by the lack of a hardware memory management unit. This alters the memory model, since the entire system, including the operating system and all application threads, runs in a single address space.

There are many techniques for reducing memory consumption, such as zero-copy network stacks, lightweight thread management, and compile-time optimizations. Some of these techniques impose limits on the services that the operating system can provide.

The lack of hardware memory management makes proper software engineering critically important on a sensor node. There is no way to prevent threads from writing over each other's memory (although it may be possible to detect after the fact). Even without multiple threads, an application could write over memory locations that hold operating system state information, or even memory-mapped registers. For example, dereferencing NULL on a Mica2 node will overwrite one of the microcontroller's internal general-purpose registers, resulting in unpredictable behavior without ever raising an exception.

6.3.3.2 *Program Memory*

Program memory is not as seriously constrained as data memory, since flash memory is cheaper; storing a complete program image is generally not a problem. It is also difficult (sometimes impossible) to write to a flash bank by accident. The only unusual constraint with program memory is that flash memory has a limited number of write–erase cycles. Of course, program memory does not have to be changed often, so this constraint may not be of practical significance.

6.3.3.3 *CPU Bandwidth*

At first glance, it would seem that the common tasks on a sensor node are I/O-bound: waiting for sensor events and transmitting/receiving data are the primary two. However, many research directions involve tasks that may be CPU-intensive, such as data aggregation [9,10] and/or transformation, in-network processing [11], and cryptography [12]. On an 8-bit microcontroller, operations such as encryption with split-key methods may take a long time to complete.

As mentioned in Section 6.3.2, a sensor operating system may or may not manage the allocation of CPU cycles between tasks. If it does handle CPU scheduling, the operating system must be capable of intelligently interrupting CPU-bound tasks in order to service I/O-driven tasks within a reasonable time frame. Of course, if the operating system does not manage scheduling, the application programmer must ensure that the CPU-bound tasks pause periodically to give cycles to the I/O-bound tasks.

6.3.4 Power Management

Power constraints on sensor nodes are severe. While Moore's law applies to CPU speed and memory size, it does not apply to battery capacity. Thus, it is not surprising the majority of the volume and mass of a sensor node such as the Mica2 is occupied by the battery.

Because battery technology is not improving very fast, the lifetime of a sensor node is increased primarily by reducing power consumption, rather than increasing supply. Low-power components help, but even very low-power microcontrollers and radios will drain a pair of AA batteries in a matter of days. Large-scale increases in node lifetime are obtained by turning components off during times when they are not needed. The "duty cycle" of a node is the fraction of the time that the node's high-power components are active, and may be on the order of 1%. This extends the lifetime of a node from several days to several months (Fig. 6.2).

From the perspective of an operating system, power management comes in two flavors: implicit and explicit. Implicit power management involves techniques that do not require cooperation from the user or from applications. Implicit power management is common in PCs. For example, laptop operating systems may dim the screen when running off battery power, and switch to full brightness when connected to an AC outlet. Many systems turn off components that have not been used for a long time (e.g., spinning down disks), and leave them off until needed. In both cases, neither the users nor the applications are involved in the power management decisions.

Explicit power management is user- or application-directed. In an explicit power management scheme, the application uses a system call to give instructions or hints to the operating system. These calls may indicate which resources the application is going to use in the near future so that the operating system can determine which components need to be powered on. Explicit systems are inherently more efficient than implicit ones, since the operating system has more information to work with. However, there is more work to be done by each application.

In sensor nodes, one of the largest power consumers is the radio. The cost of transmitting one bit is on the order of a thousand times the cost of processing one bit. Despite power optimizations in other modules, the power consumed by radio transmission will continue to remain a concern because the laws of physics

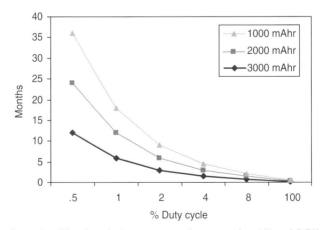

Figure 6.2 Operating life of a wireless sensor node, comparing 1% and 0.5% duty cycles. Period = 300 seconds.

fundamentally limit the amount that can be conserved. Electromagnetic transmission requires a certain transmit power in order to achieve a desired signal-to-noise ratio (SNR) at the receiver. Radios present a difficult power management problem, because they generally must be powered on in order to detect incoming packets. To successfully transmit a packet, both the transmitter and receiver nodes must power up their radios before the start of transmission. This means that power management in a sensor network is not a local issue, but rather a network-wide one. In order to move data across a multihop network, all of the nodes on the path from data source to data sink must activate their radios in a coordinated manner that is agreed upon in advance. This may involve integration with other subsystems, such as the network routing system and time synchronization.

6.3.5 Networking

As implied by the term "sensor network," the network stack is the primary application running on a sensor node. The actual application software may be simple and lightweight in comparison to the network stack. Of particular importance is the memory consumption of the network stack: while a typical sensor network packet might be only 32 bytes, 32 bytes is not that small compared to the total data memory available.

Since memory is so limited and the network may be the primary memory consumer, it is important to define cross-layer interfaces that avoid copying buffers. It also may make sense to integrate network memory management closely with the operating system, to make maximum use of the data memory that is available.

6.3.6 Sensing

As the name sensor network implies, sensing is a key requirement that must be smoothly supported by a sensor operating system. A variety of input sensors should be supported, ranging from simple analog/digital (A/D) sampling and resistive sensors to complex sensors requiring calibration and warm-up procedures, including global positioning system (GPS) receivers. Calibrating sensors is an especially difficult problem in situ. Some approaches seek to provide distributed autocalibration to improve ease of in situ deployment and management while retaining the accuracy of sensor readings [13]. A given sensor node may house multiple sensors in order to provide multimodal sensing, for example, temperature, pressure, and relative humidity. Some types of sensing behavior may be periodic in nature; that is, the node wakes up every T seconds and takes a sensor reading. A sensor operating system should provide primitives when possible to support such periodic duty cycle–based sensing. Other types of sensing behavior are more adaptive and data-driven, for example, target tracking [14].

6.3.7 Applications

Sensor network applications are structured differently from traditional PC applications. Individual sensor nodes do not run complete applications—they do not even have user interfaces onboard! Instead of a complete user-driven application, an individual sensor node runs a small piece of a distributed application, which may be connected to the user via some other back-end system, perhaps connected to the sensor network through a gateway. Thus, it makes sense to optimize the operating system for interactions with other machines, rather than direct operation by a human user.

6.3.8   Maintenance

Since sensor nodes are intended to be deployed in large numbers [15], it is infeasible to perform maintenance on each node manually. Thus, if any sort of software maintenance is required, the operating system must provide a remote maintenance interface. For example, dynamic reprogramming (also called *retasking*) features allow a network administrator to change the software on every node in a network after the network is deployed. Such features require operating system support and must be planned for in advance of deployment.

6.4   FEATURES OF SENSOR OPERATING SYSTEMS

In order to be of practical use, sensor operating systems must include features that are uncommon in PC operating systems, or significantly different from their analogs in the PC world. Several of those features are described below.

6.4.1   Dynamic Network Reprogramming

Two of the main features of wireless sensor networks are that they are unattended and long-lived. Sensor networks could be installed in inaccessible and hostile environments that cannot be accessed easily. Even if the individual nodes are accessible, the scale of a network may be so large that manually accessing each node in the network is highly impractical. Management of the network under such conditions is a daunting task, and a sensor node operating system should have features that enable network management to be handled remotely. The network administrator should be able to address subsections of the network, as well as individual nodes. Over time, due to changing conditions of the environment, it may be necessary to retask some or all of the nodes in order to modify the behavior of the network.

Dynamic network reprogramming is defined as the process of programming the sensor nodes by disseminating code over a multihop network [16,17]. At present, there are two main techniques that may be used to reprogram a sensor network. The first technique is to transmit the entire code image as raw binary data. The main drawback of this approach is that it requires synchronizing large amounts of data across a multihop wireless network: the size of the updated code image is often much larger than main memory (a typical Mica2 node has only 4 kb of RAM) and dwarfs the size of a 30–64-byte packet.

An alternative approach is to use a virtual machine (VM) on each node in the network [18]. The use of a VM allows nodes to be retasked through small virtual programs—high-level program specifications that can be interpreted by the VM—that can be disseminated throughout the network at a lower cost than an entire code image. Upon receiving an updated virtual program, a node's VM can interpret the instructions into byte code, at which point the new program can replace the existing code. While the use of VMs can provide a low-cost approach to dynamic network reprogramming, it is still necessary to support raw binary retasking, since the VM itself may require changes.

A sensor node operating system will need to provide the basic facilities that are required to support both reprogramming paradigms. First, the network infrastructure must be capable of disseminating code updates to all nodes in the network. The operating system must also allow access to nonvolatile data storage; since a sensor node will need to be rebooted in order to replace old code, the updated code cannot be stored in main memory. A file system is also required to accumulate the data as packets arrive over the network, and to organize data in memory (Section 6.4.2). The final requirement is a boot loader program that is capable of replacing the old code with the updated code.

To support either of these reprogramming paradigms, the operating system requires a mechanism by which data can be incrementally added to persistent storage until all the updated code has been received. This generally requires a file system, as discussed in Section 6.4.2. Once the complete code image has been stored, the operating system will set a flag that can be read by the boot loader, indicating that the sensor node is ready to be retasked. Since the entire code image cannot be replaced during run-time, the operating system must then restart the sensor node, at which point the boot loader takes over. Upon rebooting, the flag indicates to the boot loader that the node should be reprogrammed, at which point the boot loader overwrites the existing code with the updated code image that was stored in persistent memory.

Depending on the scale of the network and the nature of the application, it may not be feasible for the reprogramming operation to stall the currently running task. If continuous operation is required while the sensor node accumulates and organizes the new code image, a portion of memory must be reserved that cannot be used by the running task. Prior to reprogramming the node, the resources that are allocated to the currently running task must be freed, after which memory can be reflashed with the verified update. Fine-grained retasking of a network is an ongoing area of research, since current mechanisms only allow retasking of the entire code image of a node. New techniques are required in order to replace a single thread while the operating system continues to run. In such cases, resource management will play an important role. The operating system will need to ensure that no dependencies exist between threads that could lead to a deadlock, since any deadlock, while freeing resources, would require manual intervention to reboot the node.

6.4.2 File System

It may not always be possible to transmit sensor data frequently, due to hardware failures, environmental conditions, and the presence of physical layer jamming. Furthermore, frequent transmissions over the radio are an expensive operation in terms of energy consumption. Due to these and other factors, a sensor node must be able to store data locally over a long period of time while multiple sensor readings are collected.

At present, most sensor nodes only have around 4–12 kb of RAM, enough to accommodate data processing, but insufficient for long-term logging of data. However, sensor nodes also provide a relatively large amount of persistent storage through external flash memory. To take advantage of this persistent storage, an abstraction layer must be provided that enables applications to access this memory in a manner that is simple, efficient, and reliable. While a simple circular buffer may suffice for the most rudimentary programs, this approach cannot satisfy the needs of more complex applications, such as the Mate virtual machine [18]. In order to support complex applications, as well as to provide a reliable logging mechanism for sensor data, a sensor network operating system must include a file system.

A file system tailored for sensor nodes needs to take into consideration a multitude of resource constraints, and it should be optimized for the common operations and data types observed in a sensor network. The three most commonly encountered data sources in a sensor network are configuration data (sampling rate, various thresholds, etc.), sequentially appended sensor data, and binary code images. Clearly, the most common operation on files would be sequential appending and reading of sensor data. Random access of file data is very rare, and is generally only performed during dynamic reprogramming.

The most common persistent storage medium for sensor nodes is flash memory. The characteristics of flash memory are significantly different from those of most magnetic media, and it is important to take these differences into account when designing a file system for a sensor network operating system. Flash memory has a limited write lifetime—on the order of 10,000 write operations. The process of writing data to flash is a multistep process that consumes a great deal of energy: A write operation onto a flash page requires reading the entire page into RAM, modifying the data as needed, erasing the target flash page, and finally, writing the entire flash page.

The extremely low-frequency (ELF) flash-file system [19] is an example of a file system designed for microsensor nodes. ELF employs a log-structured paradigm and write caching to achieve wear leveling. ELF is adapted to the most common operations of sensor data logging, namely, write-appends, but supports a full set of file I/O features, namely, sequential reads and writes as well as random reads and writes. ELF provides additional features such as best-effort reliability of designated files. A traditional log-structured file system creates a new sequential log entry for each write operation that occurs. This type of operation causes very good wear leveling, since the flash memory can be used sequentially all the way through, only returning to previously used blocks after all of the blocks in flash have been written to at least once. Creating a log entry for each write-append operation is not feasible, as it would cause the run-time memory representation of a traditional log-structured file with many small appends to grow to an unwieldy size. Caching individual writes to the same page reduces the number of write-append operations, and therefore improves the wear leveling of the flash pages. Furthermore, a log-structured file system makes the file system resilient to crashes, which is critical if sensor data are stored locally. Logging the operations performed by the file system provides a reliable form of data recovery.

6.4.3 Bridging to IP Networks

Today's architecture for interconnecting wireless sensor networks (WSNs) and the Internet is based on treating the WSNs as a separate entity. In most existing deployments, a sensor network is treated as a large-scale distributed database, logically separated from the Internet by a database application program interface (API), and physically partitioned by a gateway that resides at the border between the sensor network and an Internet-connected machine. The gateway's primary task is to collect data from the sensor network, then store the data in a local or remote database. This architecture makes it difficult to introduce and integrate sensor networks with new Internet-based applications. In order to add new features or new Internet-based services, such as monitoring and management, each application developer must either develop their own application-specific gateway or modify an existing gateway's API. The gateway is therefore a bottleneck when multiple applications require access to the sensor network.

One solution to this problem is to *overlay* sensor networks over portions of the Internet [20]. The gateway could encapsulate the sensor network packets within transmission control protocol/Internet protocol (TCP/IP) (or user datagram protocol/IP (UDP/IP)) packets, which would then be sent over the Internet to the appropriate application end point on a remote IP host. With this approach, the gateway only needs to understand the lowest level (i.e., the network layer) of the sensor network in order to handle the translation between IP and sensor network routing. To realize this architecture, conscious design decisions must be made in the sensor network operating system. The network stack needs to be extensible enough to allow seamless, easy translation between IP and sensor network packets. The design of the gateway needs to take into consideration both data-driven routing protocols [21] (which are typically addressless) and the integration with IP-based routing. Lightweight data-driven routing protocols should be extensible via application-level overlays over the Internet; since the Internet is relatively high-bandwidth in comparison to sensor networks, this should cause minimal impact in terms of overhead. Further discussion on the design of a sensor network network stack is presented in Section 6.7.3.

6.5 SENSOR NETWORK MANAGEMENT

Sensor networks have a fundamentally different architecture from that of wired data networks. Sensor networks are highly resource-constrained, communicate over low-bandwidth wireless links, and are often deployed in physically inaccessible environments. The protocols and applications designed for sensor networks must therefore be highly optimized for low resource consumption as compared to protocols developed for wired data networks [22]. Managing problems in a sensor network requires querying the network for information, diagnosing the information to determine the faults, and taking the necessary corrective measures to mitigate the problems.

Querying the network for diagnostic information needs to be performed in such a way that it does not overload the network [23–25]. One commonly encountered problem is known as the *response implosion problem* [26,27], which occurs when a diagnostic query triggers a high volume of incoming replies, causing the central gateway node to become the bottleneck. A sensor network operating system must also export a rich set of interfaces to support troubleshooting; applications should have access to the underlying system parameters so that optimization and troubleshooting can be performed as needed. Interfaces should be exported at each layer of the operating system so that the higher-layer applications can make full use of them to optimize performance. For example, the medium-access control (MAC) layer of the network stack should export variables like transmit power, preamble length, frequency channel, enable or disable request to send/clear to send (RTS/CTS), and enable or disable reliable link-layer acknowledgments (ACKs). Based on information from network diagnosis, a command server could use these interfaces to refine the operation of the MAC; for example, disabling ACKs when the link quality is excellent, or disabling RTS/CTS in the absence of hidden terminals in the network.

6.6 TINYOS

The TinyOS operating system utilizes a unique software architecture that was designed specifically for resource-constrained sensor nodes [1]. Primarily based on the concept of wiring together components to create an application, the paradigm strives to use as little memory as possible, while still handling multiple roles in a sensor networking environment. Initially written in standard C, the project has since moved to a custom language, nesC. This "Network Embedded Systems C" uses C-like syntax, but adds some new features to support the structure and execution model of a TinyOS application. We will discuss nesC and the component design further, but first let us consider the execution model.

6.6.1 Execution Model

During the execution of a typical task in any operating system, stack memory is allocated for storing activation records and local variables. This typically leads to the allocation of a separate stack for each running task; however, since most low-power microcontrollers have a small amount of system memory, the designers of TinyOS chose to adopt a new execution model that is well suited for limited memory scenarios.

In order to minimize the amount of memory used during execution, TinyOS applications consist of multiple tasks that all share a single stack. Because of this design, a task must run to completion before giving up the processor and stack memory to another task. These tasks can be preempted by hardware event handlers, which also run to completion, but before giving up the shared stack a task must store any required state in global memory.

6.6.2 TinyOS Components

Another unique aspect of the TinyOS system is its usage of components to create applications, rather than developing libraries of functions that would be called by user programs. These components are separate blocks of code that have clearly defined interfaces for both input and output. In order to "provide" an interface, a component must implement a set of commands defined by the interface. In order to "use" an interface, a component implements a different set of functions, called *events*. Using this structure, a component that wants to utilize the commands of a specific interface must also implement the events for that same interface.

Once a variety of components have been developed they must be organized in an application-specific way to implement the desired application functionality. This is done by using configuration components. These components essentially connect, or "wire," the functional components together. In wiring components, any component that implements an interface can use or provide multiple interfaces as well as multiple instances of a single interface.

When completed, a TinyOS application can be represented as a directed graph in which the wiring of commands and events between components dictates the edges of the graph. This structure is best visualized as an upside-down tree where the root node is the single executing task, and the bottom-tier leaf nodes are hardware event handlers. In this model, events propagate from the bottom of the tree up through various components where they are either handled directly by intermediate components, or post a new task to run when the stack and processing time is next available. From the executing task, commands typically run down the stack through various protocol and driver components before actually reaching the point of hardware manipulation.

6.6.3 The nesC Language

The first generation of TinyOS was implemented in the C language, which forced its creators to use a large number of macros to reduce the amount of extra code necessary to implement the unique component design and execution model. There were four primary reasons for the large number of macros: simplifying access to microcontroller features, accessing global variables specific to a given task (the task frame), calling commands in another component, and signaling events from a component. In many applications one of these four operations was occurring on almost every line of source code, which made developing TinyOS components quite cumbersome. It was decided that a new language, nesC, was the best way to help ease the development of TinyOS applications.

The main goals of nesC were to allow for strict checking at compile time, while also easing the development of TinyOS components. Two of the primary constraints in nesC allow for the majority of its optimization. First, nesC programs are intended to have all of their components compiled simultaneously. This allows for a large number of in-line functions and streamlined execution, and it also lets the compiler check for possible data race conditions. Second, dynamic memory allocation is not supported in nesC or TinyOS. While this limitation lets the developer know exactly what the resource requirements are for a given application, it also forces the developer to determine these requirements in advance, rather than letting them be dictated by run-time characteristics.

6.7 MANTIS OPERATING SYSTEM (MOS)

The MANTIS Operating System (MOS) [28] is a larger sensor operating system that is designed to behave similarly to UNIX and provide more built-in functionality than TinyOS. MOS applications are written in standard C and executed as threads; the MOS kernel includes scheduling and synchronization mechanisms. MOS also includes a framework to allow a high degree of integration between device drivers, network protocols, and the operating system.

6.7.1 Kernel

MOS is based on a multithreaded kernel, with Portable Operating System Interface for UNIX (POSIX)-like semantics [29]. Scheduling is priority-based; round-robin scheduling is used for multiple threads at the same priority level. Since all threads must coexist in the same address space, the maximum stack size of each thread must be specified when the thread is spawned. This allows the kernel to allocate a block of data memory for the thread's stack. The need to keep multiple stacks in memory at the same time makes MOS (and multithreaded systems in general) more resource-intensive than single-threaded systems like TinyOS.

The MOS kernel also provides counting semaphores and mutual-exclusion semaphores, as well as timers and sleep functions. These create a multiprogramming model similar to that seen in conventional operating systems.

6.7.2 Memory Management

The layout of data memory in MOS is shown in Figure 6.3. Text does not appear, as it is stored in program memory. Statically allocated memory, whose size is known at start-up, begins at low addresses. When the node first starts up, the stack pointer is at the high end of memory, and therefore the INIT thread stack is located in the top block of memory. After start-up, the INIT thread becomes the idle thread, and it keeps the same stack.

All of memory between the data section and the INIT/idle stack is managed as a heap. When a thread is spawned, its stack space is allocated out of the available heap

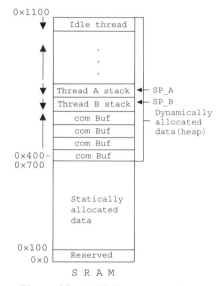

Figure 6.3 MOS data memory layout.

space. The space is reclaimed when the thread exits. Since communications buffers are relatively large (and may be dynamic), they are also allocated out of the heap space. It is also technically legal (albeit discouraged) for an application to allocate memory dynamically out of the heap.

MOS fills a thread's stack with a flag byte (0xEF) when the thread is spawned, so stack usage can be determined while a thread is running by counting the number of flag bytes still present from the end of the thread's stack space. While this makes it possible to detect stack overruns after the fact, it does not prevent them. Determining the amount of stack space necessary for a thread—and not overrunning it—is still the application programmer's responsibility.

6.7.3   Com, Dev, and Net Layers

MOS separates hardware devices into two broad categories, each managed by its own interface layer. The first category includes synchronous, unbuffered devices and is associated with the Dev layer. The second category includes devices that receive data asynchronously, requiring buffering, and is associated with the Com layer. The abstraction layers for devices are shown in Figure 6.4.

Examples of Dev layer devices are most sensors, the file system, the onboard sounder, and the random-number generator. Any number of such devices may exist in a single system. Devices are all accessed through the same set of read, write, mode, and ioctl functions, which are similar to UNIX stream functions. The mode function is used to turn devices on and off. All Dev-layer functions are synchronous, so they return only after the operation is completed.

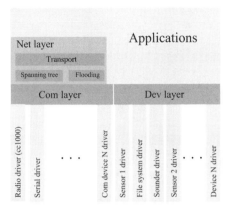

Figure 6.4   MOS network and device abstraction layers.

Examples of Com layer devices are the radio and serial port. These devices are handled separately from Dev-layer devices because they require the ability to receive data in the background during times when there is no application thread currently blocked on a receive call. The interface to Com-layer devices is similar to that of Dev-layer devices: sends are synchronous and receives block until a packet is present. However, packets are being received in the background and buffered whenever the device is turned on. The Com layer also provides the ability to perform a select on multiple devices, returning when a packet is available on any selected device. Select also has a nonblocking option and a time-out option.

It is important to note that device drivers in MOS are not associated with threads. Rather, they are interrupt-driven state machines. Simple interrupt handlers shift bytes from buffers to the hardware or vice versa while the application thread is blocked, and the thread is woken up after the entire operation is complete. Thus, other application threads may be running while a thread is waiting for a Com-layer or Dev-layer operation.

One important break from UNIX is that MOS uses zero-copy mechanisms at all levels of the network stack, including the Com layer [30]. There is a preallocated pool of packet buffers that is "owned" by the Com layer. Device drivers may request buffers, which they then fill with incoming data. When a buffer is full, the driver swaps the filled buffer for another empty buffer. The filled buffer is queued until an application performs a receive operation on the associated device.

When an application performs a receive, it is passed a pointer to the first filled buffer on the specified device. At this point, the application "owns" a buffer that originally was allocated by the Com layer. When the application is finished with the buffer, it must return it to the Com layer through a "free" function call. When the application frees the buffer, it is returned to the buffer pool.

This interface decision puts additional responsibility on the application, since it must free every packet it receives. If the application fails to do so, it will eventually consume all of the buffers in the Com-layer buffer pool, and the Com-layer devices will no longer be able to receive packets. The trade-off for this added responsibility and loss of isolation is a more efficient use of memory and CPU time.

To allow easy development of network algorithms, MOS also includes an integrated, modular network layer. A user can elect to activate the Net layer instead of dealing with the Com layer directly. The Net layer has its own thread, which is necessary because network protocols may perform significant processing, including generating and sending packets. The Net thread performs a select on all Com-layer devices. Upon receiving a packet, it looks for a protocol ID in the packet and then invokes the appropriate routing module. This allows multiple routing algorithms to coexist and share the same communications devices.

The Net layer includes a notion of ports. Applications may perform a receive operation on a port at the Net level, and routing modules may deliver packets to ports. The Net level will buffer packets at each port until they are received by an application, similar to the Com layer. During a send operation, the application specifies a destination port number and a routing module, followed by arguments specific to the routing algorithm.

6.7.4 Power Management

MOS includes both explicit and implicit forms of power management. Explicit power management is performed through the mode functions in the Com and Dev layers, which activate and deactivate hardware devices such as the radio and sensors. Besides on and off, the mode functions also support intermediate states (such as standby), which have device-specific meanings.

To encourage power-efficient programming, all devices in MOS are set to the off state initially, where they consume the minimum amount of power possible. Com-layer devices will not receive packets until they are turned on. However, all of the device drivers currently implemented for MOS include logic to turn the device on and then off again if the user attempts to perform a synchronous operation while in the off mode. In effect, the user is explicitly turning the device on by using it in a synchronous call, since the user probably does not intend to block forever. Even with this logic present, the mode functions are still useful because there may be a performance advantage to keeping a given device on; for example, when power-up is slow and the user knows the device will be needed again soon.

While device power management is handled explicitly, MOS handles CPU power management implicitly. The MOS kernel may be in one of three modes: active, idle, and power-save. In active mode, the CPU is fully powered and running instructions. In idle mode, the CPU is not running instructions, but interrupts and peripherals are enabled. Power consumption in idle mode is roughly half that of active mode. In power-save mode, the CPU is completely powered down except for external interrupts and the watchdog timer, and power consumption is reduced by three orders of magnitude.

The MOS kernel uses a simple algorithm, executed as part of the idle thread, to determine the CPU power mode. If at least one thread is ready to run, then the CPU is left in active mode. If no thread is ready (that is, the idle thread is executed), but at least one thread is blocked waiting for an interrupt, then the CPU is put into idle mode. If no thread is running and no thread is waiting for an interrupt, then the CPU is put into power-save mode.

The kernel sleep function is the main noninterrupt mechanism that threads use to block. When switching to power-save mode, the kernel checks the sleep time on every thread and computes the proper time to wake up so as not to miss any deadlines. The watchdog timer is used to wake the CPU back up at the proper time.

6.8 COMPARISON SUMMARY

In this section, we summarize the trade-offs between event-driven run-to-completion sensor operating systems such as TinyOS, and preemptively time-sliced multithreaded sensor operating systems such as MANTIS OS. Table 6.1 compares the advantages and disadvantages of each approach.

An event-driven run-to-completion operating system is well suited to highly memory-constrained devices such as the original Rene motes, which had only 512 bytes of SRAM [1]. Indeed, it is unlikely that a multithreaded system would be capable of being implemented in such limited memory while also being able to support multiple threads of execution. As a result, event-driven designs represent the primary option, given severe memory constraints. An event-driven system also offers the advantage of being well-suited to achieve energy efficiency. When there are no events to be handled, for example, sensing events or radio packet events, then the system need not execute, and can sleep itself. As a result, event-driven systems were perfectly adapted to the first generation of wireless sensor nodes.

The drawbacks of an event-driven system relate to its run-to-completion paradigm. Each task must run to completion before another task can execute. As a

TABLE 6.1 **Comparing Event-Driven and Multithreaded Sensor Operating Systems**

Operating System	Advantages	Disadvantages
Event-driven Run-to-completion OS (TinyOS)	• Very compact memory footprint • Energy-efficient—sleeps system when no events to process (no tasks in task queue) • OS code is simplified	• Application developer is exposed to complexity—must determine when to relinquish control • No fault-tolerant isolation between applications • Porting of existing C code base becomes complicated
Preemptively Time-sliced Multithreaded OS (MOS)	• Programmer is hidden from complexity of control and timing issues • Fault-tolerant isolation of applications • Leveraging existing C code base is straightforward	• Context switch time • Memory overhead of a stack per thread

result, a programmer must be very careful to relinquish control in a timely manner, to avoid blocking other tasks that need to run. The application programmer is thus exposed to complexity in terms of needing to determine when to give up control of the CPU. If a programmer is implementing a complex task, for example, in-network processing or aggregation that employs onboard statistical analysis and/or a compression algorithm, then implementing such a task as a single monolithic long-lived execution module or component would prevent other application tasks from executing in a timely manner; for example, radio packets could overflow their network buffer because the networking component was unable to execute and process radio packets fast enough. Programmers are forced to alter their programming practice and "slice-and-dice" their program into sufficiently small execution components to avoid blocking the CPU. This further imposes a requirement on the programmer to achieve a detailed semantic understanding of the program to be decomposed, so that each component is sufficiently small. For example, porting a compression algorithm or encryption algorithm to an event-driven system would require that the programmer understand compression or encryption in order to properly decompose the algorithm. Even worse, understanding when the components are "sufficiently" small depends on the delay tolerances of other tasks. An application developer will not know a priori the other tasks to be run on the system, nor their latency tolerances. In the worst case, the developer is forced to decompose a program into the finest granularity. In addition, there is no isolation of faults between application tasks. If one task executes a conditional while( ) loop whose condition is logically never satisfied, then that task will execute in an infinite loop, blocking all other tasks.

In contrast, a multithreaded preemptively time-sliced system seeks to offer services to the programmer that hide complexity, at the cost of additional complexity in the operating system. Because each application thread operates in its own execution environment, and the system handles automatic switching between threads of execution, then the application developer need not be concerned with such complexities as when to relinquish control of the CPU. This is automatically done for the programmer by the system's preemptive time-slicing. Moreover, fault isolation between applications is accomplished by having multiple execution environments. A single thread that operates in an infinite loop or that crashes will not prevent other threads from executing. As a result, the complexities of control and timing issues are hidden from the programmer, who can then concentrate on the correctness of the program. A further benefit is that typical programmers already trained in writing code for multithreaded systems will not need to change their programming practice. This has important implications for easing the porting from the vast C code base developed for multithreaded systems to a microsensor node. The programmer who is porting an algorithm need not have a detailed semantic understanding, and can focus on memory usage of the ported algorithm, which also must be dealt with in an event-driven system. For example, porting of C implementations of encryption standards, compression algorithms, stop-and-wait reliability protocols, and so forth, have all been accomplished with relative ease to the multithreaded MOS.

The costs of offering these multithreading services are a more complex operating system, context switch time, and extra memory overhead for one stack per thread. The advent of second-generation wireless microsensor nodes such as the MICA2 motes, which have 4 kB of SRAM, has enabled the development of complete multithreaded systems such as MOS capable of supporting four to eight application threads simultaneously. The RAM in current systems is sufficient to support multiple stacks, with the expectation that from this point forward that Moore's law as applied to RAM will make multithreading increasingly affordable on microsensor nodes. Context switch time will continue to remain an issue, though at present our experience has not found this to be a significant problem in the sensing and forwarding applications that have been developed on MOS. Multithreaded systems additionally must be adapted to be energy-efficient. MOS implements power management such that when there are no more threads with meaningful execution, the scheduler sleeps the system rather than have the idle thread spin in a continuous loop. The duration of sleep is determined from the hints provided by each application thread, namely, the argument it provided to the sleep( ) call. The complexity of power managmenet is largely hidden from the user except through the sleep( ) API.

Both multithreaded and event-driven systems also must handle synchronization issues introduced by concurrency. When there is concurrent execution in a multithreaded system, it is important that code be synchronized in a thread-safe manner. MOS provides mutual exclusion primitives, which can be used to build thread-safe code. Provided the programmer is furnished with reentrant libraries, then synchronization complexity can be reduced for the application programmer. However, the designer in charge of the multithreaded operating system must then assume the burden and complexity of realizing thread safety while avoiding deadlock and race conditions. An event-driven system such as TinyOS also faces similar synchronization issues. TinyOS introduced atomic operations and the "async" keyword to address race conditions [31]. This is limited to hardware event handlers. These operations seek to guarantee correct concurrency in an event-driven model.

As sensor networks evolve, we expect to see increasing heterogeneity, with nodes of different capabilities; for example, aggregators differ from leaf nodes in purpose and capabilities. Thus, future sensor networks may consist of mixtures of nodes with multithreaded and event-driven sensor operating systems. Recent research has also investigated combining the best features of multithreaded and event-driven systems in a new sensor operating system [32].

6.9 CONCLUSION

There are many unique challenges in designing a sensor node operating system, ranging from limitations on power consumption, to flash-based file systems, to remote network maintenance. This chapter has provided an overview of the principles of sensor node operating system design. We showed how sensor hardware characteristics must influence the decisions about the operating system architecture, covering the problems of task coordination, management of constrained memory and CPU resources, power management, network infrastructure, application design, and remote maintenance. Three of the more unusual features of sensor node operating systems include dynamic network reprogramming, sensor node file systems, and the integration of sensor networks with traditional IP networks. Two implementations of sensor operating systems were discussed, namely, the event-based TinyOS, and the multithreaded MANTIS OS, and a comparison of the two systems was provided. TinyOS is compact and energy efficient, but exposes the application programmer to the complexity of control, as well as to the faulty behavior of other applications. MOS adds multithreading capability, which shields the application developer from control issues and fault isolation, but comes at the cost of context switching and stack memory overhead.

REFERENCES

1. J. Hill, R. Szewczyk, A. Woo, S. Hollar, D. Culler, and K. Pister, System architecture directions for networked sensors. In *Proceedings of the 9th International Conference on Architectural Support for Programming Languages and Operating Systems*, November 2000.

2. H. Abrach, S. Bhatti, J. Carlson, H. Dai, J. Rose, A. Sheth, B. Shucker, J. Deng, and R. Han. MANTIS: System support for MultimodAl NeTworks of In-situ Sensors. In *Proceedings of the 2nd ACM International Workshop on Wireless Sensor Networks and Applications (WSNA)* pages, 50–59. San Diego, California, September 2003.

3. Atmel AVR 8-bit RISC processor, from http://www.atmel.com/products/AVR.

4. M. Leopold, M. B. Dydensborg, and P. Bonnet. Bluetooth and sensor networks: A reality check. In *Proceedings of the 1st ACM Conference on Sensor Systems (SenSys'03)*, pages 103–113, Los Angeles, California, November 2003.

5. The Smart-Its project, from http://www.smart-its.org/.

6. The Eyes project, from http://eyes.eu.org/.

7. Crossbow motes, from http://www.xbow.com.

8. Single chip ultra low power RF transceiver, from http://www.chipcon.com/files/CC1000_Data_Sheet_2_1.pdf, 2001.

9. J. Zhao, R. Govindan, and Estrin. Computing aggregates for monitoring wireless sensor networks. In *Proceedings of the 1st IEEE International Workshop on Sensor Network Protocols and Applications*, Anchorage, Alaska, April–May 2003.

10. S. Madden, M. Franklin, J. Hellerstein, and W. Hong. TAG: A Tiny AGgregation service for ad-hoc sensor networks. In *Proceedings of the 5th Annual Symposium on Operating Systems Design and Implementation (OSDI)*, Boston, Massachusetts, December 2002.

11. R. Kumar, V. Tsiatsis, and M. Srivastava. Computation hierarchy for in-network processing. In *Proceedings of the 2nd ACM International Workshop on Wireless Networks and Applications (WSNA 2003)*, San Diego, California, September 2003.

12. A. Perrig, R. Szewczyk, V. Wen, D. Culler, and J. Tygar, SPINS: Security suite for sensor networks. In *Proceedings of ACM MobiCom 2001*, pages 189–199, July 2001.

13. V. Bychkovskiy, S. Megerian, D. Estrin, and M. Potkonjak. A collaborative approach to in-place sensor calibration. In *Proceedings of the 2nd International Workshop on Infor-*

mation Processing in Sensor Networks (IPSN'03), volume 2634 of Lecture Notes in Computer Science, pages 301–316, Springer-Verlag.

14. J. Liu, P. Cheung, L. Guibas, and F. Zhao. A dual-space approach to tracking and sensor management in wireless sensor networks. In *Proceedings of the 1st ACM International Workshop on Wireless Sensor Networks and Applications (WSNA)*, Atlanta, Georgia, September 2002. Also, Palo Alto Research Center Technical Report P2002-10077, March 2002.

15. A. Mainwaring, J. Polastre, R. Szewczyk, D. Culler, and J. Anderson Wireless Sensor Networks for Habitat Monitoring. In *Proceedings of the 1st ACM Workshop on Wireless Sensor Networks and Applications (WSNA)*, pages 88–97, Atlanta, Georgia, September 2002.

16. J. Hui and D. Culler. The dynamic behavior of a data dissemination protocol for network programming at Scale. In *Proceedings of the 2nd ACM Conference on Embedded Networked Sensor Systems (SenSys)*, November 2004.

17. N. Reijers and K. Langendoen. Efficient code distribution in wireless sensor networks, In *Proceedings of the 2nd ACM International Conference on Wireless Sensor Networks and Applications (WSNA)*, pages 60–67, 2003.

18. P. Levis and D. Culler. Mate: A virtual machine for Tiny networked sensors. In *Proceedings of the ACM Conference on Architectural Support for Programming Languages and Operating Systems (ASPLOS)*, October 2002.

19. H. Dai, M. Neufeld, and R. Han. ELF: An efficient log-structured flash file system for wireless micro sensor nodes. In *Proceedings of the 2nd ACM Conference on Embedded Networked Sensor Systems (SenSys)*, pages 176–187, November 2004.

20. H. Dai and R. Han. Unifying Micro Sensor Networks with the Internet via Overlay Networking, Paper presented at the First IEEE Workshop on Embedded Networked Sensors (EmNetS-I), Tampa, Florida, 2004.

21. C. Intanagonwiwat, R. Govindan, and D. Estrin. Directed diffusion. In *Proceedings of ACM MobiCom 2000*, pages 56–67, Boston, Massachusetts, August 2000.

22. J. D. Case, M. Fedor, M. L. Schostall, and C. Davin. Simple Network Management Protocol (SNMP). Internet RFC 1157, May 1990.

23. J. Elson, S. Bien, N. Busek, V. Bychkovskiy, A. Cerpa, D. Ganesan, L. Girod, B. Greenstein, T. Schoellhammer, T. Stathopoulos, and D. Estrin. *EmStar: An Environment for Developing Wireless Embedded Systems Software.* CENS Technical Report 0009, March 24, 2003.

24. L. Girod, J. Elson, A. Cerpa, T. Stathopoulos, N. Ramanathan, and D. Estrin. EmStar: A software environment for developing and deploying wireless sensor networks. In *Proceedings of the USENIX Technical Conference 2004.*

25. P. Levis, N. Lee, and M. Welsh, TOSSIM: Accurate and scalable simulation of entire TinyOS applications. In *Proceedings of the 1st ACM Conference on Embedded Networked Sensor Systems (SenSys)*, pages 126–137, 2003.

26. C. Jaikaeo, C. Srisathapornphat, and C. Shen, Diagnosis of sensor networks. In *Proceedings of the IEEE International Conference on Communications (ICC)*, June 2001.

27. B. Deb, S. Bhatnagar, and B. Nath. *A Topology Discovery Algorithm for Sensor Networks with Applications to Network Management.* DCS Technical Report DCS-TR-441, Rutgers University, May 2001.

28. S. Bhatti, J. Carlson, H. Dai, J. Deng, J. Rose, A. Sheth, B. Shucker, C. Gruenwald, A. Torgerson, and R. Han, MANTIS OS: An embedded multithreaded operating system for wireless micro sensor platforms. *Mobile Networks and Applications*, forthcoming.

29. Portable Operating System Interface (POSIX)—Part 1: System Application Programming Interface (API)[C Language]. ISO/IEC 9945-1:1996, IEEE standards, 1996.

30. H. K. Jerry Chu. Zero-copy TCP in Solaris. In *Proceedings of the USENIX 1996 Annual Technical Conference*, San Diego, California, January 1996.

31. Race conditions, from http://www.tinyos.net/tinyos-1.x/doc/changes-1.1.html.

32. A. Dunkels, B. Gronvall, and T. Voight, Contiki—A Lightweight and Flexible Operating System for Tiny Networked Sensors. Paper presented at the First IEEE Workshop on Embedded Networked Sensors (EmNetS-I), Tampa, Florida, 2004.

■■■■■■ **CHAPTER 7**

Time Synchronization and Calibration in Wireless Sensor Networks

KAY RÖMER, PHILIPP BLUM, and LENNART MEIER
Swiss Federal Institute of Technology (ETH), Zurich, Switzerland

This chapter reviews time synchronization and calibration for wireless sensor networks. First, time synchronization is considered, and then calibration. Time synchronization can be considered as a calibration problem and many observations about time synchronization can be transferred to calibration. Wireless sensor networks present a number of novel challenges to time synchronization, which many traditional approaches fail to meet. We classify common approaches for synchronization in sensor networks and discuss underlying models, synchronization techniques, and algorithms. In addition, common techniques for evaluating synchronization algorithms and selected evaluation results are presented.

7.1 INTRODUCTION

Sensor networks are used to monitor real-world phenomena. For such monitoring applications, physical time often plays a crucial role. We discuss these applications of time in Section 7.1.1. Providing synchronized physical time is a complex task due to various challenging characteristics of sensor networks. In Section 7.1.2, we present these challenges and discuss why synchronization algorithms for traditional distributed systems often do not meet these challenges.

7.1.1 The Need for Synchronized Time

Physical time plays a crucial role for many sensor network applications. While many traditional applications of time also apply to sensor networks, we will focus here on

Handbook of Sensor Networks: Algorithms and Architectures, Edited by Ivan Stojmenović
Copyright © 2005 John Wiley & Sons, Inc.

areas specific to sensor networks. Figure 7.1 illustrates a rough classification of applications of physical time: (a) at the interface between the sensor network and an external observer, (b) among the nodes of the sensor network, and (c) at the interface between the sensor network and the observed physical world. The following paragraphs discuss applications of time in these three domains. Note that some applications are hard to assign to a single domain. In such cases, we picked the most appropriate domain.

7.1.1.1 *Sensor Network: The Observer*

In many applications, a sensor network interfaces to an external observer for tasking, reporting results, and management. This observer may be a human operator or an autonomous computing system. Tasking a sensor network often involves the specification of time windows of interest such as "only during the night." Since a sensor network reports observation results to an external observer, temporal properties of observed physical phenomena may be of interest. For example, the times of occurrence of physical events are often crucial for the observer to associate event reports with the originating physical events. Physical time is also crucial for determining properties such as speed or acceleration.

7.1.1.2 *Sensor Network: The Real World*

In sensor networks, many sensor nodes may observe a single physical phenomenon. One of the key functions of a sensor network is hence the assembly of those distributed observations into a coherent estimate of the original phenomenon: this process is known as data fusion. Time is a key ingredient for data fusion. For example, if sensors can only detect the proximity of an object, then higher-level information (such as speed, size, or shape) can be obtained by correlating data from multiple sensor nodes. The velocity of a mobile object, for example, can be estimated by the quotient of the spatial and temporal distances between two consecutive sightings of the object by different sensor nodes.

Since many instances of a physical phenomenon can occur within a short time, one of the tasks of a sensor network is the separation of sensor samples, that is,

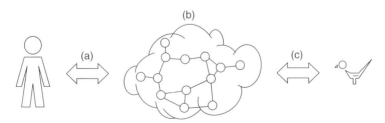

Figure 7.1 Applications of physical time: (a) interaction of an external observer with the sensor network, (b) interaction among sensor nodes, and (c) interaction of the sensor network with the real world.

the partitioning of sensor samples into groups so that each represents a single physical phenomenon. Temporal relationships (e.g., distance) among sensor samples are a key element for separation.

Temporal coordination among sensor nodes may also be necessary to ensure correctness and consistency of distributed measurements [1]. For example, if the sampling rate of sensors is low compared to the frequency of an observed phenomenon, it may be necessary to ensure that sensor readout occurs concurrently at all sensor nodes in order to avoid false observation results (e.g., for calibration, see Section 7.7.5.2).

It is anticipated that large-scale, complex actuation functions will be implemented by the coordinated use of many simple actuator nodes. This requires temporal coordination.

7.1.1.3 *Within the Sensor Network* Time is also a valuable tool for intranetwork coordination among different sensor nodes. Many applications of time known from traditional distributed systems also apply to wireless sensor networks. Reference [2] points out a number of applications of time in distributed systems, such as concurrency control (e.g., atomicity, mutual exclusion), security (e.g., authentication), data consistency (e.g., cache consistency, consistency of replicated data), and communication protocols (e.g., at-most-once message delivery).

One particularly important example for concurrency control is the use of time-division multiplexing in wireless communication, where multiple access to the shared communication medium is typically achieved by assigning time slots to the communicating nodes. This requires the participating sensor nodes to share a common view of physical time.

A number of approaches intend to improve energy efficiency by frequently switching sensor nodes or components thereof into power-saving sleep modes (e.g., ref. [3]). In order to nonetheless ensure seamless operation of the sensor network, temporal coordination of the sleep periods among sensor nodes may be required.

Another important service for sensor network applications is temporal message ordering (e.g., ref. [4]). Many data-fusion algorithms have to process sensor readings ordered by the time of occurrence (e.g., the approach for velocity estimation just sketched). However, the highly variable message delays in sensor networks imply that messages from distributed sensor nodes may often not arrive at a receiver in the order in which they were sent. Reordering messages according to the time of sensor readout requires temporal coordination among sensor nodes.

Methods for localization of sensor nodes based on the measurement of time of flight or difference of arrival time of certain signals also require synchronized time (e.g., ref. [5]).

7.1.2 Revisiting Time Synchronization for Sensor Networks

Time synchronization is a research area with a very long history. Over time, numerous algorithms have been proposed and have been in large-scale use. The network time protocol (NTP) [6] is perhaps one of the most advanced and time-tested systems. However, several unique characteristics of sensor networks often preclude the use of existing synchronization techniques in this domain.

In the following, we discuss sensor network challenges that impact the design of time synchronization approaches. Using NTP as an example, we will outline why traditional approaches often do not meet the requirements of sensor networks (see also ref. [7]). Note that some of the illustrated shortcomings of NTP are relatively easy to fix, while others are not. To provide the necessary background, we will first give an overview of NTP.

NTP was designed for large-scale networks with a rather static topology (such as the Internet). Nodes are externally synchronized to a global reference time that is injected into the network at many places via a set of master nodes (so-called "stratum 1" servers). These master nodes are synchronized out of band, for example, via global positioning system (GPS) (which provides global time with a precision significantly below 1 μs). Nodes participating in NTP form a hierarchy: leaf nodes are called clients, inner nodes are called stratum L servers, where L is the level of the node in the hierarchy. The parents of each node must be specified in configuration files at each node. Nodes frequently exchange synchronization messages with their parents and use the obtained information to adjust their clocks by regularly incrementing them.

7.1.2.1 *Energy and Other Resources*   Sensor-network applications often require sensor nodes to be small and cheap. This has a number of important implications. First of all, the amount of energy that can be stored in or scavenged by small devices is typically very limited due to the low-energy density of available and foreseeable technology. To ensure longevity despite this limited energy budget, energy-efficient design both in hardware and software becomes a dominating goal. Additionally, computing, storage, and communication capabilities of individual sensor nodes are rather limited due to size and energy constraints.

These constraints may preclude the use of GPS or other technologies for out-of-band synchronization of NTP master nodes. NTP is also not optimized for energy efficiency, simply because this is not an issue in infrastructure-based distributed systems. Energy overhead in NTP results from several sources. First, the service provided by NTP typically cannot be dynamically adapted to the varying needs of an application. Hence, with NTP all nodes would be continuously synchronized with maximum precision, even though only subsets of nodes might occasionally need synchronized time with less-than-maximum precision.

Second, NTP uses the processor and the network in ways that would lead to significant overhead in energy expenditure in sensor networks. For example, NTP maintains a synchronized system clock by regularly adding small increments to the system-clock counter. This behavior precludes the processor from being switched to a power-saving idle mode. In addition, NTP servers must be prepared to receive synchronization requests at any point in time. However, constantly listening is an energywise costly operation in sensor networks; many sensor network protocols therefore switch off the radio whenever possible.

7.1.2.2 Network Dynamics Due to their deployment in the physical environment, sensor networks are subject to a high degree of network dynamics. Sensor nodes can be mobile, die due to depleted batteries or due to environmental influences, and new sensor nodes may be added at any point in time. This results in relatively frequent and unpredictable changes in the network topology and possibly even in (temporary) network partitions. Mobile nodes can transport messages across partition boundaries by storing a received message and forwarding it as soon as a new partition is entered. The end-to-end delay of such message paths is very unstable and hard to predict.

The operation of NTP is independent of the underlying physical network topology. In the NTP overlay hierarchy, a master and a client can be separated by many hops in the physical network, even though they are neighbors in the overlay hierarchy. Due to the previously mentioned effects, multihop paths may be very unstable and unpredictable in a sensor network. NTP, however, depends on the ability to accurately estimate the delay characteristics of network links.

NTP implicitly assumes that network nodes that shall be synchronized are a priori connected by the network. However, this assumption may not hold in dynamic sensor networks with mobile nodes. Consider, for example, an application where mobile sensor nodes with sporadic network connectivity time-stamp sensor readings and deliver these records to an observer as they pass by a base station (e.g., ref. [8]). The base station may then want to compare time stamps generated by different sensor nodes in order to evaluate the collected data. However, in the preceding scenario, there might not be a network connection between the various originators of the time-stamped messages at any point in time. Hence, NTP cannot be applied in such settings.

7.1.2.3 Infrastructure In many applications, sensor networks have to be deployed in remote, unexploited, or hostile regions. Sensor networks therefore often cannot rely on sophisticated hardware infrastructure. For example, under dense foliage or inside buildings, GPS cannot be used, since there is no free line of sight to the GPS satellites.

In order to improve the precision and availability of synchronization in large networks, NTP injects the reference time into the network at many points. Hence, any node in the network is likely to find a source of reference time in a distance of only a few hops. Note that shorter paths tend to be more reliable and more predictable, since they include fewer sources of error and unpredictability.

However, such an approach requires an external infrastructure of reference-time sources that have to be synchronized with some out-of-band mechanism. Where this is not feasible, NTP would have to operate with a single master node, which uses its local time as the reference time. In large sensor networks, the average path length from a node to this single master is long, leading to reduced precision. This is particularly problematic when collocated sensor nodes require very precise mutual synchronization, for example, to cooperate in observing a nearby physical event. With a single master node, the collocated nodes might end up using different synchronization paths, which results in different synchronization errors (i.e., time offsets) with respect to the master node.

7.1.2.4 Configuration   After initial deployment, it is often infeasible to physically access the sensor nodes for hardware or software maintenance. The large number of nodes also precludes manual configuration of individual nodes. While traditional networks such as the Internet do also consist of a large number of nodes, there is an accordingly large number of human network administrators, such that each one takes care of a manageable number of computers. With sensor networks, however, half a dozen human operators may be responsible for thousands of sensor nodes.

NTP requires the specification of one or more potential synchronization masters for each node. This is an appropriate solution for networks with a rather static topology, where configurations remain valid for extended periods of time. In sensor networks, however, network dynamics necessitate a frequent adaptation of configuration parameters.

7.2 SYSTEM MODEL

In the following subsections we analyze various synchronization approaches. We now specify the system model for time synchronization, which we use as the foundation of our analysis. First, we describe how we model clocks. We then specify the characteristics of communication between nodes in a sensor network.

All our modeling is done in terms of discrete time and events. An event can represent communication between nodes, a sensor measurement, the injection of time information at a node, and so on. We denote the real time at which event a occurs as t_a, and the local time of node N_i at that time as h_a^i. Note that our model does not explicitly contain node mobility or network dynamics; these aspects are included implicitly by the absence or existence of corresponding communication events.

7.2.1 Clock Models

Digital clocks measure time intervals. They typically consist of a counter h (which we will also refer to as "the (local) clock") that counts time steps of an ideally fixed length; we denote the reading of the counter at real time t as $h(t)$. The counter is incremented by an oscillator with a rate (or frequency) f. The rate f at time t is given as the first derivative of $h(t)$: $f(t) = dh(t)/dt$. An ideal clock would have rate 1 at all times, but the rate of a real clock fluctuates over time due to changes in supply voltage, temperature, and so on. If the fluctuation were allowed to be arbitrary, the clock's reading would obviously give no information at all. Fortunately, it is limited by known bounds. Different types of bounds on the rate fluctuation lead to different clock models.

7.2.1.1 *Constant-Rate Model*   The rate is assumed to be constant. This is reasonable if the required precision is small compared to the rate fluctuation.

7.2.1.2 *Bounded-Drift Model*   The deviation of the rate from the standard rate 1 is assumed to be bounded. We call this deviation the clock's *drift* $\rho(t) = f(t) - 1 = dh(t)/dt - 1$, and denote the corresponding bound with $\rho_{\max}$:

$$-\rho_{\max} \leq \rho(t) \leq \rho_{\max} \quad \forall t \tag{7.1}$$

A reasonable additional assumption is $\rho_i(t) > -1$ for all times t. This means that a clock can never stop ($\rho_i(t) = -1$) or run backward ($\rho_i(t) < -1$). Thus, if two events a, b with $t_a < t_b$ occur at a node N_i whose clock's drift ρ_i is bounded according to equation (7.1), then node N_i can compute lower and upper bounds $\Delta_i^l[a, b], \Delta_i^u[a, b]$ on the real-time difference $\Delta[a, b] := t_b - t_a$ as:

$$\Delta_i^l[a, b] := \frac{h^i(t_b) - h^i(t_a)}{1 + \rho_{\max}} \qquad \Delta_i^u[a, b] := \frac{h^i(t_b) - h^i(t_a)}{1 - \rho_{\max}} \tag{7.2}$$

This model is typically reasonable, since bounds on the oscillator's rate are given by the hardware manufacturer. Sensor nodes usually contain inexpensive oscillators, and thus we have $\rho_{\max} \in [10\,\text{ppm}, 100\,\text{ppm}]$.[1] Note that in this model, the drift can jump arbitrarily within the bounds specified in equation (7.1). The next model limits the variation of the drift.

7.2.1.3 *Bounded-Drift-Variation Model*   The variation $\vartheta(t) = d\rho(t)/dt$ of the clock drift is assumed to be bounded:

$$-\vartheta_{\max} \leq \vartheta(t) \leq \vartheta_{\max} \quad \forall t \tag{7.3}$$

This assumption is reasonable if the drift is influenced only by gradually changing conditions such as temperature or battery voltage. It makes drift compensation possible: A node can estimate its current drift and compute bounds on its drift for future times.

We can also assume both equation (1.1) and equation (1.3).

7.2.2 Software Clocks

A synchronization algorithm can either directly modify the local clock h or otherwise construct a software clock c. A software clock is a function taking a local clock value $h(t)$ as input and transforming it to the time $c(h(t))$. This time is the final result of synchronization, and we therefore call it the synchronized time. For example, $c(h(t)) = t_0 + h(t) - h(t_0)$ is a software clock that starts with the correct

---

[1]Parts per million, that is, 10^{-6}. A clock with a drift of 100 ppm drifts 100 seconds in a million seconds, or 100 µs in one second.

real time t_0 and then runs with the same speed as the local clock h. In general, we require that a software clock be a piecewise continuous, strictly monotonically increasing function.

7.2.3 Communication Models

Communication is needed to obtain and maintain synchronization. In the following, we identify different communication parameters that affect time synchronization.

7.2.3.1 Unicast versus Multicast If a message is sent by one network node and is received by at most one other network node, we call this unicast or point-to-point communication. Multicast communication occurs when a message is sent by one network node and is received by an arbitrary number of other network nodes. The case where all nodes within transmission range are recipients is called *broadcast*. Wireless sensor networks typically use simple broadcast radios, such that a sensor node's transmission is overheard by all nodes within its transmission range.

7.2.3.2 Symmetrical versus Asymmetrical Links If we assume that node A can receive messages sent by node B if and only if node B can receive messages sent by node A, we say that the link between these two nodes is symmetrical. Otherwise, it is asymmetrical. An example for an asymmetrical link is the link between a base station with high transmit power and a mobile device with low transmit power: Beyond a certain distance between the two, only communication in direction from the base station to the mobile device is possible. In wireless sensor networks, it is reasonable to assume that there is a large number of small sensor nodes, and a small number of more powerful (regarding energy, memory, processing power, and transmit power) nodes. The links between these two types of nodes would clearly be asymmetrical.

7.2.3.3 Implicit versus Explicit Synchronization When comparing clock synchronization approaches, it is important to distinguish whether synchronization information can be sent only with the messages that the sensor network application transmits ("piggyback"), or whether additional communication (i.e., messages sent only for the sake of synchronization) is allowed. There is a trade-off between the amount of additional communication and the achievable synchronization quality. Additional communication incurs additional energy consumption and can reduce the bandwidth available for application data. Piggybacked time information does typically not reduce available bandwidth significantly, since there are no additional message headers to be transmitted or transmission slots to be occupied, and the time information is small in size.

7.2.3.4 Delay Uncertainty As far as synchronization is concerned, the goal of communication is to convey time information. The delay of the messages sent between nodes has to be taken into account when extracting this time information; we explore this in Section 7.4.1. The message delay consists of

- The send time, lasting from when the application issues the send command to when the node actually starts trying to send; it is caused by kernel processing, context switches, and system calls, and hence varies with the current system load.
- The (medium) access time, lasting from when the node is ready to send to when it actually starts the transmission; this is the time that is spent waiting for access to the wireless channel, and hence depends on the current network load.
- The propagation time, which is the time it takes for the radio signal to travel from the sender to the receiver; it is constant for any pair of nodes with constant distance, and is negligible compared to the other delay components in wireless sensor networks (since distances are small and radio signals travel very fast).
- The receive time, lasting from the reception of the signal to the arrival of the data at the application.

The send and receive time (and especially the uncertainty about them) can be reduced by implementing the time-stamping of outgoing and incoming messages at a very low level, for instance, in the MAC layer. As a general rule, message-delay uncertainties in typical wireless sensor networks are rather large compared to those in wired networks. This is due to the lower link reliability and bandwidth (see Section 7.1.2).

7.2.4 Sources of Synchronization Errors

Clock synchronization algorithms face two problems: the information a node has about the local time of another node degrades over time due to clock drift (the two clocks "drift apart"), and its improvement through communication is hindered by message-delay uncertainty.

The influence of drift and delay uncertainty on the quality of synchronization can to a large extent be studied separately. The influence of the clock drift may dominate over that of the message delays. This is the case in those sensor networks where communication is *infrequent*. The reason for this is that with decreasing frequency of communication, the uncertainty due to clock drift increases, while the uncertainty due to message delays remains constant. *A numeric example:* Suppose the message delay contributes 1 millisecond to a node's uncertainty, and the clock drift is bounded by $\rho_{\max} = 10$ ppm. After 50 seconds, the drift's contribution to the uncertainty equals that of the delay. After one hour, it is 72 times larger. In this setting, neglecting the delay uncertainty is acceptable.

The time information that is obtained through communication has to be processed to achieve synchronization. As we show in Sections 7.4.1 and 7.4.2, the computation power and memory size required to do this in a timely fashion can increase (even nonlinearly) with the amount of communication and thus become very large. There is a trade-off between computational power and storage capacity spent and achievable synchronization.

7.3 CLASSES OF SYNCHRONIZATION

Synchronization is commonly understood as "making clocks show the same time," but there are actually many different types of synchronization. In the following, we give an overview of the various choices available for synchronization. When choosing the synchronization approach for a given sensor network application, the maxim is to fulfill the application's requirements with the smallest possible effort in terms of computation, memory, and especially energy.

7.3.1 Internal versus External

The synchronization of all clocks in the network to a time supplied from outside the network is referred to as *external* synchronization. NTP performs external synchronization, and so do sensor nodes synchronizing their clocks to a master node. Note that it makes no difference whether the source of the common system time is also a node in the network or not.

Internal synchronization is the synchronization of all clocks in the network, without a predetermined master time. The only goal here is consistency among the network nodes. External synchronization requires consistency within the network *and* with respect to the externally provided system time.

In everyday life, we are mostly faced with external synchronization, namely, with keeping wristwatches and clocks in computers, cell phones, personal digital assistants (PDAs), cars, microwave ovens and so on, synchronized to the legal time.

7.3.2 Lifetime: Continuous versus On-Demand

The *lifetime* of synchronization is the period of time during which synchronization is required to hold. If time synchronization is *continuous*, the network nodes strive to maintain synchronization (of a given quality) at all times. For some sensor network applications, *on-demand* synchronization can be as good as continuous synchronization in terms of synchronization quality, but much more efficient. During the (possibly long) periods of time between events, no synchronization is needed, and communication, and hence energy consumption, can be kept at a minimum. As the time intervals between successive events become shorter, a break-even point is reached where continuous and on-demand synchronization perform equally well. There are two kinds of on-demand synchronization: event-triggered and time triggered.

Event-triggered on-demand synchronization is based on the idea that in order to time-stamp a sensor event, a sensor node needs a synchronized clock only immediately after the event has occurred. It can then compute the time stamp for the moment in the recent past when the event occurred. Post facto synchronization [9] is an example for event-triggered synchronization.

We use *time-triggered* on-demand synchronization if we are interested in obtaining sensor data from multiple sensor nodes for a specific time. This means that there is no event that triggers the sensor nodes, but the nodes have to take a sample at precisely the right time. This can be achieved via *immediate* synchronization (where sensor nodes receive the order to immediately take a sample and time-stamp it) or *anticipated* synchronization (where the order is to take the sample at some future time, the *target time*). Anticipated synchronization is necessary if it cannot be guaranteed that the order can be transmitted rapidly and simultaneously to all involved sensor nodes. This is especially the case if sensor nodes are more than one hop away from the node giving the order.

Note that for successful anticipated synchronization, it is sufficient to maintain a synchronization quality that guarantees that the target time is not missed. This means that the required synchronization quality grows as the real time approaches the target time. There is no need to synchronize with maximum quality right from the beginning.

Analogously to the event-triggered post facto synchronization, we might refer to time-triggered synchronization as pre facto synchronization.

7.3.3 Scope: All Nodes versus Subsets

The *scope* of synchronization defines which nodes in the network are required to be synchronized. Depending on the application, the scope comprises all or only a subset of the nodes (Fig. 7.2). Event-triggered synchronization can be limited to the collocated subset of nodes that observe the event in question.

7.3.4 Rate Synchronization versus Offset Synchronization

Rate synchronization means that nodes measure identical time-interval lengths. In a scenario where sensor nodes measure the time between the appearance and disappearance of an object, rate synchronization is a sufficient and necessary condition for comparing the duration of the object's presence within the sensor range of different nodes (but not for ordering the observations chronologically).

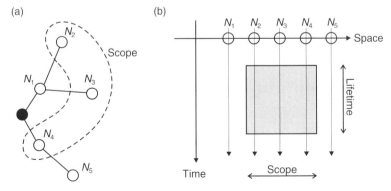

Figure 7.2 Scope and lifetime define where and when synchronization is required. (a) Shows the topology of some network, (b) illustrates the scope and lifetime of the synchronization: Only nodes N_2, N_3, and N_4 need synchronization.

Offset synchronization means that nodes measure identical points in time, that is, at some time t, the software clocks of all nodes in the scope show t. Offset synchronization is needed for combining time stamps from different nodes.

The difference between rate and offset synchronization is illustrated in Figure 7.3. Node N_2 can compute the bird's speed all by itself by dividing the distance between the bird's positions at events a and b by the corresponding local-time difference. For this, the node's clock must be rate synchronized to the real-time rate 1. Alternatively, data from nodes N_2 and N_3 can be combined to compute the bird's speed; here, we would use events b and c. The nodes' clocks have to be offset synchronized for this.

7.3.5   Timescale Transformation versus Clock Synchronization

Time synchronization can be achieved in two fundamentally different ways. We can synchronize clocks, that is make all clocks display the same time at any given moment. To achieve this, we have to perform rate and offset synchronization (or continuous offset synchronization, which, however, is costly in terms of energy and bandwidth and requires reliable communication links). The other approach is to transform timescales, that is, to transform local times of one node into local times of another node.

Both approaches are equal in the sense that if we have either perfect clock synchronization or perfect timescale transformation, the distributed sensor data can be combined as if it had been collected by a single node. The approaches differ in that clock synchronization requires either communication across the whole network (for internal synchronization) or some degree of global coordination (for external synchronization). This calls for communication over multiple hops (which, however, tends to degrade synchronization quality), or well-distributed infrastructure, which, for instance, guarantees that every sensor node is only a few hops away from a node equipped with a GPS receiver. Timescale transformation does not have these drawbacks, but may instead incur additional computation and memory overhead.

We illustrate the difference between clock synchronization and timescale transformation using the example shown in Figure 7.3. If the clocks of all three nodes are

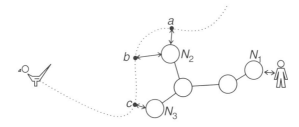

Figure 7.3   At events a, b, and c, nodes N_2 and N_3 measure the position of the bird and time-stamp these data with their current local time. Rate or offset synchronization is needed, depending on how the data from the three events are to be combined.

synchronized, node N_1 can directly combine the sensor data from nodes N_2 and N_3, since the time stamps refer to the same timescale. If the clocks are not synchronized, a timescale transformation on the received time stamps is necessary. The final result is identical to that of using synchronized clocks.

7.3.6 Time Instants versus Time Intervals

Time information can be given by specifying time instants (e.g., "$t = 5$") or time intervals ("$t \in [4.5, 5.5]$"). In both cases, the time information can be refined by adding a statement about its quality. For instance, the time information may be guaranteed to be correct with a certain probability, or even probability distributions for the time can be given. A measure for the quality of the time information can then be defined; we will speak of its inverse, the *time uncertainty*.

For sensor networks, the use of guaranteed time intervals can be very attractive. Interestingly, this approach has not received much attention, although it has a number of advantages over using time instants: (1) Guaranteed bounds on the local times at which sensor events occurred allow guaranteed bounds from sensor-data fusion to be obtained. (2) The concerted action (sensing, actuating, communicating) of several nodes at a predetermined time always succeeds, each node can minimize its uptime while guaranteeing its activity at the predetermined time. (3) The combination of several bounds for a single local time is unambiguous and optimal, while the reasonable combination of time estimates requires additional information about the quality of the estimates.

7.4 SYNCHRONIZATION TECHNIQUES

In this section, building blocks and fundamental mechanisms of time synchronization algorithms are presented. The section is organized by increasing complexity: In Section 7.4.1, various approaches for obtaining a single reading of the clock of a remote node are presented. In Section 7.4.2, techniques for maintaining synchronization are discussed. In Sections 7.4.3 and 7.4.4, it is shown how multiple samples can improve synchronization between two nodes. Finally, various approaches to organize the synchronization process in larger networks are discussed in Section 7.4.5.

7.4.1 Taking One Sample

We start with the simple model shown in Figure 7.4(a), with two nodes N_i and N_j that can exchange messages. Synchronization between these nodes means that the nodes establish some relationship between their local clocks h^i and h^j.

7.4.1.1 Unidirectional Synchronization   The conceptionally simplest solution is illustrated in Figure 7.4(b). Node N_i sends a message containing a local time stamp h_a^i to node N_j, where it is received at local time h_b^j. The node N_j

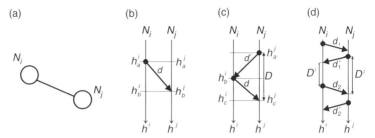

Figure 7.4   Uni- and bidirectional synchronization. (a) A node N_j determines the offset of its local clock relative to that of another node N_i, using (b) unidirectional communication or using (c, d) bidirectional communication. In contrast to (c), scheme (d) allows both nodes to measure a round-trip time.

cannot determine the delay d of the message. It only knows that the local clock of node N_i showed h_a^i before its own local clock shows h_b^j. Thus its local time when the message was sent is $h_a^j < h_b^j$, and local time at node N_i when the message is received is $h_b^i > h_a^i$. Time synchronization consists of estimating either h_b^i or h_a^j.

If a priori bounds on the message delay are known, that is, $d_{\min} \le d \le d_{\max}$, then the estimation $h_a^j \approx h_b^j - 1/2(d_{\min} + d_{\max})$ (or alternatively $h_b^i \approx h_a^i + 1/2(d_{\min} + d_{\max})$)) minimizes the synchronization error in the worst case. Alternatively, $h_b^j - d_{\max}$ and $h_b^j - d_{\min}$ are lower and upper bounds on h_a^j (and $h_a^i + d_{\min}$ and $h_a^i + d_{\max}$ are bounds on h_b^i).

7.4.1.2 Round-Trip Synchronization

A slightly more complex solution is illustrated in Figure 7.4(c). Node N_j sends a query message to node N_i, asking for the time stamp h_b^i. Node N_j measures the round-trip time $D = h_c^j - h_a^j$, that is, the length of the time interval between sending the request and receiving the reply. Without having a priori knowledge, node N_j now knows that the delay d is bounded by 0 and D. If a priori bounds on the message delay are known, that is, $d_{\min} \le d \le d_{\max}$, the node N_j knows that d is bounded by $\max(D - d_{\max}, d_{\min})$ and $\min(d_{\max}, D - d_{\min})$.

The estimation $h_b^j \approx h_c^j - D/2$ minimizes the worst-case synchronization error: $h_c^j - (D - d_{\min})$ and $h_c^j - d_{\min}$ are lower and upper bounds on h_b^j. Similarly, an estimation and bounds for h_c^i can be determined.

In comparison with the unidirectional approach, round-trip synchronization has the advantage of providing an upper bound on the synchronization error. The mechanism known as *probabilistic* time synchronization, first presented in ref. [10], uses this to decrease the synchronization error as follows: After receiving the reply message, N_j checks whether the worst-case synchronization error $D/2 - d_{\min}$ is below a specified threshold. If not, it sends a new request message to N_i. This procedure is repeated until a pair of request and reply messages occurs that achieves the required synchronization error. The smaller the chosen threshold, the more messages have to be exchanged on average.

The main disadvantage of round-trip synchronization is that the number of messages increases linearly with the number of nodes that communicate with N_i, while in the unidirectional case, a single broadcast message sent by N_i can serve an arbitrary number of nodes. A combination of the advantages of both approaches is known as *eavesdropping* or *anonymous synchronization*, and was first described in ref. [11]. The basic idea is the following: Node N_j sends a broadcast message to N_i and some additional node N_k; and N_i replies with a broadcast message to N_j and N_k. Node N_k assumes that the second message was produced after it had received the first message, thus node N_k can do round-trip synchronization with the two local receive time stamps and the send time stamp from N_i without ever producing any messages itself.

In Figure 7.4(d), two modifications of round-trip synchronization are illustrated. First, it is not necessary that N_i replies immediately to query messages. Node N_i can instead measure the duration D^i between receiving the query message and sending the reply, and the node N_j can then account for this duration in its calculations. Second, the message exchange shown in Figure 7.4(c) is asymmetrical, that is, only N_j can do round-trip synchronization. Therefore, at least one additional message from N_j to N_i is required, such that also N_i can estimate or bound remote time stamps.

7.4.1.3 *Reference Broadcasting*

A third approach is shown in Figure 7.5. In addition to nodes N_i and N_j, a so-called *beacon* node N_k is involved. The beacon sends a broadcast message to the other nodes. The delays d (to N_i) and d' (to N_j) are almost equal. N_i then sends the time stamp h_a^i to N_j. Node N_j measures the length of the time interval $D = h_b^j - h_{a'}^j$ between the arrivals of the two messages and can then estimate $h_b^i \approx h_a^i + D$.

This approach was first proposed in ref. [12] under the name *a posteriori agreement*. It became more widely known in the sensor network community as *reference broadcast* synchronization (RBS) [9]. Its main advantage is that a broadcast message is received almost concurrently (even though its delay is largely variable), and thus the synchronization error typically is smaller than with unidirectional or round-trip synchronization.

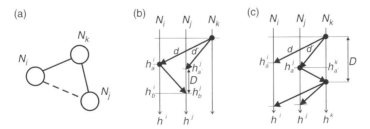

Figure 7.5 Reference broadcast synchronization. A node N_i determines the offset of its local clock relative to that of another node N_j with the help of a third node N_k. In (c), a variant of reference broadcast synchronization is shown that can be used if N_i and N_j cannot directly communicate with each other (dashed link in (a)).

The reference broadcast technique can be used in many variations. For example, Figure 7.5(c) shows a solution presented in ref. [13] for the case that nodes N_i and N_j, while being able to receive messages from N_k, cannot communicate with each other directly. Node N_j replies to N_k, which then can estimate its own local time $h_{a'}^k$ and send this information in another broadcast message to N_i and N_j. In ref. [9], yet another version is described: All nodes report their time stamps to a single node, which then broadcasts all information.

The disadvantage of the reference broadcast approach is that physical broadcasts and a beacon node are required.

7.4.2 Synchronization in Rounds

Typically, two local clocks do not run at exactly the same speed. Therefore time synchronization has to be refreshed periodically, the duration of the round depending on the error budget and the amount of relative drift between the two clocks. Let the length of a round be τ_{round}. Assume a round consists of a first period with length τ_{sample}, where one or more samples are taken according to one of the methods described in Section 7.4.1, and a second period where the nodes do nothing. Let us assume that an application allows for a total error of E_{total}, the maximum error after taking the samples is E_{sample}, and the maximal drift rate is ρ_{max}. Then the maximum length of a round τ_{round} has to satisfy

$$\tau_{round} \leq \frac{E_{total} - E_{sample}}{\rho_{max}}$$

This relation implies that rounds can be longer if E_{sample} and ρ_{max} are small. For example, algorithms that use the round-trip technique can bound E_{sample} according to the measured round-trip time, and thus can dynamically increase τ_{round} if the round-trip time was small. Other algorithms compensate the drift of the local clock and therefore can compute a smaller effective ρ_{max}, which also allows an increase in τ_{round}.

In some applications, E_{total} is smaller than what can be guaranteed by taking a single sample. In such a case, multiple samples can be taken to achieve $E_{sample} < E_{total}$. Taking multiple samples increases τ_{sample}. At the limit, $\tau_{sample} \approx \tau_{round}$; in this case, synchronization in rounds becomes a continuous process, where in rounds follow each other seamlessly.

7.4.3 Combining Multiple Time Estimates

We now discuss techniques for combining multiple estimates of the local time of a remote node. Figure 7.6(a) illustrates the situation: Every circle stands for a single estimate of node N_j's local time h_a^j at some event a, which occurs at N_i's local time h_a^i.

(a)

(b)

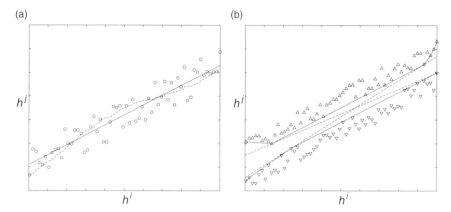

Figure 7.6 Multiple samples improve on the synchronization error. (a) Every point represents a sample, that is a local time h^i of node N_i and an estimated local time h^j of node N_j. Using interpolation techniques improves on the synchronization error. The solid line results from a linear regression on the samples, the dashed line is the result of a phaselocked loop. (b) The same idea can be used for lower ($\triangledown$) and upper ($\triangle$) bounds on the local time of N_j. Also here, interpolation can considerably improve on the synchronization error (i.e., on the uncertainty in this case). The solid lines are determined by the convexhull approach, the dashed lines according to ref. [14].

7.4.3.1 *Linear Regression*

The most widely used technique is linear regression. A linear relation $h^j = \alpha + \beta \cdot h^i$ is postulated and the coefficients α and β are determined by minimizing the square of the difference between the fitted h^j and the actual samples. This technique has a single parameter, that is, the number of samples that are accounted for when computing the coefficients. A large number of samples can improve the regression quality, but requires a large amount of memory.

The coefficient β can be interpreted as an estimation of h^j's drift relative to h^i. Linear regression thus implicitly compensates for clock drift. If the drift is variable, the postulated linear relationship between h^j and h^i does not describe reality very well. In such a situation, the number of samples accounted for should be small.

The linear regression can be computed on-line, that is, incrementally whenever a new sample is taken. An efficient on-line implementation can be found in ref. [15]. A disadvantage of the linear-regression technique is that it weighs data points by the square of their error against the fitted line. Outliers thus have a particularly strong influence on the resulting coefficients α and β.

7.4.3.2 *Phase-Locked Loops*

Another method for processing a continuous sequence of samples is based on the principle of phase-locked loops (PLL) [16]. The PLL controls the slope of the interpolation using a proportional-integral (PI) controller. The output of a PI controller is the sum of a component that is proportional to the input and a component that is proportional to the integral of the input. The input of the controller is the difference between the actual sample and the interpolated value. If the interpolation is smaller than the sample, its slope is increased, otherwise it is decreased. The main advantage of the PLL-based approach is that it requires far less memory than the linear-regression technique (in essence only the current state of the integrator sum). The main disadvantage is that PLLs require a long convergence time to achieve a stable rate [17]. The NTP algorithm uses a PLL [18].

7.4.4 Combining Multiple Time Intervals

The techniques of Section 7.4.1 can also be used to derive lower and upper bounds on the local time of a remote node. Figure 7.6(b) shows a sequence of lower and upper bounds on the local times h^j of a remote node N_j on the y-axis and the corresponding local times h^i of a node N_i on the x-axis. In the previous section, the samples formed a single cloud and the interpolation was a line "through the middle of this cloud." Here we have two clouds, one formed by the lower-bound samples, the other by the upper-bound samples.

The *convex-hull* technique [19,20] interpolates the two clouds separately. One curve is drawn above all lower bounds, a second below all upper bounds. While linear-regression and PLL techniques tend toward the average of the individual samples, the convex-hull technique ignores average values and accounts for the samples with minimal or maximal error. This can result in improved robustness: While the current average message delay can be very unstable, the minimal message delay remains stable, though it may occur more or less frequently.

In ref. [14], it is proposed to interpolate lower- and upper-bound samples by a single line as follows: First, the steepest and flattest lines that do not violate any lower or upper bound are determined. The slopes of these lines represent bounds on the drift of clock h^j relative to h^i. The "average"-line of these two extremal solutions is used as the final interpolation; for a more detailed description, see Section 7.5.3.

7.4.5 Synchronization of Multiple Nodes

Sensor networks most often have a much more complicated topology than the simple examples shown in Figures 7.4 and 7.5, and not all sensor nodes can communicate with each other directly. Thus, multihop synchronization is required, which adds an additional layer of complexity. Clearly, this could be avoided by using an overlay network that provides virtual, single-hop communication from every sensor node to a single master node. But as we saw in Section 7.4.1, the synchronization error directly depends on the message delay, which is very difficult to control on a logical link that is composed of many physical hops. Therefore, performant synchronization schemes have to deal with the multihop problem explicitly.

Figure 7.7 illustrates various approaches to multihop synchronization. We now describe these four schemes and use them as examples to discuss the main problems of multihop synchronization.

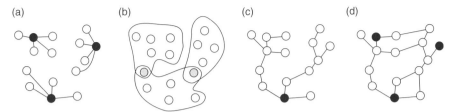

Figure 7.7 Organizing synchronization in multihop networks. (a) Single-hop synchronization with a set of master nodes that are synchronized out of band (e.g., using GPS). (b) Single-hop synchronization in overlapping clusters, gateway nodes translate time stamps. (c) Tree hierarchy with a single master node at the root. (d) Unstructured.

7.4.5.1 *Out-of-Band Synchronization*

The conceptually simplest solution is to avoid the problem: A large number of master nodes is distributed in the network such that every node has a direct connection to at least one of these masters (e.g., ref. [21]). The master nodes are synchronized among each other using some out-of-band mechanism. The GPS is well suited to this purpose, as it provides time information with submicrosecond accuracy. However, GPS receivers are still relatively costly, consume a considerable amount of energy, and require a direct line of sight to a number of satellites, and thus cannot operate inside buildings.

7.4.5.2 *Clustering*

The authors of the RBS algorithm proposed to partition the network into clusters [9]. All nodes within a cluster can broadcast messages to all other members of the cluster, and thus the reference broadcast technique can be used to synchronize the cluster internally. Some nodes are members of several clusters and participate independently in all corresponding synchronization procedures. These nodes act as time gateways to translate time stamps from one cluster to the other. There is a trade-off in choosing the size of the clusters. On the one hand, a small number of large clusters reduces the number of translations, and thus improves the synchronization error; on the other hand, energy consumption grows quickly with increasing transmission range; this makes choosing many small clusters attractive. This trade-off has been examined in ref. [22].

7.4.5.3 *Tree Construction*

The most common solution of the multihop synchronization problem is to construct a synchronization tree with a single master at the root [14,23,24,31]. Single-hop synchronization is applied along the edges of the tree. Various well-known algorithms can be used to construct such a tree [24]. Since the accuracy degrades with the hop distance from the root, a tree with minimum depth is preferable. On the other hand, a small depth implies that the root has to serve many clients, and thus consumes far more energy than the other nodes.

Tree construction faces two main problems: First, in sensor networks, the network topology may be dynamic; nodes may be mobile and repeatedly join or leave the network. The multihop synchronization algorithms have to explicitly deal with such events. In particular, if the root node fails, a new root has to be elected

[31]. Second, two neighboring (in terms of physical location) nodes may have a large hop distance in the synchronization tree. In consequence, the accuracy of synchronization between these nodes is not as good as if they would synchronize directly with each other.

7.4.5.4 Unstructured As illustrated in the tree-construction approach, the multihop synchronization problem can be interpreted as the problem of determining the links and directions over which time information is disseminated. In contrast to tree-construction approaches, *unstructured* approaches do not first explicitly solve this problem and then perform pairwise synchronization. Instead, time information is exchanged between any pair (or group) of nodes that communicate. Whereas in the tree-construction approach every pairwise synchronization is asymmetrical (i.e., between a client and a local master), it is symmetrical in the unstructured approach (i.e., between two equal peers). In ref. [25], such an approach has been presented for interval-based synchronization. Two nodes combine their bounds on real time by selecting the larger lower bound and the smaller upper bound. A similar approach for point estimates is *asynchronous diffusion* proposed in ref. [26]. Here, nodes that communicate adjust their synchronized clocks to the average of their synchronized times. Like the interval-based solution from ref. [25], this approach is completely local. Since these approaches do not maintain any global configuration, node mobility does not cause particular problems. In contrast, clustering and tree-construction schemes require that the global configuration has to be updated whenever nodes move or fail or when new nodes are added to the system.

Because algorithms that follow the unstructured approach do not attempt to communicate with a particular node (e.g., the parent node in a synchronization tree), some of these algorithms piggyback time stamps on messages that are sent for some other, not synchronization-related reason (e.g., refs. [25] and [27]). It could be argued that these algorithms have virtually no communication overhead, as no messages are generated exclusively for time synchronization.

7.5 CASE STUDIES

In the following subsections, we discuss a number of concrete synchronization algorithms from the literature (ordered by publication date). The goal here is to give an overview of the approaches (with reference to the techniques and classes discussed earlier in this chapter), rather than to discuss all the details. In addition, for each algorithm we will give some experimental results. Table 7.1 summarizes the underlying assumptions of the various protocols and classifies the approaches according to the criteria discussed in Section 7.3.

7.5.1 Time-Stamp Synchronization

Time-stamp synchronization (TSS) [27] provides internal synchronization on demand. Node clocks run unsynchronized, that is, time stamps are valid only in

TABLE 7.1 Synchronization Classes and Assumptions of Time-Synchronization Protocols

	RBS	TPSN	TS/MS	LTS	TSS	IBS	TSync	FTSP	TDP	AD
Classes										
Internal vs. external	I	E	I	E	I	E	E	I	I	I
Cont. vs. on-demand	O	C	C	O	O	C	C	C	C	C
All nodes vs. subsets	S	A	S	A/S	S	A	A	A	A	A
Rate vs. offset	RO	O	RO	O	O	O	O	RO	O	O
Transform vs. clocksync	T	C	—	C	T	C	C	C	C	C
Instants vs. intervals	S	S	TS	S	T	T	S	S	S	S
Assumptions										
Broadcast	X	X	X	X			X	X	X	X
Bidirectional communication		X	X	X	X		X		X	X
Constant rate			X							
Bounded drift				X	X	X				
Multichannel							X			
MAC access		X						X		

Abbreviations: RBS = reference broadcast synchronization; TPSN = Timing-Sync Protocol for Sensor Networks; TS/MS-Tmy-Sync/Mini-Sync; LTS-lightweight tree-based synchronization; TSS = time-stamp synchronization; IBS = interval-based synchronization; FTSP = flooding time synchronization protocol; TDP = time diffusion synchronization; AD = asynchronous diffusion.

219

the node that generated them. However, when a time stamp is sent to another node as part of a message, the time stamp is transformed to the timescale of the receiver. For messages sent over multiple hops, the transformation is repeated for each hop.

Time-stamp transformation is achieved by determining the age of each time stamp from its creation to its arrival at a sensor node. On a multihop path, the age is updated at each hop. The time stamp can then be transformed to the receiver's local timescale by subtracting the age from the time of arrival. The age of a time stamp consists of two components: (1) the total amount of time the time stamp resides in nodes on the path, and (2) the total amount of time needed to transfer the time stamp from node to node. The first component is measured using the local, unsynchronized clocks; the second component can be bounded by the round-trip time of the message and its acknowledgment.

For the round-trip measurement, the technique depicted in Figure 7.4(d) is used, where the sender is N_i and the receiver is N_j. Message d_2 is a data message containing the time stamp, and message d'_2 is an acknowledgment. Using the previous message exchange (d_1, d'_1), the receiver can use $D^j - D^i$ as an upper bound for the delay of message d_2. If a minimum delay is known, it can be used as a lower bound (otherwise, 0 is used). Using storage time and the bounds on transmission delay just given, lower and upper bounds of the time-stamp age can be determined. Additionally, ρ_{max} is used to transform time intervals between node clocks as in equation (7.2).

With this approach, synchronization information is piggybacked to existing (acknowledged) messages. There are no additional synchronization messages, except when two nodes exchange a message for the first time. In this case, an additional initialization message must be sent and acknowledged in order to enable round-trip measurement. An acknowledgment is not needed if the sender can overhear the receiver forwarding the message to the next hop, which is typically the case in broadcast networks.

Measurements in a wired network with $\rho_{max} = 1$ ppm showed that the average uncertainty of the time-stamp interval is about 200 μs for adjacent nodes. It increases by an additional 200 μs for each additional hop, and by about 2.5 μs per age second.

7.5.2 Reference Broadcast Synchronization

RBS [9] provides synchronization for a whole network. The basic synchronization primitive is a reference broadcast to a set of client nodes in the one-hop neighborhood of a beacon node, as illustrated in Figure 7.5(b). The beacon node broadcasts synchronization pulses. The clients then exchange their respective reception times and use linear regression to compute relative offsets and rate differences to each other. Using offset and rate difference, each client can transform a local clock reading to the local timescale of any other client.

To extend this scheme to multihop networks, the network is clustered such that a single beacon can synchronize all nodes in its cluster. Gateway nodes that participate in two or more clusters independently take part in the reference broadcast procedure of all their clusters. By knowing offsets and rate differences to nodes in all adjacent clusters, gateway nodes can transform time stamps from one cluster to another.

Time synchronization across multiple hops is then provided as follows. Nodes time-stamp sensor data using their local clocks. Whenever time stamps are exchanged among nodes, the time stamps are transformed to the receiver's local time using offset and rate difference.

In experiments it has been shown that adjacent Berkeley Motes can be synchronized with an average error of 11 μs by using 30 broadcasts. Over multiple hops, the average error grows with $O(\sqrt{n})$, where n is the number of hops.

7.5.3 Tiny-Sync and Mini-Sync

Tiny-Sync and Mini-Sync (TS/MS) [14] are methods for pairwise synchronization of sensor nodes. Both TS and MS use multiple round-trip measurements and a line-fitting technique to obtain the offset and rate difference of the two nodes. For this, a constant-rate model (see page 205) is assumed. To obtain data points for line fitting, multiple round-trip synchronizations are performed, as depicted in Figure 7.4(c), where the client is N_j and the reference is N_i. Each round-trip measurement results in a data point $(h_b^i, [h_a^j, h_c^i])$. Then, the line-fitting technique depicted in Figure 7.6(b) is used to calculate two lines with minimum and maximum slope. The slope and axis intercept of these two lines then gives bounds for the relative offset and rate difference of the two nodes. The line with average slope and intercept of the two lines is then used as the offset and rate difference between the two nodes.

Note that each of the two lines is unambiguously defined by two (a priori unknown) data points. The same results would be obtained if the remaining data points could be eliminated. Since the computational and memory overhead depends on the number of data points, it is a good idea to remove as many data points as possible before the line fitting. TS and MS only differ in this elimination step. Essentially, TS uses a heuristic to keep only two data points for each of the two lines. However, the selected points may not be the optimal ones. MS uses a more complex approach to eliminate exactly those points that do not change the solution. Hence, TS achieves a slightly suboptimal solution with minimal overhead, and MS gives an optimal solution with increased overhead.

Measurements on a 802.11 network with 5000 data points resulted in an offset bound of 945 μs (3230 μs) and a rate bound of 0.27 ppm (1.1 ppm) for adjacent nodes (nodes five hops away).

7.5.4 Lightweight Time Synchronization

Lightweight time synchronization (LTS) [24] is a synchronization technique that provides a specified precision with little overhead, rather than striving for maximum precision as do many other techniques.

Two algorithms are proposed: one that operates on demand for nodes that actually need synchronization, and one that proactively synchronizes all nodes. Both algorithms assume the existence of one or more master nodes that are synchronized out-of-band to a reference time. The proactive algorithm proceeds to construct spanning trees with the masters at the root by flooding the network. In a second phase, nodes synchronize to their parent in the tree by means of round-trip synchronization. The synchronization frequency is calculated from the requested precision, from the depth of the spanning tree, and from the drift bound ρ_{max}.

The on-demand version also assumes the existence of one or more master nodes. When a node needs synchronization, it sends a request to one of the masters using any routing algorithm (this is not further specified). Then, along the reverse path of the request message, nodes synchronize using round-trip measurements. The synchronization frequency is calculated as in the proactive version just described. In order to reduce synchronization overhead, each node may ask its neighbors for pending synchronization requests. If there are any such requests, the node synchronizes with the neighbor, rather than executing a multihop synchronization with a reference node.

The overhead of the algorithms was examined in simulations with 500 nodes uniformly placed in a 120-m × 120-m area, a target precision of 0.5 s, and a duration of 10 h. The centralized algorithm performed an average of 36 pairwise synchronizations per node. The distributed algorithm executed 4–5 synchronizations per node on average, if 65% of all nodes request synchronization.

7.5.5 Timing-Sync Protocol for Sensor Networks

The timing-sync protocol for sensor networks (TPSN) [23] provides synchronization for a whole network. First, a node is elected as a synchronization master (details for this are not specified), and a spanning tree with the master at the root is constructed by flooding the network. In a second phase, nodes synchronize to their parent in the tree by means of round-trip synchronization. Synchronization is performed in rounds and initiated by the root broadcasting a synchronization-request message to its children. Each child then performs a round-trip measurement to synchronize with the root. Nodes further down in the tree overhear the messages of their parents and start synchronization when their parents have synchronized. To eliminate message-delay uncertainties, time-stamping for the round-trip synchronization is done in the medium-access control (MAC) layer. In the case of node failures and topology changes, master election and tree construction must be repeated.

Measurements showed that two adjacent Berkeley Motes can be synchronized with an average error of 16.9 μs, which is a worse figure than the 11 μs given for RBS in ref. [9]. However, the authors of ref. [23] claim that a reimplementation of RBS on their hardware resulted in an average error of 29.1 μs between adjacent nodes, effectively claiming that TPSN is about twice as precise as RBS.

7.5.6 TSync

TSync [13] provides two protocols for external synchronization: the hierarchy referencing time synchronization protocol (HRTS) for proactive synchronization of the whole network, and the individual-based time request protocol (ITR) for on-demand synchronization of individual nodes. Both protocols use an independent radio channel for synchronization messages in order to avoid inaccuracies due to variable delays introduced by packet collisions. In addition, the existence of one or more master nodes with access to a reference time is assumed.

With HRTS, a spanning tree with the master at the root is constructed. Then, the master uses the reference broadcasting technique illustrated in Figure 7.5(c) to synchronize its children. Each child node now repeats the procedure for its subtree.

Measurements in a network of MANTIS sensor nodes showed a mean synchronization error of 21.2 μs (29.5 μs) for two adjacent nodes (nodes three hops away). For comparison, RBS was also implemented, giving an average error of 20.3 μs (28.9 μs).

7.5.7   Interval-Based Synchronization

Interval-based synchronization (IBS) was first proposed in ref. [28], where a bounded-drift model (see page 205) is assumed. The network nodes perform external synchronization by maintaining a lower and upper bound on the current time. During communication between two nodes, the bounds are exchanged and combined by choosing the larger lower and the smaller upper bound. This amounts to intersecting the time intervals defined by each pair of bounds. Between communications, each node advances its bounds according to the elapsed real time and the known drift bounds. In ref. [29], the model was refined by including bounded drift variation and fault tolerance.

In ref. [25], the simple approach from ref. [28] was shown to be worst-case optimal, where the worst case is the one where all clocks run with maximal drift. A considerable improvement in the synchronization quality can be achieved by having each node store, maintain, communicate, and use the bounds from its last communications with other nodes. In ref. [30], it was shown that optimal IBS can only be achieved by having nodes store and communicate their entire history. Obviously, this becomes prohibitive with growing network size and lifetime. In realistic settings, the value of a piece of history data decreases rapidly with its age. Therefore, efficient average-case optimal synchronization can be obtained by using only recent data.

7.5.8   Flooding Time-Synchronization Protocol

The flooding time-synchronization protocol (FTSP) [31] can be used to synchronize a whole network. The node with the lowest node ID is elected as a leader that serves as a source of reference time. If this node fails, then the node with the lowest ID in the remaining network is elected as the new leader. The leader periodically floods the network with a synchronization message that contains the leader's current time. Nodes that have not yet received this message record the contained time stamp and the time of arrival, and broadcast the message to their neighbors after updating the time stamp. Time-stamping is performed in the MAC layer to minimize delay variability, and hence uncertainty.

Each node collects eight (time stamp, time of arrival) pairs and uses linear regression on these eight data points to estimate offset and rate difference to the leader.

Measurements were performed in an eight-by-eight grid of Berkeley Motes, where each Mote has a direct radio link to its eight closest neighbors. With this setup, the network synchronized in 10 minutes to an average (maximum) synchronization error of 11.7 μs (38 μs), giving an average error of 1.7 μs per hop.

7.5.9 Asynchronous Diffusion

Asynchronous diffusion (AD) [26] supports the internal synchronization of a whole network. The algorithm is very simple: each node periodically sends a broadcast message to its neighbors, which reply with a message containing their current time. The receiver averages the received time stamps and broadcasts the average to the neighbors, which adopt this value as their new time. It is assumed that this sequence of operations is atomic, that is, the averaging operations of the nodes must be properly sequenced.

Simulations with a random network of 200 static nodes showed that the synchronization error decreases exponentially with the number of rounds.

7.5.10 Time Diffusion Synchronization

Time diffusion synchronization (TDP) [32] supports the synchronization of a whole network. Initially, a set of master nodes — so-called leaders — is elected. For external synchronization, these nodes must have access to a global time. This is not required for internal synchronization, where masters are initially unsynchronized.

Master nodes then broadcast a request message containing their current time, and all receivers send back a reply message. Using these round-trip measurements, a master node calculates and broadcasts the average message delay and standard deviation. Receiving nodes record these data for all leaders. Then they turn themselves into so-called "diffused leaders" and repeat the procedure. The average delays and standard deviations are summed up along the path from the masters. The diffusion procedure stops at a given number of hops from the masters.

All nodes have now received from one or more masters m the time h_m at the initial leader, the accumulated message delay Δ_m, and the accumulated standard deviation β_m. A clock estimate is computed as $\Sigma_m \omega_m (h_m + \Delta_m)$, where the weights ω_m are inversely proportional to the standard deviation β_m. After all nodes have updated their clocks, new masters are elected and the procedure is repeated until all node clocks have converged to a common time.

In a simulation with 200 static nodes with 802.11 radios and a delay of 5 seconds between consecutive synchronization rounds, the deviation of time across the network dropped to 0.6 second after about 200 seconds.

7.6 EVALUATION STRATEGIES

Evaluating the precision of time synchronization in wireless sensor networks is not a trivial task. For example, the authors of the RBS algorithm report 11μs precision on the Berkeley Motes platform [9], while the authors of the TPSN algorithm report

29 µs for RBS on the same platform, concluding that TPSN is better, as it achieves 17 µs [23]. Which numbers are correct? Probably all of them, but the evaluation was done slightly differently.

In this section, we discuss different evaluation strategies that have been applied to time-synchronization algorithms for wireless sensor networks. There are various aspects of the performance achieved by an algorithm than can be evaluated, for example, the energy consumption or the message and memory overhead. The discussion in this section concentrates on various alternatives for the evaluation of the *precision* of time-synchronization algorithms.

7.6.1 What Is Precision?

Figure 7.2(b) illustrates the scope and lifetime of synchronization in a sensor network. The scope defines which nodes have to be synchronized, and the lifetime defines when these nodes have to be synchronized. Thus, it is natural to evaluate the precision in the shaded area of Figure 7.2(b). The precision is a metric that is closely related to the synchronization error. While the precision is a single scalar value for a whole network, the synchronization error is a function of time for a single node. In the following, we discuss several alternatives to map such functions to a single scalar precision value P.

7.6.1.1 Combining the Synchronization Error of Many Nodes At some
time t within the lifetime of a sensor network, every node N_i within the scope has a synchronized time $c^i(h^i(t))$. In the case of internal synchronization, the *instantaneous precision* $p(t)$ is often defined as the maximal difference between any two synchronized times, that is,

$$p(t) = \max_{i,j}\left\{c^i(h^i(t)) - c^j(h^j(t))\right\}$$

for any nodes N_i and N_j within the scope. Some authors (e.g., ref. [32]) use the standard deviation among all $c^i(h^i(t))$ as a measure for the instantaneous precision at time t.

In the case of external synchronization, the instantaneous precision is more often defined as the maximal synchronization error, that is,

$$p(t) = \max_{i}\left\{c^i(h^i(t)) - t\right\}$$

for any node N_i within the scope. This variant of precision is sometimes called *accuracy*. Alternatively, the precision can be defined as the average synchronization error within the scope or the maximal synchronization error among the 90% (or 99%, etc.) nodes in the scope with the smallest synchronization error.

7.6.1.2 Steady State and Convergence Time The instantaneous precision
$p(t)$ obviously varies during the synchronization lifetime. The final precision metric

P can be derived by taking the maximum of $p(t)$ for all t in the lifetime. Alternatively, the average of $p(t)$ can be used.

It is clear that the precision P improves in proportion to the time the synchronization process is active, and that at some point, the improvement stops. Usually, the precision P is evaluated after this point, that is, the lifetime of synchronization starts after the synchronization process, and the precision P describes the *steady state*.

Some authors evaluate the *convergence time*, which is the length of the interval from the start of the synchronization process to the point in time where the precision P stops to improve or reaches a specific value. If the lifetime is defined, the convergence time indicates when the synchronization process has to be started such that the desired precision P is achieved before the start of the lifetime and is maintained until the end of the lifetime.

7.6.2 Goals of Performance Evaluation

There can be different reasons why the performance of an algorithm has to be evaluated, and different goals lead to different solutions.

The actual performance of a given synchronization algorithm strongly depends on properties of the target platform. It is difficult to identify and model all the influence factors explicitly. A *realistic estimation* of the achievable precision is thus best obtained by using *measurements on the actual target platform*, rather than using simulation of a simplified target platform.

Sometimes, realistic estimation of the performance is less important than *fairness and repeatability* of the evaluation. This is the case if several competing algorithms have to be compared. Also in the optimization process of the parameters of a particular algorithm, it is important that differences in the performance are due to differences in the algorithm and not due to different conditions (e.g., message delays, clock drift). Here, *simulation based on recorded or generated traces* is more appropriate than direct measurements.

If the goal of analyzing a particular synchronization algorithm is to give worst-case guarantees on its performance, neither measurements nor simulation based on recorded traces can be used, since both strategies only evaluate a finite number of instances. Instead, the worst case has to be identified and the worst-case performance has to be determined analytically.

7.6.3 Measurements

7.6.3.1 Measurement Techniques Three fundamentally different measurement strategies, which are illustrated in Figure 7.8, have been used in recent publications.

Consider Figure 7.8(a). Every sensor node executes two synchronization procedures, synchronizing two different clocks. The first procedure is the actual synchronization algorithm under test, using only the means of the platform on which it is executed. The second procedure is another algorithm, which achieves a far

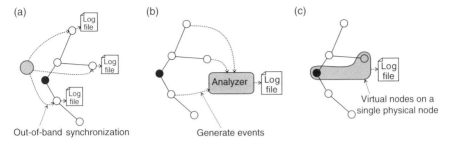

Figure 7.8 Precision-measurement techniques. (a) Every node is synchronized out of band and measures its own precision. (b) Every node generates events, the evaluation is centralized. (c) Some nodes are virtual nodes on the same hardware platform as the master node.

better precision than the first. This is possible since this second synchronization uses resources that are not offered by the target platform, but which are introduced for the measurements. A GPS receiver for every sensor node can serve this purpose. Alternatively, cable connections can be used as an out-of-band mechanism with very low delay variability to provide a reference time (e.g., refs. [9,33,34]). In ref. [31], a single-hop RBS scheme is used to measure the precision achieved by the FTSP multihop algorithm. This approach has the advantage that every node can evaluate and log its own precision, and these values can be collected at the end of the experiment (or even on-line), providing complete information.

An alternative is shown in Figure 7.8(b). All sensor nodes generate some directly observable event, for example, a rising edge on a particular I/O pin, when their synchronized time reaches a particular value X. An external analyzer device then records the time interval between the instance when a node's synchronized time is X and the instance when it really is X. Such a procedure has been used, for example, in ref. [23]. Its advantage is that the precision of the measurement is not limited by the resolution of the nodes' clocks or the performance of a second synchronization procedure.

As illustrated in Figure 7.8(c), ref. [22] proposes to measure the precision achieved by *one* client node as follows: A client node synchronizes over several hops to a master node. Master and client nodes are virtual nodes emulated on a single physical node, and the intermediate nodes are all separate physical nodes. Since the master and the client share a single hardware clock, the precision of the client can easily be evaluated.

7.6.3.2 *Systems and Topologies*

All three approaches do not scale well. Therefore, only small networks have been used so far for measurements. The largest experiment is described in ref. [31], where an 8-by-7 grid of Mica2 Motes is evaluated. In ref. [23], a chain of six Mica Motes is used, ref. [13] evaluates five MANTIS Nymph nodes, ref. [27] evaluates a chain of seven standard PCs with 100 Mbit/s wired Ethernet, and ref. [9] evaluates IPAQ nodes communicating over 802.11-b WLAN and Mica Motes.

How the synchronization error of hundreds of nodes should be measured is an open question. Current evaluations of such large networks are all based on simulation.

7.6.3.3 Results We will now give some measurement results from recent publications. Our intention is to give an idea about the order of magnitude of the achievable precision and to illustrate that although all results are about precision, they are difficult to compare. In ref. [31], the convergence time of the FTSP algorithm in a 7-by-8 grid is reported to be 10 minutes. A maximal error of 38 μs and an average error (over all nodes) of 12 μs is reported. For the TPSN algorithm, ref. [23] reports a maximal error of 45 μs for one hop and 74 μs for five hops. Average errors (over time) are 17 μs for one hop and 38 μs for five hops. The authors also provide the percentage of the time when the synchronization error was below the average error (>60%). The authors of RBS present in ref. [9] the distribution of the synchronization error (over time) for one hop and the mean, median, 95%, and 99% values over 300 trials for one to four hops.

Some authors evaluate the distribution of the synchronization quality in the system. At some time t, either the synchronized times $c^i(h^i(t))$ of all nodes i [32], or alternatively the corresponding synchronization errors $e^i(t)$ [13,23], are shown in a histogram.

7.6.4 Simulation

Performance evaluation through simulation has the advantage that the resulting precision or accuracy of all nodes does not have to be measured, but is directly accessible. Thus, much larger systems can be evaluated.

7.6.4.1 Systems and Topologies In ref. [32], systems with 200 nodes are evaluated, and in refs. [26] and [32] systems with up to 500 nodes, always randomly placed in a square area. The transmission range of the nodes is 10 m in a square of length 80 m [33] or 120 m [24]; in ref. [26], various transmission ranges from 0.4 m to 1 m are used in a square of length 10 m. In ref. [25], the transmission range is varied between 0.1 and 0.5 times the width of the square area. In ref. [14], a chain of 5 nodes is simulated.

7.6.4.2 Message Delays For simulation, a number of assumptions about the behavior of the system have to be made (e.g., about message delays). In ref. [14], measured delay traces from an 802.11 wireless local area network (LAN) are used, and the authors of refs. [24] and [32] generate delay traces according to a normal distribution. In ref. [32], an additional offset is added that increases when the medium is saturated, that is, when more than 75% of the channel capacity is used. The authors of ref. [25] assume zero message delay, arguing that the synchronization errors induced by delay uncertainty and drift can be studied separately.

7.6.4.3 Clock Drift In refs. [26] and [33], every node is assigned an arbitrary but constant drift rate between -100 ppm and $+100$ ppm. In ref. [24], all nodes have a drift rate of 50 ppm.

7.6.4.4 Results The main concern of ref. [24] is to compare centralized and distributed versions of the LTS algorithm in terms of required messages and achieved synchronization error. The average error (over all nodes) is evaluated as a function of the hop distance to the master node. Reference [14] evaluates the synchronization error and the drift-compensation error achieved by the TS/MS algorithms as a function of time. A node one hop away from the master has an error of 1 ms after 83 minutes. A node with five hops distance achieves 3 ms. In ref. [25], the average synchronization error (over time and over all nodes) is evaluated as a function of the number of messages exchanged between the nodes. Also the impact of the transmission range and of the number of master nodes is evaluated. The authors of ref. [26] mainly evaluate how quickly (number of rounds) a network synchronizes using the AD algorithm. This is evaluated as a function of the transmission range and as a function of the number of nodes in the system. It is also shown that the synchronization error decreases exponentially with the number of rounds. The speed of convergence is also evaluated in ref. [32], for the TDP and TPSN algorithms; the standard deviation of the nodes' synchronization error is shown as a function of time. It is argued that node mobility makes convergence slower. In addition, histograms and three-dimensional plots of the distribution of the synchronization error after convergence are presented.

7.6.5 Challenges of a Benchmark

So far, we have presented how synchronization algorithms are evaluated in the current literature. We have seen that results of different authors are quite incomparable due to widely differing goals, assumptions, and techniques. On the one hand, there is not yet a common understanding about the requirements on synchronization in sensor networks. On the other hand, there is also disagreement about available resources and platforms.

A benchmark for comparing the various algorithms on common ground has not yet been presented. In the following, we discuss why it is difficult to devise a benchmark that can be used with a large number of algorithms. Ideally, the comparison of algorithms is based on simulation using system traces. Such traces should contain the system and communication model (How many nodes are there? How many of them are master nodes? Which node communicates with which other node at which time?), and they should characterize the "adversary" of synchronization, namely, all message delays and the drift rates of the nodes. But this would require determining all communications before executing the algorithms. This is not possible for most of the algorithms, since they actively decide to generate messages, depending on previous events. Furthermore, some algorithms require broadcast communication, while others do not.

7.7 CALIBRATION

In the previous sections, we have considered the problem of time synchronization, where the output of a hardware clock had to be mapped to a timescale. Sensor calibration is the problem of mapping the output of a sensor to a well-defined scale. In this section, we take a step back from time synchronization and consider the more general problem of calibration. As we will show, there is a close relationship between calibration and time synchronization, since the latter can be considered a special case of the former. It might be somewhat unfamiliar to consider a hardware clock as a sensor, but as we will show in Section 7.7.1, the difference is rather subtle. The remainder of this section is structured in a similar way as the discussion of time synchronization in the previous sections. Section 7.7.2 explains why new approaches are required for calibration in sensor networks, Section 7.7.3 presents our system model for sensors, Section 7.7.4 discusses various classes of calibration, and Section 7.7.5 presents concrete calibration algorithms from current literature.

7.7.1 Time Synchronization as Calibration

Sensors are hardware devices that have an input and an output. The input is a certain physical quantity in the real world, such as temperature, light intensity, acceleration, radio signal strength, and so on. The output typically is a variable electrical signal, such as a voltage or current. An analog-to-digital (A/D) converter converts it to a digital number.

A hardware clock typically consists of four components: a physical system that has a periodic behavior (e.g., an oscillating quartz, decaying cesium, a pendulum), a sensor that converts the physical phenomenon to an electrical signal, an A/D converter (e.g., threshold detector) that converts the output of the sensor to a one-bit number, and a counter that counts the number of rising (or falling) edges seen so far in the digital output stream. Hence, a hardware clock contains, among other things, a sensor *and* the physical phenomenon to observe.

Despite this analogy, time as a physical quantity has some unique characteristics. For example, in many practical settings, observed physical quantities have a rather limited range of values (e.g., temperature $\in [-30°C, +30°C]$). Given a bounded relative sensor error (i.e., bounded drift), the absolute measurement error is then also bounded. However, physical time eventually grows beyond all bounds. Hence, the absolute error of a software clock is unbounded unless synchronization messages are exchanged. This explains the need for precise drift compensation and for repeating synchronization after a certain amount of time.

7.7.2 Revisiting Calibration for Sensor Networks

Calibration is a very old problem, since it is needed for almost all measurement instruments. Despite this, calibration in sensor networks has so far not received much attention by researchers, at least when compared to time synchronization. However, a number of challenges to calibration in sensor networks pose interesting questions.

A large number of sensor nodes often cannot be calibrated manually and individually. This is particularly true for pairwise calibration, where a sensor measures a quantity emitted by another device (i.e., an actuator). One example for this is measuring the radio signal strength to infer the distance between devices. In such a scenario, every sensor would have to be calibrated against every transmitter, resulting in a quadratic number of calibration steps.

Sensors may be exposed to significant changes of environmental parameters (e.g., temperature, humidity) during the lifetime of a sensor network. Since the commonly used low-cost sensors are rather sensitive to such changes, a one-time factory calibration may not be sufficient. In this case, periodical calibration during the lifetime of the sensor network is necessary.

7.7.3   System Model

The physical quantity q that is observed by a sensor is mainly a function of the size of the sensor, of its orientation, of its location, and of time t. Since the size of a sensor is nonzero, sensors can typically only observe the accumulation (e.g., weighted average) of a physical quantity over a certain area or volume. If we assume that size, orientation, and location are constant properties of a sensor i, we can denote the time-dependent physical quantity observed by the sensor as $q^i(t)$. Often, q is a real-valued scalar function (e.g., for temperature sensors), but may also be more complex (e.g., for a location sensor, q might return triples (x, y, z)).

The output of a sensor i under stimulus $q^i(t)$ is denoted as $h^i(t)$. Note that $h^i(t)$ for a given $q^i(t)$ typically depends on a number of parameters, for example, fabrication tolerance of the sensor, environmental parameters such as temperature and humidity, and wear of the sensor.

In analogy to software clocks, we introduce a *software sensor* as a function c that maps a sensor output $h(t)$ to $c(h(t))$. Software sensors are typically introduced to map a sensor output to a standard scale (e.g., the Celsius scale). Here, the goal of calibration is to find a suitable c for a given scale. Often, q does refer to such a standard scale, in which case the goal of calibration could be to find a c that approximates $c^i(h^i(t)) = q^i(t)$ for all t.

As mentioned in the previous section, calibration may also be applied to actuators. An actuator can be considered a reverse sensor that accepts a digital value $h(t)$ as input and produces a physical quantity $q(t)$ as output. For example, a heater may accept a temperature specification as input and heat until this temperature is reached. A *software actuator* c^{-1} then maps a given value v to $c^{-1}(v)$, which can be used as an input to the actuator to produce a certain physical quantity (e.g., such that $q(t) = v$).

7.7.4   Classes of Calibration

In this subsection, we adapt the classes of time synchronization introduced in Section 7.3 to calibration.

- *Internal versus External*   For internal calibration, all software sensors i should output the same value $c^i(h^i(t))$ if they are exposed to an identical stimulus $q(t)$

(note that if for instance $q(t) = 25°C$, then $c^1(h^1(t)) = c^2(h^2(t)) = 10°C$ would mean that sensors 1 and 2 are internally calibrated). For external calibration, the output of all software sensors must conform to a specified scale (e.g., if $q(t) = 25°C$, then $c^1(h^1(t)) = c^2(h^2(t)) = 25°C$ is required).

- *Lifetime: Continuous versus On-Demand*   Because some of the parameters that influence h may change over time, calibration may have to be repeated to adapt to these parameters. Calibration may be performed continuously or on demand.

- *Scope: All Nodes versus Subsets*   All nodes or only subsets of nodes might participate in calibration. For example, only some nodes might be equipped with a certain type of sensor, or the sensor might only be used by some nodes.

- *Rate versus Offset*   Sometimes it is sufficient if differences $c^i(h^i(t_1)) - c^i(h^i(t_2))$ (e.g., temperature differences) obtained from different sensor instances can be compared. In this case, rate calibration is sufficient. If, however, absolute values $c^i(h^i(t))$ (e.g., absolute temperature values) originating from different sensor nodes are to be compared, offset calibration is needed.

- *Scale Transformation versus Global Scale*   Rather than having all software sensors adhere to a global scale, it might be advantageous (e.g., in terms of overhead) to maintain local scales and transform sensor readings as they are sent to nodes with a different scale (e.g., if node 1 uses the Celsius scale and node 2 uses the Fahrenheit scale, then the transformation function for transforming from node 1 to node 2 is $c^{12}(h^1(t)) = 1.8h^1(t) + 32$).

- *Point Estimates versus Bounds*   Software sensors may either output point estimates (in analogy to time instants for time synchronization) or bounds (analogous to intervals for time synchronization).

7.7.5 Case Studies

In this subsection, we present two calibration algorithms for sensor networks. As in Section 7.5, we outline the algorithm and give an idea of its performance.

7.7.5.1 *Calibration as Parameter Estimation*   Calibration as parameter estimation (CPE) [35] provides a framework for external calibration where sensors measure a quantity emitted by an actuator. Both the behavior of the actuator and the sensor are unknown and must be taken into account for calibration. The general approach here is the joint calibration of sensors and actuators such that the overall system response is optimized.

The algorithm is illustrated by pairwise distance measurements between sensor nodes using the time of flight of an ultrasound signal. Each node is equipped with a speaker (the actuator) and a microphone (the sensor). Assuming the nodes have synchronized clocks, one node emits an acoustic signal containing a time stamp and the other receives the acoustic signal, computes the time of flight, and multiplies with the speed of sound to obtain a distance estimate.

For our discussion, we consider a virtual distance sensor that directly outputs a distance estimate $h(t)$ given the true distance as the physical stimulus $q(t)$. Due to reasons discussed later, distance estimates include various systematic errors. Hence, the goal of calibration is to find a function c that maps the distance estimate $h(t)$ to the correct distance $q(t)$.

For CPE, c must now be parameterized, such that it is expressed as a function of h and of parameters that describe the various error sources. These parameters are related to both the sensor and to the actuator. For example, the distance $h(t)$ output by the sensor can be expressed in terms of the true distance $q(t)$:

$$h(t) = B_T + B_R + G_T q(t) + G_R q(t)$$

where B_T and B_R refer to constant distance offsets caused by startup times for diaphragm oscillation in the transmitter and the receiver, respectively, and G_T and G_R represent the distance-dependent influence of the transmitter volume and the receiver sensitivity, respectively. Solving for $c(h(t)) := q(t)$, we obtain

$$c(h(t)) = \frac{h(t) - B_T - B_R}{G_T + G_R}$$

Although there is one instance of the previously given joint-calibration function for every transmitter–receiver pair, there is only one set of parameters (B_T, B_R, G_T, G_R) for each of the N nodes. Hence, we can formulate a linear equation system with $4N$ variables (i.e., the four parameters for each node just given) and $4N$ equations, which requires $4N$ pairs $(h(t), c(h(t)) = q(t))$ to solve. Alternatively, more samples can be collected and least-squares optimization can be used to obtain more accurate estimates for the parameters. Once the parameters (B_T, B_R, G_T, G_R) for each node are known, the calibration functions are also known.

The authors performed an experiment with an 8×4 square grid of Berkeley Motes with a node distance of 30 cm, such that the true distances between pairs of nodes can be easily calculated. Each node emitted an ultrasound beacon, which allows all other nodes to estimate their distance to the transmitting node. The average error of the uncalibrated distance estimates is 74.6%. With the preceding calibration procedure, the average error could be reduced to 10.1%.

7.7.5.2 Collaborative In-Place Calibration
Collaborative in-place calibration (CIC) [37] supports internal calibration under a number of assumptions: the sensor nodes should be densely deployed, sensor orientation should have a negligible impact on the sensor output, spatial frequency of the observed physical quantity should be low, temporal frequency of the quantity should be high. Essentially, these assumptions ensure that collocated sensor nodes will see very similar stimuli $q(t)$ that change quickly over time. Additionally, it is assumed that q, h, c are real-valued scalar functions. Calibration functions c are assumed to be linear functions, although the method could also be adapted to nonlinear functions. The algorithm consists of two phases. In the first phase, pairwise calibration among collocated

nodes is performed. In the second phase, calibration among remote nodes is performed.

In the first phase, collocated pairs of nodes are calibrated against each other. Let us assume node 1 has to be calibrated against node 2. Then the goal is to derive a linear calibration function c^{12} with $c^{12}(h^1(t)) = h^2(t)$. First, both sensor nodes record sensor readings $h^i(t_x)$, such that both nodes read out the sensor concurrently at t_x, which requires clock synchronization. The sensor nodes exchange these readings, so that node 1 obtains a set of data points $(t_x, h^1(t_x), h^2(t_x))$. Since c^{12} is assumed to be linear, it can be derived by linear regression from this set of data points. However, even though sensor readout is synchronized, it is possible that the two sensors perceived different physical stimuli $q^1(t_x) \neq q^2(t_x)$ due to their different locations. Hence, such data points have to be eliminated before regression.

For this, confidence values are assigned to the data points, and the data points with low confidence values are eliminated. The confidence values are obtained by shifting a time window with a given size over the set of data points. For each window position, the linear correlation coefficient $r \in [-1, 1]$ for the contained data points is calculated. The larger r is, the better the data points fall on a line (i.e., the closer $q^1(t_x)$ and $q^2(t_x)$ are). If r is positive, then each data point in the window is further examined. If the data point contributes a positive addend to r, then the confidence of this data point is increased by r. The initial confidence of each data point is zero.

After this procedure, all data points with a confidence below a certain threshold value are eliminated. With the remaining set, linear regression is performed to derive a linear calibration function c^{12}. This function is also assigned a confidence that equals the linear correlation coefficient of the used data points.

Calibration functions for remote nodes are obtained by concatenating multiple calibration functions for collocated nodes. For example, for two remote nodes 1 and 3, the calibration functions c^{12} and c^{23} may be concatenated to give a calibration function $c^{13}(h^1(t)) = c^{23}(c^{12}(h^1(t)))$. However, there are typically many ways to construct a remote calibration function from many local function. Due to inconsistencies, it must be expected that for an alternate calibration function c'^{13}, we have $c^{13}(h^1(t)) \neq c'^{13}(h^1(t))$. Hence, the second phase of the algorithm computes more consistent nonlocal calibration functions $\hat{c}$.

The algorithm generates a new set of data points and uses linear regression to compute $\hat{c}^{ij}$. To obtain the data points, the algorithm enumerates all concatenated paths up to a specified maximum length. The kth path $\{c_k^{ij}\}$ is assigned a confidence r_k^{ij} by multiplying the confidences of the path segments. Using a set of random data values x_i, the data points are calculated as $(x_i, (1/N)\Sigma_k r_k^{ij} c_k^{ij}(x_l))$, where N is the number of concatenated paths c_k^{ij}.

The accuracy of the derived calibration functions for collocated sensors was measured in an experiment, where nine Berkeley Motes with temperature sensors were placed in a 3×3 square grid with a node distance of 5 cm. A slowly moving hair dryer was used as a heat source. About 70% of the pairwise calibration functions deviated by less than 5°C, while more than 10% were off by more than 10°C.

7.8 SUMMARY

In this chapter, we discussed various aspects of time synchronization and calibration in sensor networks. We outlined the applications of physical time and discussed why existing algorithms for time synchronization have to be revisited. We also presented common classes of and techniques for synchronization, reviewed time-synchronization algorithms from the literature, and discussed evaluation strategies. Time synchronization was identified as a special case of calibration, and many of the observations about time synchronization could be transferred to calibration.

While time synchronization for sensor networks is an established field of research, calibration has not received that much attention yet. However, we expect that calibration becomes a more active field as sensor networks move beyond the lab and small field experiments. Unfortunately, calibration is a much more general and complex problem than time synchronization. Hence, it is likely that research will first focus on more specific calibration problems. An interesting question is, can techniques developed for time synchronization be adapted to calibration problems?

The case studies of time synchronization algorithms and the discussion of evaluation techniques illustrated the very real problem of evaluating and comparing synchronization algorithms. Note that these difficulties also apply to calibration and many other distributed algorithms. Hence, one of the challenges for future research is the development of methods and tools for the evaluation of time synchronization and calibration in large-scale sensor networks.

Current application-oriented projects (e.g., ref. [8]) indicate that many simplifying assumptions about sensor networks (e.g., immobile nodes, fixed-network topology) may not hold in practice. Hence, future work might have to revisit existing approaches for time synchronization and calibration under updated assumptions.

REFERENCES

1. D. Ganesan, S. Ratnasamy, H. Wang, and D. Estrin. Coping with Irregular Spatio-Temporal Sampling in Sensor Networks. *Computer Communication Review*, **34**(1):125–130, 2004.

2. B. Liskov. Practical uses of synchronized clocks in distributed systems. In *Proceedings of the 10th Annual ACM Symposium on Principles of Distributed Computing (PODC '91)*, pages 1–10, August 1991.

3. W. Ye, J. Heidemann, and D. Estrin. An energy-efficient MAC protocol for wireless sensor networks. In *Proceedings of the 21st Annual Joint Conference of the IEEE Computer and Communications Societies (INFOCOM)*, Volume 3, pages 1567–1576, New York, June 2002.

4. K. Römer. Temporal message ordering in wireless sensor networks. In *Proceedings of the IFIP Mediterranean Workshop on Ad-Hoc Networks (Med-Woc-Net 2003)*, pages 131–142, Madhia, Tunisia, June 2003.

5. L. Girod, V. Bychkovskiy, J. Elson, and D. Estrin. Locating tiny sensors in time and space: A case study. In *Proceedings of the International Conference on Computer Design (ICCD 2002)*, Freiburg, Germany, September 2002.

6. D. L. Mills. Internet time synchronization: The network time protocol. *IEEE Transactions on Communications*, **39**(10):1482–1493, October 1991.

7. J. Elson and K. Römer. Wireless sensor networks: A new regime for time synchronization. In *Proceedings of the 1st Workshop on Hot Topics in Networks (HotNets-I)*, Princeton, New Jersey, October 2002.

8. P. Juang, H. Oki, Y. Wang, M. Martonosi, L. S. Peh, and D. Rubenstein. Energy-efficient computing for wildlife tracking: Design tradeoffs and early experiences with ZebraNet. In *Proceedings of the 10th International Conference on Architectural Support for Programming Languages and Operating Systems (ASPLOS-X)*, pages 96–107, San Jose, California, October 2002.

9. J. Elson, L. Girod, and D. Estrin. Fine-grained network time synchronization using reference broadcasts. In *Proceedings of the Fifth Symposium on Operating Systems Design and Implementation (OSDI 2002)*, Boston, Massachusetts, December 2002.

10. F. Cristian. Probabilistic clock synchronization. *Journal of Distributed Computing*, **3**:146–158, 1989.

11. D. Dolev, R. Reischuk, R. Strong, and E. Wimmers. *A Decentralized High Performance Time Service Architecture*. Technical Report 95/26, Institute for Computer Science, University of Lübeck, November 1995.

12. J. Y. Halpern and I. Suzuki. Clock synchronization and the power of broadcasting. *Distributed Computing*, **5**(2):73–82, 1991.

13. H. Dai and R. Han. Tsync: A lightweight bidirectional time synchronization service for wireless sensor networks. *Mobile Computing and Communications Review*, **8**(1):125–139, January 2004.

14. M. L. Sichitiu and C. Veerarittiphan. Simple, accurate time synchronization for wireless sensor networks. In *Proceedings of the IEEE Wireless Communications and Networking Conference (WCNC 2003)*, Volume 2, pages 1266–1273, New Orleans, Louisiana, March 2003.

15. W. H. Press, S. A. Teukolsky, W. T. Vetterli, and B. P. Flannery. *Numerical Recipes in C*, Second Edition. Cambridge University Press, 1992.

16. F. M. Gardner. *Phaselock Techniques*. John Wiley & Sons, 1979.

17. R. Noro. Synchronization over Packet-Switched Networks: Theory and Applications. Ph.D. thesis, EPFL, Lausanne, Switzerland, 2000.

18. D. L. Mills. Improved algorithms for synchronizing computer network clocks. *IEEE/ACM Transactions on Networks*, **3**(3):245–254, June 1995.

19. J.-M. Berthaud. Time synchronization over networks using convex closures. *IEEE/ACM Transactions on Networking*, **8**(2):265–277, 2000.

20. L. Zhang, Z. Liu, and C. Honghui Xia. Clock synchronization algorithms for network measurements. In *Proceedings of the 21st Annual Joint Conference of the IEEE Computer and Communications Societies (INFOCOM)*, Volume 1, pages 160–169, New York, June 2002.

21. P. Verissimo, L. Rodrigues, and A. Casimiro. Cesiumspray: A precise and accurate global time service for large-scale systems. *Real-Time Systems*, **3**(12):243–294, 1997.

22. S. Mitra and J. Rabek. Power efficient clustering for clock synchronization in dynamic multihop networks, unpublished. See at http://theory.lcs.mit.edu/mitras/courses/6829/project/final_report.ps, 2003.

23. S. Ganeriwal, R. Kumar, and M. B. Srivastava. Timing-sync protocol for sensor networks. In *Proceedings of the 1st International Conference on Embedded Networked Sensor Systems (SenSys)*, pages 138–149, November 2003.

24. J. van Greunen and J. Rabaey. Lightweight time synchronization for sensor networks. In *Proceedings of the 2nd ACM International Workshop on Wireless Sensor Networks and Applications (WSNA)*, pages 11–19, San Diego, California, September 2003.

25. P. Blum, L. Meier, and L. Thiele. Improved interval-based clock synchronization in sensor networks. In *Proceedings of the 3rd International Symposium on Information Processing in Sensor Networks (IPSN)*, pages 349–358, Berkeley, California, April 2004.

26. Q. Li and D. Rus. Global clock synchronization in sensor networks. In *Proceedings of IEEE INFO COM* 2004, Hong Kong, China, March 2004.

27. K. Römer. Time synchronization in ad hoc networks. In *Proceedings of the 2nd ACM Symposium on Mobile Ad Hoc Networking and Computing (MobiHoc)*, pages 173–182, Long Beach, California, October 2001.

28. K. Marzullo and S. Owicki. Maintaining the time in a distributed system. In *Proceedings of the 2nd Annual ACM Symposium on Principles of Distributed Computing*, pages 295–305, ACM Press, 1983.

29. U. Schmid and K. Schossmaier. Interval-based clock synchronization. *Real-Time Systems*, **12**(2):173–228, 1997.

30. L. Meier, P. Blum, and L. Thiele. Internal synchronization of drift-constraint clocks in ad-hoc sensor networks. In *Proceedings of the 5th ACM International Symposium on Mobile Ad Hoc Networking and Computing*, pages 90–97, Tokyo, Japan, May 2004.

31. M. Maroti, B. Kusy, G. Simon, and A. Ledeczi. The flooding time synchronization protocol. In *Proceedings of the 2nd ACN Conference on Embedded Networked Sensor Systems (SenSys)*, pages 39–49, Baltimore, Maryland, November 2004.

32. W. Su and I. F. Akyildiz. Time-diffusion synchronization protocol for sensor networks. *IEEE/ACM Transactions on Networking*, **13**(2): 384–397, 2005.

33. M. Mock, R. Frings, E. Nett, and S. Trikaliotis. Clock synchronization in wireless local area networks. In *Proceedings of the 12th Euromicro Conference on Real Time Systems*, pages 183–189, June 2000.

34. P. Blum and L. Thiele. Clock synchronization using packet streams. In *Brief Announcements of the 16th International Symposium on DIStributed Computing (DISC 2002)*, pages 1–8, Toulouse, France, October 2002.

35. K. Whitehouse and D. Culler. Calibration as Parameter Estimation in Sensor Networks. In *Proceedings of the 1st ACM International Workshop on Wireless Sensor Networks and Applications (WSNA)*, pages 59–67, Atlanta, Georgia, September 2002.

36. V. Bychkovskiy, S. Megerian, D. Estrin, and M. Potkonjak. A collaborative approach to in-place sensor calibration. In *Proceedings of the 2nd International Workshop on Information Processing in Sensor Networks (ISPN)*, pages 301–306, Palo Alto, California, April 2003.

CHAPTER 8

The Wireless Sensor Network MAC

EDGAR H. CALLAWAY, Jr.

Motorola Labs, Plantation, Florida

Wireless sensor networks usually employ performance metrics that are different from those of more conventional data networks, emphasizing low power consumption and low cost rather than data throughput or channel efficiency. Since power is consumed every time a networked device accesses the channel (either transmitting or receiving), the method by which the device accesses the channel can have a large effect on its power consumption, and therefore the valuation of the wireless sensor network as a whole. Should the channel-access method require specialized or additional hardware (e.g., a second transceiver, or more processor memory), the cost incurred can similarly affect network valuation. The open systems interconnection (OSI) stack places the responsibility for channel access in the medium-access control (MAC) sublayer of the data link layer, the second layer of the stack. This chapter will discuss the function of the MAC sublayer, identify some of the problems that are addressed (or avoided) in a good MAC design, and review the relationships between the performance goals of the network and tradeoffs that can be made in selecting and designing a MAC protocol. The chapter will then review the major categories of MAC protocols used in data networks of all types, describe a selection of MAC protocols designed for wireless sensor networks, and conclude with a look at directions for future research.

8.1 INTRODUCTION

Wireless sensor networks usually employ performance metrics that are different from those of more conventional data networks, emphasizing low-power consumption and low cost rather than data throughput or channel efficiency [1]. Since power is consumed every time a networked device accesses the channel (either transmitting or receiving), the method by which the device accesses the channel can have a large

Handbook of Sensor Networks: Algorithms and Architectures, Edited by Ivan Stojmenović
Copyright © 2005 John Wiley & Sons, Inc.

effect on its power consumption, and therefore the valuation of the wireless sensor network as a whole. Should the channel-access method require specialized or additional hardware (e.g., a second transceiver or more processor memory), the cost incurred can similarly affect network valuation. The open systems interconnection (OSI) stack [2] places the responsibility for channel access in the medium-access control (MAC) sublayer of the data-link layer, the second layer of the stack. This chapter discusses the function of the MAC sublayer, identify some of the problems that are addressed (or avoided) in a good MAC design, and review the relationships between the performance goals of the network and trade-offs that can be made in selecting and designing a MAC protocol. The chapter then reviews the major categories of MAC protocols used in data networks of all types, describes a selection of MAC protocols designed for wireless sensor networks, and concludes with a look at directions for future research.

8.1.1   Function of the MAC Layer

The wireless communication medium is a resource that must be shared by all network devices. Therefore, a scheme must be devised to provide access to it in some way that meets the needs of the network application. This problem is termed the MAC problem.

The communication medium has several dimensions that can be exploited for access control, including time, frequency, and coding. In addition, several physical parameters of the network devices themselves can also be employed, including spatial separation, antenna directionality, transmitter power output, and receiver sensitivity. Gummalla and Limb [3] provide a good survey of wireless MAC protocols in general. Murthy and Manoj [4] provide an encyclopedic collection of MAC protocols for ad hoc wireless networks.

For wireless sensor networks, the MAC problem assumes a collection of quasi-stationary network devices. Here we define "quasi-stationary" to mean that the network devices move slowly compared to the speed of network operation. Note that, unlike many other types of networks, the assumption that all network devices generate new frames at the same average rate is often not a good one for wireless sensor networks, which may have very asymmetric data-generation patterns. The transmissions of some devices, in fact, may consist entirely of acknowledgment frames (devices that are data sinks), while other devices may generate the majority of network traffic (devices that are data sources). This network heterogeneity can be used to advantage in MAC algorithms.

There is a subtle yet important distinction between *medium*-access protocols and *multiple*-access protocols. Multiple-access protocols solve the MAC problem, but in addition attempt to service multiple simultaneous communication links [5]. This, of course, requires multiple logical channels, so the MAC problem becomes one of optimally sharing multiple communication media in a way that meets the needs of the network application. Multiple-access protocols are therefore a subset of the set of medium-access protocols.

The structure of this chapter is as follows. The remainder of Section 8.1 discusses features of the MAC problem, emphasizing those of special importance to wireless sensor networks. Section 8.2 discusses several popular MAC methods employed today. Section 8.3 focuses this discussion on MAC methods proposed for use in wireless sensor networks. Section 8.4 concludes by considering some future directions for research.

8.1.2 Problems to Be Solved or Avoided by the MAC Layer

There are several important issues that arise during the solution of the MAC problem. Many are specific to the particular type of solution selected; however, a few are more generic and concern all MAC algorithms.

8.1.2.1 *Fairness*

Most MAC algorithms are designed to be fair, that is, to provide equal access to the channel for all network devices that desire to use it. In most applications it is undesirable to give some devices preferential treatment, allowing them better access to the channel than other devices. "Better access to the channel" in this context is usually defined as "faster access to the channel," although in general it may be best defined as "access to the channel leading to better quality of service," since the network application may value performance metrics other than data throughput. This situation may occur, for example, in wireless sensor networks that value long operational life over message latency. Such networks may elect to distribute channel access in a manner that will tend to equalize the remaining battery life of network devices, regardless of message latency; the goal of these schemes is to have the batteries of all network devices reach depletion at the same time, in a manner analogous to that of Oliver Wendell Holmes' "one-hoss shay" ("It ran a hundred years to the day") [6]. It is also important to consider that often a wireless sensor network performs a single application, as opposed to a wireless local area network (WLAN), which has many independent users performing independent applications competing for the available communication medium. Having collective or atomic network functionality can reduce the value of MAC fairness; it may not be important, for example, that an individual message is sent if others carrying identical information reach the same destination.

8.1.2.2 *Latency*

Message latency, the time it takes for a message to travel from an application on a source network device to an application on a destination device, is important in most wireless data networks; in some, message latency (or its variation) can be of primary importance (most multimedia applications fall into this category). MAC latency is a significant contributor to message latency in most networks; the design of the MAC therefore can have a significant effect on the suitability of the network for a proposed application. Increased MAC latency is frequently traded for reduced energy consumption in the design of wireless sensor network protocols [7].

8.1.2.3 Deadlock, Livelock, and Other Undesirable States

Like most algorithms, it is possible for MAC algorithms to exhibit unexpected (and therefore undesirable) behavior under certain conditions, behavior that usually takes the form of a drop in message throughput, often to zero. The algorithm reaches a state in which communication is not possible, and from which it cannot extract itself (at least in a given amount of time). The entry conditions to these states can be particularly difficult to identify since, from an ad hoc network perspective, the MAC algorithm is usually a distributed algorithm, and the variables involved may include particular combinations of the physical location of the network devices, the network topology, the offered message load, traffic history, and other factors difficult to model a priori.

As the name suggests, *MAC deadlock* is a logical condition that can arise in which a channel is available, network devices are operating properly, yet no attempt to communicate is made. Deadlock can arise, for example, if each device in a network is simultaneously expecting to receive a frame from another as a condition for future transmissions. Since none of them receives the frame it expects, all stay in a state of suspended animation, and communication halts.

A somewhat less common occurrence is *receive livelock*, in which there is so much MAC control message (message setup) overhead (perhaps from a large number of neighbors with much traffic to send) that the processing power of the individual devices is exceeded, and no data transfer is possible. This can occur in a so-called "broadcast storm," when a broadcast message is distributed in an ad hoc network without controlling the number of message retransmissions [8].

Another undesirable condition occurs in carrier sense multiple-access (CSMA) algorithms (see Subsection 8.2.3.2) when the binary exponential backoff exponent is allowed to increase without limit, a condition that forces network devices to self-censor themselves to silence. Several issues with CSMA algorithms employing the request-to-send/clear-to-send (RTS/CTS) exchange have been identified by Ray, Carruthers, and Starobinski [9].

Deadlock, livelock, and other logical cul-de-sacs are often quite subtle, and must be carefully considered with any MAC algorithm—along with the network topology, network device density, and other assumptions about the environment in which the MAC algorithm must perform. M.A. Youssef et al. suggest that analyses of a formal model for MAC algorithms are useful [10].

8.1.3 Important Factors in Wireless Sensor Network MAC Design

Due to the differing performance metrics applied to them, a MAC protocol suitable for wireless sensor networks is often significantly different than one designed for other applications, such as WLANs.

8.1.3.1 Emphasis on Power Consumption

As we have said, a primary concern of wireless sensor networks is power consumption. It is desirable to place the network devices in a low-power sleep mode as much as possible, to minimize average power consumption. This means that a MAC protocol that requires network devices to monitor the channel constantly would be a poor choice for wireless sensor networks, since their receivers would have to be constantly active and drawing current. (Due to their low transmitter output power, the receivers of many wireless sensor network devices dissipate more power than their transmitters, exacerbating this situation.) Any energy expended monitoring a silent channel, or listening to a network device that does not have a message to send, is wasted energy that could better be used for actual communication.

It is not just the active time that costs. Transceivers transitioning from a sleep state to an active state require a finite amount of time to lock synthesizers, regulate voltages, program registers, and a host of other tasks. These take time and consume energy, too, and if the MAC design requires frequent, short periods of activity, a significant amount of energy will be consumed just waking the transceiver up [1]. This wake-up problem can occur, for example, in a time-division multiple-access (TDMA) system (see Section 8.2.1.2) sending frequent control frames.

8.1.3.2 Deemphasis on Throughput, Message Latency, and Fairness
As previously noted, wireless sensor networks emphasize low power consumption and low cost over more traditional network performance metrics such as data throughput, message latency, and even fairness. This change in emphasis leads to the selection of different trade-offs during the design of the wireless sensor network MAC [7]. For example, as discussed before, the successful wireless sensor network MAC is unlikely to employ a scheme requiring constant monitoring of the channel. However, the decision to allow the network devices to sleep will almost certainly result in an increase in message latency, since messages must wait for devices to awaken before communication can be established. (The concept of a low-power wake-up radio has been proposed as a way to have the best of both worlds, i.e., the low message latency of an always-awake network device with low average power consumption [11–13]. Such a receiver would draw only a few tens of microwatts, and serve only to wake up the main receiver to receive the message.)

Fairness, to both network devices and messages, may also be traded for improved power consumption. In many cases, wireless sensor networks are designed with a tree topology, with a central device that is the primary data source, or primary data sink, in the network. This is often the case in environmental sensing, for example, where the purpose of the network is to send sensor data to a gateway device for collection. In this case, it has been proposed that the MAC be optimized for message transmission in the direction most of the messages are traveling, at the expense of those that may travel in other directions, for an improvement in energy efficiency and message latency [14]. It is also important to realize that there is often a philosophical difference between wireless sensor networks and more conventional data networks: Since wireless sensor networks often support a single application, the information from multiple sensors can be redundant. Redundant information can be delayed or even dropped from the network without affecting the common task of the network devices. MAC protocols can take advantage of this to reduce message fairness in certain situations, especially if a lower-cost or lower-power network results. Conventional data networks, on the other hand, more often function as "radio common carriers," supporting applications of all types. In this case, the loss of a single frame can be catastrophic, and link-level message fairness is of greater importance. In these cases, energy may be spent to enhance fairness.

8.1.3.3 *Low Channel Occupancy*

In the design of a MAC for wireless sensor networks, it should be kept in mind that the most common state of the network is that the channel is unused, that is, due to the low throughput of most wireless sensor networks, the majority of the time they are in operation there will be no traffic on the channel within range of a given device. For lowest energy consumption, therefore, it is incumbent on the designer to ensure that the most efficient state of the MAC is that in which no traffic is present.

8.1.3.4 *Self-Organization and Self-Maintenance*

Wireless sensor networks are typically designed to be ad hoc networks, installed by nonspecialists, and must be self-organizing and self-maintaining. That is, there is no system administrator available to identify and correct problems. The wireless sensor network MAC must therefore be utterly stable under a wide range of real-world conditions, including a wide array of network topologies and data generation patterns.

8.1.3.5 *Scalability*

Scalability is also an important factor in a wireless sensor network MAC. Since wireless sensor networks are self-organizing, and have an almost unlimited variety of applications, the MAC must be capable of operation in networks of both large and small order, and with a wide range of device densities.

8.1.3.6 *Quasi-Stationary Assumption*

It is often assumed that the devices in wireless sensor networks are quasi-stationary, that is, any movement they may make is slow relative to the speed at which the network may respond to such movement. This assumption is often employed to increase the sleep periods of the network devices, to improve overall energy efficiency, but it should be carefully considered in light of the proposed application. For example, if an asset tracking application is considered, it is important that the MAC be able to supply channel access fast enough for the asset to be tracked through the network; otherwise, since it will be unable to transmit a data frame, it will seem to have simply disappeared from the network once its motion begins.

8.1.3.7 *Use of Unlicensed Frequency Bands*

An often-overlooked issue with the wireless sensor network MAC is the fact that they are nearly always employed on unlicensed frequency bands—bands that are shared with other services. The fact that the channel is being shared with noncooperative devices performing other services can greatly complicate MAC design. For example, the MACs employed by the two services may interact in undesirable ways, producing deadlock in one or both of the services, or greatly reducing quality of service (QoS). This behavior may be hard to predict in advance, since there are many different services operating on the unlicensed bands, and many potential interactions between them.

Coexistence between services can be an undocumented requirement in many wireless sensor network applications, but it can be an extremely important one, especially if the wireless sensor network is not the first service in the band. Being second means that it is the responsibility of the wireless sensor network designer to coexist with the existing service. However, it is unlikely that the existing service has been designed to coexist with the incoming wireless sensor network, the existence of which the designers of the existing service may not have expected. The correct policy for the wireless sensor network to follow could perhaps be, "first, do no harm," but this philosophy can have unintended consequences.

For example, consider the use of CSMA on an unlicensed band. Should a device back off if it detects *any* signal during its channel-sensing period, or just a signal from a member of its network? If it desires to be a "good neighbor," perhaps the policy should be to back off upon detection of any signal energy, but this policy is unproductive if the other signal is leakage from a microwave oven on the 2.4-GHz industrial, scientific, medical (ISM) band; in fact, all that will be accomplished is that the device's own communication will be delayed. However, if a device is required to not only detect signals in the channel but characterize them as well, additional time must be taken to attempt to demodulate the signal, determine baud rate, and perhaps even decode a frame header. Simply detecting energy in a channel can be performed in a few microseconds, but characterizing a signal can take much longer—and waste significant energy.

8.2 COMMON MEDIUM-ACCESS CONTROL METHODS

The following is a brief review of some of the most popular medium-access control methods employed today. It is important to keep in mind that these methods may be combined in a single application; for example, a system may employ a polling protocol on a given channel, but have multiple frequencies available for use (frequency-division multiple access (FDMA)).

Numerous ways have been proposed to organize the various medium-access control methods; the taxonomy of Figure 8.1 is but one method. Classification of medium-access control methods is made difficult not only by the fact that schemes may be combined in various ways, but also because each method has many variations, so that even a precise definition of some methods can be elusive, and different practitioners may reasonably disagree in some areas. In addition, it should be noted that Figure 8.1 is not exhaustive; there are many other medium access control methods that have been proposed but are not included here. Among these are polarization-division multiple access (PDMA) [15] and space-division multiple access (SDMA) [16,17], in which orthogonally polarized and highly directive antennas, respectively, are employed to transmit to multiple receivers at the same time, on the same frequency. Despite these limitations, it is instructional to consider some broad classifications; Figure 8.1 divides medium-access protocols into fixed-assignment, demand-assignment, and contention-access protocol categories.

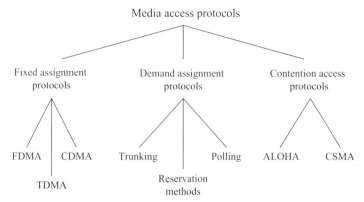

Figure 8.1   Medium-access protocol taxonomy. Fixed-assignment and demand-assignment protocols are often collectively referred to as "contention-free" protocols.

Fixed-assignment protocols are those for which, as the name implies, channel assignments are fixed, regardless of need. Demand-assignment protocols schedule channel access based on the demand of users having message traffic to transmit. Those without traffic are not given channel access. In both of these types, users transmitting messages are assured by the medium-access algorithm that their messages will not collide with messages from other devices in the network. In the third category, contention-access (also called random access) protocols, this assurance is not made; the protocols must include a recovery mechanism for message collisions.

8.2.1 Fixed-Assignment Protocols

As previously noted, fixed-assignment protocols make fixed-channel assignments, without consideration of any variation in communication needs that may exist among devices in the network. This makes fixed-assignment protocols often the easiest to implement, but also the most inflexible in response to changing network conditions. This last characteristic is often a large disadvantage for wireless sensor networks, due to their ad hoc, self-organizing, and self-maintaining nature, and fixed-assignment protocols are rarely proposed for them. However, most fixed-assignment protocols can be modified into demand-assignment variants, some of which have been proposed for wireless sensor networks, so it is important to understand their characteristics and limitations.

8.2.1.1 *Frequency-Division Multiple Access*   Possibly the oldest method of medium-access control is that of frequency-division multiple access (FDMA), which can trace its roots to Marconi's famous "four sevens" British patent of 1900 [18]. Marconi's patent disclosed the use of tuned circuits to enable multiple wireless stations to operate simultaneously without interference, by employing different frequencies of operation. Prior to this time, only untuned systems were in use and, as the range of wireless systems improved, interference was becoming a serious problem that threatened to limit the utility of the nascent wireless industry. Dividing the frequency spectrum into bands to be utilized by individual stations (much later to be assigned and enforced by government regulation) greatly reduced this problem, and FDMA was the primary multiple-access method used in all types of wireless services for many decades to follow. The concept of FDMA is shown in Figure 8.2.

Despite its popularity, FDMA is not without its weaknesses. Primary among these, like all fixed-assignment protocols, is the need to plan ahead for the "worst case" maximum-use scenario. Assigning frequencies for all potential users of the spectrum, when far fewer users will in fact be present most of the time, leads to poor spectrum utilization. Much of the time, a significant fraction of the available spectrum will be unoccupied. While this is a minor or nonexistent problem for some services, such as broadcasting, that occupy their assigned frequencies almost continuously, it is a significant problem for other services, such as public safety communications, that seldom require use of their frequencies (but must have them when required). Just being able to predetermine the maximum number of users may be impractical for some ad hoc systems, such as wireless sensor networks. Another weakness is the spacing between users (guard bands) that must be employed. A trade-off must be made between spectrum utilization efficiency (higher when the guard bands are small) and the cost of filtering needed to select the desired user (unfortunately, also higher when the guard bands are small). Guard bands may be reduced, and less-expensive filtering used, by physically spacing users so that the received adjacent-channel energy is attenuated; however, this limits the attainable density of networked devices, and may not be practical with mobile users.

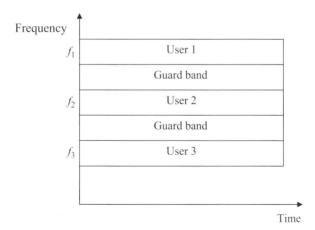

Figure 8.2 Frequency-division multiple access. A frequency band is reserved at all times for each user.

8.2.1.2 *Time-Division Multiple Access*

A second method of fixed-assignment multiple access is time-division multiple access (TDMA). In the basic TDMA scheme, shown in Figure 8.3, a single channel is time-shared, that is, use of the channel is divided among several users by allowing each user to access the channel periodically, but only for a small period of time (a "time slot"). After this time slot, the user must relinquish the channel to another user. Since the channel is only available to each user for a fraction of time, the raw (over-the-air) data rate used by each user must be proportionally higher to maintain a given throughput.

TDMA first came into practical use in satellite communication systems [19], an application for which FDMA has certain disadvantages, including the development of intermodulation products in the satellite transponder [20]. Since it employs only one channel, the use of TDMA eliminates the intermodulation problem in that application. It has since come in to wide use in cellular telephone systems, including Global System for Mobile Communications (GSM) [21], and the American National Standards Institute/Telecommunications Industry Association/Electronics Industry Association ANSI/TIA/EIA-136-B digital cellular standard [22].

Difficulties with TDMA largely center on the problem of synchronizing a number of independent users. Since the time-base references of the users are independent, the transmission of each user typically begins with a preamble that includes a bit synchronization pattern, followed by a "start of burst delimiter" (SBD), a synchronization code word indicating that data are to immediately follow. To account for the finite accuracy of synchronization, the possibility of clock drift during the slot, and differences in propagation delay between users, guard bands (periods of unassigned time) between slots are typically employed. These features are shown in Figure 8.4.

TDMA is frequently employed as a demand-assignment protocol, rather than a fixed-assignment protocol, by allowing users to request and receive multiple slots. As a demand-assignment protocol, it has been proposed as the underlying structure for a variety of wireless sensor network channel access methods [23,24]. Clare, Pottie, and Agre [25] note that the network time synchronization inherent in TDMA may also be used to synchronize sensor sampling and signal processing, enabling coherent beam formation of the sensors. The achievement and maintenance of time synchronization across a large multihop network, however, is quite difficult, especially if low-power operation is required. Due to the unavoidable communication delay across the network, it is possible for instabilities to develop; the

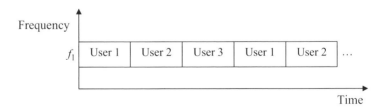

Figure 8.3 Time-division multiple access. A periodic time slot is reserved for each user on a single frequency.

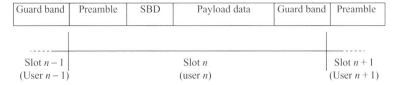

Guard band	Preamble	SBD	Payload data	Guard band	Preamble

Slot $n - 1$ Slot n Slot $n + 1$
(User $n - 1$)   (user n)   (User $n + 1$)

Figure 8.4 Typical structure of a TDMA slot (not to scale).

delays themselves vary, of course, due to the changing routes messages may take over the life of the network.

8.2.1.3 Code-Division Multiple Access

As a type of spread-spectrum communication, code-division multiple access (CDMA) is a more recent development than FDMA and TDMA, and relies on the observation that coding, as well as frequency and time, can be used to separate simultaneous transmissions in a frequency band and thereby achieve multiple access. If the codes are orthogonal, or nearly so, so that any bit errors caused by cochannel interference (interference from other users on the same frequency, but employing different codes) can be handled by forward error correction, multiple users may occupy the same band.

There are two fundamental types of spread-spectrum communication, frequency hopping and direct sequence, and each may be used for CDMA. Frequency hopping, as the name suggests, switches the carrier frequency of the modulated signal from channel to channel in the frequency band in a pseudorandom pattern (the "code"). The device may linger on a given channel for the duration of an entire message ("slow hopping"), or it may stay on a given channel only for the duration of a few symbols ("fast hopping"). In either case, multiple devices attempting to send message traffic on the network may do so simultaneously, since it is unlikely that they will be on the same carrier frequency at the same time. The sequence of transmission frequencies is, of course, known at both the transmitter and the receiver, but to other devices the sequence seems random. Direct sequence, on the other hand, keeps the carrier frequency constant, but instead multiplies the transmitted binary data by a predetermined, high-frequency, pseudorandom spreading code prior to modulation of the carrier. The elements of the spreading code are called *chips*, and there are typically 10 or more chips per transmitted bit of data. Since the rate at which chips are sent (the *chip rate*) is greater than the bit rate, the direct sequence signal occupies more bandwidth than the unspread signal. At the receiver, the received signal is multiplied by an identical spreading code, producing the recovered data. Multiple devices therefore may send message traffic on a given frequency simultaneously, using orthogonal spreading codes.

CDMA employing direct-sequence spread spectrum has been widely used in cellular telephone systems [26], including the ANSI/TIA/EIA-95-B [27] and TIA/EIA/IS-2000 (Interim Standard) [28] standards. Frequency-hopping CDMA has been less popular, although it has seen use in cordless telephone applications [29]. While the use of CDMA has greatly increased the system capacities of cellular telephone systems, it is not without its limitations; for example, in order to employ CDMA at a receiver the power received from the multiple transmitters must be substantially equal. This "near–far problem" creates a requirement for power control at the transmitters, so that their transmit powers vary inversely with respect to the path loss to the receiver. Not only does this requirement make the network devices more expensive and complicated, it makes CDMA useful only for networks having a single coordinator or base station, since it is impossible in general to solve the near–far problem for multiple receivers simultaneously.

8.2.2 Demand-Assignment Protocols

The economic viability of a wireless service is often a strong function of how efficiently it uses available spectrum [30]. Spectral efficiency, in turn, is greatly affected by the channel access method employed and, as we have said, fixed-assignment protocols are not particularly efficient. Demand-assignment protocols attempt to improve on the channel inefficiencies of fixed-assignment protocols, by reassigning unused channel assets to users that can use them. In general, this requires the use of a controller to arbitrate between users, and makes demand assignment protocols more complex than their fixed counterparts, since needy users and available channel assets must be matched. It also generates the need for a logical control channel, separate from the logical data channel over which messages are passed. In addition, the requirement to request channel access before actually having the channel assigned implies that there will be a setup delay between the time a user identifies a need to communicate and the time a channel is assigned to do so. This time, which is not present in fixed-assignment protocols, must be considered in the design of the protocol to ensure adequate QoS for the desired application.

8.2.2.1 Polling The most straightforward way to perform demand-based channel access is to have a controlling device in the network (a controller) repetitively ask all other network devices, one by one, if they need channel access. Devices that do not need channel access decline, while those that do need channel access inform the controller of that fact. The controller then assigns channel access for the requesting device. (In the degenerate case, devices merely begin transmitting upon receiving the query.) Polling is inherently fair, in the sense that the controller is able to ensure that all devices can have access to the channel with the same QoS.

Possibly the most well-known system that employs the polling technique is the Bluetooth™ (IEEE 802.15.1) Wireless Personal Area Network standard [31]. Bluetooth is ideally suited for a polling protocol, since it uses a star network topology, with one master device (the controller) and a maximum of seven slave devices. Time is divided into slots in the Bluetooth system; the master device can begin its transmissions only on odd-numbered slots, while slave devices can begin their transmissions only on even-numbered slots. Slave devices can only transmit in response to a query from the master device. Should a slave device have a message to transmit, it must wait until the master device polls it before it can transmit.

While useful in some applications, it is clear that polling is not appropriate for many types of networks. For example, star networks that have a large number of slaves are poor candidates for the polling technique, since an individual slave would be infrequently polled, leading to message transmission delays and poor quality of service. Further, if message generation is not evenly distributed among network devices, much network time will be spent uselessly polling devices with no message traffic, while devices with large amounts of traffic stand idly by—unless the polling device tracks message generation rates and modifies its polling pattern accordingly, a complex undertaking. Probably the biggest liability for the polling technique, however, is that it sets a power-consumption floor for devices in the network. Even if the network has no message traffic to exchange, the controller must continue to query each network device, and each network device must wake up, receive the query, and transmit a reply. The power penalty is worst for the controller, which must periodically contact all devices in the network, but is also a problem for the other network devices. This problem can be ameliorated somewhat by allowing the network device to miss a certain number of polls, in order to lower its power consumption. Bluetooth, in fact, has established three lower-power modes—HOLD, SNIFF, and PARK—to eliminate the polling overhead from network devices and allow them to reduce their average power consumption [32].

8.2.2.2 *Reservation Methods*

As the name suggests, reservation methods require a device to reserve a communication channel prior to transmission. A typical method is shown in Figure 8.5.

In a network of n devices, in which all devices can receive transmissions from all other devices, time is divided into superframes, and each superframe is then further divided into a reservation period and a data-transmission period. The reservation period is divided into frames, with one frame assigned to each device in the network. In a device's reservation frame, the device transmits a code word, indicating whether or not it has message traffic to send and, if it does, how many of data-transmission slots A, B, and C it needs. All other devices do the same in turn. At the end of the reservation period, all devices know which devices will be transmitting during the data-transmission period.

As just described, the protocol is an example of an unfair algorithm: There are a finite number of available data-transmission slots, and the devices request them in a preferred order. Device 0 will always find slots available, and will therefore always be able to transmit, while device n will find slots available only if all other network

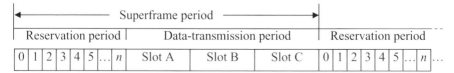

Figure 8.5 A reservation channel-access protocol. Network devices 0–7 reserve time in data transmission slots A, B, and C.

devices have left some for it. This unfairness may be acceptable (and even desirable) if the devices are organized by expected QoS; for example, if device 0 transmits emergency video for the fire department, and device *n* transmits, say, daily weather reports. However, if a fairer algorithm is desired, there are a number of ways this can be accomplished, for example, each device can include an aging value in its reservation code word, indicating how long its data has been waiting for transmission. Messages with the higher values can then be selected for transmission during the data-transmission period.

A second example of a reservation protocol is packet-reservation multiple access (PRMA) [33]. PRMA is an example of a combination MAC protocol, as it is a reservation protocol with features of TDMA and ALOHA (see Section 8.2.3.1). In PRMA, a star network is assumed. Time is divided into frames, each of which has many numbered slots. The network controller transmits an acknowledgment message at the end of each slot, which identifies that slot as being "reserved" or "unavailable." When a network device has message traffic (for the controller), it uses the ALOHA protocol to contend for an available slot. When the controller successfully receives the message, it replies with a "reserved" acknowledgment message, indicating receipt and indicating that the network device has reserved that slot for future frames. The network device now has an assigned slot in the frame, similar to TDMA, and can transmit without fear of frame collision in that slot in all future frames, since other network devices within range also detect the acknowledgment message. When the network device has completed its traffic, the slot reservation is released by the simple expedient of not transmitting in it. The network controller then transmits an "available" acknowledgment message at the end of the slot, informing the rest of the network.

8.2.2.3 *Trunking*

Trunking is a multiple-access scheme that dynamically assigns communication requests to available logical channels. Any fixed-assignment MAC protocol—FDMA, TDMA, CDMA—can be employed, with the goal to substantially improve the channel efficiency without causing the QoS to any user to degrade.

The earliest trunked systems were wired telephone systems, in which multiple lines between points were installed, and calls were routed to a line with available capacity. This greatly increased the reliability of the network, since a single line outage would be unlikely to result in a loss of service, and also improved infrastructure economy, since any peaks in call volume could be rerouted over other, less busy, lines, and every line did not have to be designed for the peak call volume requested over that route.

The first wireless systems to employ trunking methods (other than the microwave systems employed by the telephone system itself) were FDMA land-mobile radio (LMR) systems. These systems employed repeaters to provide communication links among mobile devices, normally organized into groups, and between mobile devices and wireline infrastructure such as telephone interconnect. Devices on LMR systems typically have very low average data throughput, but very high peak throughput—the worst-case scenario for channel efficiency using FDMA.

To install an FDMA trunking system, the system operator amasses a collection of 5 to 40 FDMA frequency pairs (inbound/outbound), using a repeater for each of them. One repeater is designated the control channel, while the others are used for message traffic. When not engaged in sending or receiving message traffic, network devices monitor the outbound control channel. When a device has a message for a particular group, it sends the request on the inbound control channel. The trunking system identifies an available repeater (channel) and transmits a command on the outbound control channel for all devices in the requested group to change to the available channel, where the requesting device transmits and the rest of the group receives [34]. Since a much larger number of users can be served with the existing spectrum allocation, this scheme greatly improves the economics of LMR. In the United States, FDMA trunking has since been expanded into the Associated Public Safety Communications Officials (APCO) Project 25 Advanced Narrowband Digital Communications (ANDC) standard [35].

Trunking principles can also be applied to TDMA systems. The TErrestrial Trunked RAdio (formerly the Trans-European Trunked RAdio) (TETRA) standard [36] employs TDMA with four slots per frame. The control frame is the last frame in a series of eighteen consecutive frames, called a TETRA "multiframe." Operation is analogous to FDMA trunked systems; mobile devices monitor the (outbound) control frame transmitted by the base station, and are assigned communication resources, in the form of identified slots in identified frames, to communicate. TETRA is designed to transmit both voice and data as separate services; the use of TDMA engenders great flexibility in channel access for this purpose, since voice services can be assigned frequent, repetitive slots, while data transfers can be assigned larger blocks of time and interrupted for the more latency-critical voice transmissions.

8.2.3 Contention-Access Protocols

As we have seen, demand-assignment protocols can improve the channel efficiency of fixed-assignment protocols. However, most demand-assignment schemes require the existence of an entity from which to demand an assignment, that is, a network controller. In many networks, for example, wireless sensor networks, such a controller does not exist. Multihop ad hoc networks, in which the network architecture (and even the order of the network) is not known a priori, are another difficult application for both fixed and demand assignment protocols. To make matters worse, many types of multihop ad hoc networks generate traffic patterns that have a low average message rate, but a high peak rate—as noted earlier, a difficult type of traffic pattern for a channel access protocol to support.

The solution to this dilemma is the third class of assignment protocols, contention (random)-access protocols. In these protocols, devices contend (compete) among each other for channel access; devices that lose access to the channel merely try again later. Since frame collisions are not prohibited by contention-access protocols, a method for detecting collisions (or at least determining a posteriori that they must have occurred) and recovering from them must be included in the protocol.

8.2.3.1 **ALOHA** The ALOHA channel-access protocol [37] is generally considered the first channel-access protocol for wireless digital communications to employ random access. The ALOHA communication system was part of a wireless time-sharing system used to connect a mainframe computer near Honolulu with remote users on other Hawaiian islands. The network therefore had a star topology; it was assigned two radio channels, one for inbound traffic and one for outbound traffic. All traffic was in the form of fixed-length frames, each 704 bits long (including identification, control, payload, and parity bits). Since there was only one source of outbound traffic, the mainframe computer, there was no medium-access control issue on that channel. In the other direction, however, a method was needed to assign the single inbound channel among the multiple remote users. This was a nontrivial problem, especially since there was no guarantee that the remote users could hear each other.

The ALOHA channel-access scheme that was developed to address this issue is elegant in its simplicity, and operates as follows: Remote devices simply transmit each frame as soon as it is generated, on the inbound channel. The transmissions are completely asynchronous and independent of those that may (or may not) be transmitted by other remote devices. If and only if a frame is received without error at the mainframe, an acknowledgment frame is sent to the remote device via the outbound channel. If a remote device does not receive the acknowledgment frame within a time-out period, it waits a random length of time and then retransmits the entire frame.

Assuming that message generation follows Poisson statistics, it can be shown that the ALOHA system becomes unstable (i.e., the number of retransmissions grows without bound) when the fraction of time the channel is utilized exceeds $1/(2e) \approx 0.184$. This relatively low value is the major drawback to ALOHA channel access, and modifying ALOHA to increase its channel utilization (and therefore channel capacity) has been of much research interest. One of the first approaches taken was to realize that a single transmitted frame could collide with *two* frames—one starting before it and one starting after it had started—and that the frame collision rate could be halved, and therefore the channel capacity doubled, by quantizing time into slots, then synchronizing the remote users so that frame transmission could only start at the beginning of each slot. Frames would not be transmitted as soon as they were generated, as in ALOHA, but would be held in queue until the beginning of the next slot. This became known as "slotted ALOHA" [38], and has a capacity of $1/e \approx 0.368$ (see Fig. 8.6).

Slotted or nonslotted, the ALOHA protocol is quite simple, and is often used as part of more complex medium-access methods (e.g., PRMA, described in Section 8.2.2.2). One advantage is in asymmetrically powered star networks, in which the controller is mains powered but the remote devices have limited power resources, and in which communication is initiated solely by the remote devices. In these applications, using ALOHA, the initiating device(s) may stay asleep until a message is generated, then wake up, transmit the message, receive the acknowledgment, and return to sleep. The responding device must keep its receiver constantly active, but the initiating device may have an extremely low duty cycle, since it need only

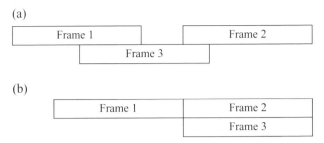

Figure 8.6 (a) Nonslotted ALOHA: Frame 3 collides with frames 1 and 2. All three are lost. (b) Slotted ALOHA: Frame 3 collides only with frame 2. Frame 1 survives.

be active for transmission and reception of a single packet. An application fitting this description is that of wireless light switches, in which the lamp itself is mains powered and the switches are battery powered (or powered by the toggling of the switch itself [39]). The power consumption of a receiver is trivial compared to that of the lamp itself, so it remains on at all times; however, the battery-powered switches benefit from the low duty cycle of the ALOHA protocol.

8.2.3.2 *CSMA* Carrier sense multiple-access (CSMA) algorithms attempt to improve upon the relatively poor channel capacity of ALOHA, by requiring that a device attempting to transmit first sense (monitor) the channel for any ongoing activity prior to transmission [40]. There are many variations of CSMA, and they have become popular for wireless personal area networks (WPANs), largely because of their distributed nature (no controller is needed), a requirement for ad hoc networks, and for their overall adequate performance.

The fundamental principle of CSMA—that one should check to see that the channel is idle prior to transmission—is an old idea, and probably an anthropomorphic one. A natural part of human conversation is to pause before speaking, to ensure that the recipient is not already engaged. Early manual radiotelegraphic networks operated in a similar fashion, even going so far as to employ what is now called the request-to-send/clear-to-send (RTS/CTS) protocol, to minimize the effect of simultaneous transmissions [41–43]. Its use in packet data networks, however, is relatively recent; Kleinrock and Tobagi [ref. 40, p. 1401 fn] give credit to D. Wax of the University of Hawaii, in an internal memorandum dated March 4, 1971. In a manner analogous to the ALOHA protocols, time in CSMA protocols may be considered to be continuous (*unslotted* CSMA) or broken into discrete intervals (*slotted* CSMA).

CSMA protocols can be further divided into two types, the *nonpersistent* CSMA protocol and a number of *persistent* CSMA protocols. In nonpersistent CSMA, a network device with a message to transmit operates as follows:

- The device senses the channel.
- If the channel is idle, the message is transmitted immediately.

- If the channel is busy, the device waits a random period of time (the "backoff period"), senses the channel again, and repeats the process.

The protocol attempts to ensure that frame collisions do not occur by sensing the channel, then waiting until sometime later if the channel is busy. Nonpersistent CSMA is simple, and is used in several popular network standards, including the IEEE 802.15.4 low-rate wireless personal area network (LR-WPAN) standard [44]. However, in applications that value message throughput it is less than optimal, because it is possible that the channel may become idle during the backoff time, when the device is not monitoring the channel. Waiting until the backoff period expires before attempting retransmission is therefore a waste of the channel resource.

As a first attempt to overcome this weakness, one can consider a modification, the so-called *1-persistent* protocol. 1-Persistent CSMA operates as follows:

- The device senses the channel.
- If the channel is idle, the message is transmitted immediately.
- If the channel is busy, the device continues to sense the channel (it is *persistent*). When the channel becomes idle, the device immediately transmits (with probability one, hence the name *1-persistent*).

The goal of 1-persistent CSMA is to make maximum use of the channel, by avoiding the "dead air" during the backoff period of nonpersistent CSMA. However, consider the situation of two network devices generating a message at the same time, and finding the channel busy. Under 1-persistent CSMA the two devices will wait until the channel is idle, then both transmit simultaneously—with disastrous results!

One method to avoid this undesired behavior, while still improving channel efficiency over nonpersistent CSMA, is, upon sensing an idle channel, to transmit the message with some probability p, where $p < 1$. This variant, *p-persistent* CSMA, operates as follows:

- The device senses the channel.
- If the channel is idle, the message is transmitted with probability p. With probability $(1 - p)$, the device waits a fixed time (a single slot in slotted CSMA, a predetermined time in unslotted CSMA). At the end of this new time, the device senses the channel again, and repeats the process.
- If the channel is busy, the device continues to sense the channel. When the channel becomes idle, the device proceeds as just outlined.

The optimal value of p for maximum throughput depends on the offered traffic rate (and the propagation delay, usually neglected in wireless sensor networks); for a detailed analysis, the interested reader is referred to Kleinrock and Tobagi [40].

A drawback to all CSMA protocols is the so-called hidden and exposed terminal problems. Consider the linear network of Figure 8.7. Devices A and C are each

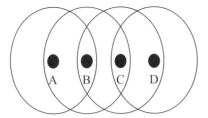

Figure 8.7   The hidden- and exposed-terminal problems. The circles indicate the communication range of each device.

within range of device B, but they are not in range of each other. Suppose device C generates a message for device B while device A is transmitting to device B. Using CSMA, device C will sense the channel, but find it idle—even though device A is transmitting, it is out of range of device C. Device C will then begin to transmit (perhaps after a few backoff periods if employing p-persistent CSMA), causing a frame collision at device B. Device A is a "hidden terminal" to device C; it is in range of a device that device C desires to contact, but is out of range of device C itself. The hidden-terminal problem leads to a reduction in network throughput.

Now consider a second scenario. Device B transmits a frame to device A. Device C generates a message for device D. Device C will sense the channel, but find it occupied by device B. Since it finds the channel busy, device C will delay transmission, even though, were it to transmit, it would not cause interference at either device B (because it is transmitting) or device D (because it is out of range of device B). This is the "exposed terminal" problem for CSMA; it, too, leads to a reduction in network throughput, since it prohibits transmission by devices that would otherwise be able to safely do so. Much of the research into CSMA protocols has been directed at ways to eliminate or reduce the severity of the hidden- and exposed-terminal problems [45].

The first solution proposed for the hidden-terminal problem was the use of busy tones [46]. This solution rests on the realization that the hidden-terminal problem, and frame collisions in general, occur at the receiving device, while the CSMA algorithm is being performed at the transmitting device. The busy-tone solution requires each network device receiving a frame to simultaneously transmit a "busy tone" on another signaling channel, indicating that its receiver is busy. Devices desiring to transmit are required to check for the presence of busy tones prior to transmission. If present, they delay transmission, since the channel (at the receiving device, where it matters) is busy.

An implementation difficulty with the use of busy tones is the need for simultaneous transmission and reception (duplex operation). While busy tones have been proposed as part of other medium-access protocols for wireless sensor networks (e.g., power aware multiaccess protocol with signaling (PAMAS), in Section 8.3.2; see also Haas, Deng, and Tabrizi [47]), duplex operation greatly increases the complexity, cost, and power consumption of the network devices, and so alternative solutions are frequently desired.

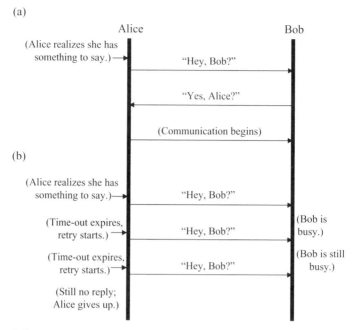

Figure 8.8   Anthropomorphic view of CSMA/CA. (a) Successful. (b) Unsuccessful.

One way to attack these problems is to continue the anthropomorphic analogy a step further (see Figure 8.8), by recognizing that something common in the initial stages of a human conversation is a request to speak ("Hey, Bob?"), followed by a grant of permission to speak ("Yes, Alice?"). This short exchange warns Bob that Alice is attempting to speak, and ensures Alice that Bob is ready to receive her statement. In particular, it confirms to Alice that Bob is not listening to someone else speak—perhaps someone that Alice cannot hear. Should Bob not reply to Alice, after a moment's delay Alice may retry, repeating until the affirmative reply is received or Alice gives up. Should Alice hear either a request to speak from someone else, Carol ("Hey, Bob?"), or a grant of permission to speak from Bob to someone else ("Yes, Carol?"), prior to her own request to speak, Alice will wait a moment for that conversation to complete before speaking (see Figure 8.9).

An analogous procedure can be performed on wireless data networks; this procedure is called carrier sense multiple-access with collision avoidance (CSMA/CA). In CSMA/CA, after sensing that the channel is idle, the initiating device first sends a short RTS frame to the responding device. If the responding device is, in fact, idle, it returns a CTS frame to the initiating device, indicating that it can begin transmission of its queued data frame. If the initiating device does not receive a CTS frame after a predetermined period of time, it waits a further random time (to avoid collisions with a potential competing device on the channel in the same state), then retransmits the RTS frame.

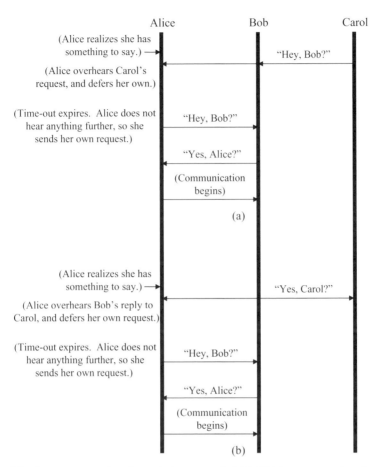

Figure 8.9 Request to speak and permission to speak. (a) Inhibition upon receipt of a request to speak. (b) Inhibition upon receipt of permission to speak.

When network devices overhear RTS and CTS frames transmitted by others, they are prohibited from transmitting for a period of time—long enough for other frames to be successfully communicated. (This is the "collision avoidance" part of CSMA/CA.) It is worth noting, however, that not all collisions are avoided in CSMA/CA. For example, simultaneous RTS transmissions by two devices, both of which detected an empty channel, are still possible. The recovery mechanism is the random backoff they employ after they do not receive the expected CTS frame in reply. Although never zero, the probability of RTS and other possible types of frame collisions is made small by making the RTS and CTS frames very short.

Another influential single-channel solution to the hidden-terminal problem of CSMA was MACA [48]. MACA solves these problems, and avoids the implementation complexity associated with channel sensing, by not sensing the channel at all. (The name "MACA" is derived from "CSMA/CA," by deleting the "CS.") Rather, it relies on the effect RTS and CTS frames have on eavesdropping devices.

MACA is based on the insight that, if a data field is placed in RTS and CTS frames indicating the amount of data that is queued to be transmitted at the initiator, any device that can hear either the RTS or CTS has sufficient information to avoid frame collisions. An idle network device using MACA constantly monitors the channel for RTS or CTS frames. If it receives an RTS frame addressed to itself, it replies with a CTS frame and communication begins. If, however, it receives an RTS frame addressed to another network device, it inhibits all potential transmissions of its own for the period of time needed for the device sending the RTS frame to (1) receive a CTS frame, and (2) send its data frame. This time is known, since the length of the data frame is included in the RTS frame. Similarly, if it receives a CTS frame, it inhibits all potential transmissions of its own, for the period of time needed for the device sending the CTS frame to receive its data frame. This behavior is shown in Figure 8.10.

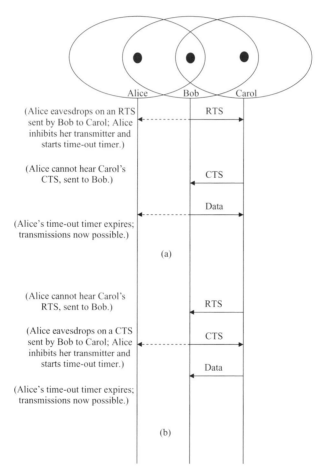

Figure 8.10 MACA operation. (a) Overheard RTS. (b) Overheard CTS. Carol is a hidden device to Alice.

MACA was very influential, and led to many variants, including MACAW [49], floor-acquisition multiple-access (FAMA) [45], and the medium-access control methods used by WLAN standards, such as IEEE 802.11 [50]. Most variants attempted to address identified weaknesses in MACA, such as the still-nonzero probability of frame collisions [51] and its backoff algorithm. MACA proposed the use of a simple binary exponential backoff, in which the backoff time is doubled after every collision and returned to the minimal value after a successful RTS/CTS exchange. This algorithm was shown [49] to be unfair, in that over time one network device would "win" the channel and have a low backoff value (with frequent channel access), while all remaining network devices would have very large backoff values and be effectively frozen out of the network. MACA's backoff problem has been addressed in a number of ways; MACAW, for example, shares backoff values between network devices.

One issue with CSMA schemes when applied in wireless sensor networks is the active time of the receiver [52]. All persistent CSMA schemes require the receiver to be active for relatively long periods to sense the channel, spending a considerable amount of energy while receiving nothing but noise. This produces an average power-consumption floor for the network device, even if it does not transmit or receive a single frame. Wireless sensor networks are particularly sensitive to this, since not only is power consumption a primary performance metric but data throughput is typically low, making the CSMA channel monitoring all the more inefficient. The IEEE 802.15.4 LR-WPAN standard, which employs a CSMA algorithm for channel access, attempts a compromise for beaconing networks by offering an optional battery-life extension (BLE) mode. In BLE mode, the number of slots following each beacon available for slotted CSMA channel-access is greatly restricted. This improves device battery life, since devices can return to sleep quickly, but also greatly limits channel capacity.

8.3 SOME MAC METHODS PROPOSED FOR WIRELESS SENSOR NETWORKS

This section reviews a number of MACs proposed for wireless sensor networks. This collection is by no means exhaustive, and is meant only to convey to the reader the wide variety of designs possible, and how the assumptions and priorities of the designers affected their final designs.

8.3.1 Self-Organizing Medium-Access Control for Sensor Networks and Eavesdrop-and-Register Protocols

Katayoun Sohrabi et al. proposed the self-organizing medium-access control for sensor networks (SMACS) and eavesdrop-and-register (EAR) protocols in 1999 [23,53]. SMACS is a distributed protocol that incorporates features of FDMA, TDMA, and CDMA. It is an excellent example of a wireless sensor network MAC that trades something usually dear in data networks, bandwidth, for increased energy efficiency. EAR expands the utility of the wireless sensor network by enabling quasi-stationary network devices to support roaming devices. EAR allows roaming devices to connect to the network without the large energy cost usually associated with the tracking of mobile devices.

As shown in Figure 8.11, devices under SMACS begin their operation by monitoring a previously determined common frequency X for a random period of time. At the end of its random monitoring period, a device transmits a Type 1 message to establish a communication link if it has not heard any Type 1 messages from other network devices. In Figure 8.11, this occurs with device B. A Type 1 message is an invitation for other devices in range to establish a joint communication link. Devices A and C receive the Type 1 message and, after waiting for a random backoff period, each replies with a Type 2 message. Device B receives both messages (assuming they did not collide), and selects one with which to form a communication link. It may make this selection on received signal strength, order in which the messages were received, or upon some information included in the Type 2 messages, such as the number of attached devices. Device B then sends a Type 3 message at the conclusion of its receiving period, to notify all devices in range of its decision. In this example, device C was selected; device A, which was not selected, then turns off its receiver for a random period of time before restarting the procedure with another receiving period.

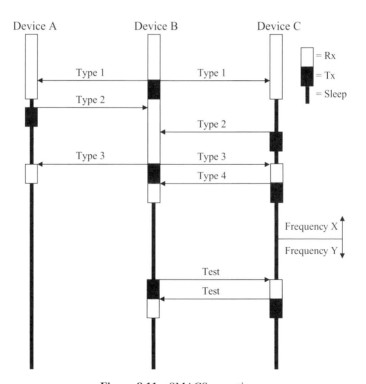

Figure 8.11   SMACS operation.

In addition to the identity of the selected device, the Type 3 message also includes timing information. SMACS and EAR employ a superframe of length T_{frame}; however, unlike conventional TDMA, the superframes are not synchronous between devices. Rather, the phase of their superframes are independent (i.e., the superframe of each device begins at an independent time). To establish the communication link, device B sends device C in its Type 3 message its schedule of existing links in its superframe, plus the time until the start of its next superframe. Device C receives the Type 3 message, and compares the link schedule of device B with its own device schedule, taking into consideration the differing times at which their superframes start. It then identifies two (ideally sequential) time slots that are available to both devices. It transmits the location of these time slots, together with a proposed frequency for the new communication link (frequency Y), in a final Type 4 message. Device B receives the Type 4 message and, at the appropriate time, moves to frequency Y and exchanges a pair of test messages with device C to ensure that the wireless link on that frequency is in fact open.

Devices B and C now have a communication link established with each other. This link establishment process can be repeated a number of times, creating a dense network among many devices. After the network is formed, as part of network maintenance the devices periodically transmit a broadcast invitation (BI) message (a beacon) to announce their presence to other network devices, encouraging them to form connections. These invitation messages need not be sent every T_{frame}, but may be sent on some multiple of T_{frame}, to trade off connection formation latency with energy efficiency. Following the BI messages there is a small period (collection of slots) during which the device monitors the frequency for the replies of any invitees.

Since coordination is not performed with all devices within range of devices B and C to establish a noninterfering pair of time slots for this communication link, if the SMACS procedure were performed in an ad hoc TDMA network on a single frequency, frame collisions with neighboring devices would be unavoidable. By employing multiple frequencies for its communication links, SMACS employs a distributed form of dynamic FDMA to avoid these otherwise inevitable frame collisions.

Sohrabi et al. [53] note that SMACS can be generalized to define the communication links to be specific frequency-hopping patterns, rather than fixed frequencies. This extension moves SMACS from a TDMA/FDMA hybrid to a TDMA/CDMA hybrid, and offers the advantage of protection against channel degradations, such as multipath flat fading and the presence of fixed interfering signals, at the cost of increased complexity. Since SMACS already requires a frequency-agile transceiver, the additional complexity is due largely to the increased complexity of the protocol, which must identify, select, and synchronize the orthogonal hopping patterns.

EAR extends SMACS for use with mobile devices. The assumptions made by the EAR protocol are that there are only a few mobile devices in a randomly distributed, much larger collection of stationary devices, and that, as before, energy consumption, rather than connectivity, is of primary importance.

A mobile device begins the EAR algorithm by searching (monitoring) for BI messages sent by stationary devices. It may receive several; if so, it selects one based on its signal quality, identification, or other features. It then replies to the selected BI message with a mobile invite (MI) message, requesting a connection to the selected stationary device. If it elects to do so, the stationary device then accepts the MI request by the transmission of a mobile response (MR) message that includes the suggested slots for communication. Later, as the signal quality of the connected stationary device falls below an acceptable threshold, a mobile disconnect (MD) message is sent by the mobile device to the stationary device, informing of the disconnection.

No acknowledgments are sent in the EAR algorithm, and the stationary devices need send only one specialized message (the MR message). Instead of acknowledgments, time-outs are used to reduce state misunderstandings between devices. This simplifies the protocol and speeds connection establishment.

Since it establishes reserved, periodic communication links among neighboring devices, SMACS is a good choice for wireless sensor networks supporting multimedia sensor applications like real-time video security systems. Message latency in its TDMA-like structure should have much less variance than if, say, a CSMA channel-access mechanism were employed, with its random backoff periods prior to transmissions; this meets a critical need of most multimedia applications. A weakness of SMACS is the relatively high duty cycle imposed by its TDMA-like structure. Since separate slots are reserved for communication with each device in range, the duty cycle degrades as the network density (or communication range of the device) increases. While CSMA protocols, by comparison, monitor the channel for messages from all neighboring devices at once, performing a parallel operation, SMACS reserves separate receive slots for each device individually, performing the function in series. This increases energy expenditure accordingly in a network that is lightly loaded. In addition, energy is expended during each transceiver warm-up period prior to each communication slot; energy consumed during warm-up can have a significant effect on overall energy efficiency [54,55].

8.3.2 PAMAS

As previously noted, power consumption of wireless sensor network devices is of critical importance, and energy should be conserved whenever possible. As it happens, the power consumption of the receivers typically used in such networks often approaches or even exceeds the power consumption of the transmitters (due to their low output power), and if receivers are operated indiscriminately their energy use can represent a significant fraction of the total energy used by network devices. It is therefore productive to evaluate techniques to minimize the use of receivers.

Consider the network of four devices shown in Figure 8.12. In this network, device B is transmitting a frame to device A. However, device C is also within range of device B, and it overhears device B's transmission. The key insight is that reception of device B's transmission to device A represents a waste of energy by device C: Device B's transmission is not for it, and it cannot receive any other transmissions (for example, from device D) during device B's transmission, since the channel is occupied. To save energy, device C should then turn off its receiver

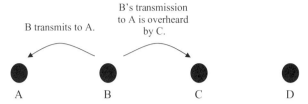

Figure 8.12 Unnecessary reception in an ad hoc network.

and go to sleep for the duration of device B's transmission. When one considers that many wireless sensor networks are quite dense, with many receiving devices within range of a single transmitting device, it becomes apparent that, networkwide, significant energy can be saved.

This receiver power-off technique was applied in an early MAC protocol for multihop ad hoc networks called PAMAS [51]. PAMAS is derived from the MACA CSMA protocol (see Section 8.2.3.2), with several interesting modifications. Principal among these is the use of a separate signaling channel: The RTS and CTS messages in PAMAS are transmitted on the signaling channel, rather than on the data channel, as done in MACA. To prevent collisions of RTS and CTS messages sent by hidden devices (rare but possible in MACA), PAMAS includes the use of receiver busy tones, also sent on the signaling channel. If a network device is receiving a message on the data channel, and receives an RTS message on the signaling channel, it responds with a busy tone on the signaling channel, with a length twice that of a CTS message. Should the neighboring target device of the RTS send a CTS message, it would collide with the busy tone and appear to the requestor as noise. The requesting device then performs the MACA binary exponential backoff and retries the RTS at a later time.

The receiver power-off feature of PAMAS is employed in a network device when either of the following two conditions exists:

1. The transmit message queue (messages generated but not yet sent) is empty and a neighbor begins transmitting on the data channel.
2. The transmit message queue is not empty, but at least one neighbor is transmitting a data message (detected by monitoring the data channel) and one is receiving (detected by monitoring the signaling channel for the busy tone sent at the start of each received frame). In this case, the device goes to sleep because it is unable to transmit or receive a message.

An interesting feature of PAMAS's receiver power-off scheme is that, unlike many other energy-saving MAC techniques, message latency is not affected. To see why, suppose device C in Figure 8.12 generates a message for device B while device A is transmitting to device B. Device C is assumed to be out of the range of device A; it is a "hidden device" (see Section 8.2.3.2) to device A. Since device C detects the receive busy tone sent by device B, it goes to sleep; however, it would be unable to transmit to device B in any case, since device B is already occupied. Conversely, suppose device C again has a message for device B, but this time device B is transmitting to device A. Again, device C goes to sleep; also again, it would be unable to transmit its message anyway, since device B is occupied. Hence, sleeping does not affect the latency of the message.

The fundamental difficulty with PAMAS is one of implementation: The requirement for a second (control) channel, in addition to the data channel, adds significantly to the cost of the network device, since a second wireless transceiver, plus duplexer, is required.

8.3.3 Sensor-MAC

Sensor-MAC (S-MAC—not to be confused with SMAC, discussed in Section 8.3.1), was designed to address the following sources of energy waste in wireless sensor networks [56,57]:

- *Frame Collisions*   When frames collide, they must be retransmitted, at additional energy cost.
- *Overhearing*   When a network device receives a frame destined for another device, it is a waste of energy, since it could have been sleeping instead. This was an insight gleaned from PAMAS (Section 8.3.2).
- *Control Frame Overhead*   Since they require an expenditure of energy, but do not directly result in the communication of information, the transmission and reception of control packets represents a waste of energy.
- *Idle Listening* (*fruitless channel monitoring*)   Monitoring the channel for the possible reception of messages that are not, in fact, sent represents another type of wasted energy. As noted in Section 8.2.3.2, this type of wasted energy is especially problematic in wireless sensor networks employing CSMA protocols, since their devices monitor the channel during their CSMA contention-based channel-access periods, yet network data throughput on such networks is low.

S-MAC trades some message fairness and latency for reduced power consumption by network devices. It assumes that the wireless sensor network is composed of a large number of devices, that it employs multihop routing, and that message destinations will be uniformly distributed throughout the network (i.e., there is no single gateway device acting as a data sink). It also assumes that, as discussed in Section 8.1.3.2, message-level fairness can be sacrificed as long as application-level fairness is maintained, and that message latency on the order of a few seconds is tolerable.

To reduce energy consumption, S-MAC limits device reception and transmission to periodic active periods, interspersed by sleep periods. A complete active-sleep cycle is called a *frame*, not to be confused with a data frame (i.e., data packet) as defined in other services. (To avoid confusion, in the rest of this section a data "frame" will be termed a "packet.") Frames are synchronized with neighboring devices by periodically broadcasting SYNC packets, which are very short, and contain the address of the sender and the relative time at which it will return to sleep. From these values, recipient devices can construct a table of their neighbors' schedules, so that future synchronization to exchange messages is possible. As the multihop network forms, differing clusters of devices may synchronize differently; devices bordering several asynchronous clusters may elect to either synchronize to all neighboring clusters (waking up in all of their active periods and thereby suffering an energy expenditure penalty) or select and follow one cluster, recalling the timing of the other(s) should that connectivity be needed.

As shown in Figure 8.13(a), unless transmissions are needed, the active portion is spent constantly receiving. The active period is divided into two parts, one for SYNC packets and one for data packets; the entire active period is also divided into a large number of time slots. Slotted CSMA with RTS/CTS is used to access the channel; each of the synchronization and data portions of the active period therefore can be considered to be a separate contention access period.

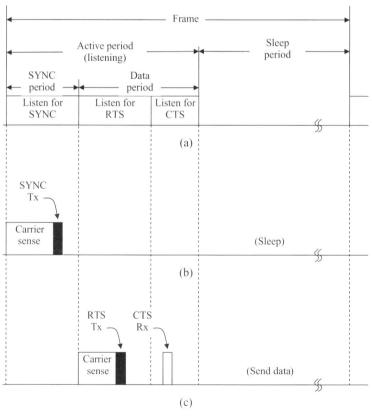

Figure 8.13 S-MAC frame. (a) Reception. (b) Transmission of a SYNC packet. (c) Transmission of a data packet.

When a device has no pending message traffic, it monitors the SYNC period for possible SYNC transmissions from its neighbors, and then monitors the RTS portion of the data period for possible message traffic for it. Following the RTS portion, it then listens to the CTS portion—even though it has not transmitted an RTS of its own—for reasons to be explained shortly. When a device is to transmit a SYNC packet, Figure 8.13(b), it senses the channel for a random period of time during the SYNC portion of the active period. After this time, if nothing is heard, it transmits the SYNC packet. When a device is to transmit a data packet, Figure 8.13(c), it senses the channel for a random period of time during the RTS portion of the data period. After this time, if nothing is heard, it transmits its RTS packet, and then awaits the CTS reply packet in the corresponding portion of the data period. The data packet itself is transmitted at the beginning of what would otherwise be the sleep period. This is deemed acceptable, since data packets are assumed to be relatively rare events in a wireless sensor network and so do not significantly affect overall energy consumption.

S-MAC incorporates an intriguing technique to limit message latency while achieving significant energy efficiency, called *adaptive listening*. Adaptive listening is based on the observation that overhearing a neighbor's transmissions can give a device early warning that a message may be arriving shortly. Adaptive listening requires a device that overhears a neighbor's transmission of an RTS or CTS packet to wake up for a short period of time at the end of the transmitted data packet (the packet length is included in RTS and CTS frames). If the listening device is in fact the next hop for the data packet, it may receive it from its neighbor immediately, rather than at the next scheduled active time. (The use of active listening is why all devices monitor the CTS period even if they have not sent RTS packets.) If it is not the next hop for the data packet, it merely goes to sleep and returns to its regularly scheduled operation. Adaptive listening enables two communication links to be traversed per frame, rather than just one.

A final feature of S-MAC is the use of message passing. Consider the transmission of a relatively large amount of data. They could be transmitted in a single packet but, since wireless links have a nonzero bit error rate, it is likely that a small number of bit errors would occur in such a long packet. The packet likely would have to be retransmitted, at considerable energy cost, and perhaps more than once. On the other hand, the data could be fragmented into many smaller, independent packets, and sent via conventional CSMA. One then incurs a significant control overhead, as many RTS/CTS pairs would be sent—one per packet—even though it is not desirable to lose the channel to other devices before the data transmission is complete and the transmission of so many control packets is a waste of energy.

Message passing is the transmission of a series of small, related packets, each fully acknowledged after their transmission, with only one RTS/CTS exchange at the beginning of the sequence. (In S-MAC, the packet length field in the RTS and CTS packets is modified in message passing to indicate the length of the transmission of the entire series of packets, including acknowledgment times.) Message passing can occupy a channel, "freezing out" devices with single packets to send, and is therefore inherently unfair at the communication link level; however, as discussed in Section 8.1.3.2, in wireless sensor networks it is often more important to prioritize operation of the network as a whole, rather than individual messages.

S-MAC is a very promising MAC for wireless sensor networks. An area for future evaluation is its performance in dense networks, where asynchronous frames from neighboring device clusters may interfere with data transmissions. It would also be interesting to investigate its performance in networks employing a gateway device, where the assumption of evenly distributed message destinations does not apply.

8.3.4 The IEEE 802.15.4/ZigBee™ MAC

The Institute of Electrical and Electronics Engineers (IEEE) 802.15.4 low-rate wireless personal area network (LR-WPAN) standard [44,58] is the first open standard designed for wireless sensor networks. Promoted by the ZigBee™ Alliance [59], an industry consortium, it is a flexible standard suitable for many network topologies and wireless sensor applications, and includes many features designed to enable low power consumption and low-cost implementation.

The IEEE 802.15.4 MAC supports both beaconing and nonbeaconing modes. The nonbeaconing mode is especially useful for star networks in which there is one central device that may be mains powered, with essentially unlimited power resources available, surrounded by other network devices that may be battery-powered, with more austere power budgets. The pedagogical example is that of a wireless light switch controlling a lamp. The lamp, attached to the mains, can monitor the channel constantly, while the switch remains idle unless it is toggled, when it transmits this information to the lamp. Since it is rarely active, the switch may have an almost unlimited battery life. The nonbeaconing mode also supports multihop networks, in which a collection of always-active relaying devices transport messages perhaps generated by another collection of very-low-duty cycle devices.

In the nonbeaconing mode, IEEE 802.15.4 specifies the use of unslotted, nonpersistent CSMA. Due to the small size of IEEE 802.15.4 frames (the maximum size, including physical-layer preamble, is 133 bytes), an RTS/CTS exchange is not used. To avoid the state of perpetual backoff, in which the CSMA backoff exponent grows without bound, the number of backoffs that can be performed is limited. When the maximum value is reached, a channel-access failure report is generated by the MAC and sent to an upper layer of the communication stack.

The IEEE 802.15.4 standard also supports an optional superframe structure, incorporating beacons. As shown in Figure 8.14, two exponent parameters, the superframe order (SO) and beacon order (BO), with SO $\leq$ BO, define the superframe. The length of the active portion of a superframe is 15.36 ms $\times 2^{SO}$, while the time between beacon starts is 15.36 ms $\times 2^{BO}$. (Numerical timing values given in this section are for the 2.4-GHz band physical layer. There is another physical layer specified for operation below 1 GHz for which the logical operation is the same, but the numerical timing values are different.) When SO < BO, an inactive period exists prior to the next beacon, and may be used to sleep. The maximum value of

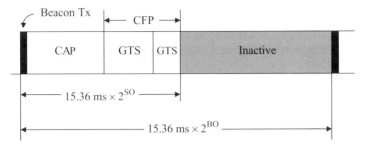

Figure 8.14 The IEEE 802.15.4 superframe. Time values shown are for the 2.4-GHz physical layer. *Abbreviations*: CAP = contention access period. CFP = contention-free period. GTS = guaranteed time slot.

SO and BO in beacon mode is 14, defining a beacon period of 251.65824 seconds, or more than 4 minutes. Application and network designers can use these parameters to trade off message latency, channel capacity, and battery life.

A beacon is transmitted at the start of a superframe. Following the beacon, the rest of the active portion is a CSMA contention-access period (CAP). In star networks, however, a portion of the CAP can be reserved for specific devices, to guarantee them access to the channel. These reservations are called *guaranteed time slots* (GTSs), and a maximum of seven of them, of varying length, can be used to avoid the message latency jitter associated with the CSMA process. This is useful, for example, in wireless game controllers and mice. When no GTSs have been allocated, there is no contention-free period (CFP) and the entire active portion of the superframe (excluding the beacon transmission) is the CAP.

The active portion of the superframe is divided into 16 slots. Each slot is further divided into three backoff periods. When SO = 0, each slot is 960 μs long and each backoff period 320 μs long.

A device attempting to contact a beaconing device begins by synchronizing with its beacon. It then performs a slotted, nonpersistent CSMA algorithm, again without an RTS/CTS exchange, to gain access to the channel. The algorithm uses the backoff periods as the slotted structure, rather than the superframe slots, to speed the algorithm. To account for the nonzero receive-to-transmit turnaround time in practical hardware, the algorithm must find the channel clear during the first 128 μs of two consecutive backoff periods before it declares the channel idle. If only a single sampling of the channel were performed, it would be possible to sample the channel during the turnaround time of an exchange between neighboring devices, determine that the channel was idle, and then produce a frame collision upon transmission.

For some applications, even the minimum length of the CAP (15.36 ms, less the beacon transmission time) is much longer than required for the low activity of the network. Keeping the receiver active is, as previously noted, a weakness of CSMA, and receiver activity should be minimized wherever possible to optimize energy efficiency. To this end, the IEEE 802.15.4 standard incorporates a BLE mode. To employ the BLE mode, the beaconing device sets a BLE flag in its beacon. It then limits its monitoring of the CAP to only six backoff periods. If it hears no activity by the end of this time, it will return to sleep. When listening devices detect the BLE flag, devices attempting to contact the beaconing device set the initial value of their CSMA backoff exponent to a value of two or less. While use of the BLE mode greatly increases the likelihood of frame collisions due to the much shorter channel-sensing period, for low-activity networks it can greatly reduce network device duty cycle: Employing BLE, with BO = 14, a device can have a total duty cycle (inclusive of transmit, receive, and warm-up periods) of less than 50 parts per million.

In addition to ZigBee, the IEEE 802.15.4 standard has been proposed for use in a number of networks, including the neuRFon[TM] netform [60]. Independent performance evaluations of the standard are becoming available; Lu, Krishnamachari, and Raghavendra [61], for example, point out the significant trade-offs made between energy saving and message latency.

8.4 FUTURE DIRECTIONS

MAC development for wireless sensor networks has really just begun. Since there are a wide variety of network applications (from multimedia distribution to the transmission of daily weather reports), many different network topologies, and many performance metrics from which to choose, the wireless sensor network MAC is of much research interest.

The standard methods of TDMA and CSMA are subject to refurbishment when old assumptions are reevaluated. For example, the assumption of random traffic patterns is probably not realistic; many types of traffic occur in bursts in so-called event-driven applications. Recent work [62,63] has shown that existing wireless sensor network MAC designs can be improved by adapting them to this reality.

The assumption of omnidirectional communication, in which the range of a given device is the same in all directions, can also be modified, with the use of directional antennas. By improving spatial reuse, MAC algorithms employing these antennas can enable improved QoS, specifically throughput and reduced message latency [64,65]. While these parameters are usually not of primary interest in wireless sensor network applications, it is interesting to speculate on the performance of a MAC protocol that combines the use of directional antennas with the use of power control [66,67]. Such a protocol may be very energy-efficient.

In addition to minimizing energy expenditures while maximizing QoS (however it is defined for the network in question), there are a few other areas that deserve investigation. One is the study of how different MACs perform when they are placed in the same channel, as often happens in unlicensed wireless bands. Is it possible to establish some global rules for MAC operation that can aid coexistence between services competing for the same channel? Is it possible to predict, without a special-purpose, event-driven simulator, the performance of two (or more) coexisting services?

Continuing further, there is the issue of cognitive radio, or radio systems that dynamically adapt their behavior to their existing electromagnetic environment.

Is it possible to conceive of a wireless sensor network employing cognitive radio techniques? It would seem that the channel sensing needed for effective cognitive radio operation would be incompatible with the energy-expenditure requirements of wireless sensor networks, but perhaps this problem awaits only a sufficiently clever researcher for a solution.

REFERENCES

1. Anantha Chandrakasan et al. Design considerations for distributed microsensor systems. In *Proceedings of the IEEE Custom Integrated Circuits Conference*, pages 279–286, May 1999.

2. Hubert Zimmermann. OSI reference model—the ISO model of architecture for open systems interconnection. *IEEE Transactions on Communications*, **COM-28**(4):425–432, April 1980.

3. Ajay Chandra V. Gummalla and John O. Limb. Wireless medium-access control protocols. *IEEE Communications Surveys*, **3**(2):2–15, 2000.

4. C. Siva Ram Murthy and B. S. Manoj. Chapter 6 in *Ad Hoc Wireless Networks–Architectures and Protocols*. Prentice Hall, 2004.

5. Dimitri Bertsekas and Robert Gallager. *Data Networks*, Second Edition. Prentice Hall, 1992.

6. Henry Petroski. *To Engineer Is Human: The Role of Failure in Successful Design*, pages 35–39. St. Martin's Press, 1985.

7. Woo Chool Park et al. Trade-off energy and delay between MAC protocols for wireless sensor networks. In *Proceedings of the 6th International Conference on Advanced Communication Technology*, Volume 1, pages 157–160, Phoenix Park, Republic of Korea, February 2004.

8. Sze-Yao Ni et al. The broadcast storm problem in a mobile ad hoc network. In *Proceedings of the 5th Annual ACM/IEEE International Conference on Mobile Computing and Networking (MobiCom)*, pages 151–162, Seattle, Washington, August 1999.

9. Saikat Ray, Jeffrey B. Carruthers, and David Starobinski. RTS/CTS-induced congestion in ad hoc wireless LANs. In *Proceedings of the IEEE Wireless Communications and Networking Conference (WCNC 2003)*, Volume 3, pages 1516–1521, New Orleans, Louisiana, March 2003.

10. Moustafa A. Youssef, Arunchandar Vasan, and Raymond E. Miller. Specification and analysis of the DCF and PCF protocols in the 802.11 standard using systems of communicating machines. In *Proceedings of the 10th IEEE International Conference on Network Protocols*, pages 132–141, Paris, France, 2002.

11. Frazer Bennett et al. Piconet: Embedded mobile networking. *IEEE Personal Communications*, **4**(5):8–15, October 1997.

12. Chunlong Guo, Lizhi (Charlie) Zhong, and Jan M. Rabaey. Low power distributed MAC for ad hoc sensor radio networks. In *Proceedings of the IEEE Global Telecoms Conference*, Volume 5, pages 2944–2948, 2001.

13. Jan M. Rabaey et al. Picoradios for wireless sensor networks: The next challenge in ultra-low power design. In *IEEE International Solid State Circuits Conference Digest of Technical Papers*, Volume 1, pages 200–202; Volume 2, pages 156–157, 444–445, 2002.

14. Gang Lu, Bhaskar Krishnamachari, and Cauligi S. Raghavendra. An adaptive energy-efficient and low-latency MAC for data gathering in wireless sensor networks. In *Proceedings of the 18th International Parallel and Distributed Processing Symposium*, pages 224–231, 2004.

15. Bernard Sklar. A structured overview of digital communications—A tutorial review—Part II. *IEEE Communications Magazine*, **21**(7):6–21, October 1983.

16. King-Tim Ko and Bruce R. Davis. A space-division multiple-access protocol for spot-beam antenna and satellite-switched communication network. *IEEE Journal Selected Areas in Communications*, **SAC-1**(1):126–132, January 1983.

17. Soheila V. Bana and Pravin Varaiya. Space division multiple access (SDMA) for robust ad hoc vehicle communication networks. In *Proceedings of the IEEE Intelligent Transportation Systems Conference*, pages 962–967, 2001.

18. Guglielmo Marconi. Improvements in Apparatus for Wireless Telegraphy. British patent 7777. The Patent Office, Newport, South Wales, April 26, 1900.

19. Tadahiro Sekimoto and John G. Puente. A satellite time-division multiple-access experiment. *IEEE Transactions on Communications Technology*, **COM-16**(4):581–588, August 1968.

20. Adel A. M. Saleh. Intermodulation analysis of FDMA satellite systems employing compensated and uncompensated TWTs. *IEEE Transactions on Communications*, **COM-30**(5):1233–1242, May 1982.

21. GSM Technical Specifications. European Telecommunication Standards Institute (ETSI), Sophia Antipolis, France. For a tractable entrance to GSM, see also Siegmund M. Redl, Matthias K. Weber, and Malcolm W. Oliphant, *An Introduction to GSM*, Artech House, 1995.

22. TDMA Third Generation Wireless, Rev B (ANSI/TIA/EIA-136-B-99), Telecommunications Industry Association, Arlington, Virginia, 1999.

23. Katayoun Sohrabi et al. A self organizing wireless sensor network. In *Proceedings of the 37th Annual Allerton Conference on Communication, Control, and Computing*, pages 1201–1210, 1999.

24. Saurabh Mishra and Asis Nasipuri. An adaptive low power reservation based MAC protocol for wireless sensor networks. In *Proceedings of the IEEE International Conference on Performance Computing and Communications*, pages 731–736, 2004.

25. Loren P. Clare, Gregory J. Pottie, and Jonathan R. Agre. Self-organizing distributed sensor networks. In *Proceedings of the SPIE Conference on Unattended Ground Sensor Technologies and Applications*, Volume 3713, pages 229–237, 1999.

26. William C. Y. Lee. Overview of cellular CDMA. *IEEE Transactions on Vehicular Technology*, **40**(2):291–302, May 1991.

27. Mobile Station–Base Station Compatibility Standard for Wideband Spread Spectrum Cellular Systems (ANSI/TIA/EIA-95-B-99), Telecommunications Industry Association, Arlington, Virginia, 1999. For a good reference for CDMA system design, see also Jhong Sam Lee and Leonard E. Miller, *CDMA Systems Engineering Handbook*, Artech House, 1998.

28. CDMA 2000® Series, Release A (2000), Telecommunications Industry Association, Arlington, Virginia, 2000.

29. André Noll Barreto, Jürgen Deißner, and Gerhard Fettweis. A frequency hopping algorithm for cordless telephone systems. In *Proceedings of the IEEE International Conference on Universal Personal Communications*, Volume 2, pages 1273–1277, 1998.

30. Richard N. Lane. Spectral and economic efficiencies of land mobile radio systems. *IEEE Transactions on Communications*, **COM-21**(11):1177–1187, November 1973.

31. Institute of Electrical and Electronics Engineers, Inc. *IEEE Standard for Information Technology—Telecommunications and Information Exchange between Systems—Local and Metropolitan Area Networks—Specific requirements—Part 15.1: Wireless Medium access Control (MAC) and Physical Layer (PHY) Specifications for Wireless Personal Area Networks (WPANs)*, IEEE Std 802.15.1-2002. IEEE Press, 2002.

32. Jaap C. Haartsen and Sven Mattisson. Bluetooth—A new low-power radio interface providing short-range connectivity. In *Proceedings of the IEEE*, **88**(10):1651–1661, October 2000.

33. D. J. Goodman et al. Packet reservation multiple-access for local wireless communications. *IEEE Transactions on Communications*, **37**(8):885–890, August 1989.

34. Arthur Chrapkowski and Gary Grube. Mobile trunked radio system design and simulation. In *Proceedings of the IEEE Vehicular Technology Conference*, pages 245–250, 1991.

35. Gregory M. Stone and Karen Bluitt. Advance digital communications system design considerations for law enforcement and internal security purposes. In *Proceedings of the IEEE 29th Annual International Carnahan Conference on Security Technology*, pages 402–408, 1995.

36. European Telecommunication Standards Institute. *Terrestrial Trunked Radio (TETRA); Voice plus Data (V + D); Part 2: Air Interface (AI)*, Document ETSI EN 300 392-2 V2.4.2 (2004-02). European Telecommunication Standards Institute, 2004.

37. N. Abramson. The ALOHA system—Another alternative for computer communications. In *Proceedings of the AFIPS Fall Joint Computer Conference*, Volume 37, pages 281–285, 1970.

38. Lawrence G. Roberts. *ALOHA Packet System With and Without Slots and Capture*, ARPANET Satellite System Note 8 (NIC Document 11290). ARPA Network Information Center, Stanford Research Institute, Menlo Park, California, June 26, 1972. Reprinted in *Computer Communications Review*, **5**(2):28–42, April 1978.

39. Fast wie ein Perpetuum Mobile (Almost like perpetual motion). Markt & Technik, **47**:45–47, November 15, 2002. See at http://www.elektroniknet.de.

40. Leonard Kleinrock and Fouad A. Tobagi. Packet switching in radio channels: Part I—Carrier sense multiple-access modes and their throughput-delay characteristics. *IEEE Transactions on Communications*, **COM-23**(12):1400–1416, December 1975.

41. William Walker. How a C.W. traffic net operates. *QST*, **36**(4):48–49, 128, 130, April 1952.

42. George Hart. Message handling. In *The ARRL Operating Manual*, Second Edition, Robert Halprin (ed.), Chapter 4. American Radio Relay League, Newington, Connecticut, 1985.

43. Edgar H. Callaway, Jr. *Wireless Sensor Networks: Architectures and Protocols*, page 28. CRC Press, Boca Raton, Florida, 2004.

44. Institute of Electrical and Electronics Engineers, Inc. *IEEE Standard for Information Technology—Telecommunications and Information Exchange between Systems—Local and Metropolitan Area Networks—Specific requirements—Part 15.4: Wireless Medium Access Control (MAC) and Physical Layer (PHY) Specifications for Low Rate Wireless Personal Area Networks (WPANs)*, IEEE Std 802.15.4-2003, IEEE Press, 2003.

45. Chane L. Fullmer and J. J. Garcia-Luna-Aceves. Solutions to hidden terminal problems in wireless networks. *ACM SIGCOMM Computer Communications Review (Proceedings of the ACM SIGCOMM '97 Conference on Applications, Technologies, Architectures, and Protocols for Computer Communications)*, **27**(4):39–49, October 1997.

46. Fouad A. Tobagi and Leonard Kleinrock. Packet switching in radio channels—Part II: The hidden terminal problem in carrier sense multiple-access and the busy tone solution. *IEEE Transactions on Communications*, **COM-23**(12):1417–1433, December 1975.

47. Zygmunt J. Haas, Jing Deng, and Siamak Tabrizi. Collision-free medium access control scheme for ad-hoc networks. In *Proceedings of the IEEE Military Communications Conference*, Volume 1, pages 276–280, 1999.

48. Phil Karn. MACA—A new channel-access method for packet radio. In *Proceedings of the ARRL/CRRL Amateur Radio 9th Computer Networking Conference*, pages 134–140, 1990.

49. Vaduvur Bharghavan et al. MACAW. *ACM SIGCOMM Computer Communication Review (Proceedings of the Conference on Communications Architectures, Protocols and Applications)*, **24**(4):212–225, October 1994.

50. Institute of Electrical and Electronics Engineers, Inc. *IEEE Standard for Information Technology—Telecommunications and Information Exchange Between Systems—Local and Metropolitan Area Networks—Specific Requirements—Part 11: Wireless LAN Medium Access Control (MAC) and Physical Layer (PHY) Specifications*, IEEE Std 802.11-1999 (ISO/IEC 8802-11: 1999). IEEE Press, 1999.

51. Suresh Singh and C. S. Raghavendra. PAMAS—Power Aware Multi-Access protocol with Signalling for ad hoc networks. *ACM Sigcomm Computer Communication Review*, **28**(3):5–26, July 1998.

52. Alec Woo and David E. Culler. A transmission control scheme for media access in sensor networks. In *Proceedings of the 7th Annual Conference on Mobile Computing and Networking*, pages 221–235, 2001.

53. Katayoun Sohrabi et al. Protocols for self-organization of a wireless sensor network. *IEEE Personal Communications*, **7**(5):16–27, 2000.

54. Eugene Shih et al. Physical layer driven protocol and algorithm design for energy-efficient wireless sensor networks. In *Proceedings of the 7th Annual ACM/IEEE International Conference on Mobil Computing and Networking (MobiCom)*, pages 272–287, Rome, Italy, July 2001.

55. Andrew Y. Wang et al. Energy efficient modulation and MAC for asymmetric RF micro-sensor systems. In *Proceedings of the IEEE International Symposium on Low Power Electronics and Design*, pages 106–111, Huntington Beach, California, 2001.

56. Wei Ye, John Heidemann, and Deborah Estrin. *An Energy-Efficient MAC Protocol for Wireless Sensor Networks*. USC/ISI Technical Report ISI-TR-543, September 2001.

57. Wei Ye, John Heidemann, and Deborah Estrin. Medium access control with coordinated adaptive sleeping for wireless sensor networks. *IEEE/ACM Transactions on Networking*, **12**(3):493–506, June 2004.

58. Jose A. Gutierrez, Edgar H. Callaway, Jr., and Raymond L. Barrett, Jr. *Low-Rate Wireless Personal Area Networks ... Enabling Wireless Sensors with IEEE 802.15.4^{TM}*. IEEE Press, 2003.

59. See at http://www.zigbee.org.

60. L. Hester et al. neuRFon™ Netform: A self-organizing wireless sensor network. In *Proceedings of the 11th International Computer Communication and Networks Conference*, pages 364–369, 2002.

61. Gang Lu, Bhaskar Krishnamachari, and Cauligi S. Raghavendra. Performance evaluation of the IEEE 802.15.4 MAC for low-rate low-power wireless networks. In *Proceedings of the IEEE International Conference on Performance, Computing, and Communications*, pages 701–706, 2004.

62. Y. C. Tay, Kyle Jamieson, and Hari Balakrishnan. Collision-minimizing CSMA and its applications to wireless sensor networks. *IEEE Journal on Selected Areas in Communications*, **22**(6):1048–1057, August 2004.

63. Jing Li and Georgios Y. Lazarou. A bit-map-assisted energy-efficient MAC scheme for wireless sensor networks. In *Proceedings of the 3rd International Symposium on Information Processing in Sensor Networks (IPSN)*, pages 55–60, Berkeley, California, April 2004.

64. Young-Bae Ko, Vinaychandra Shankarkumar, and Nitin H. Vaidya. Medium access control protocols using directional antennas in ad hoc networks. In *Proceedings of the 19th Annual Joint Conference of the IEEE Computer and Communications Societies (INFOCOM)*, Volume 1, pages 13–21, Tel-Aviv, Israel, March 2000.

65. Yu Wang and J. J. Garcia-Luna-Aceves. Collision avoidance in single-channel ad hoc networks using directional antennas. In *Proceedings of the 23rd International Conference on Distributed Computing Systems (ICDCS)*, pages 640–649, May 2003.

66. Shu-Lin Wu, Yu-Chee Tseng, and Jang-Ping Sheu. Intelligent medium access for mobile ad hoc networks with busy tones and power control. In *Proceedings of the 8th International Conference on Computer Communications and Networks*, pages 71–76, 1999.

67. Maciej Zawodniok and Sarangapani Jagannathan. A distributed power control MAC protocol for wireless ad hoc networks. In *Proceedings of the IEEE Wireless Communications and Networking Conference (WCNC 2004)*, Volume 3, pages 1915–1920, Atlanta, Georgia, March 2004.

▄▄▄▄▄▄▄ CHAPTER 9

Localization in Sensor Networks

JONATHAN BACHRACH and CHRISTOPHER TAYLOR

Massachusetts Institute of Technology, Cambridge, Massachusetts

Location, Location, Location

— Anonymous

In emerging sensor network applications it is necessary to accurately orient the nodes with respect to a global coordinate system in order to report data that is geographically meaningful. Furthermore, basic middle ware services such as routing often rely on location information (e.g., geographic routing). Application contexts and potential massive scale make it unrealistic to rely on careful placement or uniform arrangement of sensors. Rather than use globally accessible beacons or expensive GPS to localize each sensor, we would like the sensors to self-organize a coordinate system. This chapter reviews localization hardware, discusses issues in localization algorithm design, present the most important localization techniques, and finally suggests future directions in localization. The goal of this chapter is to outline the technical foundations of today's localization techniques and present the tradeoffs inherent in algorithm design. No specific algorithm is a clear favorite across the spectrum. For example, some algorithms rely on pre-positioned nodes while others are able to do without. Other algorithms require expensive hardware capabilities. Some algorithms need a way of performing off-line computation, while other algorithms are able to do all their calculations on the sensor nodes themselves. Localization is still a new and exciting field, with new algorithms, hardware, and applications being developed at a feverish pace; it is hard to say what techniques and hardware will be prevalent in the end.

9.1 INTRODUCTION

Advances in technology have made it possible to build ad hoc sensor networks using inexpensive nodes consisting of a low-power processor, a modest amount of

Handbook of Sensor Networks: Algorithms and Architectures, Edited by Ivan Stojmenović memory, a wireless network transceiver, and a sensor board; a typical node is comparable in size to two AA batteries [1]. Many novel applications are emerging: habitat monitoring, smart building failure detection and reporting, and target tracking. In these applications it is necessary to accurately orient the nodes with respect to a global coordinate system in order to report data that are geographically meaningful. Furthermore, basic middleware services such as routing often rely on location information (e.g., geographic routing).

Ad hoc sensor networks present novel trade-offs in system design. On the one hand, the low cost of the nodes facilitates massive scale and highly parallel computation. On the other hand, each node is likely to have limited power, limited reliability, and only local communication with a modest number of neighbors. These application contexts and potential massive scale make it unrealistic to rely on careful placement or uniform arrangement of sensors. Rather than use globally accessible beacons or expensive global positioning systems (GPSs) to localize each sensor, we would like the sensors to self-organize a coordinate system.

In this chapter, we review localization hardware, discuss issues in localization algorithm design, present the most important localization techniques, and, finally, suggest future directions in localization. The goal of this chapter is to outline the technical foundations of today's localization techniques and present the trade-offs inherent in algorithm design. No specific algorithm is a clear favorite across the spectrum. For example, some algorithms rely on prepositioned nodes (Subsection 9.2.1) while others are able to do without. Other algorithms require expensive hardware capabilities. Some algorithms need a way of performing off-line computation, while other algorithms are able to do all their calculations on the sensor nodes themselves. Localization is still a new and exciting field, with new algorithms, hardware, and applications being developed at a feverish pace; it is hard to say what techniques and hardware will be prevalent in the end.

9.2 LOCALIZATION HARDWARE

The localization problem gives rise to two important hardware problems. The first, the problem of defining a coordinate system, is covered in Subsection 9.2.1. The second, which is the more technically challenging, is the problem of calculating the distance between sensors (the ranging problem), which is covered in the balance of this section.

9.2.1 Anchor/Beacon Nodes

The goal of localization is to determine the physical coordinates of a group of sensor nodes. These coordinates can be global, meaning they are aligned with some externally meaningful system like GPS, or relative, meaning that they are an arbitrary "rigid transformation" (rotation, reflection, translation) away from the global coordinate system.

Beacon nodes (also frequently called *anchor nodes*) are a necessary prerequisite to localizing a network in a global coordinate system. Beacon nodes are simply ordinary sensor nodes that know their global coordinates a priori. This knowledge could be hard-coded, or acquired through some additional hardware like a GPS receiver. At a minimum, three noncollinear beacon nodes are required to define a global coordinate system in two dimensions. If three-dimensional coordinates are required, then at least four noncoplanar beacons must be present.

Beacon nodes can be used in several ways. Some algorithms (e.g., multidimensional sealing–mobile application part (MDS–MAP), Subsection 9.4.2) localize nodes in an arbitrary relative coordinate system, then use a few beacon nodes to determine a rigid transformation of the relative coordinates into global coordinates (see Appendix B). Other algorithms (e.g., approximate point in triangle (APIT) Subsection 9.5.4) use beacons throughout, using the positions of several beacons to "bootstrap" the global positions of nonbeacon nodes.

Beacon placement can often have a significant impact on localization. Many groups have found that localization accuracy improves if beacons are placed in a convex hull around the network. Locating additional beacons in the center of the network is also helpful. In any event, there is considerable evidence that real improvements in localization can be obtained by planning beacon layout in the network.

The advantage of using beacons is obvious: the presence of several prelocalized nodes can greatly simplify the task of assigning coordinates to ordinary nodes. However, beacon nodes have inherent disadvantages. GPS receivers are expensive. They also cannot typically be used indoors, and can also be confused by tall buildings or other environmental obstacles. GPS receivers also consume significant battery power, which can be a problem for power-constrained sensor nodes. The alternative to GPS is preprogramming nodes with their locations, which can be impractical (for instance, when deploying 10,000 nodes with 500 beacons) or even impossible (for instance, when deploying nodes from an aircraft).

In short, beacons are necessary for localization, but their use does not come without cost.

The remainder of Section 9.2 focuses on hardware methods of computing distance measurements between nearby sensor nodes (i.e., ranging).

9.2.2 Received Signal-Strength Indication

In wireless sensor networks, every sensor has a radio. The question is: How can the radio help localize the network? There are two important techniques for using radio information to compute ranges. One of them, hop count, is discussed in Subsection 9.2.3. The other, received signal-strength indication (RSSI), is covered here.

In theory, the energy of a radio signal diminishes with the square of the distance from the signal's source. As a result, a node listening to a radio transmission should be able to use the strength of the received signal to calculate its distance from the transmitter. RSSI suggests an elegant solution to the hardware ranging problem: all sensor nodes are likely to have radios, so why not use them to compute ranges for localization?

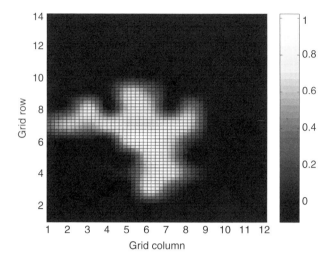

Figure 9.1 Diagram by Alec Woo that shows the probability of successful packet transmission with respect to distance from the source. It shows that the fixed-radius disk approximation of radio connectivity is quite inaccurate. It also demonstrates the difficulties inherent in retrieving distance information from signal strength. (From ref. [3], with permission.)

In practice, however, RSSI ranging measurements contain noise on the order of several meters [2]. This noise occurs because radio propagation tends to be highly nonuniform in real environments (see Fig. 9.1). For instance, radio propagates differently over asphalt than over grass. Physical obstacles such as walls, furniture, and the like, reflect and absorb radio waves. As a result, distance predictions using signal strength have been unable to demonstrate the precision obtained by other ranging methods, such as time difference of arrival (Subsection 9.2.4).

However, RSSI has garnered new interest recently. More careful physical analysis of radio propagation may allow better use of RSSI data, as might better calibration of sensor radios. Whitehouse [3] did an extensive analysis of radio signal strength, which he was able to parlay into noticeable improvements in localization. Thus, it is quite possible that a more sophisticated use of RSSI will eventually prove to be a superior ranging technology, from a price/performance standpoint. Nevertheless, the technology is not there today.

9.2.3 Radio Hop Count

Even though RSSI is too inaccurate for many applications, the radio can still be used to assist localization. The key observation is that if two nodes can communicate by radio, their distance from each other is less than R with high probability, where R is the maximum range of their radios, no matter what their signal strength reading is. Thus, simple connectivity data can be useful for localization purposes.

In particular, many groups have found "hop count" to be a useful way to compute internode distances. The local connectivity information provided by the radio defines an unweighted graph, where the vertices are sensor nodes and edges represent direct radio links between nodes. The hop count h_{ij} between sensor nodes s_i and s_j is then defined as the length of the shortest path in the graph between s_i and s_j.

Naively, if the hop count between s_i and s_j is h_{ij}, then the distance between s_i and s_j, d_{ij}, is less than R^*h_{ij}, where R is again the maximum radio range.

It turns out that a better estimate can be made if we know n_{local}, the expected number of neighbors per node. Then, as shown by Kleinrock and Silvester [4], it is possible to compute a better formula for the distance covered by one radio hop:

$$d_{hop} = R\left(1 + e^{-n_{local}} - \int_{-1}^{1} e^{-(n_{local}/\pi)\arccos t - t\sqrt{1-t^2}}\, dt\right) \qquad (9.1)$$

Then, $d_{ij} \approx h_{ij} * d_{hop}$. Experimentally [5], equation (9.1) has been shown to be quite accurate when n_{local} grows above 5. However, when $n_{local} > 15$, d_{hop} approaches R, so equation (9.1) becomes less useful.

There are two problems with using hop count as a measurement of distance. First, distance measurements are always integral multiples of d_{hop}. This inaccuracy corresponds to a total error of about $0.5R$ per measurement, which can be too high for some applications. Second, environmental obstacles can prevent edges from appearing in the connectivity graph that otherwise would be present. As a result, hop count–based distances can be substantially too high, for example, as in Figure 9.2.

Nagpal et al. [5] demonstrate by algorithm that even better hop-count distance estimates can be computed by averaging distances with neighbors. This benefit does not begin to appear until $n_{local} \geq 15$; however, it can reduce hop-count error down to as little as $0.2R$.

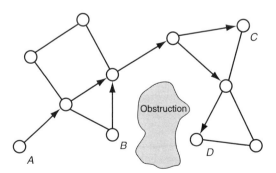

Figure 9.2 Examples of hop count. In this diagram, $h_{AC} = 4$. Unfortunately, h_{BD} is also four, due to an obstruction in the topology. This is one of the ways that hop-count distance metrics can experience dramatic error.

9.2.4 Time Difference of Arrival

Time difference of arrival (TDoA) is a commonly used hardware ranging mechanism. In TDoA schemes, each node is equipped with a speaker and a microphone. Some systems use ultrasound while others use audible frequencies. However, the general mathematical technique is independent of particular hardware.

In TDoA, the transmitter first sends a radio message. It waits some fixed interval of time, t_{delay} (which might be zero), and then produces a fixed pattern of "chirps" on its speaker.

When listening nodes hear the radio signal, they note the current time, t_{radio}, then turn on their microphones. When their microphones detect the chirp pattern, they again note the current time, t_{sound}. Once they have t_{radio}, t_{sound}, and t_{delay}, the listeners can compute the distance d between themselves and the transmitter using the fact that radio waves travel substantially faster than sound in air.

$$d = (s_{radio} - s_{sound}) * (t_{sound} - t_{radio} - t_{delay}) \tag{9.2}$$

See Fig. 9.3 for an illustration.

TDoA methods are impressively accurate under line-of-sight conditions; however, they perform best in areas that are free of echoes and when the speakers and microphones are calibrated to each other. Several groups are working to compensate for these issues, which will likely lead to even better field accuracy.

Nevertheless, rather good results can already be obtained, even in subpar conditions. The Cricket ultrasound ranging system [6] can obtain close to centimeter accuracy without calibration over ranges of up to 10 meters in indoor environments, provided the transmitter and receiver have line-of-sight.

The downside of TDoA systems is that they inevitably require special hardware to be built into sensor nodes, specifically a speaker and a microphone. TDoA systems perform best when they are calibrated properly, since speakers and microphones never have identical transmission and reception characteristics. Furthermore, the speed of sound in air varies with air temperature and humidity, which introduce

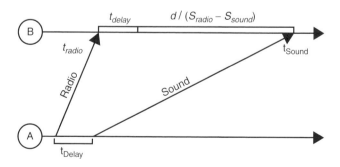

Figure 9.3 Time difference of arrival (TDoA) illustrated. Sensor A sends a radio pulse followed by an acoustic pulse. By determining the time difference between the arrival of the two pulses, sensor B can estimate its distance from A.

inaccuracy into equation (9.2). Finally, the line-of-sight constraint can be difficult to meet in some environments.

It is possible to use additional constraints to identify and prune bad ranging data ("outliers") [7]. Representative constraints include:

- The range from node A to node B should be approximately equal to the range from node B to node A ($r_{AB} \approx r_{BA}$).
- The pairwise ranges between nodes A, B, and C should obey the triangle inequality ($r_{AB} + r_{AC} \geq r_{BC}$).

In the end, many localization algorithms use time difference of arrival ranging simply because it is dramatically more accurate than radio-only methods. The actual reason why TDoA is more effective in practice than RSSI is due to the difference between using signal travel time and signal magnitude, where the former is vulnerable only to occlusion while the latter is vulnerable to both occlusion and multipath.

9.2.5 Angle of Arrival, Digital Compasses

Some algorithms depend on angle of arrival (AoA) data. These data are typically gathered using radio or microphone arrays, which allow a listening node to determine the direction of a transmitting node. It is also possible to gather AoA data from optical communication methods.

In these methods, several (3–4) spatially separated microphones hear a single transmitted signal. By analyzing the phase or time difference between the signal's arrival at different microphones, it is possible to discover the AoA of the signal.

These methods can obtain accuracy to within a few degrees [8]. Unfortunately, AoA hardware tends to be bulkier and more expensive than TDoA ranging hardware, since each node must have one speaker and several microphones. Furthermore, the need for spatial separation between speakers is difficult to accommodate as the form factor of sensors shrinks.

AoA hardware is sometimes augmented with digital compasses. A digital compass simply indicates the global orientation of its node, which can be quite useful in conjunction with AoA information.

In practice, few sensor localization algorithms absolutely require AoA information, though several are capable of using it when it is present.

9.3 ISSUES IN LOCALIZATION ALGORITHM DESIGN

9.3.1 Resource Constraints

Sensor networks are typically quite resource-starved. Nodes have rather weak processors, making large computations infeasible. Moreover, sensor nodes are typically battery powered. This means communication, processing, and sensing actions are all expensive, since they actively reduce the lifespan of the node performing them.

Not only that, sensor networks are typically envisioned on a large scale, with hundreds or thousands of nodes in a typical deployment. This fact has two important consequences: nodes must be inexpensive to fabricate, and trivially easy to deploy. Nodes must be inexpensive, since fifty cents of additional cost per node translates to $500 for a one-thousand node network. Deployment must be easy as well: 30 seconds of handling time per node to prepare for localization translates to over 8 man-hours of work to deploy a 1000-node network.

Localization is necessary to many functions of a sensor network; however, it is not the purpose of a sensor network. Localization must cost as little as possible while still producing satisfactory results. That means designers must actively work to minimize the power cost, hardware cost, and deployment cost of their localization algorithms.

9.3.2   Node Density

Many localization algorithms are sensitive to node density. For instance, hop-count–based schemes generally require high node density so that the hop count approximation for distance is accurate (Subsection 9.2.3). Similarly, algorithms that depend on beacon nodes fail when the beacon density is not high enough in a particular region. Thus, when designing or analyzing an algorithm, it is important to notice the algorithm's implicit density assumptions, since high node density can sometimes be expensive if not totally infeasible.

9.3.3   Nonconvex Topologies

Localization algorithms often have trouble positioning nodes near the edges of a sensor field. This artifact generally occurs because fewer range measurements are available for border nodes, and those few measurements are all taken from the same side of the node. In short, border nodes are a problem because less information is available about them and that information is of lower quality. This problem is exacerbated when a sensor network has a nonconvex shape: Sensors outside the main convex body of the network can often prove unlocaliz-able. Even when locations can be found, the results tend to feature disproportion-ate error.

9.3.4   Environmental Obstacles and Terrain Irregularities

Environmental obstacles and terrain irregularities can also wreak havoc on localiz-ation. Large rocks can occlude line of sight, preventing TDoA ranging, or interfere with radios, introducing error into RSSI ranges and producing incorrect hop-count ranges. Deployment on grass versus sand versus pavement can affect radios and acoustic ranging systems. Indoors, natural features like walls can impede measurements as well. All of these issues are likely to come up in real deployments, so localization systems should be able to cope.

9.3.5 System Organization

This Subsection defines a taxonomy for localization algorithms based on their computational organization.

Centralized algorithms (Section 9.4) are designed to run on a central machine with plenty of computational power. Sensor nodes gather environmental data and pass them back to a base station for analysis, after which the computed positions are transported back into the network. Centralized algorithms circumvent the problem of nodes' computational limitations by accepting the communication cost of moving data back to the base station. This trade-off becomes less palatable as the network grows larger, however, since it unduly stresses nodes near the base station. Furthermore, it requires that an intelligent base station be deployed with the nodes, which may not always be possible. This scaling problem can be partially alleviated by deploying multiple base stations (forming a multitier network).

In contrast, distributed algorithms are designed to run in the network, using massive parallelism and internode communication to compensate for the lack of centralized computing power. Often distributed algorithms use a subset of the data to solve for each position independently, yielding an approximation of a corresponding centralized algorithm where all the data are considered and used to solve for all the positions simultaneously.

There are two important approaches to distributed localization. The first group, beacon-based distributed algorithms (Section 9.5), typically starts with some group of beacons (Subsection 9.2.1). Nodes in the network obtain a distance measurement to a few beacons, then use these measurements to determine their own location. In some algorithms, these newly localized nodes become beacons to help other nodes localize.

The second group approaches localization by trying to optimize a global metric over the network in a distributed fashion. This group splits out into two substantially different approaches. The first approach, relaxation-based distributed algorithms (Section 9.6) is to use a coarse algorithm to roughly localize nodes in the network. This coarse algorithm is followed by a refinement step, which typically involves each node adjusting its position to optimize a local error metric. By doing so, these algorithms hope to approximate the optimal solution to a networkwide metric that is the sum of the local error metric at each of the nodes.

Coordinate system stitching (Section 9.7) is the second approach to optimizing a networkwide metric in a distributed manner. In these algorithms, the network is divided into small overlapping subregions, each of which creates an optimal local map. Finally, the subregions use a peer-to-peer process to merge their local maps into a single global map. In theory, this global map approximates the global optimum map.

The next four sections treat each of these groups in turn.

9.4 CENTRALIZED ALGORITHMS

This section is devoted to centralized localization algorithms. Centralization allows an algorithm to undertake much more complex mathematics than is possible in a distributed setting. However, as we said in the previous section, centralization requires the migration of internode ranging and connectivity data to a sufficiently powerful central base station and then the migration of resulting locations back to respective nodes. The main difference between centralized algorithms is the type of processing they do at the base station. We will discuss two types of processing: semidefinite programming and multidimensional scaling.

9.4.1 Semidefinite Programming

The semidefinite programming (SDP) approach to localization was pioneered by Doherty et al. [9]. In this algorithm, geometric constraints between nodes are represented as linear matrix inequalities (LMIs). Once all the constraints in the network are expressed in this form, the LMIs can be combined to form a single semidefinite program. This is solved to produce a bounding region for each node, which Doherty et al. simplify to be a bounding box. See Figure 9.4 for some sample LMI constraints.

Unfortunately, not all geometric constraints can be expressed as LMIs. In general, only constraints that form convex regions are amenable to representation as an LMI. Thus, AoA data can be represented as a triangle and hop-count data can be represented as a circle, but precise range data cannot be conveniently represented, since rings cannot be expressed as convex constraints. This inability to accommodate precise range data may prove to be a significant drawback.

Solving the linear or semidefinite program must be done centrally. The relevant operation is $O(k^2)$ for angle of arrival data, and $O(k^3)$ when radial (e.g., hop count)

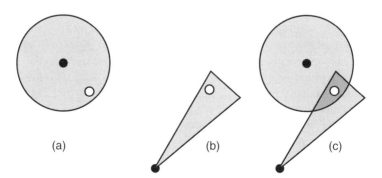

Figure 9.4 Semidefinite program constraints. (a) A radial constraint, for example, from radio connectivity. (b) A triangular constraint, for example, from angle of arrival data. (c) Location estimate derived from intersection of two convex constraints.

data is included, where k is the number of convex constraints needed to describe the network. Thus, running time is something of an Achilles' heel for this algorithm. A hierarchical version of this algorithm might have better scaling properties, but no relevant performance data have been published to our knowledge.

The real advantage of this algorithm is its elegance. Given a set of convex constraints on a node's position, SDP simply finds the intersection of the constraints. However, SDP's poor scaling and inability to effectively use range data will likely preclude the algorithm's use in practice.

9.4.2 MDS–MAP

MDS–MAP is a centralized algorithm due to Shang et al. [10]. Instead of using semidefinite programming, however, MDS–MAP uses a technique from mathematical psychology called multidimensional scaling (MDS).

The intuition behind MDS is simple. Suppose there are n points, suspended in a volume. We do not know the positions of the points, but we do know the distance between each pair of points. MDS is an $O(n^3)$ algorithm that uses the law of cosines and linear algebra to reconstruct the relative positions of the points based on the pairwise distances. The mathematical details of MDS are in Appendix C of this chapter.

MDS–MAP is almost a direct application of the simplest kind of MDS: classic metric MDS. The algorithm has four stages, which are as follows:

Step 1. Gather ranging data from the network, and form a sparse matrix R, where r_{ij} is the range between nodes i and j, or zero if no range was collected (for instance, if i and j are physically too far apart).

Step 2. Run a standard all-pairs shortest-path algorithm (Dijkstra's, Floyd's) on R to produce a complete matrix of internode distances D.

Step 3. Run classic metric MDS on D to find estimated node positions X, as described in Appendix C.

Step 4. Transform the solution X into global coordinates using some number of fixed anchor nodes using a coordinate system registration routine B.

MDS–MAP performs well on RSSI data alone, getting performance on the order of half the radio range when the neighborhood size n_{local} is higher than 12. As expected, MDS–MAP estimates improve as ranging improves. MDS–MAP also does not use anchor nodes very well, since it effectively ignores their data until stage 4. As a result, its performance lags behind other algorithms as anchor density increases. The main problem with MDS–MAP, however, is its poor asymptotic performance, which is $O(n^3)$ on account of stages 2 and 3. It turns out that this problem can be partially ameliorated using coordinate system stitching: see Section 9.7 for details.

9.5 BEACON-BASED DISTRIBUTED ALGORITHMS

In this section, we talk about beacon-based distributed algorithms. These algorithms all extrapolate unknown node positions from beacon positions. Thus, they localize nodes directly into the global coordinate space of the beacons. These algorithms are also all distributed, so that all the relevant computation is done on the sensor nodes themselves. We will present four beacon-based distributed algorithms: diffusion, bounding box, gradient multilateration, and APIT.

9.5.1 Diffusion

Diffusion arises from a very simple idea: the most likely position of a node is at the centroid of its neighbors' positions. Diffusion algorithms require only radio connectivity data. We describe two different variants below.

Bulusu et al. [11] localize unknown nodes by simply averaging the positions of all beacons with which the node has radio connectivity. Thus, Bulusu et al. assume that nodes have no way of ranging to beacons. This method is attractive in its blinding simplicity; however, the resulting positions are not very accurate, particularly when beacon density is low, or nodes fall outside the convex hull of their audible beacons.

Fitzpatrick and Meertens [12] describe a more sophisticated variant: each node is at the centroid of its neighbors, including nonbeacons. The algorithm is as follows:

Step 1. Initialize the position of all nonbeacon nodes to (0, 0).

Step 2. Repeat the following until positions converge:

Step 2a. Set the position of each nonbeacon node to the average of all its neighbors' positions.

This variant requires fewer beacons than Bulusu et al.'s algorithm; nevertheless, its accuracy is poor when node density is low, nodes are outside the convex hull of the beacons, or node density varies across the network. In all of these cases, a more sophisticated algorithm would improve accuracy dramatically. Fitzpatrick and Meertens's variant also uses substantially more computation than Bulusu et al.'s approach, since positions must be exchanged between adjacent nodes during step 2.

However, this algorithm is quite useful in networks where nodes are capable of very little computation, but the network topology can be selectively changed to improve localization. In particular, Savvides et al. [13] recommend placing some beacons around the edges of the sensor network field. Selectively adding additional beacons can also help resolve pathologies in the diffusion estimates. Bulusu et al. [11] describe an approach for adaptive beacon placement to improve diffusion-based localization.

9.5.2 Bounding Box

The bounding-box algorithm [14,15] is a computationally simple method of localizing nodes given their ranges to several beacons. See Figure 9.5 for an example. Essentially, each node assumes that it lies within the intersection of its beacons' bounding boxes. The bounding box for a beacon b is centered at the beacon position (x_b, y_b), and has height and width $2d_b$, where d_b is the node's distance to the beacon measurement.

The intersection of the bounding boxes can be computed without use of floating-point operations:

$$[\max(x_i - d_i), \ \max(y_i - d_i)] \times [\min(x_i + d_i), \ \min(y_i + d_i)] \tag{9.3}$$
$$i = 1 \cdots n$$

The position of a node is then the center of this final bounding box, as shown in Figure 9.5.

Whitehouse [3] analyzes a distributed version of this algorithm [15], showing that unfortunately this version is highly susceptible to noisy range estimates, especially small estimates that tend to propagate.

The accuracy of the bounding-box approach is best when the nodes' actual positions are closer to the center of their beacons. Simic and Sastry [15] prove results about convergence, errors, and complexity.

In any event, bounding box works best when sensor nodes have extreme computational limitations, since other algorithms may simply be infeasible. Otherwise, more mathematically rigorous approaches such as gradient multilateration (Subsection 9.5.3), may be more appropriate.

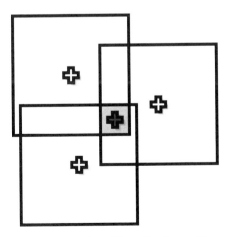

Figure 9.5 An example of the intersection of bounding boxes. The center of the intersection is the position estimate for the unknown node. The size of the boxes is based on hop count radio range from the beacons to the unknown node.

9.5.3 Gradient

The principal mathematical operation of the gradient method is called *multilateration*. Multilateration is a great deal like triangulation, except that multilateration can incorporate ranges from more than three reference points. Formally, given m beacons with known Cartesian positions $b_i, i = 1 \cdots m$ and possibly noisy range measurements r_i from the known nodes to an unknown sensor node s, multilateration finds the most likely position of s. The mathematics of multilateration are outlined in Appendix 1.9.

Using gradients to compute ranges for multilateration has been proposed by a number of researchers [11,16–19]. These algorithms all assume that there are at least three beacon nodes somewhere in the network (though probably more). Each of these beacon nodes propagates a gradient through the network, which is the distributed equivalent of computing the shortest path distance between all the beacons and all the unlocalized nodes. The gradient propagation is as follows:

Step 1. For each node j and beacon k, let d_{jk} (the distance from j to k) be 0 if $j = k$, and ∞ otherwise.

Step 2. On each node j, perform the following steps repeatedly:

Step 2a. For each beacon k and neighbor i, retrieve d_{ik} from i.

Step 2b. For each beacon k and neighbor i, apply the following update formula:

$$d_{jk} = \min(d_{ik} + \hat{r}_{ij}, \ d_{jk})$$

where $\hat{r}_{ij}$ is the estimated distance between nodes i and j. These internode distance estimates can be either unweighted (one if there is connectivity, zero otherwise) or measured distances (e.g., using RSSI or TDoA).

After some amount of settling time, each value d_{jk} will be the length of the shortest path between node j and beacon k. Figure 9.6 shows the results of running the gradient propagation algorithm with one beacon.

The gradient-based distance estimate to a beacon must be adjusted, since even given perfect internode distance estimates, gradient distance estimates will always be longer than (or exactly equal to) corresponding straight-line distances. Of course, given imperfect internode distance estimates, gradient-based distance estimate can actually be shorter than straight distances. In fact, Whitehouse [3] shows that it is actually more likely that they are shorter, since underestimated internode distances skew all subsequent gradient-based estimates. Niculescu and Nath [20] suggest using a correction factor calculated by comparing the actual distance between beacons to the shortest path distances computed during gradient propagation. Each unlocalized node simply applies the correction factor from its closest beacon to its gradient distance estimate.

As an alternative, Nagpal et al. [5] in their Amorphous algorithm suggest correcting this distance based on the neighborhood size n_{local}, as we previously discussed in Subsection 9.2.3.

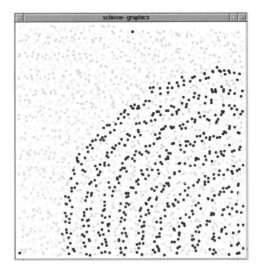

Figure 9.6 Gradients propagating from a beacon (in the lower right corner). Each dot represents a sensor node. Gray levels are based on their gradient value.

Once final distance estimates to beacons have been computed, the actual localization process simply uses multilateration directly on the beacon positions k and the distance measurements d_{jk}.

Like the other beacon-based distributed algorithms, this algorithm has the virtue of being direct and easy to understand. It is also scales well (provided the density of beacons is kept constant; otherwise, the communication cost can be prohibitive). It is also quite effective in homogeneous topologies where there are few environmental obstructions. However, even when using high-quality range data, this algorithm is subject to the deficiencies described in Subsection 9.2.3 and demonstrated in Figure 9.2, so it behaves badly in obstructed settings. It also requires substantial node density before its accuracy reaches an acceptable level.

A number of variations to the multilateration approach have been suggested. Niculescu and Nath [21] suggest propagating AoA information along links. Nagpal [22] proposes refining the hop-count estimates by averaging values among neighbors. This turns out to greatly increase the accuracy of gradient multilateration.

9.5.4 APIT

APIT [23] is quite a bit different from the beacon-based distributed algorithms described so far. APIT uses a novel area-based approach, in which nodes are assumed to be able to hear a fairly large number of beacons. However, APIT does not assume that nodes can range to these beacons. Instead, a node forms some number of "beacon triangles," where a beacon triangle is the triangle formed by three arbitrary beacons. The node then decides whether it is inside or outside a given triangle by comparing signal-strength measurements with its nearby

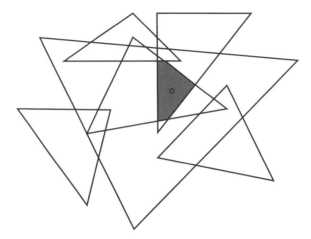

Figure 9.7 Node position estimated as the center of mass of the intersection of a number of beacon triangles for which a given node is inside.

nonbeacon neighbors. Once this process is complete, the node simply finds the intersection of the beacon triangles that contains it. The node chooses the centroid of this intersection region as its position estimate. Figure 9.7 shows an example of this process: each of the triangles represents a triple of beacons and the intersection of all the triangles defines the position of the unknown node.

The actual algorithm is as follows:

Step 1. Receive beacon positions from hearable beacons.

Step 2. Initialize *inside-set* to be empty.

Step 3. For each triangle T_i in possible triangles formed over beacons, add T_i to *inside-set* if node is inside T_i. Goto *Step 4* when accuracy of *inside-set* is sufficient.

Step 4. Compute position estimate as the center of mass of the intersection of all triangles in *inside-set*.

The point in triangle (PIT) test is based on geometry. For a given triangle with points A, B, and C, a given point M is outside triangle ABC, if there exists a direction such that a point adjacent to M is further/closer to points A, B, and C simultaneously. Otherwise, M is inside triangle ABC. Unfortunately, given that typically nodes cannot move, an approximate APIT test is proposed that assumes sufficient node density for approximating node movement. If no neighbor of M is further from/ closer to all three anchors A, B, and C simultaneously, M assumes that it is inside triangle ABC. Otherwise, M assumes it resides outside this triangle.

This algorithm is described as being range-free, which means that RSSI range measurements are required to be monotonic and calibrated to be comparable but are not required to produce distance estimates. It could be that the effort put into RSSI calibration would produce an effective enough ranging estimate to be useful for gradient techniques described in Subsection 9.5.3, making the range-free distinction potentially moot. The APIT algorithm also requires a relatively high ratio of beacons to nodes, requires longer-range beacons, and is susceptible to erroneously low RSSI readings. On the other hand, He et al. [23] show that the algorithm requires smaller amounts of computation and less communication than other beacon-based algorithms. In short, APIT is a novel approach that is a potentially promising direction that requires further study.

9.6 RELAXATION-BASED DISTRIBUTED ALGORITHMS

The class of relaxation-based distributed algorithms starts with nodes estimating their positions with any of a variety of methods, such as gradient distance propagation. These initial positions are then refined from position estimates of neighbors.

Savarese et al. [24] refine the initial gradient-derived positions using local neighborhood multilateration. Each node adjusts its position by using its neighbors as temporary beacons. Convex optimization can also be used to find an improved position for situations where beacon distance estimates are unavailable.

An equivalent formulation to local multilateration is presented in ref. [25] and is generally referred to as a spring model. This description considers edges between nodes as springs, with resting lengths being the actual measured distances. The algorithm involves iteratively adjusting nodes in the direction of their local spring forces. The optimization stops when all nodes have zero forces acting on them. If the magnitude of all the forces between nodes is also zero, then the final positions form a global minimum.

Unfortunately, these relaxation techniques are quite sensitive to initial starting positions. Bad starting positions will result in local minima. Priyantha et al. [25] describe a technique for producing starting positions for nodes that nearly always avoid bad local minima. The insight is that the network gets tangled and that, using the spring model style, optimization is unable to fully untangle the network. Their approach starts the network in a "fold-free" state.

The fold-free algorithm works by choosing five reference nodes, one in the center n_0 and four on the periphery, n_1, n_2, n_3, n_4. The four on the periphery are chosen so that the two pairs n_1, n_2 and n_3, n_4 are roughly perpendicular to each other. The choice of these nodes is performed using a hop-count approximation to distance. The node positions (x_i, y_i) are calculated using polar coordinates (θ_i, ρ_i):

$$\theta_i = h_{0,i} R$$

$$\rho_i = \arctan \frac{h_{1,i} - h_{2,i}}{h_{3,i} - h_{4,i}}$$

$$x_i = h_{0,i} R \frac{h_{3,i} - h_{4,i}}{l_i} \tag{9.4}$$

$$y_i = h_{0,i} R \frac{h_{1,i} - h_{2,i}}{l_i}$$

$$l_i = \sqrt{(h_{3,i} - h_{4,i})^2 + (h_{1,i} - h_{2,i})^2}$$

where $h_{j,i}$ is the hop count to reference node j and R is the maximum radio range.

These relaxation algorithms have the virtue that they are fully distributed and concurrent and operate without beacons. While the computations are modest and local, it is unclear how well these algorithms scale to much larger networks. Furthermore, there are no provable means for avoiding local minima and local minima problems could worsen at larger scales. To date, researchers have avoided local minima by starting optimizations at favorable starting positions, but another alternative would be to utilize optimization techniques, such as simulated annealing [26], which tend to fall into fewer local minima.

9.7   COORDINATE SYSTEM STITCHING

In section 9.6, we showed one method of fusing the precision of centralized schemes with the computational advantages of distributed schemes. Coordinate system stitching is a different way of approaching the same problem. It has received a great deal of recent work [20,27–29]. Coordinate system stitching works according to the following algorithm:

Step 1. Split the network into small overlapping subregions. Very often each subregion is simply a single node and its one-hop neighbors.

Step 2. For each subregion, compute a "local map," which is essentially an embedding of the nodes in the subregion into a relative coordinate system.

Step 3. Finally, merge the subregions using a coordinate system registration procedure. Coordinate system registration finds a rigid transformation that maps points in one coordinate system to a different coordinate system. Thus, step 3 places all the subregions into a single global coordinate system. Many algorithms do this step suboptimally, since there is a closed-form, fast, and least-square optimal method of registering coordinate systems. We describe this optimal method in Appendix B.

Steps 1 and 2 tend to be unique to each algorithm, whereas step 3 tends to be the same in every algorithm. We will describe three different methods of performing steps 1 and 2, and finally explain the typical method of performing step 3.

Meertens and Fitzpatrick [28] form subregions using one-hop neighbors. Local maps are then computed by choosing three nodes to define a relative coordinate system and using multilateration (Subsection 9.5.3) to iteratively add additional nodes to the map, forming a "multilateration subtree."

Moore et al. [29] outline an approach that they claim produces more robust local maps. Rather than use three arbitrary nodes to define a map, Moore et al. use "robust quadrilaterals" (robust quads), where a robust quad is a fully connected set of four nodes, where each subtriangle is also "robust." A robust subtriangle must have the property that:

$$b \sin^2 \theta > d_{\min}$$

where b is the length of the shortest side, θ is the size of the smallest angle, and d_{min} is a predetermined constant based on average measurement error. The idea is that the points of a robust quad can be placed correctly with respect to each other (i.e., without "flips"). Moore et al. demonstrate that the probability of a robust quadrilateral experiencing internal flips given zero mean Gaussian measurement error can be bounded by setting d_{min} appropriately. In effect, d_{min} filters out quads that have too much positional ambiguity to be localized with confidence. The appropriate level of filtering is based on the amount of uncertainty σ^2 in the distance measurements.

Once an initial robust quad is chosen, any node that connects to three of the four points in the initial quad can be added using simple multilateration (Subsection 9.5.3). This preserves the probabilistic guarantees provided by the initial robust quad, since the new node forms a new robust quad with the points from the original. By induction, any number of nodes can be added to the local map, as long as each node has a range to three members of the map.

These local maps (which Moore et al. call "clusters") are now ready to be stitched together. Optionally, an optimization pass such as those in Section 9.6 can be used to refine the local maps first.

Ji and Zha [30] use MDS to form local maps. We discussed MDS with MDS–MAP in Subsection 9.4.2, and cover the mathematics of MDS in Appendix C. Ji and Zha use an iterative variant of MDS to compensate for missing internode distances. This iterative variant turns out to be intimately related to standard iterative least-square algorithms, though it is somewhat more sophisticated. Ji and Zha focus on RSSI for range data. Once again, subregions are defined to be one-hop neighborhoods.

The stitching phase (step 3 in previous algorithm), uses coordinate system registration (described in Appendix B) in a peer-to-peer fashion to shift all the local maps into a single coordinate system. One way of performing this stitching is now described:

Step 1. Let the node responsible for each local map choose an integer coordinate system ID at random.

Step 2. Each node communicates with its neighbors; each pair performs the following steps:

Step 2a. If both have the same ID, then do nothing further.

Step 2b. If they have different IDs, then register the map of the node with the lower ID with the map of the node with the higher ID. Afterward, both nodes keep the higher ID as their own.

Step 3. Repeat step 2 until all nodes have the same ID; now all nodes have a coordinate assignment in a global coordinate system.

Limited work has been done on the mathematical properties of this scheme. Moore et al. prove the probability of their algorithm constructing correct local maps and prove error lower bounds on the local map positions. Meertens and Fitzpatrick [28] devote some discussion to the topic of error propagation caused by local map stitching. They point out that registering local maps iteratively can lead to error propagation and perhaps unacceptable error rates as networks grow. Furthermore, they argue that in the traditional communication model, where nodes can communicate only with neighbors, this algorithm may converge quite slowly, since a single coordinate system must propagate from its source across the entire network. Future work is needed to curb this error propagation.

Furthermore, these techniques have a tendency to orphan nodes, either because they could not be added to a local map or because their local map failed to overlap sufficiently with neighboring maps. Moore et al. argue that this is acceptable because the orphaned nodes are the nodes most likely to display high error. However, this answer may not be satisfactory for some applications, many of which cannot use unlocalized nodes for sensing, routing, target tracking, or other tasks.

Nonetheless, coordinate system stitching techniques are quite compelling. They are inherently distributed, since subregion and local map formation can trivially occur in the network and stitching is easily formulated as a peer-to-peer algorithm. Furthermore, they enable the use of sophisticated local-map algorithms which are too computationally expensive to use at the global level. For example, map formation using robust quadrilaterals is $O(n^4)$, where n is the number of nodes in the subregion; however, in networks with fixed neighborhood size n_{local}, map formation is $O(1)$. Likewise, coordinate system stitching enables the realistic use of $O(n^3)$ multidimensional scaling in sensor networks.

9.8 FUTURE DIRECTIONS

The sensor network field and localization in particular are in their infancy. Much work remains in order to address the varied localization requirements of sensor network services and applications. Many future directions stand out as important areas to pursue in order to meet both current and future needs.

Localization hardware will always involve fallible and imperfect components; thus, calibration is imperative [3]. For example, raw measurements from RSSI vary wildly from node to node, while most algorithms expect measurements to be at minimum monotonic and comparable. If calibration can bridge this gap, a wide variety of algorithms would become practical on cheap hardware.

Even with accurate calibration, localization hardware produces noisy measurements due to occlusion, collisions, and multipath effects. This mandates an improvement in measurement outlier rejection algorithms. Early work [7] has suggested that outlier rejection can greatly improve the performance of localization algorithms. Some early ideas [7] involve using consistency checks such as symmetry and geometric constraints to reject improbable measurements, as discussed in Subsection 9.2.4. Other possibilities involve using statistical error models to identify outliers.

Future sensor networks will involve movable sensor nodes. New localization algorithms will need to be developed to accommodate these moving nodes. Some algorithms can tolerate a certain amount of movement, but more experiments and algorithm development is required. Some researchers [11,31] have touched on this issue with adaptive beacon placement, but much more work is needed.

No current localization algorithm adequately scales for ultrascale sensor networks (i.e., 10,000 nodes and beyond). It seems likely that such networks will end up being multitiered, and will require the development of more hierarchical algorithms.

9.9 CONCLUSION

In this chapter we presented the foundations of sensor network localization. We discussed localization hardware, issues in localization algorithm design, major localization techniques, and future directions. In this section, we summarize the trade-offs and provide guidelines for choosing different algorithms based on context and available hardware.

The first primary distinction between algorithms is those that require beacons (described in Section 9.5) and those that do not (described in Sections 9.4, 9.6, and 9.7). Beaconless algorithms necessarily produce relative coordinate systems that can optionally be registered to a global coordinate system by positioning three (or four) nodes. Often sensor network deployments make the use of beacons prohibitive and furthermore many applications do not require a global coordinate system. In these situations beaconless algorithms suffice. Finally, some algorithms (such as APIT from Subsection 9.5.4) require a higher beacon-to-node ratio than others to achieve a given level of accuracy.

The next distinction between localization algorithms is their hardware requirements. All sensor nodes have radios and most can measure signal strength, thus, algorithms that rely on hop count or RSSI require the least hardware. Varying degrees of ranging precision can be achieved from RSSI, with hop count being at the low end, with one bit precision. Gradient algorithms (from Subsection 9.5.3 such as DV-hop and Amorphous can often produce quite accurate results using only hop counts and sufficient node density. Sometimes, a microphone and speaker are required for other reasons, making the use of more accurate TDoA ranging possible. Sometimes nodes lack sufficient arithmetic processing, making certain algorithms impractical. Algorithms such as bounding-box and APIT make the least demands on processors (although APIT makes some demands on memory).

Finally, certain algorithms are centralized while others are distributed. Centralized algorithms typically compute more exact positions and can be competitive in situations where accuracy is important and the exfiltration of ranging data and dissemination of resulting location data is not prohibitively time-consuming nor error-prone. Centralized algorithms could actually be a viable option in many typical deployments where a base station is already needed for other reasons. Distributed algorithms are often local approximations to centralized algorithms, but have the virtue that they do not depend on a large centralized computer and potentially have better scalability.

Other issues to consider are battery life and communication costs. Often these two are intertwined, as typically communication is the most battery-draining sensor node activity. Consult He et al. [23] for a comparison of communication costs (and other metrics) of a number of localization algorithms.

The development of localization algorithms is proceeding at a fast pace. While the task appears simple, to compute positions for each node in a sensor network, the best algorithm depends heavily on a variety of factors such as application needs and available localization hardware. Future algorithms will address new sensor network needs such as mobile nodes and ultrascale sizes.

APPENDIX A: MULTILATERATION

This Appendix derives a solution to the multilateration problem (Subsection 9.5.3). See Figure 9.8 to see an example of this solution in practice.

Multilateration is a simple technique, but the specific mathematics of its implementation vary widely, as do its application in sensor networks. The purpose of multilateration is simple: given m nodes with known Cartesian positions $b_i, i = 1 \cdots m$ and possibly noisy range measurements r_i from the known nodes to an unknown node s, multilateration finds the most likely position of s.

Multilateration is typically done by minimizing the squared error between the observed ranges r_i and the predicted distance $\|s - b_i\|$:

$$s = \operatorname*{argmin}_{s} E(s)$$

$$E(s) = \sum_{i=1}^{m} (\|s - b_i\| - r_i)^2$$

(9.5)

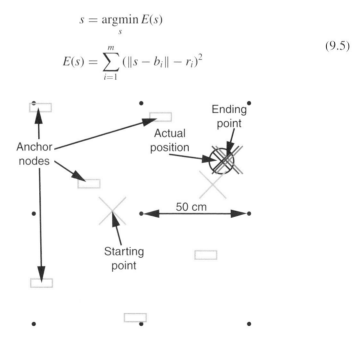

Figure 9.8 In this diagram, a single unknown node with ranges to six different beacons localizes itself using multilateration. The ground truth position of the unknown node is circled. The X's mark the best estimate after each iteration of least squares, with darker shades indicating higher iterations.

This minimization problem can be solved using Newton–Raphson/least squares as follows. First, approximate the error function $e(s, b_i) = \|s - b_i\| - r_i$ in equation (9.5) with a first-order Taylor series about s_0:

$$e(s, b_i) \approx e(s_0, b_i) + \nabla e(s_0, b_i)(s - s_0)$$
$$= \nabla e(s_0, b_i)s - (-e(s_0, b_i) + \nabla e(s_0, b_i)s_0)$$
$$\nabla e(s, b_i) = \frac{s - b_i}{\|s - b_i\|}$$

Plug this approximation back into equation (9.5):

$$s \approx \underset{s}{\operatorname{argmin}} \sum_{i=1}^{m} (\nabla e(s_0, b_i)s - (-e(s_0, b_i) + \nabla e(s_0, b_i)s_0))^2$$

Stacking terms:

$$s \approx \underset{s}{\operatorname{argmin}} \|As - b\|^2 \tag{9.6a}$$

$$A = \begin{bmatrix} \nabla e(s_0, b_1) \\ \nabla e(s_0, b_2) \\ \vdots \\ \nabla e(s_0, b_m) \end{bmatrix} \tag{9.6b}$$

$$b = \begin{bmatrix} -e(s_0, b_1) + \nabla e(s_0, b_1)s_0 \\ -e(s_0, b_2) + \nabla e(s_0, b_2)s_0 \\ \vdots \\ -e(s_0, b_m) + \nabla e(s_0, b_m)s_0 \end{bmatrix} \tag{9.6c}$$

The right side of equation (9.6a) is in exactly the right form to be solved by an off-the-shelf iterative least-square solver. The resulting s is a good estimate of the unknown sensor's position, provided b_i and r_i are accurate. Here is a summary of the multilateration method:

Step 1. Choose s_0 to be a starting point for the optimization. The choice is somewhat arbitrary, but the centroid $\bar{b}$ is a good one:

$$\bar{b} = \frac{1}{m} \sum_{i=1}^{m} b_i$$

Step 2. Compute A and b using s_0 and equations (9.6b) and (9.6c).

Step 3. Compute $s_0' = \operatorname{argmin}_x \|Ax - b\|^2$ using a least-square solver.

Step 4. If $E(s_0) - E(s_0') < \epsilon$, then s_0' is the solution, otherwise set $s_0 = s_0'$ and return to *Step 2*.

There are many ways to solve the multilateration problem. The one presented here is equivalent to Newton–Raphson descent on the error function E (equation (9.5)). Most alternate methods also attempt to minimize squared error using some form of iterative optimization. To see a prototypical example of an algorithm that uses multilateration, see Subsection 9.5.3.

APPENDIX B: COORDINATE SYSTEM REGISTRATION

Many localization algorithms compute a relative coordinate assignment for a group of sensors and later transform this local coordinate assignment into a different coordinate system. To do this, the algorithm must compute a translation vector, a scale factor, and an orthonormal rotation matrix that define the transformation from one coordinate system to the other. The process of finding these quantities is known as "coordinate system registration." Registration can be performed for two dimensions as long as three points have known coordinates in both systems. The three-dimensional version naturally requires four points.

We present Horn et al.'s method of solving the coordinate system registration problem [32]. It has many advantages over commonly used registration methods:

- It has provable optimality over the canonical least-square error metric (equation (9.7)).
- It uses *all* the data available, though it can compute a correct result with as few as three (or four) points.
- It can be computed quickly, since its running time is proportional to the number of common points n.

There is one caveat: even after a rigid transformation, it is unlikely that the known points will precisely align, since the measurements used to localize the points are likely to have errors. Thus, the best that can be done is a minimization of the misalignment between the two coordinate systems. Let $x_{l,i}$ and $x_{r,i}$ be the known positions of node $i = 1 \cdots n$ in the left-hand and right-hand coordinate systems, respectively. The goal of registration is to find a translation t, scale s, and rotation R that transform a point x in the left-hand coordinate system to the equivalent point x' in the right-hand coordinate system using the formula:

$$x' = sRx + t \qquad (9.7)$$

Horn et al. approach this problem using squared error; they look for a t, s, and R that meet the following condition:

$$(t, s, R) = \underset{t,s,R}{\operatorname{argmin}} \sum_{i=1}^{n} \|e_i\|^2 \qquad (9.7a)$$

$$e_i = x_{r,i} - sRx_{l,i} - t \qquad (9.7b)$$

In ref. [32], Horn et al. derive a closed form for equation (9.7) that can be computed in $O(n)$ time. The method is outlined in the following with emphasis on the precise steps required to perform the computation. For more detail on the mathematical underpinnings, see ref. [32]. To see the method in action, see Figure 9.9.

Step 1. Compute the centroids of x_l and x_r:

$$\bar{x}_l = \frac{1}{n} \sum_{i=1}^{n} x_{l,i} \qquad \bar{x}_r = \frac{1}{n} \sum_{i=1}^{n} x_{r,i}$$

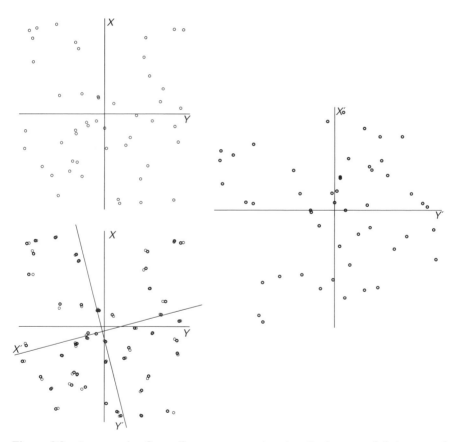

Figure 9.9   An example of coordinate system registration. In the upper left is a set of reference points (X,Y). On the right, the reference points have been moved into a new coordinate system by a linear transformation $(X',Y') = L(X,Y)$ and then jittered to simulate position error. Finally, in the lower left the (X',Y') coordinate system is brought into registration with the reference coordinate system (X,Y).

Step 2. Shift the points so that they are defined with respect to the centroids:

$$x'_{l,i} = x_{l,i} - \bar{x}_l \qquad x'_{r,i} = x_{r,i} - \bar{x}_r$$

Now the error term in equation (9.7b) can be rewritten as

$$e_i = x'_{r,i} - sRx'_{l,i} - t'$$

$$t' = t - \bar{x}_r + sR\bar{x}_l$$

As it turns out, the squared error from equation (9.7) is minimized when $t' = \mathbf{0}$, independent of s and R. Therefore:

$$t = \bar{x}_r - sR\bar{x}_l \tag{9.8}$$

So after s and R have been computed, equation (9.8) can be used to compute t. Since $t' = \mathbf{0}$, the error term can be rewritten as:

$$e_i = x'_{r,i} - sRx'_{l,i} \tag{9.9}$$

Now that t is out of the way, we can focus on finding s and R. Equation (9.9) can be rewritten as

$$e_i = \frac{1}{\sqrt{s}}x'_{r,i} - \sqrt{s}Rx'_{l,i} \tag{9.10}$$

So now we need only find

$$(s, R) = \underset{s,R}{\text{argmin}} \sum_{i=1}^{n} \|e_i\|^2$$

$$= \underset{s,R}{\text{argmin}} \frac{1}{s} \sum_{i=1}^{n} \|x'_{r,i}\|^2 + s \sum_{i=1}^{n} \|r_{l,i}\|^2 \tag{9.11}$$

$$-2 \sum_{i=1}^{n} x'_{r,i} \cdot (Rx'_{l,i})$$

By completing the square in s, it can be shown that equation (9.11) (and thus equation (9.7)) is minimized when

$$s = \sqrt{\sum_{i=1}^{n} \|x'_{r,i}\|^2 \bigg/ \sum_{i=1}^{n} \|x'_{l,i}\|^2} \tag{9.12}$$

Step 3. Use equation (9.12) to compute the optimal scale factor s. Now equation (9.11) can be simplified to

$$R = \underset{R}{\text{argmin }} 2 \left(\sqrt{\left(\sum_{i=1}^{n} \|x'_{r,i}\|^2 \right) \left(\sum_{i=1}^{n} \|x'_{l,i}\|^2 \right)} - \sum_{i=1}^{n} x'_{r,i} \cdot (Rx'_{l,i}) \right) \tag{9.13}$$

Equation (9.13) is minimized when the following is true:

$$R = \underset{R}{\text{argmax}} \sum_{i=1}^{n} x'_{r,i} \cdot (Rx'_{l,i})$$

This is the same as

$$R = \underset{R}{\text{argmax}} \ \text{Trace}(R^T M) \tag{9.14a}$$

$$M = \sum_{i=1}^{n} x'_{r,i}(x'_{l,i})^T \tag{9.14b}$$

M is a 2×2 or 3×3 matrix, depending on whether the points $x_{l,i}$ and $x_{r,i}$ are two- or three-dimensional. For the remainder of this discussion, assume M is 3×3; the results are similar for the two-dimensional case.

Step 4. Compute M using equation (9.14b).

Step 5. Compute the eigen decomposition of $M^T M$. That is, find eigenvalues λ_1, λ_2, λ_3 and eigenvectors $\hat{u}_1, \hat{u}_2, \hat{u}_3$ such that

$$M^T M = \lambda_1 \hat{u}_1 \hat{u}_1^T + \lambda_2 \hat{u}_2 \hat{u}_2^T + \lambda_3 \hat{u}_3 \hat{u}_3^T$$

Step 6. Compute $S = (M^T M)^{1/2}$ and $U = MS^{-1}$. That is,

$$S = \sqrt{\lambda_1} \hat{u}_1 \hat{u}_1^T + \sqrt{\lambda_2} \hat{u}_2 \hat{u}_2^T + \sqrt{\lambda_3} \hat{u}_3 \hat{u}_3^T$$

$$U = MS^{-1} = M\left(\frac{1}{\sqrt{\lambda_1}} \hat{u}_1 \hat{u}_1^T + \frac{1}{\sqrt{\lambda_2}} \hat{u}_2 \hat{u}_2^T + \frac{1}{\sqrt{\lambda_3}} \hat{u}_3 \hat{u}_3^T \right)$$

Note that $M = US$, and that U is orthonormal, since $U^T U = I$. We can now write $\text{Trace}(R^T M)$ from equation (9.14a) as:

$$\text{Trace}(R^T US) = \sqrt{\lambda_1} \text{Trace}(R^T U \hat{u}_1 \hat{u}_1^T)$$

$$+ \sqrt{\lambda_2} \text{Trace}(R^T U \hat{u}_2 \hat{u}_2^T)$$

$$+ \sqrt{\lambda_3} \text{Trace}(R^T U \hat{u}_3 \hat{u}_3^T)$$

$\text{Trace}(R^T U \hat{u}_i \hat{u}_i^T)$ can be rewritten as $(R\hat{u}_i \cdot U\hat{u}_i)$. Since $\hat{u}_i$ is a unit vector, and since U and R are orthonormal transformations, $(R\hat{u}_i \cdot U\hat{u}_i) \leq 1$, with equality only when $R\hat{u}_i = U\hat{u}_i$. Therefore:

$$\text{Trace}(R^T US) \leq \sqrt{\lambda_1} + \sqrt{\lambda_2} + \sqrt{\lambda_3} = \text{Trace}(S)$$

The maximum value of Trace $(R^T US)$ occurs when $R^T U = I$, that is, when $R = U$. Therefore, the rotation R necessary to minimize the error in equation (9.13) is given by

$$R = U = M\left(\frac{1}{\sqrt{\lambda_1}} \hat{u}_1 \hat{u}_1^T + \frac{1}{\sqrt{\lambda_2}} \hat{u}_2 \hat{u}_2^T + \frac{1}{\sqrt{\lambda_3}} \hat{u}_3 \hat{u}_3^T \right) \tag{9.15}$$

Step 7. Compute R using equation (9.15). R is an orthonormal matrix that encapsulates the rotation and possible reflection necessary to transform $x_{l,i}$ into $x_{r,i}$.

Step 8. Now we have R and s, so use equation (9.8) to compute t. R, s, and t form a complete linear transformation between the two coordinate systems that minimizes equation (9.7).

Step 9. For each point x in the left-hand coordinate system, compute the corresponding position x' in the right-hand coordinate system using

$$x' = t + sRx$$

Even though this math may look imposing, it is straightforward to implement, and gives provably optimal results. As you will see shortly, many algorithms depend on coordinate system registration, either to shift a completely localized relative topology into global coordinates, or to "stitch together" small local topologies into a single consistent coordinate assignment. This Appendix described a powerful closed-form method of performing the necessary registration operations.

APPENDIX C: MULTIDIMENSIONAL SCALING

Multidimensional scaling (MDS) was originally developed for use in mathematical psychology. It comes in many variations, but all the variations share a common goal. Given a set of points whose positions are unknown and measured distances between each pair of points, multidimensional scaling determines the underlying dimensionality of the points, and finds an embedding of the points in that space that honors the pairwise distances between them.

Clearly, MDS has potential in the sensor localization domain. Using only ranging data, without anchors or GPS, MDS can solve for the relative coordinates of a group of sensor nodes with resilience to measurement error and rather high accuracy.

This Appendix focuses on a type of multidimensional scaling called *classical metric MDS*, classical because it uses only one matrix of "dissimilarity" or distance information, and metric because the dissimilarity information is quantitative (e.g., distance measurements), as opposed to ordinal. There are many other types, but they are not common in sensor networks, so they are omitted for brevity.

Let there be n sensors in a network, with positions $X_i, i = 1 \cdots n$, and let $X = [X_1, X_2, \ldots, X_n]^T$. Here, X is $n \times m$, where m is the dimensionality of X. For now, consider m to be an unknown. Let $D = [d_{ij}]$ be the $n \times n$ matrix of pairwise distance measurements, where d_{ij} is the measured distance between X_i and X_j for $i \neq j$, and $d_{ii} = 0$ for all i. The distance measurements d_{ij} must obey the triangular inequality: $d_{ij} + d_{ik} \geq d_{jk}$ for all (i, j, k).

The goal of MDS is to find an assignment of X in low-dimensional space that minimizes a "stress function," defined as:

$$X = \underset{x}{\text{argmin}} \, \text{Stress}(X) \qquad (9.16)$$

$$\text{Stress}(X) = \sqrt{\frac{\sum_{i=1}^{n}\sum_{i=1}^{i-1}(d_{ij} - \delta_{ij})^2}{\sum_{i=1}^{n}\sum_{j=1}^{i-1}\delta_{ij}^2}} \tag{9.17}$$

In equation (9.16), δ_{ij} is the distance between X_i and X_j. Thus, the metric MDS stress function is closely related to the squared error function we have seen in other techniques, such as multilateration (Subsection 9.5.3).

Classic metric multidimensional scaling is derived from the law of cosines, which states that given two sides of a triangle d_{ij}, d_{ik}, and the angle between them θ_{jik}, the third side can be computed using the formula:

$$d_{jk}^2 = d_{ij}^2 + d_{ik}^2 - 2d_{ij}d_{ik}\cos\theta_{jik} \tag{9.18}$$

Rewriting:

$$d_{ij}d_{ik}\cos\theta_{jik} = \frac{1}{2}(d_{ij}^2 + d_{ik}^2 - d_{jk}^2) \tag{9.19}$$

The left side of equation (9.19) can be rewritten as a dot product:

$$(X_j - X_i)\cdot(X_k - X_i) = \frac{1}{2}(d_{ij}^2 + d_{ik}^2 - d_{jk}^2) \tag{9.20}$$

If all measurements are perfect, then a good zero-stress way to solve for the positions X is to choose some X_0 from X to be the origin of a coordinate system, and construct a matrix $B_{(n-1)x(n-1)}$ as follows:

$$b_{ij} = \frac{1}{2}\left(d_{0i}^2 + d_{0j}^2 - d_{ij}^2\right) \tag{9.21}$$

Matrix B is known as the matrix of scalar products. As we know from equation (9.20), we can write B in terms of X. Call $X'_{(n-1)xm}$ the matrix X where each of the X'_i is shifted to have its origin at X_0: $X'_i = X_i - X_0$. Then, using equations (9.20) and (9.21),

$$X'X'^T = B$$

We can solve for X' by taking an eigen decomposition of B into an orthonormal matrix of eigenvectors and a diagonal matrix of matching eigenvalues:

$$B = X'X'^T = UVU^T$$
$$X' = UV^{1/2} \tag{9.22}$$

The problem is that X' has too many columns: we need to find X in 2-space or 3-space. To do this, we throw away all but the two or three largest eigenvalues from V, leaving a 2×2 or 3×3 diagonal matrix, and throw away the matching eigenvectors (columns) of U, leaving $U_{(n-1)x2}$ or $U_{(n-1)x3}$. Then X' has the proper dimensionality.

Note that this method produces a coordinate system that is a linear transformation from the coordinate system of the true X_i. Reconciling the two requires a registration procedure like that of Appendix B.

Remember, though, that we said this method only works when the data d_{ij} is perfect, which is an unrealistic assumption. In practice, there is some error, which ends up in the stress value of the final coordinate assignment. Fortunately, the classic metric MDS method generalizes to gracefully cover measurement errors. In the preceding, we chose a single point from our data to be the origin. This choice gives X_0 an undue influence on the error of X. Thus, real MDS does not use a point from the data; rather, it uses a special point in the center of the X_i. This point is found by "double centering" the squared distance matrix. The squared distance matrix $D^2 = [d_{ij}^2]$. To double center a matrix, subtract the row and column means from each element. Then, add the grand mean to each element. Finally, multiply by $-1/2$. The elementwise formula for double centering is below:

$$
b_{ij} = -\frac{1}{2}\left(d_{ij}^2 - \frac{1}{n}\sum_{k=1}^{n} d_{kj}^2 - \frac{1}{n}\sum_{k=1}^{n} d_{ik}^2 + \frac{1}{n^2}\sum_{k=1}^{n}\sum_{l=1}^{n} d_{kl}^2\right)
$$

$$
= \sum_{a=1}^{m} x_{ia}x_{ja}
$$

(9.23)

Reformulating equation (9.23) in matrix notation:

$$
B_{n\times n} = -\frac{1}{2}JD^2J = XX^T
$$

(9.24a)

$$
J_{n\times n} = I_{n\times n} - \frac{1}{n}e^T e
$$

(9.24b)

$$
e_{1\times n} = [1, 1, 1, \ldots, 1]
$$

(9.24c)

Equation (9.24) is an expression for X in terms of D, in m-dimensional space. If $m = n - 1$, then there is a trivial assignment of $X_1 \cdots X_n$ that makes Stress $(X) = 0$. As m decreases, it turns out that Stress (X) must increase or stay the same; it cannot decrease. We know that the measurements D originate from a two- or three-dimensional space. If the measurements from D are perfect, then there is a zero-stress assignment of X when $m = 2$ or 3. However, measurement error makes it unlikely that such an assignment really exists. Thus, some stress is inevitable as we reduce the dimensionality from n to 2 or 3.

As before, this dimensionality reduction is done by taking an eigen decomposition of B, then removing eigenvalues and eigenvectors. This is a safe operation because B is symmetric positive definite, and therefore has n positive eigenvalues.

$$
B = XX^T = UVU^T
$$

$$
X = UV^{-1/2}
$$

(9.25)

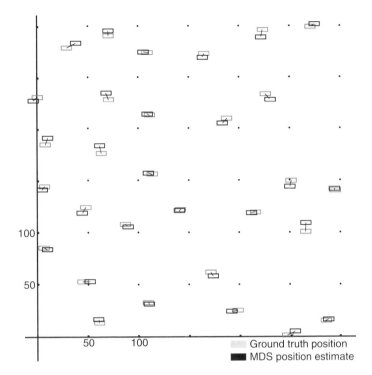

Figure 9.10 Topology constructed by multidimensional scaling. Each internode range measurement has zero-mean Gaussian error with a standard deviation of 10 units.

Thus, multidimensional scaling provides a method of converting a complete matrix of distance measurements to a matching topology in 2-space or 3-space. This conversion is quite resilient to measurement error, since increased measurement error simply becomes an increase in the stress function. To see an example of MDS in action, look at Figure 9.10.

Unfortunately, multidimensional scaling has some disadvantages. First, the main computation of MDS, the eigen decomposition of B (equation (9.25)) requires $O(n^3)$ time. As a result, a single pass of multidimensional scaling cannot operate on a large topology, particularly in the constrained computational environment of sensor networks. Second, classic MDS requires that D contain a distance measurement for *all* pairs of nodes. This requirement is impossible to meet with ranging hardware alone in large networks; thus, implementations of MDS in sensor networks must do preprocessing on measured data to generate D (Subsection 9.4.2) or use coordinate system stitching to distribute the computation (9.7).

To conclude, here are the steps of classic metric multidimensional scaling:

Step 1. Create the symmetric matrix $D = [d_{ij}]$, with $d_{ii} = 0$ and $d_{ij} + d_{ik} \geq d_{jk}$.

Step 2. Create the symmetric matrix J (equation 9.24b).

Step 3. Compute B using $D^2 = [d_{ij}^2]$ and J (equation (9.24a)).

Step 4. Take an eigen decomposition UVU^T of B.

Step 5. Let V_d be the diagonal matrix of the d largest eigenvalues in V, where d is the desired dimensionality of the solution.

Step 6. Let U_d be the d eigenvectors from U that match the eigenvalues in V_d.

Step 6. Compute $X_d = [X_1, X_2, \ldots X_n]^T$ using $X_d = U_d V_d^{1/2}$. Here, $V_d^{1/2}$ can be computed by taking the square root of each of V_d's diagonal elements.

Step 7. (Optional) Transform the X_i from X_d into the desired global coordinate space using some coordinate system registration algorithm (Appendix B). These transformed X_i are the solution.

REFERENCES

1. J. Hill, R. Szewczyk, A. Woo, S. Hollar, D. Culler, and K. Pister. System architecture directions for networked sensors. In *Proceedings of the 9th International Conference on Architectural Support for Programming Language and Operating Systems (ASPLOS-IX)*, pages 93–104, Cambridge, Massachusetts, November 2000.

2. P. Bahl and V. Padmanabhan. Radar: An in-building rf-based user location and tracking system. In *Proceedings of the 19th Annual Joint Conference of the IEEE Computer and Communication Societies (INFOCOM 2000)*, pages 775–784, Tel-Aviv, Israel, March 2000.

3. Cameron Whitehouse. The Design of Calamari: An Ad-Hoc Localization System for Sensor Networks. Master's thesis, University of California at Berkeley, 2002.

4. L. Kleinrock and J. A. Silvester. Optimum transmission radii for packet radio networks or why six is a magic number. In *Proceedings of the IEEE National Telecommunications Conference*, pages 4.3.1–4.3.5, Birmingham, Alabama, December 1978.

5. R. Nagpal, H. Shrobe, and J. Bachrach. Organizing a global coordinate system from local information on an ad hoc sensor network. In *Proceedings of the 2nd International Workshop on Information Processing in Sensor Networks (ISPN '03)*, Palo Alto, California, April 2003.

6. H. Balakrishnan, R. Baliga, D. Curtis, M. Goraczko, A. Miu, N. Priyantha, A. Smith, K. Steele, S. Teller, and K. Wang. Lessons from developing and deploying the cricket indoor location system. Preprint, November 2003.

7. Y. Kwon, K. Mechitov, S. Sundresh, W. Kim, and G. Agha. *Resilient Localization for Sensor Networks in Outdoor Environments*. Technical Report UIUCDCS-R-2004-2449, University of Illinois at Urbana-Champaign, June 2004.

8. N. Priyantha, A. Miu, H. Balakrishnan, and S. Teller. The cricket compass for context-aware mobile applications. In *Proceedings of the 7th Annual ACM/IEEE International Conference on Mobile Computing and Networking (MobiCom)*, pages 1–14, Rome, Italy, July 2001.

9. L. Doherty, L. El Ghaoui, and K. S. J. Pister. Convex position estimation in wireless sensor networks. In *Proceedings of the 20th Annual Joint Conference of the IEEE Computer and Communications Societies (INFOCOM 2001)*, Volume 3, pages 1655–1663, Anchorage, Alaska, April 2001.

10. Y. Shang, W. Ruml, Y. Zhang, and M. P. J. Fromherz. Localization from mere connectivity. In *Proceedings of the 4th ACM International Symposium on Mobile Ad Hoc Networking and Computing (MobiHoc 2003)*, pages 201–212, Annapolis, Maryland, June 2003.

11. N. Bulusu, V. Bychkovskiy, D. Estrin, and J. Heidemann. Scalable, ad hoc deployable rf-based localization. In *Proceedings of the Grace Hopper Celebration of Women in Computing Conference 2002*, Vancouver, British Columbia, Canada, October 2002.

12. S. Fitzpatrick and L. Meertens. Diffusion based localization. Private communication, 2004.

13. A. Savvides, H. Park, and M. Srivastava. The bits and flops of the n-hop multilateration primitive for node localization problems. In *Proceedings of the 1st ACM International Workshop on Wireless Sensor Networks and Applications (WSNA)*, pages 112–121, Atlanta, Georgia, September 2002.

14. A. Savvides, C.-C. Han, and M. B. Srivastava. Dynamic fine-grained localization in ad-hoc networks of sensors. In *Proceedings of the 7th Annual Conference on Mobile Computing and Networking*, pages 166–179, Rome, Italy, July 2001.

15. S. Simic and S. Sastry. Distributed Localization in Wireless Ad Hoc Networks. Memorandum No. UCB/ERL M02/26, University of California, Berkeley, 2002.

16. W. J. Butera. Programming a Paintable Computer. Ph.D. thesis, Massachusetts Institute of Technology, 2002.

17. D. Niculescu and B. Nath. Localized positioning in ad hoc networks. *Ad Hoc Networks*, 1:247–259, 2003.

18. J. D. McLurkin. Algorithms for Distributed Sensor Networks. Master's thesis, UCB, December 1999.

19. R. Nagpal, H. Shrobe, and J. Bachrach. Organizing a global coordinate system from local information on an ad hoc sensor network. In *Proceedings of the 2nd International Workshop on Information Processing in Sensor Networks (IPSN '03)*, Palo Alto, California, April 2003.

20. D. Niculescu and B. Nath. Ad hoc positioning system (APS). In *Proceedings of GLOBECOM '01 (IEEE)*, pages 2926–2931, San Antonio, Texas, November 2001.

21. D. Niculescu and B. Nath. Ad hoc positioning system (APS) using AOA. In *Proceedings of IEEE INFOCOM 2003—The Conference on Computer Communications*, 22(1):1734–1743, March 2003.

22. R. Nagpal. Organizing a global coordinate system from local information on an amorphous computer, 1999.

23. T. He, C. Huang, B. Blum, J. Stankovic, and T. Abdelzaher. Range-Free Localization Schemes in Large Scale Sensor Networks. Paper Submitted to MobiCom 2003.

24. C. Savarese, J. Rabaey, and J. Beutel. Locationing in distributed ad-hoc wireless sensor networks. In *Proceedings of the 2001 International Conference on Acoustics, Speech, and Signal Processing (ICASSP 2001)*, pages 2037–2040, Salt Lake City, Utah, May 2001.

25. N. Priyantha, H. Balakrishnan, E. Demaine, and S. Teller. Anchor-free distributed localization in sensor networks. In *Proceedings of the 1st International Conference on Embedded Networked Sensor Systems (SenSys-03)*, pages 340–341, Los Angeles, California, November 2003.

26. S. Kirkpatrick, C. D. Gelatt Jr., and M. P. Vecchi. Optimization by simulated annealing. *Science*, no. 4598, pages 671–680, May 13, 1983.

27. S. Capkun, M. Hamdi, and J.-P. Hubaux. GPS-free positioning in mobile ad-hoc networks. In *Proceedings of the 34th Annual Hawaii International Conference on System Sciences (HICSS-34)*, Volume 9, page 9008, Maui, Hawaii, January 2001.

28. L. Meertens and S. Fitzpatrick. *The Distributed Construction of a Global Coordinate System in a Network of Static Computational Nodes from Inter-Node Distances.* Kestrel Institute Technical Report KES.U.04.04, Kestrel Inistitute, Palo Alta, California, 2004.

29. D. Moore, J. Leonard, D. Rus, and S. Teller. Robust distributed network localization with noisy range measurements. In *Proceedings of the 2nd International Conference on Embedded Networked Sensor Systems (SenSys-04)*, pages 50–61, Baltimore, Maryland, November 2004.

30. X. Ji and H. Zha. Sensor positioning in wireless ad hoc networks using multidimensional scaling. In *Proceedings of IEEE INFOCOM 2004—The Conference on Computer Communication*, **23**(1):2652–2661, March 2004.

31. N. B. Priyantha, H. Balakrishnan, E. D. Demaine, and S. Teller. Mobile-assisted localization in wireless sensor networks. In *Proceedings of the 24th Annual Joint Conference of the IEEE Communications Society on Computer Communications (INFOCOM 2005)*, Miami, Florida, March 13–17, 2005.

32. B. K. P. Horn, H. Hilden, and S. Negahdaripour. Closed-form solution of absolute orientation using orthonormal matrices. *Journal of the Optical Society of America A*, **5**(7), 1988.

CHAPTER 10

Topology Construction and Maintenance in Wireless Sensor Networks

JENNIFER C. HOU and NING LI

University of Illinois, Urbana, Illinois

IVAN STOJMENOVIĆ

University of Ottawa, Ontario, Canada

Energy efficiency and network capacity are two of the most important issues in wireless sensor networks. Topology-control algorithms have been proposed to maintain network connectivity while reducing energy consumption and improving network capacity. Several studies have also been performed to investigate critical conditions on several network parameters in order to ensure network k-*connectivity (in the asymptotic sense). In this chapter, several problems (and their corresponding solutions) related to topology construction, maintenance, and connectivity in wireless sensor networks are discussed. Specifically, topics discussed include (1) various communication models and generation of random network topologies; (2) neighbor discovery and maintenance; (3) basic connectivity properties of wireless sensor networks (with the random unit graph model as the underlying model); (4) localized topology construction algorithms, along with their associated geometric structures in both homogeneous and heterogeneous networks; and (5) how to enhance fault tolerance in topology construction and connectivity.*

10.1 INTRODUCTION

To construct and maintain an efficient network topology is a very important task in wireless sensor networks. Instead of transmitting with the maximal power, nodes in a multihop wireless network collaboratively determine their transmission power and

Handbook of Sensor Networks: Algorithms and Architectures, Edited by Ivan Stojmenović
Copyright © 2005 John Wiley & Sons, Inc.

define the network topology by forming the proper neighbor relation under certain criteria. This is in contrast to the "traditional" network, in which each node transmits with its maximal transmission power and the topology is built implicitly by routing protocols (that update their routing caches as in timely a way as possible) without considering the power issue. A desirable network topology not only reduces energy consumption and prolong network lifetime, but also improves spatial reuse (and hence the network capacity [1]) and mitigate the medium-access control (MAC) level contention.

The research in topology construction and connectivity has been approached independently along two paths. In one path, researchers aim to determine critical conditions on network parameters (such as the transmission range [2–5], the number of neighbors [5,6], the minimum total power required [7–11], or the node failure probability) to ensure network (k-)connectivity with high probability. Of particular interest is how these critical conditions scale as the number of wireless devices increases. In the other path, researchers aim to devise distributed algorithms to enable each node to choose its own transmission power in order to minimize the total transmission power of all wireless nodes, while maintaining (k-)connectivity. This problem is, in general, NP-hard in the Euclidean plane [8]. What makes the problem more complicated is that there is, in general, no central authority in a multiple-hop wireless network, and each node has to make its decision based on the information collected from the network. This implies the topology construction algorithm should be distributed. To be less susceptible to mobility, the algorithm should depend only on the information collected locally. Algorithms that depend only on local information also incur less message overhead/delay in the process of information collection, and are hence more scalable. Several researchers have leveraged elegant *graph-theoretic* structures to develop localized heuristics [12–19] or efficient algorithms with bounded approximation ratios [20,21].

In this chapter, we give an overview of research activities along these two research thrusts, and present several problems (and their corresponding solutions) related to topology construction, maintenance, and connectivity in wireless sensor networks. We consider the problem that arises in topology construction (1) when the network is heterogeneous and nodes may have different attributes (such as transmission radii); and (2) when fault tolerance has to be taken into account (in addition to the primary objectives of network connectivity and reduction in power consumption).

The rest of this chapter is organized as follows. In Section 10.2, we introduce the communication models for wireless sensor networks, including the unit-graph model and other probabilistic models. In Section 10.3, we discuss the methods for generating random unit graphs. The issue of neighbor discovery and maintenance is treated in Section 10.4. Following that, we summarize in Section 10.5 several localized geometric structures for topology construction, and discuss in Section 10.6 the problem that arises in the case of network heterogeneity. Several essential connectivity properties of wireless sensor networks (with the random unit graph as the underlying model) are presented in Section 10.7. Finally, we discuss the issue of enhancing fault tolerance in topology construction and connectivity in Section 10.8.

10.2 COMMUNICATION MODELS

In the subsequent discussion, we assume that all the communication activities of a node take place in a single wireless channel, that is, each wireless device has one transmitter and one receiver. A widely accepted basic graph-theoretical model for wireless sensor networks is the *unit-graph* model. In the unit-graph model, two nodes A and B are neighbors (and thus joined by an edge) if the Euclidean distance between their coordinates in the network is at most R, where R is the transmission radius and is assumed to be the same for all nodes in the network. Figure 10.1 gives an example of a unit graph with transmission radius as indicated. Because of limited transmission radius, the routes are normally created through several hops in such a multihop wireless network.

The unit-graph model assumes that all nodes use the same and fixed-transmission radius. Variation of this model includes unit graphs with obstacles (or subgraphs of unit graphs), and *minpower* graphs where each node has its own maximum transmission radius and links are unidirectional or allowed only when bidirectional communication is possible. Nodes in an ad hoc network may transmit with their maximum transmission radius, or may adjust their transmission range, normally selected from a discrete set of possible values.

The network is normally assumed *homogeneous*, with all the nodes possessing the same network attributes (such as computational capacity, battery power, and transmission radii). In *heterogeneous* networks, however, nodes may have different network attributes. An example is heterogeneous (and often hierarchical) sensor networks, in which a large number of low-cost lightweight wireless devices (that simply sense the environmental changes) and a few energy-rich devices (that serve as cluster heads for data aggregation and in-network processing) coexist. They scatter in a geographic region, have to dynamically organize themselves,

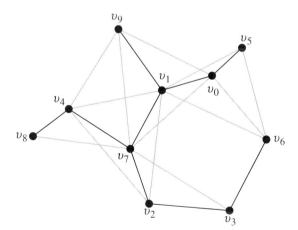

Figure 10.1 Unit-graph representation of multihop wireless network.

and convey the location/tracking information periodically or on-demand to data sinks. Each type of node has its own battery power (and hence transmission radius), computational capacity, and memory.

Credible research was conducted in the literature only for the unit-graph model, while other models are sporadically mentioned, without many results. One example is *fuzzy unit graph*, used in [22] to improve routing with guaranteed delivery [23]. In this model, there exist two transmission radii, r and R. Two nodes always communicate if their distance is $\leq r$, never communicate of their distance if $\geq R$, and may or may not communicate otherwise.

The unit-graph model is ideal in the sense that the probability of receiving a packet between two nodes does not, in reality, suddenly change from 1 to 0 when nodes move from distance $R - \varepsilon$ to $R + \varepsilon$ for a very small ε. There exist few articles that consider a more realistic physical layer, but the first one to formalize the communication model is ref. [24]. They applied the log normal shadow fading model to represent a realistic physical layer to derive the probability $p(x)$ for receiving a packet successfully as a function of distance x between two nodes, as shown in Figure 10.2. The transmission radius R is defined in ref. [23] as the distance at which $p(R) = 0.5$. Then $p(x)$ is approximated reasonably accurately by $P(x) = 1 - (x/R)^{v\beta}/2$ for $x < R$, and $((2R - x)/R)^{v\beta}/2$ otherwise, where β is the power attenuation factor (between 2 and 6), and v depends on packet length L (e.g., $v = 2$ for $L = 120$). Two nodes are considered neighbors if the distance x between them is such that $p(x) \geq w$, where w is a threshold parameter (for example, when $w = 0.05$, then $x \approx 1.4R$).

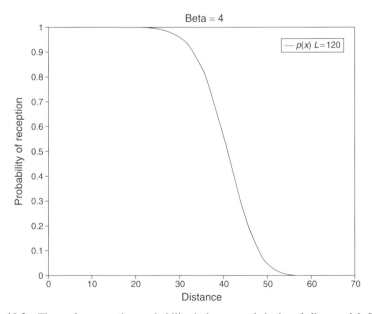

Figure 10.2 The packet reception probability in log normal shadow fading model, $R \approx 41$.

10.3 GENERATING RANDOM UNIT GRAPHS

Parameters of a random unit graph are normally the number of nodes N and the common transmission radius R. However, R may be a misleading parameter when used in simulations, since all generated graphs may be either dense or sparse, which has significant impact on the performance (for example, of routing protocols). It is therefore preferable to use another parameter that can be easily interpreted and matched directly with the graph density. That parameter is the average number of neighbors d per each node. There exists a relation between R and d, since an approximate radius R can be obtained from the formula $d = (N - 1)\pi R^2/A$, where A is the area of the region where nodes are placed. In case of a square of edge length a, $A = a^2$. This formula is obtained by finding the expected number of nodes inside a circle with radius R (the formula multiplies number of other nodes with probability of being generated inside the region).

In the literature, random unit graphs are normally generated by selecting each of N nodes at random locations inside a square or circle. In the case of a square region, this means generating x and y coordinates at random in an interval $[0, a)$. Then all $N*(N - 1)/2$ potential edges in the network among the N nodes are sorted by their length in ascending order. The radius R that corresponds to a chosen value of d is equal to the length of the $N*d/2$-th edge in the sorted order. Any edge no longer than R will remain in the graph. Other edges are eliminated from the graph. Dijkstra's *shortest-path algorithm* (from one node to all other nodes) is used to check the connectivity of the graph, if desired. This generation method is proposed in ref. [25].

Sparse random unit disk graphs have high probability of being partitioned, which increases the generation time of connected ones. In [52], Atay and Stojmenovic addressed the following two problems: fast generation of sparse connected random unit graphs and the nature of obtained graphs. They describe several new generation schemes that resemble conference scenarios. These schemes select the next node position based on the distribution of the nodes already placed. In the *Minimum Degree Proximity Algorithm (MIN-DPA)*, first a center node C is selected at random among nodes having the smallest number of neighbors. Then the new node is placed at random but constrained to be neighbor of C. The placed node might affect the degree of not only the center node but also other nodes in the proximity. Based on this observation, *Maximum Degree Proximity Algorithm (MAX-DPA)* imposes a maximum degree constraint for each node. In each round, a random position X is generated repeatedly until it passes both the *proximity test* (to be the neighbor of at least one already selected node) and the *maximum degree test*. To check the latter, the approximate degrees of X and all already selected nodes are calculated assuming that X is added to the set. If none of these degrees is greater than or equal to the maximum degree allowed *dmax*, the position is accepted. Authors [52] show that their new algorithms are significantly faster than the well-known generation scheme for sparse graphs, and also have lower standard deviation in node degree. In addition, they analyze the degree distribution and partition patterns of graphs generated by different algorithms.

10.4 NEIGHBOR DISCOVERY

It is usually assumed in the literature that each node is aware of its direct neighbors. In the case that the assumption does not hold, a node may broadcast a "hello" message, with all the nodes that receive the message being defined as the neighbors. Although the problem of neighbor discovery does not appear trivial in real networks, where transmission activities may interfere with each other and messages may collide, surprisingly, this problem did not receive much attention in this literature.

Alonso et al. [27] proposed a protocol for node discovery in single-hop ad hoc networks, where each node is within transmission radius of each other. The protocol discovers one edge in the network, which is considered to be detected after one endpoint was a single transmitter in one step, followed by the other endpoint being a single transmitter in the next step with the edge confirmation. Therefore the protocol remains short of achieving the larger goal of recognizing all edges and all nodes in the network, since repeated application of one of the protocols described may not lead to an efficient and terminating scheme for that goal.

In multihop ad hoc networks, nodes may not be within communication range of each other. We assume that node B detects edge AB if B receives a packet from A without any collision. Nodes that send "hello" messages are not able to recognize a collision, because the signal from its own transceiver is so strong that the receiver of the transmitting node cannot hear or even recognize the existence of any other signal. The problem can be considered to have two variants: with and without time synchronization of the sensors.

McGlynn and Borbash [28] assumed that time is synchronized and is divided into slots. They considered the variant of the neighbor discovery problem by enabling sensors to broadcast "hello" messages in different starting slots, and devised methods for saving energy during the neighbor discovery process over a longer period of time. They propose some variants of birthday protocols, where each sensor may be active or in the sleeping mode. In their protocol, the average number of neighbors N of each sensor is overestimated by a fixed number N', equal to all sensors. Then each sensor transmits the message with probability $1/N'$.

The variant of the problem when all sensors are active and all begin initialization in the same time slot is considered in ref. [29]. In this protocol, each node maintains its own estimate N of the number of its neighbors based on its own collision experience, and transmits with probability $1/N$. If a nontransmitting node experiences collision, N is doubled. If silence is detected, N is halved. If a nontransmitting node hears only one neighbor, N does not change. The algorithm in ref. [29] is applied to single-hop networks (complete graphs), but can be extended to multihop networks.

Neighbor discovery with all the physical layer characteristics taken into account is considered in ref. [24]. The simplest protocol is that each node transmits "hello" packets for a prespecified number of times, thus increasing the chance of being discovered by its neighbor(s). In the variable transmission-count protocol, each node transmits "hello" packets until it correctly identifies a certain number of neighbors. Note that protocol in ref. [24] does not consider the impact of collisions. The problem becomes much more difficult if both the physical-layer characteristics and collisions are taken into account; this has not been addressed in the literature.

10.5 LOCALIZED STRUCTURES FOR TOPOLOGY CONSTRUCTION

In this section, we describe some basic geometric structures used in the various topology construction protocols. Most of this section is devoted to a "chain" of planar and such connected structures: MST (minimum spanning tree), LMST (local minimum spanning tree), RNG (relative neighborhood graph), GG (Gabriel graph), PDT (partial Delaunay triangulation), and DT (Delaunay triangulation). They create a chain since MST $\subseteq$ LMST $\subseteq$ RNG $\subseteq$ GG $\subseteq$ PDT $\subseteq$ DT. MST and DT are global structures, while LMST, RNG, GG, and PDT are constructed based only on local knowledge. Some other structures exist in literatures that need only local knowledge. However, in addition to the knowledge of (mostly one-hop and sometimes k-hop) neighbors, they may also require nodes to send additional messages between them in order to construct the graph. In addition, for a complete treatment of the subject, we introduce relay-region–based and cone-based topology construction approaches.

10.5.1 Minimum Spanning Tree

MST is a subgraph of a given unit graph that is connected, contains all the nodes, and whose sum of edge lengths is minimized. It can be constructed using Dijkstra's algorithm as follows. All edges are sorted according to the increasing order by

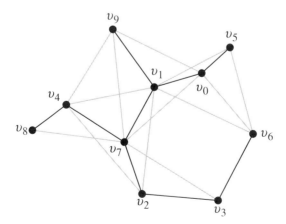

Figure 10.3 MST for a unit graph.

their lengths. Each edge is tested in that order (from shortest to longest) for inclusion in MST. If adding the edge to the already constructed graph does not create a cycle, the edge is then added to the constructed graph (that is, eventually to MST); otherwise it is not added. This construction also can be used to show that the connected unit graph contains MST as its subgraph (since edges whose length does not exceed the transmission radius will already connect the graph while constructing MST). Figure 10.3 shows an MST for a unit graph, with MST edges drawn in bold lines.

When several edges are of the same length, the MST may not be unique. In order to have a unique structure for MST, one can label edges so that all edges become distinct, and edges of the same length can be compared. This can be achieved by introducing the weight of an edge AB as a record $w(AB) = (|AB|, \min(\mathrm{id}(A), \mathrm{id}(B)),$ $\max(\mathrm{id}(A), \mathrm{id}(B)))$, where $|AB|$ is the length of AB, and $\mathrm{id}(A)$ and $\mathrm{id}(B)$ are unique identifiers of its endpoints, and $|AB|$, $\min(\mathrm{id}(A), \mathrm{id}(B))$, and $\max(\mathrm{id}(A), \mathrm{id}(B))$ serve as the primary, secondary, and ternary keys in comparisons. Two edges are compared first by their primary keys. If the primary keys are the same, secondary keys are used. If the secondary keys are also the same, the ternary keys are used. The defined edge weights can also be applied to other structures, which are described in the following subsections, thus eliminating ambiguity and leading to some desirable characteristics (e.g., degree limitation).

10.5.2   Local Minimum Spanning Tree

A localized MST-based topology control algorithm was proposed in ref. [14]. Each node u first collects positions of its one-hop neighbors $N1(u)$. Node u then computes the minimum spanning tree $\mathrm{MST}(N1(u))$ of $N1(u)$. Node u keeps a directed edge uv in LMST if and only if the edge uv is also an edge in $\mathrm{MST}(N1(v))$. If each node already has 2-hop neighboring information, the construction does not involve any message exchange between neighboring nodes; otherwise, each node contacts

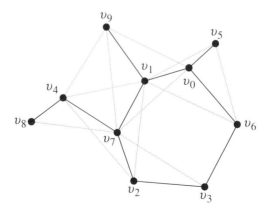

Figure 10.4   LMST for a unit graph.

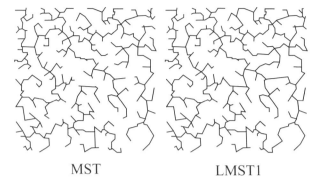

MST   LMST1

Figure 10.5   MST and LMST in two dimensions.

neighbors along its LMST link candidates, to verify the status at other nodes. The variant with the union of edge candidates rather than their common intersection is also considered in ref. [14], possibly leading to a directed graph (no message exchange is then needed even with one-hop neighbor information). Figure 10.4 shows an example of an LMST graph.

In ref. [18], Li et al. showed that LMST is a planar graph (no two edges intersect). Then they extended LMST to k-hop neighbors, that is, the same construction but with each node having more local knowledge. They also proved that MST is a subset of 2-hop–based LMST, but MST is not a subset of the one-hop–based LMST considered in this article. We observed, however, on their diagrams that LMST with 2-hop and higher local knowledge was mostly identical to the one constructed with only one-hop knowledge, and recommend that only that limited knowledge be used, thereby reducing the communication overhead needed to maintain k-hop knowledge. MST and LMST are illustrated in Figure 10.5.

Theorem 10.1   MST is a subset of LMST [30].

Proof. Let $LMST(A) = MST(n(A))$ be the minimal spanning trees constructed from $n(A)$, which is the set containing A and its one-hop neighbors. We will show that if an edge from MST has endpoints in $n(A)$, then it belongs to $LMST(A)$. Suppose that this is not true, and let e be the shortest such edge. $LMST(A)$ can also be constructed by following Kruskal's algorithm described earlier. Thus edges from A to its neighbors and between neighbors of A are sorted in increasing order. They are then considered for inclusion in $LMST(A)$. Thus, when e is considered, since it is not included in $LMST(A)$, it creates a cycle C in $LMST(A)$, with e being the longest edge in that cycle. Some of the edges from C are not in MST. Consider now expanded cycle C' constructed from C as follows. Let f be an edge from C that is not in MST. Addition of f into MST creates a cycle B, with f being the longest edge in the cycle. The cycle consists of f and a path consisting of edges from MST. Replace f in C with all the edges from that path. Each such replacement enlarges the cycle C, but does not add any edge longer than f, and consequently longer than e. At the end of this process, after replacing all non-MST edges with the corresponding paths of

MST edges, edge e remains the longest edge of C', but all the other edges of C' are now also in MST. This is a contradiction, since MST has no cycles. Therefore each edge AB from MST belongs to both LMST(A) and LMST(B), and therefore to LMST. $\qquad\square$

10.5.3 Relative Neighborhood Graph

Relative neighborhood graph (RNG) is introduced by Toussaint [31], and can be defined, in the simplest form, as follows. An edge uv is included in RNG if and only if it is not the longest edge in any triangle uvw. Figure 10.6 illustrates this definition. Consider the "lune," which is the intersection of two circles centered at u and v, both with radius uv. Edge uv is in RNG if and only if this lune has no other nodes from an ad hoc network in its interior. In Figure 10.2, uv is not in RNG because of a witness neighbor w that makes uv the longest edge in triangle uvw.

Using this definition, some edges may have very large degrees in several particular scenarios (e.g., many nodes can be located at the boundary of the lune just discussed). To obtain a degree-limited structure, the record $w(AB) = (|AB|, \min(\mathrm{id}(A), \mathrm{id}(B)), \max(\mathrm{id}(A), \mathrm{id}(B)))$ can be used instead, as described earlier. We refer to this structure in the sequel, assuming random node placement and a very low chance of any two edges being of the same length. The degree of such a structure is limited to 6 (for nodes located in a plane). This can be proved by contradiction. Assume that node A has more than 6 neighbors in RNG. Then there exist two consecutive neighbors B and C so that $\angle BAC < \pi/6$. Then BC is not the shortest edge in triangle ABC, and either AB or AC is. This contradicts the definition of RNG. Note again that such an argument cannot be used if one allows that $AB = AC$. Note also that the same proof can be used to show the degree of limitation of MST or LMST in two dimensions.

Theorem 10.2 The LMST of a unit graph is a subgraph of RNG of the same graph [32].

Proof. It suffices to show that if an edge uv belongs to LMST, then it belongs to RNG. By contradiction, suppose there exists an edge uv such that $uv \in$ LMST

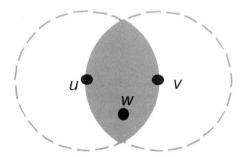

Figure 10.6 (u, v) is not in RNG graph because of a witness node w.

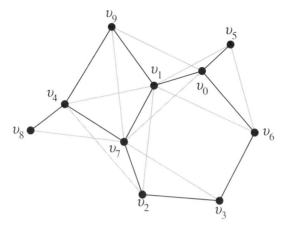

Figure 10.7 RNG in a small ad hoc network.

and $uv \notin$ RNG. Then uv belongs to MST($n(u)$) and to MST($n(v)$), and since uv does not belong to RNG, there exists a node $w \in n(u) \cap n(v)$ such that uv is the "longest" edge in triangle uvw. Either edge uw or vw is not in LMST, since LMST is a tree (from the local point of view). Without loss of generality, suppose uw is not in LMST. Then uv can be replaced by uw in MST($n(u)$), giving a spanning tree with a lower overall weight (i.e., total sum of all edge lengths) than the minimal one (MST($n(u)$)), which is a contradiction. Therefore LMST is a subgraph of RNG. □

MST, LMST, and RNG are used in a number of articles where a sparsely connected network is desirable. The average degree of a node in MST with n nodes

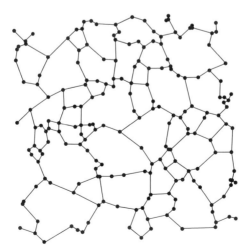

Figure 10.8 RNG in a large ad hoc network.

is $2(n - 1)/n \approx 2$. LMST has an average degree of about 2.04 (that is, about 2% more than MST), while RNG has an average degree of about 2.5.

Figure 10.7 and Figure 10.8 show RNGs in smaller and larger networks. LMST and RNG are planar graphs, which follows from the planarity of the Gabriel graph (which is their superset). Note that construction of LMST and RNG does not require that the exact positions of nodes and their neighbors be known; only the corresponding mutual distances are required. In both cases, each node requires that the distance to its neighbors be known, as well as the distances between any pair of neighbors.

10.5.4 Gabriel Graph

The Gabriel graph (GG) is proposed in ref. [33], and is defined as follows. A GG contains an edge UV if and only if the disk with diameter UV contains no other node inside it. For instance, in Figure 10.9, UV is in GG while PQ is not, because of a "witness node" W located inside the disk. This criterion can be tested in two ways. Each common neighbor W of nodes U and V should be located at a distance of at least $|UV|/2$ from the midpoint of UV for UV to be included in GG. Alternatively, one can verify the angles from neighbors to UV. As shown in Figure 10.9, if $\angle PWQ > \pi/2$ for a common neighbor W of P and Q, then PQ is not in GG. It should be observed, as in the case of LMST and RNG, that the construction of GG requires only the knowledge of the location of a node and those of its neighbors. Figure 10.10 shows an example of a GG, with GG edges drawn as thick lines.

Theorem 10.3 RNG is a subgraph of GG.

Proof. Note that the lune in the RNG definition is a subset of the disk in GG definitions. Therefore, if the lune does not contain any neighbor, then the disk also does not contain any neighbor. Therefore if an edge belongs to RNG it also belongs to GG. □

A Gabriel graph was used in ref. [23] for routing with guaranteed delivery because it was the densest known localized structure that was planar. The planarity of GG (and consequently its subgraphs RNG and LMST) can be shown from the planarity of Delaunay triangulation (see the next proof), which contains GG as its subset.

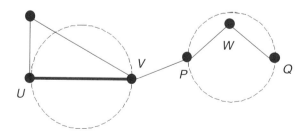

Figure 10.9 UV is included in GG, while PQ is not, because of witness W.

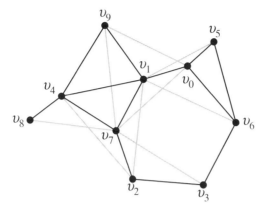

Figure 10.10 GG of a set of nodes in a unit graph.

However, in order to give a simpler proof, we concentrate on the GG case only, and prove this directly.

Theorem 10.4 The Gabriel graph is a planar graph.

Proof. Suppose that, on the contrary, GG is not a planar graph. Let UV and PQ be two of its edges that intersect (Fig. 10.11). Since UV is in GG, nodes P and Q are outside the disk with diameter UV. Therefore $\angle UPV < \pi/2$, and $\angle UQV < \pi/2$. Similarly, $\angle PUQ < \pi/2$ and $\angle PVQ < \pi/2$. Then the sum of angles in quadrilateral $UPVQ$ is $< 2\pi$. This is a contradiction, since the sum of angles in any quadrilateral is 2π. □

Huang, Lu, and Roman [34] presented some statistics on the Gabriel graph. On a random graph with 1600 nodes, they report an average face size of 4.3, 1369 faces, average node degree about 3.8, and an average size of the outer face of 248. However, they counted each dangling edge once (an edge that belongs to only one face, not to two different faces). If dangling edges are counted as two edges in the corre-

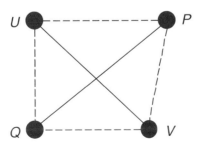

Figure 10.11 Planarity of GG, RNG, and LMST.

sponding face, then counting the average number of edges on a face of a planar graph can be done as follows. Let N, F, and E be the number of nodes, faces, and edges of a planar graph, respectively. The well-known Euler formula is $F = E - N + 2$. Let S be average number of edges on a face. Since each edge is counted twice (whether or not it belongs to same or two different faces), we have $FS = 2E$. Thus $S = 2E/F$, or $S = 2E/(E - N + 2)$.

10.5.5 Delaunay Triangulation and Partial Delaunay Triangulation

Delaunay triangulation (DT) is a well-known and frequently used applied structure in computational geometry [35]. It can be constructed in $O(n \log n)$ time for a set of n points in the plane. One definition of Delaunay triangulation is as follows: an edge uv belongs to DT if and only if there exists a circle, whose chord is uv, which does not contain any other point from the set in its interior. It immediately follows that GG is a subset of DT, since diameter uv (and the disk with diameter uv) is a special case of a chord (and the corresponding circle).

Partial Delaunay triangulation (PDT) [16] is a subgraph of DT, and contains GG as its subgraph. PDT can be constructed locally. More precisely, it is a subset of DT containing edges of DT that can be locally confirmed. Let $Disk(u, v)$ be the disk with diameter uv. Given an edge uv, we consider three cases: (1) if $Disk(u, v)$ contains no other node from the network, then $uv \in$ GG $\subseteq$ PDT; (2) if $Disk(u, v)$ contains nodes on both sides of uv, then $uv \notin DT$ and $uv \notin$ PDT; and (3) if $Disk(u, v)$ only contains nodes on one side of uv, let w be one such point that maximizes the angle $\alpha = \angle uwv \geq \pi/2$. Consider the largest angle $\beta = \angle upv$ on the other side of uv, where p is a node outside $Disk(u, v)$. If $\alpha + \beta > \pi$, then $uv \notin$ DT (and $uv \notin$ PDT); if $\alpha + \beta < \pi$ (assume that no four nodes are cocircular), we can add uv into PDT if the following two conditions are both satisfied, which depends on whether one-hop or 2-hop neighbor topology is known to each node (this defines two structures, PDT1 and PDT2, respectively). PDT definition is illustrated in Figure 10.12.

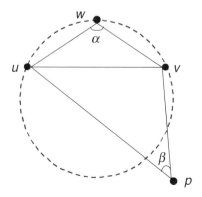

Figure 10.12 Definition of the partial Delaunay triangulation (PDT).

In the case of one-hop neighbor knowledge, assume $w \in n(u)$ is inside $Disk(u, v)$, with the largest angle $\alpha = \angle uwv \geq \pi/2$. Edge uv is added to PDT1 if (1) there is no node from $n(u)$ that lies on the different side of uv and inside the circumcircle passing through u, v, and w; and (2) $\sin \alpha > d/R$, where R is the transmission radius of each wireless node, and $d = |uv|$.

In the case of 2-hop neighbor knowledge, assume $w \in n(u) \cap n(v)$ is inside $Disk(u, v)$, with the largest angle $\alpha = \angle uwv \geq \pi/2$. Edge uv is added to PDT2 if (1) there is no node from $n(u) \cup n(v)$ that lies on the different side of uv and inside the circumcircle passing u, v, and w; and (2) $\cos \alpha/2 > d/2R$, where R is the transmission radius of each wireless node.

The construction of PDT ensures that GG $\subseteq$ PDT $\subseteq$ DT. Since DT is planar, GG and PDT are also planar. PDT1 $\subseteq$ PDT2, since more edges can be confirmed with more knowledge. PDT1 has about 2% more edges than GG. The average degree of a node of DT is about 5.9.

10.5.6 Yao Graph

The Yao$_p$ graph [36] is proposed by Yao to efficiently construct an MST in high dimensions. Any p equally separated rays originating at each node u define p cones. In each cone, u then chooses the closest node v within the transmission range, if there is any, and adds a directed link uv. This can result in a directed subgraph, as shown in Figure 10.13. Since Yao$_p$ contains MST as a subgraph, deleting all unidirectional links still preserves network connectivity. Note that Yao$_p$ is not necessarily planar.

10.5.7 Minimum Power Topology

Rodoplu and Meng [12] introduced the enclosure graph for localized power-aware topology control in ad hoc networks. The power needed for transmitting between

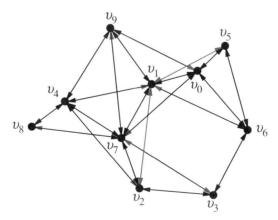

Figure 10.13 YAO$_6$ for a unit graph. The arrows indicate the direction of communication.

two nodes at distance r is proportional to $u(r) = r^\alpha + c$, where α is power attenuation factor (a number between 2 and 6), while c is a constant that accounts for the cost of running hardware at nodes and minimum reception power. Although most researchers assume $c = 0$, which enables them to prove some nice properties, in reality it is $c > 0$, which means that selecting a very close forwarding neighbor may not be the best choice when energy is the criterion. An edge AB is in the enclosure graph if and only if direct transmission between A and B is a power-optimal solution for a given set of nodes. That is, $u(|AB|) \le u(|AC|) + u(|CB|)$ for any common neighbor C of A and B. *The enclosure graph is undirectional.* In the case when $\alpha = 2$ and $c = 0$, the enclosure graph becomes equivalent to GG, illustrated in Figure 10.10.

10.5.8 Cone-Based Topology

Cone-based topology control (CBTC(α)) [13] is a two-phase localized algorithm. Each node finds the minimal power p such that transmitting with the power p ensures that the node can reach some node in every cone of degree α. The algorithm has been proved to preserve network connectivity if $\alpha < 5\pi/6$. Several optimizations to the basic algorithm have also been discussed. These include (1) a *shrink-back* operation can be added at the end to allow a boundary node to broadcast with less power, if doing so does not reduce the cone coverage; (2) if $\alpha < 2\pi/3$, asymmetric edges can be removed while maintaining the network connectivity; and (3) if there exists an edge from node u to node v_1 and from node u to node v_2, respectively, the longer edge can be removed while preserving connectivity, as long as $d(v_1, v_2) < \max\{d(u, v_1), d(u, v_2)\}$. As shown in Figure 10.14, the resulting subgraph may be directed.

An event-driven strategy was also proposed to reconfigure the network topology in the case of mobility. Each node is notified when any neighbor leaves/joins the neighborhood and/or the angle changes. The mechanism used to realize this requires the

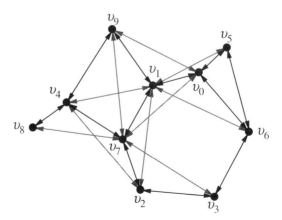

Figure 10.14   CBTC $(2\pi/3)$ of a unit graph.

state to be kept at, and message exchanges among, neighboring nodes. The node then determines whether it needs to rerun the topology control algorithm.

10.5.9 Bluetooth Scatternet Formation

Bluetooth technology is based on a medium-access frequency-hopping, where master nodes decide about a series of frequencies for communication within a piconet. A piconet consists of a master node and up to seven slave nodes. Piconets can be connected to a scatternet via common nodes that can serve as common slaves, or a master node in one piconet can serve as slave node in another piconet. Bluetooth is currently considered energy expensive for application in sensor networks. The creation of a connected degree-limited scatternet is a challenging problem that attracted research interest recently. For instance, ref. [16] proposed a solution that will optionally apply RNG, GG, or PDT to create a planar structure (if desired, e.g., for routing), and then to apply a Yao graph that will limit degree while preserving connectivity (and planarity). The last step is to decide about master–slave roles. Since the Yao graph construct selects the closest nodes in each angular range, all proposed methods tend to choose neighboring nodes that are relatively close to the current node. When routing is applied in a scatternet that is constructed in this way, the routes have relatively large hop count (although they can be power efficient). An alternative design is proposed in ref. [37]. Since LMST and RNG are degree limited, the Yao graph does not need to be applied on them. Each node has, on average, two neighbors in LMST and 2.5 neighbors in RNG, and the scatternet is already connected. Therefore the remaining links, up to seven per node, may be chosen arbitrarily. If the hop count is the optimality metric, then these links can be oriented toward the furthest neighbors in several angular ranges (to provide good forwarding neighbors in all directions). If power or other metric is considered, the best neighbors for the given metric can be added. The scatternet constructed in this way improves routing performance [36].

10.6 TOPOLOGY CONSTRUCTION IN HETEROGENEOUS NETWORKS

The homogeneity assumption usually does not hold in practice for various reasons. First, even devices of the same type may have slightly different maximal transmission power. Second, the environment settings also affect the transmission range. For instance, a wireless device usually has a larger transmission range with fewer obstacles around it. Finally, as mentioned in Section 10.2, there exist devices of dramatically different capabilities in wireless sensor networks, for example, lightweight sensor nodes versus energy-rich cluster heads. In this section, we consider the issue of topology construction in heterogeneous wireless networks, where the transmission range of each node may be different.

Most of the topology construction algorithms discussed in Section 10.5 assume homogeneous wireless nodes with uniform transmission ranges. When directly

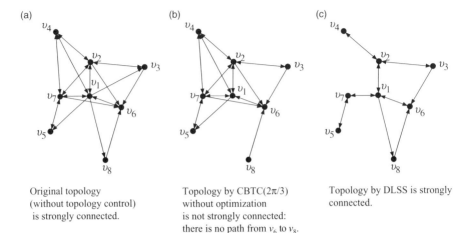

Original topology
(without topology control)
is strongly connected.

Topology by CBTC($2\pi/3$)
without optimization
is not strongly connected:
there is no path from v_6 to v_8.

Topology by DLSS is strongly
connected.

Figure 10.15 An example that shows that CBTC($2\pi/3$) may render disconnectivity in heterogeneous networks. There is no path from v_6 to v_8 due to the loss of edge (v_1, v_8), which is discarded by v_1, since v_5 and v_6 have already provided the necessary coverage.

applied to heterogeneous networks, these algorithms may render disconnectivity. For example, as shown in Figure 10.15(a) and 10.15(b) (the arrows are used to indicate the direction of the links), the network topology derived under CBTC ($2\pi/3$) without optimization may not preserve connectivity. Similarly, we show in Figure 10.16(a) and 10.16(b) that the network topology derived by RNG may be disconnected.

Since RNG is originally intended for undirected graphs only, one can tailor the definition of RNG for directed graphs. The *modified relative neighborhood graph* (MRNG) is defined as shown in Figure 10.17. An edge *uv* is included in MRNG

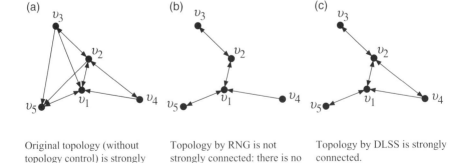

Original topology (without
topology control) is strongly
connected.

Topology by RNG is not
strongly connected: there is no
path from v_1 to v_4.

Topology by DLSS is strongly
connected.

Figure 10.16 An example that shows that RNG may render disconnectivity in heterogeneous networks. There is no path from v_1 to v_4 due to the loss of edge (v_2, v_4), which is discarded since $|(v_2, v_1)| < |(v_2, v_4)|$, and $|(v_4, v_1)| < |(v_2, v_4)|$.

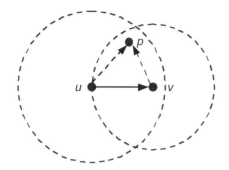

Figure 10.17   The definition of modified relative neighborhood graph (MRNG).

if and only if there does not exist a third node p such that both u and v can reach p by using the maximal transmission power, and $w(up) < w(uv)$, $w(vp) < w(uv)$. Unfortunately, the topology derived by MRNG may still be disconnected, as shown in Figure 10.18(a) and 10.18(b).

Two localized topology construction algorithms were proposed in ref. [38]. The first one, directed relative neighborhood graph (DRNG), is an extension of RNG for heterogeneous graphs. An edge uv is included in DRNG if and only if there does not exist a third node w such that both u can reach w and w can reach v by using the maximal transmission power, respectively, and $w(up) < w(uv)$, $w(pv) < w(uv)$. The second one, directed local spanning subgraph (DLSS), is an extension of LMST for heterogeneous graphs. In DLSS, each node u first builds its local spanning subgraph S_u of the one-hop neighborhood $n(u)$. The algorithm to construct S_u is similar to Kruskal's algorithm. Edges are inserted into S_u in ascending order of weight. An edge pq is kept in S_u only if p and q are not strongly connected before the inser-

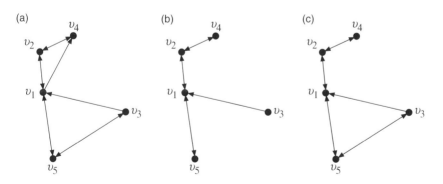

(a) Original topology (without topology control) is strongly connected.

(b) Topology by MRNG is not strongly connected: there is no path from v_1 to v_3.

(c) Topology by DLSS is strongly connected.

Figure 10.18   An example that shows MRNG may render disconnectivity in heterogeneous networks. There is no path from v_1 to v_3 due to the loss of edge (v_5, v_3), which is discarded since $|(v_5, v_1)| < |(v_5, v_3)|$, and $|(v_3, v_1)| < |(v_5, v_3)|$.

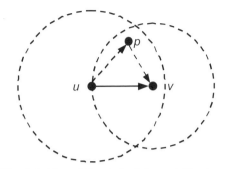

Figure 10.19   The definition of directed relative neighborhood graph (DRNG).

tion of pq. This procedure repeats until the S_u is strongly connected. Then u keeps a directed edge uv if and only if uv is also an edge in S_u. The definition of DRNG is illustrated in Figure 10.19.

The following lemmas and theorems show that both DLSS and DRNG can preserve the strong connectivity of the original graph.

Lemma 1. For any edge uv in the original graph $E(G)$, u is still connected to v in DLSS [37].

Proof. Sort all edges $uv \in E(G)$ in ascending order of weight, that is, $w(u_1v_1) < w(u_2v_2) < \cdots < w(u_lv_l)$, where l is the total number of edges. We now prove by induction.

- *Basis:* The first edge u_1v_1 satisfies $w(u_1v_1) = \min\{w(uv)|\ uv \in E(G)\}$. According to the algorithm for constructing S_u, $u_1v_1 \in E(\text{DLSS})$, that is, u_1 is connected to v_1 in DLSS.
- *Induction:* Assuming the hypothesis holds for all edges u_iv_i, $i < k$, we prove u_k is connected to v_k in DLSS. Since this is obviously true if $u_kv_k \in E(\text{DLSS})$, we only need to consider the case where $u_kv_k \notin E(\text{DLSS})$. Before edge u_kv_k was considered in the local topology construction of v_k, there must already exist a path $p = (w_0 = u_k, w_1, w_2, \ldots, w_{m-1}, w_m = v_k)$ from u_k to v_k, where $w_iw_{i+1} \in E(S_{uk})$, $w_{i+1} \in E(S_{uk})$, $i = 0, 1, \ldots, m - 1$. Since edges are inserted in ascending order of weight, we have $w(w_iw_{i+1}) < w(u_kv_k)$. Applying the induction hypothesis, w_i is connected to w_{i+1} for $i = 0, 1, \ldots, m - 1$. Therefore, u_k is connected to v_k. ☐

Theorem 10.5 (Connectivity of DLSS [37]): DLSS preserves the connectivity of G, that is, DLSS is strongly connected if G is strongly connected.

Proof. Suppose G is strongly connected. For any two nodes u, $v \in V(G)$, there exists at least one path $p = (w_0 = u, w_1, w_2, \ldots, w_{m-1}, w_m = v)$ from u to v, where $w_i w_{i+1} \in E(G)$, $i = 0, 1, \dots, m - 1$. Since w_i is connected to w_{i+1} by Lemma 10.1, u is connected to v in DLSS. □

Lemma 10.2. DLSS is a subset of DRNG [37].

Proof. We prove by contradiction. Given any edge $uv \in E(\text{DLSS})$, assume $uv \notin E(\text{DRNG})$. According to the definition of DRNG, there must exist a third node p such that $w(up) < w(uv)$, $w(pv) < w(uv)$. According to the construction of DLSS, uv should not be inserted since u is connected to p and p is connected to v, that is, $uv \notin E(\text{DLSS})$. □

Theorem 10.6 (Connectivity of DRNG [37]): If G is strongly connected, then DRNG is also strongly connected.

Proof. This is a direct result of Theorem 10.5 and Lemma 10.2. □

Figure 10.20 shows DRNG and DLSS of a large network where 50 nodes are randomly distributed in a 500 m × 500 m area, and the transmission ranges vary from 100 m to 150 m.

10.7 CONNECTIVITY

In this section, we present some basic connectivity properties of the random unit-graph model. In particular, we discuss the critical conditions, such as the transmission range, the number of neighbors, and the minimum total power of all the nodes (termed as the critical total power), to ensure network connectivity with high probability. Two of the objectives that have been most commonly considered in the literature are minimizing the maximum transmission range at each node (assuming all nodes use a common transmission radius), and minimizing the total power incurred by all the nodes.

10.7.1 Critical Transmission Range and Node Degree

One of the most fundamental questions is: Given a number of nodes n to be deployed in a region, which is the minimum value r of the transmitting range that ensures network connectivity? If node placement is known in advance, the localized structures given in Section 10.5 can be used to construct the topology and the value of r can be determined accordingly. In the case that n nodes are uniformly randomly placed on a unit disk and the transmission radius $r(n)$ satisfies $\pi r^2(n) = (\log n + c(n))/n$, it is guaranteed that the network (of large sizes) is connected with probability approaching 1 if and only if $c(n) \to \infty$ as $n \to \infty$. That is, $r = O(\sqrt{\log n / n})$ [10,39].

Penrose [2] shows that the longest edge M_n of the minimum spanning tree has the following asymptotic distribution:

$$P(n \pi M_n^2 - \log n <= \alpha) \longrightarrow \exp(-e^{-\alpha}) \qquad \text{as } n \to \infty$$

(a)

(b)

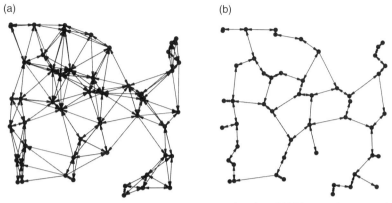

Original topology (without topology control)
is strongly connected.

Topology by DRNG is strongly connected.

(c)

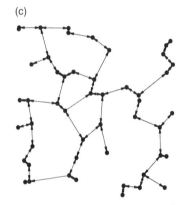

Topology by DLSS is strongly connected.

Figure 10.20   Topologies derived by DRNG and DLSS.

Thus if we let $n\pi r^2(n) = \log n + \alpha$, and $\alpha \to \infty$, the network is connected with the probability approaching one. He also showed in ref. [3] that the longest nearest neighbor and the longest MST edge have asymptotically (when $n \to \infty$) the same value. Based on this observation, Ovalle et al. [30] proposed to use the longest LMST edge to approximate the value of $r(n)$ using a wave-propagation quazi-localized algorithm. The differences between the exact and approximated values of $r(n)$ are estimated for two- and three-dimensional random unit graphs. Despite a small number of additional edges in LMST with respect to MST (under 3%), they can extend $r(n)$ by about 33% its range on networks with up to 500 nodes, which implies a 50% or more increase in energy consumption, depending on the power attenuation factor. A quazi-localized scheme to construct MST from LMST is then described in ref. [30]. The scheme needs less than seven messages per node on average (for networks up to 500 nodes). It eliminates LMST edges that are not in MST by a loop-breakage procedure, which iteratively follows dangling edges from leaves to LMST loops, and breaks loops by eliminating their longest edges, until the procedure finishes at a single node (as a by-product, this single node can also be considered as an elected leader of the network). The leader so elected also learns the longest MST edge in the process, and can broadcast it to other nodes. Note that this procedure operates only in two dimensions, since it is based on a face routing scheme (cf. ref. [23]).

Santi and Blough [4] showed that, in two and three dimensions, the transmission range can be reduced significantly if weaker requirements on connectivity are acceptable. Halving the critical transmission range, the longest connected component contains 90% of nodes approximately. This means that a considerable amount of energy is spent in order to connect relatively few nodes.

It has also been shown in ref. [40] that, as the common transmission range changes, the probability that the network is connected exhibits a sharp transition within a relatively short interval (see Fig. 10.21). Similar transition phenomena also exist for k-connectivity [19,39].

10.7.2 Critical Node Degree

Since the average number of neighbors d is approximated from $d = (n-1)r^2\pi/A$ (where A is the area of the region), it follows directly from Gupta and Kumar's result [10] (if the transmission radius $r(n)$ satisfies $\pi r^2(n) = (\log n + c(n))/n$, the network (of large sizes) is connected with probability approaching 1 if and only if

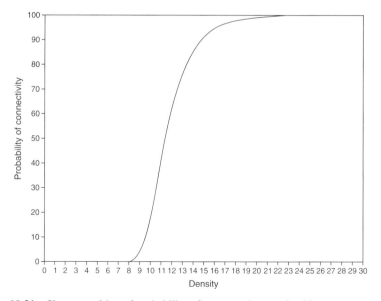

Figure 10.21 Sharp transition of probability of connected network with respect to common transmission radius.

$c(n) \to \infty$ as $n \to \infty$ that $d = \Theta(\log n)$ is the average density for critical connectivity. This means that each node needs to have, on average, $\Theta(\log n)$ neighbors for the network to be connected.

Xu and Kumar [6] went one step further and showed that in a network with n randomly placed nodes, if each node is connected to less than $0.074 \log n$ nearest neighbors, then the network is asymptotically disconnected with probability one as n increases; on the other hand, if each node is connected to more than $5.1774 \log n$ nearest neighbors, then the network is asymptotically connected with probability approaching one as n increases. It appears that the critical constant may be close to one, but that remains an open problem.

10.7.3 Critical Total Power

Blough et al. [7] studied the critical total power for one-connectivity, based on results on the asymptotic total weight for weighted minimal spanning trees [41,42]. Since the proof is based on the results on the asymptotic total weight for weighted minimal spanning trees, it cannot be easily generalized to the case of k-connectivity. Rengarajan et al. [11] gave the expectation of the critical total power for 1-connectivity.

Clementi et al. [8] studied the problem of assigning transmission ranges for wireless nodes so as to minimize the total power consumption in the special case of the path-loss exponent $\alpha = 2$, provided that any pair of nodes are within h hops. They showed that given the upper bound on the number of hops h, the total power incurred $\Theta(n^{1/h})$ with high probability. Their result cannot be readily generalized to the case of $\alpha \neq 2$. Gomez and Campbell [9] applied the results reported in ref. [41] and showed that for the n nodes that are independently, uniformly distributed in a unit d-dimensional cubic, the total length of the minimal spanning tree using the edge weight function $\psi(x) = x^\alpha$ is $\Theta(n^{1-\alpha/d})$ with probability 1 as $n \to \infty$ for $0 < \alpha < d$. Their results hold only for $0 < \alpha < 2$, which is out of the typical range for the path loss exponent in the 2-dimensional case.

The entire preceeding analysis assumes that all the nodes are subject to the same network parameter (e.g., common transmission power/range). Zhang and Hou [43] investigate the critical total power required for maintaining asymptotic k-connectivity in a heterogeneous random wireless network on a unit square $S = [0,1]^2$. In their analysis, each node is allowed to choose its own transmission power. Specifically, let $W_{t,i}$ be the critical transmission power node i uses, and $R_{t,i}$ the corresponding transmission range of node i under the power model $W_{t,i} = R_{t,i}^\alpha$, where $2 \leq \alpha \leq 4$ is the path-loss exponent. They showed that, under the assumption that wireless nodes are distributed on a unit square according to a Poisson point process with density λ and with the use of the toroidal model (torus convention), the critical total power of all the nodes, $W_\alpha = \sum W_{t,i} = \sum R_{t,i}^\alpha$, for maintaining k-connectivity is $\Theta(\Gamma(\alpha/2 + k)/(k - 1)! \lambda^{1-\alpha/2})$ with probability approaching 1 as $\lambda \to \infty$. This result suggests that the power saved using optimal, nonuniform transmission ranges is on the order of $(\log \lambda)^{\alpha/2}$ as compared to that using optimal *uniform* transmission ranges.

10.8 TOPOLOGY CONSTRUCTION AND CONNECTIVITY WITH CONSIDERATION OF FAULT TOLERANCE

Since a wireless sensor network is usually composed of a large number of unreliable sensor nodes, fault tolerance is an important requirement for topology construction. In particular, the network connectivity should be preserved even when some of the sensor nodes fail or deplete their power. With the use of smaller transmission power, most topology control algorithms actually decrease the number of possible routing paths between any pair of nodes. The topology thus derived is more susceptible to node failure. For instance, if node v_7 in Figure 10.3 fails, the network will be partitioned into three disconnected components. One way to construct fault-tolerant topology is to construct a k-vertex connected network. Note that a k-vertex connected network is $k - 1$ fault-tolerant, that is, it can tolerate failure of at most $k - 1$ nodes. A 3-vertex connected network is given in Figure 10.22. (For simplicity, the term k-connectivity is used to refer to k-vertex connectivity.)

An alternate way to enhance fault tolerance in topology construction is to make data sinks or network controllers aware of *critical nodes and links* in the network, so that additional sensors can be woken up or deployed on demand in the network. A node is defined to be critical if the subgraph of its p-hop neighbors (without the node itself) is disconnected [26]. Similar definitions were given in ref. [26] for critical links.

In this section, we first discuss the properties of k-vertex connected topologies and present several algorithms that construct such topologies. Then we discuss localized algorithms that detect such nodes and links.

10.8.1 *K*-vertex Connected Topologies

Since the problem of finding a minimum-cost k-connected subgraph has been proved to be NP-hard, many approximation algorithms have been proposed (see, for example, refs. [21] and [44] for a summary).

Penrose [3] studied k-connectivity in a geometric random graph of n nodes, each with a transmission range of r. It has been proved that the minimum value of r at

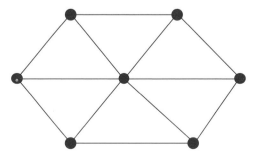

Figure 10.22 A 3-vertex connected network.

which the graph is k-connected is equal to the minimum value of r at which the graph has a minimum degree of k, with probability 1 as n goes to infinity. This result is significant, since it links k-connectivity, a global parameter of the graph, to node degree, a local parameter. However, the minimum value of r was not given in ref. [3]. Bettstetter [40] also investigated the relation between the minimum node degree and k-connectivity for geometric random graphs. The analytical expression of the required range r_0 for the almost surely k-connected network is derived and verified by simulation.

Li et al. [19] extended the work in ref. [3] and gave the lower and upper bounds on the minimum value of r at which the graph is k-connected with high probability. Yao$_{p,k}$, a localized topology control algorithm based on the Yao structure, was also proposed. Yao$_{p,k}$ is constructed by having every node u choose k closest neighbors in each of the $p \geq 6$ equal cones around u. It was proved to preserve k-connectivity and a length spanner.

Bahramgiri et al. [45] augmented the CBTC algorithm (Subsection 10.5.8) to provide fault tolerance. Specifically, let the directed subgraph of G, $D(\alpha)$, be the output of the CBTC(α) algorithm. Let $G(\alpha)$ be the result of deleting all unidirectional links in $D(\alpha)$. It was proved that $G(2\pi/3k)$ preserves k-connectivity of G.

In ref. [21], three approximation algorithms were presented to find the minimum power k-connected subgraph. The first algorithm is global and gives an $O(k\beta)$-approximation, where β is the best approximation factor for the k-UPVCS problem defined in ref. [21]. The second algorithm is also global and improves the approximation factor to $O(k)$ for general graphs. The third algorithm is distributed and gives an $k^{O(\alpha)}$-approximation, where α is the exponent in the propagation model. It first computes the MST of the input graph by using a distributed algorithm, and then adds a path among the neighbors of each node in the returned tree. Since this distributed algorithm is based on the distributed MST algorithm, it is not localized, that is, it relies on information that is multiple hops away to construct the MST. This implies more maintenance overhead, and delay will be incurred when the topology has to be changed in response to node mobility or failure.

In [15], a centralized greedy algorithm, fault-tolerant global spanning subgraph (FGSS$_k$), was first presented. FGSS$_k$ is a generalized version of the Kruskal's algorithm for MST [46]. In FGSS$_k$, different components are iteratively merged until only one k-connected component remains. FGSS$_k$ preserves k-vertex connectivity and is min–max optimal, that is, FGSS$_k$ minimizes the maximum transmission power used in the network, among all algorithms that preserve k-vertex connectivity. Based on this algorithm, fault-tolerant local spanning subgraph (FLSS$_k$) is proposed for topology control in wireless networks. In FLSS$_k$, each node u applies the FGSS$_k$ algorithm to its one-hop neighborhood, $n(u)$, and determines its neighbor set locally. It has been proved that FLSS$_k$ preserves k-vertex connectivity and maintains bidirectionality for all links in the topology, while reducing the transmission power and improving the network capacity. FLSS$_k$ has also been proved to be min–max optimal among all strictly localized algorithms.

10.8.2 Detection of Critical Nodes and Links

An alternative way to enhance fault tolerance is to make data sinks or network controllers aware of critical nodes and links in the network. Algorithms for detecting critical nodes and links based on global knowledge are well known. However, their use in sensor networks is limited, since the controllers may not be able to learn the overall network structure in a dynamic environment. It is therefore preferred that sensors themselves detect locally critical links and/or nodes and report them to the controllers.

Several localized definitions of critical nodes and links, using topological or positional information, are introduced in ref. [26]. A node is critical if the subgraph of its p-hop neighbors (without the node itself) is disconnected. Three definitions of critical links are proposed, based on verifying common p-hop neighbors, loop length, and critical status of link endpoints, respectively. The experiments with random unit graph model of ad hoc networks show high correspondence of local and global decisions. The errors mostly occur when alternative routes exist but are relatively long. Note that for a given particular report path, the reporting sensor and the network controller could be located in the same component after failure of a sensor on the route between them, and therefore the criticality of that sensor does not necessarily imply the criticality of the overall route, as an alternate route may already exist. However, in the case of monitoring and reporting by many or all sensors and multiple reporting paths, the criticality of a node is likely to require network maintenance. The notions can be generalized to the case of critical k-connectivity of the network.

Localized algorithms for testing k-connectivity are proposed in ref. [47]. In the first protocol, each node makes a criticality decision by verifying whether or not each of its p-hop neighbors has degree (number of neighbors) at least k. The second protocol also tests whether the subgraph of p-hop neighbors of a given nodes is k-connected. The third protocol also verifies whether this subgraph contains any critical nodes.

The proposed tests assume static sensors. In the case of mobile sensors (e.g., attached to some vehicles), the partition detection can be performed with the protocol [48], based on LMST structure [14]. Based on their speed and movement directions, two neighboring nodes A and B can predict when their link will break (details are first described in ref. [49]). Using movement information from all neighbors, nodes A and B construct LMST at the time their link will break. If at that time AB is in LMST (no matter whether the link is or is not currently in LMST) then the link is critical.

To the best of our knowledge, there have been no studies of connectivity issues when realistic physical-layer characteristics are taken into account to model sensor networks. What does it mean that a set of nodes is connected? There are several possible definitions as a starting point. In the simplest definition, a network can be considered q-connected if the network, consisting of edges whose probability of receiving a packet is at least q, is connected. When a message is broadcast from a source, it reaches any other node with a certain probability. However, multiple hops can drastically decrease the probability that a unicast/broadcast message reaches one or all nodes. Alternatively, one can define the network as being q-connected if the probability of a packet to be routed successfully between any two of the nodes, or broadcasting from any node to reach all of the nodes in the network, is $\geq q$ [24].

We recommend to read [50,51] for a deeper treatment of topology in ad hoc and sensor networks.

EXERCISES

10.1. Prove that MST is a subset of Yao$_p$ [36].

10.2. Show that PDT1 $\subseteq$ PDT2.

10.3. Prove that the enclosure graph for $\alpha = 2$ and $c = 0$ becomes equivalent to GG.

10.4. Prove that CBTC(α) preserves network connectivity for $\alpha < 5\pi/6$ [13].

REFERENCES

1. P. Gupta and P. R. Kumar. The capacity of wireless networks. *IEEE Transactions on Information Theory*, **46**:388–404, March 2000.

2. M. Penrose. The longest edge of the random minimal spanning tree. *Annals of Applied Probability*, **7**(2):340–361, 1997.

3. M. Penrose. On k-connectivity for a geometric random graph. *Random Structures and Algorithms*, **15**(2):145–164.

4. P. Santi and D. Blough. The critical transmitting range for connectivity in sparse wireless ad hoc networks. *IEEE Transactions on Mobile Computing*, **2**(1):1–15, 2003.

5. P.-J. Wan and C. Yi. Asymptotic critical transmission radius and critical neighbor number for k-connectivity in wireless ad hoc networks. In *Proceedings of the ACM International Symposium on Mobile Ad Hoc Networking and Computing (MobiHoc)*, Tokyo, May 2004.

6. F. Xue and P. R. Kumar. The number of neighbors needed for connectivity of wireless networks. *Wireless Networks*, **10**(2):169–181, March 2004.

7. D. M. Blough, M. Leoncini, G. Resta, and P. Santi. On the symmetric range assignment problem in wireless ad hoc networks. In *Proceedings of the 2nd IFIP International Conference on Theoretical Computer Science*, pages 71–82, Montreal, August 2002.

8. A. E. F. Clementi, P. Penna, and R. Silvestri. On the power assignment problem in radio networks. *Mobile Networks and Applications*, **9**(2), April 2004.

9. J. Gomez and A. Campbell. A case for variable-range transmission power control in wireless multihop networks. In *Proceedings of IEEE INFOCOM 2004*, Hong Kong, China, March 2004.

10. P. Gupta and P. Kumar. Critical power for asymptotic connectivity in wireless networks. *Stochastic Analysis, Control, Optimization and Applications*, W. M. McEneaney, G. Xin, and Q. Zhang (eds.), pages 547–566, Birkhauser, Boston, 1998.

11. B. Rengarajan, J. Chen, S. Shakkottai, and T. S. Rappaport. Connectivity of sensor networks with power control. In *Proceedings of the 37th Asilomar Conference on Signals, Systems and Computers*, Volume 2, pages 1691–1693, Pacific Grove, California, November 2003.

12. V. Rodoplu and T. H. Meng. Minimum energy mobile wireless networks. *IEEE Journal on Selected Areas in Communications*, **17**(8):1333–1344, August 1999.

13. L. Li, J. Y. Halpern, P. Bahl, Y.-M. Wang, and R. Wattenhofer. Analysis of a cone-based distributed topology control algorithm for wireless multi-hop networks. In *Proceedings of the ACM Symposium on Principles of Distributed Computing (PODC)*, pages 264–273, Newport, Rhode Island, USA, August 2001.

14. N. Li, J. C. Hou, and L. Sha. Design and analysis of an MST-based topology control algorithm. In *Proceedings of IEEE INFOCOM 2003*, San Francisco, California, 2003.

15. N. Li and J. C. Hou. FLSS: A fault-tolerant topology control algorithm for wireless networks. In *Proceedings of the ACM International Conference on Mobile Computing and Networking (MobiCom)*, Philadelphia, Pennsylvania, September 2004.

16. X. Li, I. Stojmenovic, and Yu Wang. Partial Delaunay triangulation and degree limited localized Bluetooth scatternet formation. *IEEE Transactions on Parallel and Distributed Systems*, **15**(4):350–361, April 2004.

17. X. Y. Li and I. Stojmenovic. Broadcasting and topology control in wireless ad hoc networks. In *Handbook of Algorithms for Mobile and Wireless Networking and Computing*, A. Boukerche and I. Chlamtac (eds.), CRC Press, to appear.

18. X. Y. Li, Yu Wang, Peng-Jun Wan, and Ophir Frieder. Localized low weight graph and its applications in wireless ad hoc networks. INFOCOM, Hong Kong, China, March 2004.

19. X. Y. Li, P. J. Wan, Y. Wang, and C. W. Yi. Fault tolerant deployment and topology control in wireless ad hoc networks. *Wireless Communications and Mobile Computing*, **4**:109–125, 2004.

20. G. Calinescu, I. L. Mandoiu, and A. Zelikovsky. Symmetric connectivity with minimum power consumption in radio networks. In *Proceedings of the 17th IFIP World Computer Congress*, pages 119–130, 2002.

21. M. Hajiaghayi, N. Immorlica, and V. S. Mirrokni. Power optimization in fault-tolerant topology control algorithms for wireless multi-hop networks. In *Proceedings of the ACM International Conference on Mobile Computing and Networking (MobiCom)*, pages 300–312, San Diego, California, September 2003.

22. L. Barriere, P. Fraigniaud, L. Narajanan, and J. Opatrny. Robust position based routing in wireless ad hoc networks with unstable transmission ranges. In *Proceedings of the 5th ACM International Workshop on Discrete Algorithms and Methods for Mobile Computing and Communications* DIALM' 01, pages 19–27, Rome, Italy, July 2001.

23. P. Bose, P. Morin, I. Stojmenovic, and J. Urrutia. Routing with guaranteed delivery in ad hoc wireless networks. In *Proceedings of the 3rd International Workshop on Discrete Algorithms and Methods for Mobile Computing and Communications*, pages 48–55, Seattle, Washington, August 1999.

24. I. Stojmenovic, A. Nayak, J. Kuruvila, F. Ovalle-Martinez, and E. Villanueva-Pena. Physical layer impact on the design and performance of routing and broadcasting protocols in ad hoc and sensor networks. *Computer Communications*, **28**(10):1138–1151, June 2005.

25. I. Stojmenovic and X. Lin. Loop-free hybrid single-path/flooding routing algorithms with guaranteed delivery for wireless networks. *IEEE Transactions on Parallel and Distributed Systems*, **12**(10):1023–1032, October 2001.

26. M. Jorgic, I. Stojmenovic, M. Hauspie, and D. Simplot-Ryl. Localized algorithms for detection of critical nodes and links for connectivity in ad hoc networks. In *Proceedings of the 3rd Annual Mediterranean Ad Hoc Networking Workshop (Med-Hoc-Net)*, pages 360–371, Bodrum, Turkey, June 2004.

27. G. Alonso, E. Kranakis, R. Wattenhofer, and P. Widmayer. Probabilistic protocols for node discovery in ad hoc, single broadcast channel networks. In *Proceedings of the IEEE International Parallel and Distributed Processing Symposium Workshops*, Nice, France, 2003.

28. M. J. McGlynn and S. A. Borbash. Birthday protocols for low energy deployment and flexible neighbor discovery in ad hoc wireless networks. In *Proceedings of the ACM International Symposium on Mobile Ad Hoc Networking and Computing (MobiHoc)*, Long Beach, California, October 2001.

29. A. Micic and I. Stojmenovic. A hybrid randomized initialization protocol for TDMA in single-hop wireless networks. In *Proceedings of the IEEE International Parallel and Distributed Processing Symposium Workshops*, Fort Lauderdale, Florida, April 2002.

30. F. J. Ovalle-Martinez, I. Stojmenovic, F. Garcia-Nocetti, and J. Solano-Gonzalez. Finding minimum transmission radii and constructing minimal spanning trees in ad hoc and sensor networks. In *Proceedings of the 3rd Workshop on Efficient and Experimental Algorithms*, Angra dos Reis, Rio de Janeiro, Brazil, May 2004.

31. G. Toussaint. The relative neighborhood graph of finite planar set. *Pattern Recognition* **12**(4):261–268, 1980.

32. J. Cartigny, F. Ingelrest, D. Simplot-Ryl, and I. Stojmenovic. Localized LMST and RNG based minimum energy broadcast protocols in ad hoc networks. *Ad Hoc Networks*, **3**(1):1–16, 2005.

33. K. R. Gabriel and R. R. Sokal. A new statistical approach to geographic variation analysis. *Systemic Zoology*, **18**:259–278, 1969.

34. Q. Huang, C. Lu, and G. C. Roman. Reliable mobicast via face-aware routing. In *Proceedings of IEEE INFOCOM 2004*, Hong Kong, China, March 2004.

35. A. Okabe, B. Boots, and K. Sugihara. *Spatial Tessellations: Concepts and Applications of Voronoi Diagrams*, John Wiley & Sons, 1992.

36. A.-C. Yao, On constructing minimum spanning trees in k-dimensional spaces and related problems. *SIAM Journal of Computing*, **11**:721–736, 1982.

37. D. Yang and I. Stojmenovic. Bluetooth Scatternet Formation for Efficient Routing in Ad Hoc Networks. In preparation.

38. N. Li and J. C. Hou. Topology control in heterogeneous wireless networks: Problems and solutions. In *Proceedings of IEEE INFOCOM*, 2004, Hong Kong, China, March 2004.

39. P. Panchapakesan and D. Manjunath. On the transmission range in dense ad hoc radio networks. In *Proceedings of the IEEE Signal Processing and Communications Conference (SPCOM)*, Bangalore, India, 2001.

40. C. Bettstetter. On the minimum node degree and connectivity of a wireless multihop network. In *Proceedings of the 3rd ACM International Symposium on Mobile Ad Hoc Networking and Computing (MobiHoc)*, pages 80–91, Lusanne, Switzerland, 2002.

41. J. M. Steele. Growth rates of Euclidean minimal spanning trees with power weighted edges. *Annals of Probability*, **16**(4), 1988.

42. J. E. Yukich. Asymptotics for weighted minimal spanning trees on random points. *Stochastic Processes and Their Applications*, **85**:123–128, 2000.

43. H. Zhang and J. C. Hou. *On the Critical Total Power for Asymptotic k-Connectivity in Wireless Networks*. Technical Report UIUCDCS-R-2004-2386, Department of Computer Science, University of Illinois at Urbana-Champaign, July 2004.

44. S. Khuller. Approximation algorithms for finding highly connected subgraphs. In *Approximation Algorithms for NP-Hard Problems,* D. S. Hochbaum, (ed.), PWS Publishing Company, Boston, Massachusetts, 1996.

45. M. Bahramgiri, M. Hajiaghayi, and V. S. Mirrokni. Fault-tolerant and 3-dimensional distributed topology control algorithms in wireless multi-hop networks. In *Proceedings of the 11th International Conference on Computer Communications and Networks (IC^{3}N)*, pages 392–397, October 2002.

46. J. B. Kruskal. On the shortest spanning subtree of a graph and the traveling salesman problem. *Proceedings of the American Mathematical Society*, **7**:48–50, 1956.

47. M. Jorgic and I. Stojmenovic. Localized algorithms for detection of k-connectivity in ad hoc networks. In preparation.

48. I. Stojmenovic. LMST-Based Partition Detection for Mobile Sensors. In preparation.

49. I. Stojmenovic. M. Russell, and B. Vukojevic. Depth first search and location based localized routing and QoS routing in wireless networks. *Computers and Informatics*, **21**(2):149–165, 2002.

50. P. Santi. Topology Control in Wireless Ad Hoc and Sensor Networks. Submitted to *ACM Computing Surveys*, 2004.

51. X. Y. Li and Y. Wang. Wireless sensor networks and computational geometry. In *Handbook of Sensor Networks*, M. Ilyas (ed.), CRC Press, 2003.

52. F. Atay and I. Stojmenovic. Fast generation of connected random unit disk graphs in ad hoc networks with reduced degree deviations, submitted for publication, 2000.

CHAPTER 11

Energy-Efficient Backbone Construction, Broadcasting, and Area Coverage in Sensor Networks

DAVID SIMPLOT-RYL

IRCICA/LIFL, Universite Lille, Villeneuve d'Ascq, France

IVAN STOJMENOVIĆ

University of Ottawa, Ontario, Canada

JIE WU

Florida Atlantic University, Boca Raton, Florida

A backbone is a subset of sensors that is sufficient for performing assigned tasks. The exact definition depends on the task or the particular desirable properties of a back-bone. We discuss two specific kinds of backbones, neighbor and area dominating sets, that we believe are the essential and perhaps only backbones required for sensor networks. A sensor is covered by a backbone if it is in the backbone or is a neighbor to a sensor in the backbone. This type of backbone is referred to here as neighbor-dominating sets, *or simply dominating sets. A point within a monitoring area is covered by a sensor backbone if it is within sensing range of at least one sensor from the backbone. This type of backbone is called* area-dominating set. *In a broadcasting (also known as* data-dissemination) *task, a message is sent from one node, which could be a monitoring center, to all nodes in the network. Sensors, which are randomly placed in an area, decide which of them should be active and monitor an area, and which of them may sleep and become active at a later time. The communication connectivity is important so that the measured data can be reported to a monitoring center. This problem is known as the* sensor-area coverage *problem, and needs to be solved efficiently to enable sensor functioning for a prolonged time. Sensors may also be placed deterministically in an area to optimize coverage and reduce power consumption. Most solutions considered in this chapter are based*

Handbook of Sensor Networks: Algorithms and Architectures, Edited by Ivan Stojmenović
Copyright © 2005 John Wiley & Sons, Inc.

on constructing area-dominating sets for sensor-area coverage. The best known solutions for backbone construction, broadcasting, and sensor-area coverage problems are based on the concept of localized connected dominating sets. For instance, one solution to the broadcasting problem is that only nodes from a connected neighbor-dominating set retransmit the message. This chapter reviews solutions to these three related problems in sensor networks.

11.1 INTRODUCTION

11.1.1 Modeling Sensor Networks

A widely accepted basic graph-theoretical model for sensor networks is the *unit-disk* graph model, defined as follows: two nodes, A and B, in the network are neighbors (and thus joined by an edge) if the Euclidean distance between their coordinates in the network is at most R, where R is the transmission radius that is equal for all nodes in the network. There are two kinds of unit-disk graphs considered in sensor networks, sensing and communication unit-disk graphs, with corresponding sensing and transmission radii, respectively. The relationship between sensing and transmission radii may vary based on the particular hardware or application. There are three basic cases: equal sensing and transmission (communication) radii, transmission radius more than twice the sensing radius, and a communication range that is between sensing and twice the sensing radii. Reasons for the three cases become apparent later in the chapter.

Some solutions make use of sensor ability to adjust transmission radius, instead of using the maximum radius, as determined by the unit-disk graph model. The unit-disk graph model is not fully realistic, but is much better for approximation of a sensor network than the random graph model (with each edge having equal probability of being selected for the graph), studied in ref. [1]. In the unit-disk graph model, the probability of receiving a packet between two nodes suddenly changes from 1 to 0 at distance R. A more realistic model is the fuzzy unit-disk graph proposed in ref. [2]. In this model, there are two transmission radii, r and R. Two nodes communicate with each other if the distance between them is $\leq r$; they do not communicate with each other if their distance is $\geq R$, and may or may not communicate if the distance is between r and R. In a realistic physical-layer model, such as the log normal shadowing model, random signal strength variations lead to a model where the packet reception probability $p(x)$ is a function of distance x between two nodes. The transmission radius R is defined in ref. [3] as the distance at which $p(R) = 0.5$. Two nodes are considered neighbors if the distance x between them is such that $p(x) \geq w$, where w is a threshold parameter (for example, when $w = 0.05$, then $x \approx 1.4R$). In the hitchhiking model [4], two transmission radii, r and R, are also used. The receiver can receive a partial packet from the sender if their distance is between r and R. The actual percentage of packet that can be decoded depends on a particular signal model. It is assumed that each receiver can assemble several partially received packets to one complete packet.

11.1.2 Localized Algorithms and Message Complexity

Among recently developed strategies for constructing small connected dominating sets, *localized protocols* offer the best prospect for achieving energy efficiency. In a localized protocol, each node makes protocol decisions based solely on some available local knowledge (to be more precise, based on the information from neighbors within *k* hops for certain *k*), without resorting to global network information. Because of the dynamic nature of sensor networks, the topology changes are frequent and unpredictable. The local information must suffice for a sensor node to make protocol decisions; otherwise, the increased communication overhead could offset the energy savings and increase latency. In a *centralized* (or *globalized*) algorithm, one or more nodes (or a central entity like a base station) need to learn global node and/or edge structure, either the whole graph (for instance, to find a route using the shortest-path algorithm), or a global structure derived from the graph (such as minimal spanning tree, which can be used for optimal energy data aggregation). Because of the huge communication overhead involved in gathering such information in dynamically changing sensor networks, such protocols cannot be energy efficient solutions in normally large sensor networks. This chapter consequently discusses primarily localized solutions (some centralized algorithms are described only for the sensor-area coverage problem).

The sensor network may operate with or without time synchronization between sensors. In an *asynchronous* protocol, there is no common clock between the sensors. Therefore, each sensor makes its own decision about being active or going to a sleep state for an arbitrary period based on the overheard communication from other sensors. In a *synchronous* protocol, sensors follow a common clock, and therefore naturally may operate in rounds. In the case of sensor-area coverage, for example, they exchange some messages (at the beginning of each round) in order to decide which of them is needed for coverage in a given round, while the remaining sensors may sleep for the rest of the round and wake up at the beginning of the next round.

We further classify localized protocols according to the amount of information required and to overhead in the construction and maintenance phases. The amount of required information is related to the *message complexity*, which can be defined as the average number of transmitted messages per sensor node in a protocol. Although some protocols appear localized, an extensive message exchange with neighbors amounts to collection and use of global information. In a *strictly localized protocol* [5], all information processed by a node is either local in nature or global in nature, but obtainable in a short constant time by querying only the node's neighbors or itself. In other words, only a small constant number of message exchanges with neighbors is allowed. Strictly localized protocols may need some information that is part of their input (such as destination position in a routing protocol) but cannot use structures that are global in nature (e.g., information about which of the outgoing links belongs to the minimum spanning tree (MST)). An interesting similar definition is given in ref. [6]. An *emergent* algorithm is any computation that achieves formally or stochastically predictable global effects by communicating directly with only a bounded number of immediate neighbors and without the use of central control or global visibility [6].

The sensors may or may not use position information in their decisions. The availability of position information for proper sensor functioning was widely recognized as highly desirable; however, it is a nontrivial problem and the precision of the location information may impact the performance of a protocol. There exists a variety of position-determination protocols [7], with a variety of message complexities. If position information is used, we will make the simple assumption that it was provided to the node message tree (here only messages transmitted by sensors are counted), which is true only if it was provided externally via a global positioning system (GPS) or similar beacons arriving from the environment.

The simplest local information required is certainly no knowledge at all about existing neighbors. The *blind flooding* protocol for broadcasting (used in a typical route discovery in reactive routing protocols), where each node will retransmit the packet after receiving it the first time, belongs to this category. The next, and commonly used, assumption is the knowledge of one-hop neighbors (direct neighbors, or nodes located within transmission radius R), and possibly their locations. To collect such knowledge, a periodic "hello" message protocol is normally assumed, where each sensor transmits one message informing neighbors about its existence. Therefore, when message complexities are compared, we assume that one message per node is needed to acquire one-hop information. A further common assumption is of 2-hop neighbors, which are obtained after each node sends a message containing the list of its one-hop neighbors. We will assume therefore that collecting this information requires two messages per node. The actual cost could be higher, since such messages in dense networks could be long and energy-consuming to transmit.

11.1.3 Does Sleeping Always Conserve Energy?

The importance of placing as many sensor nodes as possible into sleep mode is apparent from the analysis of sensor energy expenditure. A sensor's radio can be in one of three active states—transmit, receive, idle—or in the sleep state. The radio is turned off in sleep state. The power consumption for various types of sensors and ad hoc nodes [8] shows that a sensor in the sleep state consumes 7–20 times less energy than one in the idle state. The power consumption while receiving a message is up to 10% higher than in idle state. Nodes spend 10–100% more energy while transmitting than while receiving messages. For instance, the Windows Internet Naming Service (WINS) seismic sensor consumes between 0.38 W and 0.7 W in the transmit state, 0.36 W in the receive state, 0.34 W in the idle state, and 0.02 W in the sleep state [9]. Sensors in the idle state are listening to the traffic and can be "alarmed" for any action. In the sleep state, however, they cannot receive any message and cannot be alarmed to become active. The importance of placing as many sensors as possible into the sleep state in order to prolong network life is apparent. Shall sensors sleep whenever they know that they are not needed for sensing or communication? Such an assumption is made in ref. [10], which proposed an activity scheduling scheme that assumes that sensor reporting can be done at predetermined times, along predetermined routes. In the route-discovery phase, each node learns about some neighbors and receives some forwarding tasks [10]. In addition, sampling, transmissions, and receptions along the route are also scheduled. This enables sensors to sleep between two scheduled tasks [10].

Suppose, for simplicity, that sensors are changing between active and sleep states on a regular basis, in rounds. The duration of a round cannot be made arbitrary, if prolonged network life is desirable. We will demonstrate this in the case of energy efficient behavior of a single sensor. Assume that there is fixed charge C for transition between the active and sleep periods in sensor networks (this charge is not zero!). Assume also for simplicity that energy consumption in the active state remains consistent regardless of the amount of traffic handled (this, in fact, is not far from the reality [8]). Let F be the ratio of energy consumption between the active and sleep states, and let S be the energy consumed in the sleep state per unit of time (therefore, in the active state the consumption is SF per unit time). Suppose that T is the ratio of sensor reporting (active) and sleep times. If the sensor remains in active mode, its energy consumption is LSF over L time units. If the sensor decides to switch to sleep state between reporting, then there are $2L/(T + 1)$ transitions, requiring energy of $2CL/(T + 1)$. The consumption for sleep periods is $SLT/(T + 1)$, and consumption during the active states is $SLF/(T + 1)$. Thus overall consumption is $2CL/(T + 1) + SLT/(T + 1) + SLF/(T + 1) = L[2C + ST + SF]/(T + 1)$. This is compared to LSF to conclude that switching to the sleep state between reporting periods is beneficial only if $T > 2C/(S(F - 1))$. Therefore, the power needed for frequent transitions may outweigh the benefits obtained from sleeping.

Clustering can be effectively used to minimize internal transmission in a cluster with time division or frequency division. Energy awareness is not only a problem of sleep and awake but also a problem of collision avoidance.

11.2 BACKBONE CONSTRUCTION

11.2.1 Backbone Construction, Maintenance, and Analysis

Most broadcasting, activity scheduling, and sensor-area coverage algorithms rely on the concept of *backbone*. A backbone is a subset of sensor nodes that is able to perform data communication tasks and to serve nodes that are not in the backbone (because it is close to them). A backbone can also be the set of active sensor nodes, assuming then that the rest of the sensors are sleeping. There is a vast literature about backbone construction (see ref. [11] for a more comprehensive review). The primary backbone concept used in the literature is the dominating-set concept. A *dominating set* has the following property: each node is either in the dominating set or has a one-hop neighbor that is in the dominating set. Further, the connectivity property is often required for proper protocol functioning. A *connected dominating set* (CDS) is a dominating set of nodes that is also a connected set.

Figure 11.1 illustrates the CDS concept. Nodes 13, 12, 11, 10, 9, 4, 3, 1 are in CDS, and any remaining node is a neighbor of one listed. It is obvious that

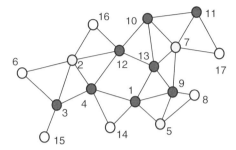

Figure 11.1 A neighbor CDS consisting of nodes 13, 12, 11, 10, 9, 4, 3, 1.

broadcasting protocol, in which all nodes belonging to a CDS retransmit the message, will reach all nodes in a sensor network (assuming an ideal medium-access layer). This does not mean that all of them indeed need to retransmit, as will be discussed later in the chapter. An activity-scheduling scheme may simply direct all sensors in the backbone to be active, and allow all others to sleep. In this chapter, we have described the two most important backbone concepts: neighbor- and area-dominating sets. They can both be applied in the sensor networks. Once the sensors for area coverage are selected (area-dominating set), their backbone (neighbor-dominating set) can be constructed. Sensors in a connected neighbor-dominating set constructed over a connected area-dominating set can be used for broadcasting in the sensor network. Thus, one can be considered a backbone of another backbone. Note that area-dominating sets provide a network of medium density, which has an impact on the selection and performance of broadcast protocols for use in sensor networks (e.g., blind flooding may be an acceptable option).

The quality of a backbone construction/maintenance protocol is normally evaluated by backbone size with respect to the minimal possible size for the same network. The problem of constructing a CDS of minimal size (with a minimal number of sensors in it) is known to be NP-complete even for centralized algorithms. Therefore it is not surprising that finding good solutions by local means is a difficult task, and one that has attracted significant interest in recent years. The *approximation ratio* of a scheme is the ratio of the number of sensors in the constructed backbone over the minimal possible number of sensors in an optimal backbone. There are other metrics that can be considered [12]: the protocol duration, message overhead, and backbone robustness (does the backbone remains connected if one node fails?). For each metric, the evaluation can be performed, analytically or experimentally, using either *average-case* or *worst-case* performances. The ultimate goal is certainly to have a winner in both categories (such as mergesort or heapsort for the sorting problem). However, so far such a winner has not emerged, and researchers have normally adopted one of the two ways for comparison. Arguably, if a sensor network designer is presented with two protocols, one with excellent average-case, but occasionally quite bad performance, and the other with firm worst case guarantees (e.g., theoretically provable constant bound for the approximation ratio) but considerably inferior in the average case, we believe that the former would be the choice. This is the "philosophy" followed in this chapter.

Before describing some backbone schemes, we discuss their determinism and cost aspects. Backbone construction schemes can be classified as *probabilistic* and *deterministic*, based on whether or not a random number generator was used to construct them. The random number usage here is limited to the network-layer decisions; the underlying medium access scheme may still use random backoff counters, for example, in a deterministic protocol.

The backbone construction protocols described in the literature normally consider only *construction cost*. However, sensor networks are dynamic and the *maintenance cost* cannot be ignored; this is the cost to update the backbone when the network changes. Both construction and localized maintenance protocols can be further divided into *quasi-local* and *local* protocols. In a quasi-local (localized) protocol, all decisions are made based on local knowledge; however, the decisions made in one part of the network may have an impact on decisions made in a distant part of the network. Clustering is a typical example of a quasi-local protocol for both the construction and maintenance phases. The construction phase starts from a few selected "seed" nodes and propagates throughout the network. While this performance is debatable, the maintenance phase of quasi-local protocols is problematic, because of possible "chain effect": a simple change in an edge or addition/deletion of nodes may trigger global backbone updates by propagation. Otherwise, a local localized clustering procedure may have a negative impact on the quality (e.g., size) of the cluster structure. This chapter is therefore inclined toward local (localized) solutions, where, in both the construction and maintenance phases, the backbone status of each node depends solely on the local network configuration, typically one-hop or 2-hop (2-hop neighbors are one-hop neighbors of one-hop neighbors).

We will now describe some localized backbone construction methods and discuss them in the light of mentioned criteria and desirable properties.

11.2.2 Backbone Construction by Clustering

The distributed clustering algorithm [13] is initiated at all nodes whose id is lowest among all their neighbors (locally lowest id nodes). All nodes are initially undecided. If all neighbors of node A, which have a lower id, sent their cluster decisions and none declared itself a clusterhead (CH), node A decides to create its own cluster and broadcasts this decision and its id as a cluster id. If a node receives a message from a neighbor that announces itself as a CH, it will send a message (to all its neighbors) declaring itself a non-CH node, to enable more clusters to be created (note that two CHs are not direct neighbors in the algorithm). Thus each node broadcasts its clustering decision after all its neighbors with lower ids have already done so. Non-CH nodes that hear two or more CHs will declare themselves as *gateway* nodes. A sophisticated maintenance procedure for cluster formation when nodes move is described in ref. [13]. To minimize the number of clusters, ref. [14] proposed that node degree be applied as the primary key in clusterhead decisions. Nodes with more neighbors are then more likely to become a CH. In the case of ties, lower id nodes have an advantage. The clustering process requires one message per node in the construction phase (after one "hello" message to find the ids of neighbors or two "hello" messages to learn their degrees). Basagni [15] proposed variants of the clustering algorithm [13], which uses a variety of weights for selecting best CHs.

In the protocol described in ref. [16], after the clustering process is completed, each CH contacts neighboring CHs (up to three hops away) in order to eliminate some gateway nodes, and to use only essential gateway nodes to preserve overall connectivity. The construction and maintenance are fully localized. The protocol in ref. [16] produces an excellent approximation ratio, but the message overhead is significant, due to the overly high complexity of the election phase of the protocol leader, which requires information to be propagated to the fragment members and to nodes in adjacent fragments every time two fragments are merged into a new one. The simulation results [12] show that approaches with nice theoretical features, such as that presented in ref. [16], may hardly be applicable in practice due to the message complexity of their operations.

Basagni, Carosi, and Petrioli [17] described such a clustering based backbone scheme where nodes with more energy have higher chances to be clusterheads. Their construction and maintenance procedures are ongoing process with decisions based on received "hello" messages from neighbors. A node declares itself a CH if it did not receive a "hello" packet from a CH with energy that differs by more than certain threshold ("older" decisions have priority).

Chan and Perrig [5] described a localized clustering algorithm. New clusters are spawned in a self-elective process, when no messages from other CHs are received. Migration of an existing cluster is controlled by its CH. Each CH will periodically poll all its followers (neighbors) to determine which is the best candidate to become the new CH. The best candidate is the node that, if it were to become a CH, would have the greatest number of nodes as followers while minimizing the amount of overlap with existing clusters. The algorithm achieves a packing, efficiency close to hexagonal packing, but is quasi-local because chain effect is not prevented. It also has significant message overhead compared to other clustering protocols.

Wu and Dai [18] proposed a simple cluster formation in a dense network. First, the neighborhood detection is done using Hello messages with shorter transmission ranges than the normal one. The regular clustering algorithm is used to find CHs. However, CHs are directly connected using the normal transmission range. There are two versions of this approach. In the first version, the range of the Hello message is $\frac{1}{3r}$, where r is the normal range. In this way, all CHs within three hops are connected, and CHs are globally connected. In the second version, the range of Hello message is $\frac{1}{4r}$. During the transmission using the normal range r, $\frac{3}{4r}$ is used to connect all CHs within three hops and $\frac{1}{4r}$ is used to cover the member in the cluster with a radius of $\frac{1}{4r}$.

In the protocol by Kuhn, Moscibroda, and Wattenhoffer [19], sensors may wake up asynchronously at any time and do not have collision detection capabilities. They only know the limit on the total number of sensors, and have no knowledge of possible neighbors. The algorithm computes asymptotically optimal clustering. The main idea is that nodes, after some initial waiting, compete to become dominators by exponentially increasing their sending probability on one channel. Two other channels are then used to guarantee that the number of further dominators emerging in the neighborhood of an already existing dominator remains small. The algorithm can be simulated to work by using only one channel.

11.2.3 Backbone Construction by Nominating and Grid Partitioning

This section will describe two very simple schemes for backbone construction. In ref. [20], the authors propose a simple method for determining the dominant set (not necessarily connected). Each node nominates the neighbor with the largest id among its neighbors to be in the dominant set (assuming that each node has a unique identifier). This can produce the $O(n)$ approximation ratio in the worst case, but works well in the average case. An example of bad performance is a linear chain of nodes with increasing identifiers. Each node needs one message to learn the identifiers of neighbors, and possibly the second message to nominate a neighbor into the dominant set. Connecting the dominant nodes is, unfortunately, a nontrivial problem (e.g., the protocol in ref. [16] could be used for it).

Xu, Heidemann, and Estrin [21] discuss the following backbone construction scheme called GAF. The given two-dimensional space is partitioned into a set of squares (called *cells*), such that any node within a square can directly communicate with any node in an adjacent square. Therefore, one representative node from each cell is sufficient for a connected backbone. Each node transmits its id (which may depend on its remaining energy) plus its coordinates (this requires one message per node). In each cell, the node with maximal id is selected for the backbone. The selected nodes in ref. [21] make a dominant set, but its average size (which depends on the selected size of the square) may be higher than for other methods considered here. Further, the dominant-set concept needs some parameters, such as the size and position of squares, which have to be propagated in the network. The method is simple, has no chain effect, and has a constant approximation ratio. When crossing a boundary, nodes need to retransmit their information to maintain the dominant set. When crossing the second boundary in a larger movement, this is not sufficient, as the moving node has no information about nodes in the new cell. This can be resolved by triggering a round of Hello messages in that cell. The most significant problem is that, for any ratio of transmission radius and grid size, the dominant set obtained may disconnect the graph [17]. Although the network topology is connected, Basagni, Carosi, and Petrioli [17] observed that, for instance, on a network with 50 nodes, GAF [21] get disconnected $\geq 40\%$ of simulation time for any grid size that produces a meaningful backbone size. An example illustrating that the partition may occur even for range transmission radii with respect to grid size is given in Figure 11.2.

11.2.4 MPR-Based Backbone

Several broadcasting schemes are based on the concept of multipoint relays (MPR) of a node S, defined as a minimal-size subset of neighbors of a given node S that will

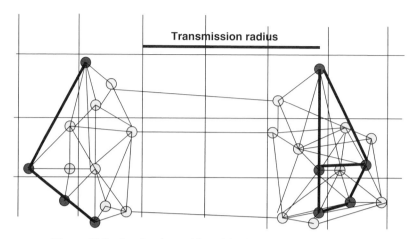

Figure 11.2   Leaders in a grid partitioning may be disconnected.

"cover" all 2-hop neighbors of S. A node is called *covered* if it can receive (directly or via retransmissions by relay nodes of S) messages originating at S. Relay points of S are one-hop neighbors of S that cover all 2-hop neighbors of S. The goal is to minimize the number of relay points of S. The computation of an MPR set with minimal size is an NP-complete problem. A heuristic algorithm, called a greedy set cover algorithm, is proposed in ref. [22]. This algorithm repeats selecting node B, which maximizes the number of neighbor nodes that are not yet covered.

Adjih, Jacquet, and Viennot [23] proposed to combine MPR and dominant-set approaches. Each node computes its set to be forwarded to its neighbors and transmits it to its neighbors. It then determines whether it belongs to the "MPR-dominating set" if it either has the smallest id in its neighborhood, or the node is a forwarding (relay) neighbor of the neighbor with the smallest id. Wu [24] observed that a node with a smaller id than all its neighbors, but without two unconnected neighbors, can be eliminated. The construction of an MPR-based backbone requires 2-hop neighbor knowledge, plus a message containing the list of relay nodes of each node. This can be treated overall as CDS construction requiring three rounds of messages, plus another round if the CDS decisions are to be communicated to neighbors.

11.2.5 Wu's Backbone

In a series of articles (the first one being ref. [25]), Wu et al. described, a lightweight backbone construction scheme. We will use a modified definition from refs. [14] and the [26] of basic concept [25], because of its reduced message overhead. A node is an *intermediate* node if it has two unconnected neighbors [25]. In the example in Figure 11.1, nodes C and K are the only nodes that are not intermediate. A node A is covered by a neighboring node B if each neighbor of A is also a neighbor of B, and key(A) < key(B). Assuming that the keys in Figure 11.1 are ordered alphabetically, node H is covered by node I, G is covered by L, while A and B are covered by E. Nodes not covered by any neighbor are *intergateway* nodes. A node A is covered by two connected neighboring nodes B and C if each neighbor of A is also a neighbor of either B or C (or both), key(A) < key(B), and key(A) < key(C). An intermediate node not covered by any neighbor becomes an intergateway node. An intergateway node not covered by any pair of connected neighboring nodes becomes a *gateway* node.

Dai and Wu [27] introduced a generalized dominant set, where coverage can be provided by an arbitrary number of connected one-hop neighbors (instead of 1 or 2 as in the original definitions). The definition was modified in ref. [11] to the following form to avoid similar message exchanges between neighbors. Node A is covered by its one-hop neighbors $B, C, D, \ldots$ if the neighbors $B, C, D, \ldots$ are connected, any neighbor of A is a neighbor of at least one of nodes $B, C, D, \ldots$, and key(A) < min (key(B), key(C), key(D), ...). It is then further computationally simplified by Carle and Symplot-Ryl [28], as follows. First, each node checks if it is an intermediate node. Then each intermediate node A constructs a subgraph G of its neighbors with higher key values. If G is empty or disconnected, then A is in the dominating set. If G is connected, but there exists a neighbor of A that is not a neighbor of any node from G, then A is in the dominant set. Otherwise A is covered and is not in the dominant set. Dijkstra's shortest-path scheme can be used to test the connectivity. This procedure is generalized since it allows coverage by any number of neighbors. It is computationally even less expensive than the two-nodes coverage case.

The CDS concept [25,27] is illustrated in Figure 11.1, where the keys are assumed to be ordered by their numerical id values: "1" < "2" < $\cdots$ < "16." Nodes 6, 15, 16, 17, 14, and 8 do not have two unconnected neighbors that are not in CDS (they are not intermediate). Node 2 is dominated by three connected neighbors (nodes 3, 4, 12), since they have higher key values, and the remaining neighbors 6 and 16 are "covered" by 3 and 12, respectively. Node 7 is covered by four connected neighbors with higher keys 9, 13, 10, and 11 (the remaining neighbor 17 is covered by 11). Node 5 is covered by its neighbor 9, since other neighbors (1 and 8) are neighbors of 9, and "5" < "9." Node 1 remains in CDS because neighbors with higher keys (4, 14, 5, 9, 13) are disconnected.

Wu's concepts require either one-hop knowledge of neighbors with their position, or 2-hop neighbor topology information. This can be obtained after one or two Hello messages from each node. Experimental data from several sources (e.g., ref. [12]) confirm that Wu's concepts provide small-size CDS on average. It was proved in ref. [27] that the generalized CDS concept has a constant approximation ratio on average, and very low probability of having an infinitely large approximation ratio. An example of a "bad" approximation ratio is the case of a linear chain of nodes with increasing keys, where almost all nodes are selected into the CDS.

Each node makes decisions about CDS membership (in Wu's concept) without communications between nodes beyond the message exchanges that nodes use to discover each other and establish neighborhood information. If knowledge of neighbors that are in the CDS is needed, then one message from these nodes suffices.

In that case, such a message can be used to further reduce the size of the backbone. As soon as one node decides to be in the CDS, it sends a packet informing neighbors about the decision. Neighbors (which did not yet decide their membership) will then consider such decided CDS neighbors as having higher key values, which may help them in withdrawal from the CDS decision [28].

11.2.6   Enhanced Dominating Sets

The number of nodes in a CDS following Wu's concepts can be reduced by applying some enhanced concepts [29,30]. The first observation [29] is that if 2-hop topological knowledge is already required, it can be used to eliminate a few more nodes from the CDS. Consider the example in Figure 11.1. Node D is in the CDS although it is actually covered by nodes E, I, and L. The later three nodes all have higher key values, and are connected. Node L, however, is not a one-hop neighbor of D. This does not prevent node D from verifying whether any of its neighbors are neighbors of L, or whether E, I, and L are connected, since L appears in the list of neighbors sent to D by its one-hop neighbors; therefore, such a conclusion can be made. The new definition therefore can be given as follows [29]. Node A is covered by its 2-hop neighbors B, C, D, ... if the neighbors B, C, D, ... are connected (according to information available to A), any neighbor of A and A itself is the neighbor of at least one of nodes B, C, D, ... and key(A) < min (key(B), key(C), key(D), ...). Note that A is not aware of possible links between its two 2-hop neighbors, and therefore may declare the set disconnected although in reality it may be connected. Note that Rule k in ref. [27] is general, allowing coverage by a set of one-hop and 2-hop "marked" neighbors that are "glued" together by other "marked" nodes ("marked" nodes are those that consider themselves to be in the dominant set), which can be at an arbitrary hop distance. However, in algorithm 2 from ref. [27], implementing Rule k, nodes send their dominating status only to their one-hop neighbors; therefore, the information about the dominating status of 2-hop neighbors and beyond is not made available for use in making a decision. While implementation [27] is based on sending messages from each node (informing about withdrawal from the dominant set), the algorithm set forth in ref. [28] does not use any message between nodes after gaining 2-hop topological knowledge. Further, the observation described in ref. [29] is based on coverage by nodes that may or may not be in the dominant set, while the definition given in ref. [27] refers to only coverage by nodes that are in the dominant set.

The second observation [29] is that key values often present obstacles to selecting proper nodes in the CDS. A definition that will allow key reversal may be beneficial. Suppose that, in Figure 11.1 node G was actually renamed node M for a reason (e.g., high energy value). Then node $M = G$ will be in the CDS, because of the highest-key value. But this does not eliminate any other node from the CDS; therefore, its inclusion is superfluous. How then can the key of $M = G$ be reversed? All the neighbors of $M = G$ are neighbors of L, and L has a neighbor that is not a neighbor of $M = G$. This is sufficient for node $M = G$ to realize that L will not declare it as a covering node, and therefore can safely withdraw from CDS. This concept can be formalized as follows. Node *u* is covered by node *v* if and only if one of the following two conditions is satisfied:

(1) $N(u) \subset N(v)$, where $N(u)$ is a proper subset of $N(v)$, that is $N(v) \neq N(u)$ is part of this condition), and

(2) $N(u) = N(v)$ and key$(u) <$ key(v).

Note that the preceding extended rule cannot be used jointly with other rules, such as Wu and Li's Rule 2 [25]. The generalization to coverage by several nodes and the corresponding algorithms for backbone construction are presented in ref. [29].

The two enhancements described can be combined into a single one, by allowing the node to be covered by either of the two ways [28].

11.2.7 Activity Scheduling in Ad Hoc Networks

In an ad hoc network that is not a sensor network, area coverage may not be required. In such a case, *activity scheduling* (deciding which nodes should be active, and which should go to sleep mode, so that the ad hoc network life is prolonged) can be performed by applying the connected-neighbor dominant set concept. Nodes in the connected-neighbor dominant set are active, while the rest of the nodes can be put to sleep. However, in order to increase network lifetime, such decisions need to be periodically reevaluated, as nodes that are saving energy need to contribute at a later time. Each node in an asynchronous ad hoc network may wake up at its predetermined time and evaluate whether it needs to be active based on a message exchange with currently active neighbors. In the case of synchronous nodes, such decisions are made in rounds. All nodes wake up at the same time, exchange Hello messages, and then decide which of them will create a backbone. Any described backbone decision process can be applied. If Wu's concept is applied, a suitably selected key value, which depends on the remaining node energy, is selected and used. The other important parameter for making decisions is the average number of neighbors (average degree) of each node. The choice of such a best metric for prolonged network life was investigated in ref. [31].

11.3 BROADCASTING IN SENSOR NETWORKS

11.3.1 Taxonomy

In addition to the taxonomy discussed for the backbone construction, the broadcasting protocols can be further classified. The next division is whether or not they are *reliable*. Reliability is the ability of a broadcast protocol to reach all the nodes in the network, assuming that the medium-access control (MAC) layer is ideal (every message sent by a node reaches all its neighbors), location update protocol provides accurate desired information to all nodes about their neighborhood, and the network is connected. The *blind-flooding* protocol, where each node receiving the packet for the first time will retransmit it, is a reliable protocol at the network layer. However, as observed in seminal work [32], due to excessive retransmissions for dense networks, collisions and contentions actually can make it very unreliable at the MAC layer, plus there exists a large amount of redundancy. The probabilistic (retransmissions with certain fixed probability), counter (retransmitting if the number of received copies does not exceed a constant), and distance (retransmitting if the distance to all senders exceeds certain threshold distance) solutions proposed in ref. [32] are not reliable at the network layer, and also have inferior rebroadcast savings (percentage of nodes that do not retransmit the packet) to the backbone-based reliable solutions reviewed here. Note that the MAC layer cannot be reliable (at least those currently considered for adoption in sensor networks), due to the hidden-terminal problem (a node simultaneously receiving messages from two other nodes that are not aware of each other's transmission) and the probabilistic nature of the protocols used.

The final classification of broadcasting schemes is determined according to the *packet content* during the broadcasting process. The broadcast message sent by the source, or retransmitted, might contain a broadcast message only. In addition, it may contain a variety of information needed for proper functioning of the broadcast protocol, such as its own id, its position, one bit about its backbone status, a list of one-hop neighbors, degree (number of its neighbors), or list of forwarding neighbors, informing them whether or not to retransmit the message.

11.3.2 Backbone and Neighbor Elimination–Based Broadcasting

In ref. [14], the following framework and general algorithm were established for reliable broadcasting. The algorithm is based on two concepts: CDS as the particular type of backbone that provides reliability, and neighbor-elimination scheme. Backbone formation was already discussed in Section 11.2. Connectivity provides propagation through the whole network, while domination assures reachability by all nodes. Excess messages in any protocol affect node power and the bandwidth available; thus, the main goal is to describe a reliable broadcast protocol with a minimal number of retransmissions, that is, to construct a connected dominating set of minimal size.

The *neighbor-elimination* scheme was independently proposed in three papers in August 2000 [33–35]. In this scheme, a node does not need to rebroadcast a message if all its neighbors have been covered by previous transmissions. After each copy of the same message is received, a node eliminates, from its rebroadcast list, neighbors that are assumed to have correctly received the same message (based on one-hop positional or 2-hop topological knowledge that the node has about its neighbors). If the list becomes empty before the node decides to rebroadcast, the rebroadcasting is canceled.

The general algorithm [14] for intelligent flooding is then the following one. The source node transmits the packet. Upon receiving the first copy of the transmitted packet intended for broadcasting, the node will not retransmit it if it is not in the CDS. If it is in the CDS, it will select a time-out based on some criteria and some random number. It will also eliminate all neighbors that received the same copy of the message from its forwarding list (originally containing all one-hop neighbors).

While waiting, more copies of the packet could be received. For each of them, all neighbors receiving it are eliminated from the forwarding list. When timeout expires, the node will retransmit if its forwarding list is non-empty, otherwise it will cancel retransmission. This framework was applied in ref. [14] using clustering based and Wu's concept based backbones. Wu and Dai [36] propose a general algorithm that unifies many neighbor elimination schemes.

Figure 11.3 illustrates the broadcasting algorithm [14], with C being the source node, and nodes F, A, G, and H being in the connected dominating set following definition [25,27], with key $=$ (degree, id). Node E is covered by node G, while node L is covered by connected neighbors with higher keys G and H. Similarly, node I is covered by A and H. Other nodes are not intermediate (do not have two unconnected neighbors). Covering relations are drawn in the dashed bolder edges. Let the time-out be defined as time-out $=$ (1/(number of uncovered neighbors), id). Note that id is added to decide which node retransmits first in case of ties. Node F then sets the time-out to $\frac{1}{3}$ (three uncovered neighbors by source transmission are A, G, E) and retransmits at the time-out expiration. Neighbors from CDS are A and G, and they set time-outs to $\frac{1}{3}$ and $\frac{1}{2}$, respectively, based on the number of neighbors not receiving that transmission (based on their knowledge; it is possible that some neighbors treated as uncovered actually already received the message from nonneighbors in the process). Node A then retransmits because of shorter timeout. After this retransmission, G changes the original time-out to $\frac{1}{1}$ (only neighbor L remains uncovered), and the remaining time-out is $\frac{1}{1} - \frac{1}{3}$, since $\frac{1}{3}$ of the time already lapsed. The time-out at H is $\frac{1}{3}$ and is shorter, so it retransmits first. Node G then cancels retransmission.

To increase reliability at the MAC layer, Stojmenovic et al. [14] proposed the retransmissions after negative acknowledgments (RANA) protocol. Collision

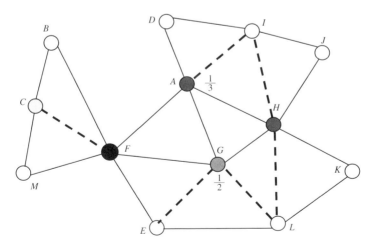

Figure 11.3 *F*, *A*, and *H* retransmit in the neighbor elimination and dominating set–based broadcasting [14].

between two packets normally occurs after the initial portion of the first packet, containing the sender's information has already been received. The receiver node can then send a negative acknowledgment back to the sender node, asking it to retransmit again.

11.3.3 MPR

Several authors [33,37–39] independently proposed reliable broadcasting schemes in which the sending node selects adjacent nodes that should relay the packet to complete the broadcast. The ids of the selected adjacent nodes are recorded in the packet as a forward list. An adjacent node that is requested to relay the packet again determines the forward list. This process is iterated until the broadcast is completed. The methods differ in the details on how a node determines its forward list. The general principle was already outlined in the section on MPR-based backbone.

The adaptation of multihop relaying presented in ref. [40] improves its performance by the following observations: the broadcasting node transmits a list of its neighbors at the time of broadcast packet transmission, not as part of any Hello message. Knowledge of the 2-hop neighbors is used to determine which neighbors also received the broadcast packet in the same transmission, and these nodes are already covered and are removed from the neighbor graph used to choose the next hop relaying nodes. Finally, if a broadcast message is received from a node that is not listed as a neighbor, the message is retransmitted to deal with high mobility issues. In connected dominant set–based broadcast algorithm [41], the sender node establishes priorities between the forwarding nodes and each forwarding node should eliminate from consideration not only neighbors of the sender node, but also neighbors of each relaying node with higher priority. Wu and Lou [43] proved several extensions of MPR to generate a smaller CDS using 2-hop neighborhood topology information to cover each node's 2-hop neighbor set. Note that 2-hop neighborhood topology includes all nodes within two hops and their connections. In addition, they extended the notion of coverage in the original MPR and showed that the extended MPR has a constant local approximation ratio compared with a logarithmic local ratio in the original MPR.

Compared to backbone-based broadcasting, MPR broadcasting has a similar or somewhat better performance in terms of rebroadcast savings, but has message overhead due to the inclusion of the forwarding list in the packet, which may be significant for energy-limited tiny sensors.

11.3.4 Broadcasting and Dominating Sets with Realistic Physical Layers

We now describe the corresponding coverage, backbone notions, and broadcasting process when the impact of the physical layer is considered. Let $A_1, \ldots, A_k$ be active neighbors of given node B, and let $x_1, \ldots, x_k$ be their respective distances to B. Then $p(x_1), \ldots, p(x_k)$ are their packet reception probability rates for packets sent by B.

The probability $q(x)$ that at least one of the packets from the active nodes is received by B is then $q = 1 - (1 - p(x_1))(1 - p(x_2)) \cdots (1 - p(x_k))$. Node B is m-covered by active nodes $A_1, \ldots, A_k$ if $q \geq m$ [3]. A set of nodes is the m-dominating set if each node is either in the set or is m-covered by nodes from the set [3]. Note that, for $m = 1$, and the unit-disk graph model, the well-known definition of dominant sets follows.

These definitions can be used as a basis to generalize some well-known types of dominating sets for the unit-disk graph to be applied under a realistic physical layer. For example, the following definition is proposed in ref. [3] as a generalization of the concept proposed by ref. [27]. Let $A_1, \ldots, A_k$ be the set of higher id neighbors of B. If the set is empty or disconnected, then B is in the dominating set. If the set is connected and each neighbor of B is m-covered by them, then B is not in the dominating set. Finally, if any neighbor of B is not m-covered by the set, then B is in the dominating set.

The broadcasting process with any notion of dominating sets and neighbor elimination [14] can proceed as follows [3]. After receiving a broadcast message, node A will set a time-out short if it is in the dominant set, and long if not. It calculates the probabilities of each neighbor for receiving the same message, and eliminates m-covered neighbors from the list. This lists is updated for any further copy of the received message. The update includes the time-out that can be extended with more received messages. At the end of time-out, if all neighbors are m-covered, retransmission is canceled. Otherwise, the node retransmits the packet.

Since the reception of any message is a probabilistic event, one retransmission by any particular node may not suffice. To learn about the existence of neighbors, each node may need to send several packets. The number of retransmissions needed for learning about the satisfactory number of neighbors depends on density. In ref. [3], it was proposed that each node retransmit Hello messages until a certain fixed number of such packets or responses is received from neighbors, as an indirect measure of density. A similar protocol also can be applied for the broadcasting task, modifying any existing protocol originally designed for the unit graph as follows. Instead of retransmitting only once, a given node can keep retransmitting until a certain fixed number of packets (carrying the same packet) has been heard from neighbors, before or after the first retransmission, or until a certain time-out expires (to handle the case of low-degree nodes). If density is known, then a fixed number of retransmissions can be replaced by a number depending on local density. Nodes that are, by original protocol, supposed not to retransmit may also contribute by retransmitting the message, but fewer times than other nodes. Further investigation and simulation is needed to find a precise description of the winning protocols, following this general design principle.

11.3.5 Minimum Energy Broadcasting

Suppose that nodes in an ad hoc network can adjust their transmission radii, and that they are aware of their own and the geographic position of their neighbors. The problem is to broadcast a packet to all the nodes in the network so that the sum of all transmission power used is minimized. The power consumption for two nodes at distance r is $r^\alpha + c$, where $\alpha \geq 2$ and c is a constant that includes signal processing and minimal reception power. It is shown in ref. [43] that, for $c > 0$ (which is a realistic assumption), it is not optimal to minimize transmission range. Furthermore, it was demonstrated that there exists an optimal radius, computed with a hexagonal tiling of the network area, that minimizes the power consumption. For $\alpha > 2$ and $c > 0$, the optimal radius is $r = (2c/(\alpha - 2))^{1/\alpha}$, which is derived theoretically and confirmed experimentally.

A localized broadcast algorithm, called *TR-LBOP* is proposed [43], which takes this optimal radius into account. This protocol is experimentally shown to be have limited energy overhead with respect to globalized algorithms for all network densities.

11.4 SENSOR AREA COVERAGE

In area-coverage problems, a set of sensors is given and distributed over a given area. Each sensor is able to cover a circle with radius centered at it. The problem is to determine a small number of sensors that still cover the same area and are connected, so that the sensor can report the detected information to a monitoring center. The maximum network lifetime is certainly a related goal. *Full coverage*, *energy efficiency*, and *connectivity* are critical requirements of any area-coverage protocol. The objective of any area-coverage protocol is to achieve full area coverage, and protocols can be classified into those that guarantee full area coverage (provided such coverage exists) and those that do not guarantee it. A set of sensor nodes that together fully cover a given area is called *area-dominating set*. Protocols can also be divided into those that guarantee connectivity of selected active sensors and those that do not.

There is a variety of problem statements, assumptions, and solution approaches for the sensor area coverage. We will review them before presenting some solutions. The problem is centered on a fundamental question: How well do the sensors observe the physical space? This chapter discusses only the *area-coverage* problem, meaning that each point in a given geographic area needs to be covered by at least one sensor. Alternative formulations include covering certain points instead of area (*point coverage*) and *barrier coverage*. Examples of barrier-coverage problems are, minimizing the probability of undetected penetration through the sensor barrier and minimal exposure path, measured as sensing time, with sensing ability diminishing with distance. A survey of point- and barrier-coverage solutions is given in ref. [44].

The area-coverage problem can be further divided into *single* and *multiple* area coverage. In single area coverage, each point in the area is required to be covered at least by one sensor. In multiple area coverage, each point needs to be covered multiple times, which could be a fixed k times coverage at a given time, or division of sensors into maximum number of layers of area coverage. These layers can then either alternate in time for coverage, or several layers can be used to cover an area simultaneously for increased reliability.

The sensor deployment mechanism can be *random* or *deterministic*. A deterministic sensor placement (placing sensors at desired locations) may be feasible in friendly and accessible environments. Random sensor distribution is generally considered in remote or inhospitable areas, or when a fast deployment of a large amount of sensors is desirable. We will consider only random placement in this chapter. An example of deterministic placement is considered in ref. [45], where the authors proposed path exposure (the likelihood of detecting a target traversing the region using a given path) as the measure of goodness of the sensors that are deployed to perform collaborative target detection. A centralized algorithm for placing sensors at selected locations to minimize path exposure is described in ref. [45].

In most articles in the literature, all sensing radii are equal, while a few articles consider coverage with different sensing radii. We will consider only the case of equal sensing radii at each node, since there is limited research done for the case of adjustable ranges [46,47]. Wu and Yang [47] considered the cases where each sensor is able to select one of two or three adjustable ranges, with the goal of minimizing the overlapped sensing area, extending results from ref. [48].

There are also several variants regarding the relation between sensing and transmission ranges. One common assumption is that sensing radius and communication radius are *equal* [28]. However, some physical measurements indicate that the communication range is normally *larger* than the reliable sensing range. This has implications on the selection of sensors for coverage, and also on the performance of other relevant protocols. For example, Xing, Lu, Pless, and Huang [49] show that greedy routing always works when the communication range is twice or more the sensing range, and the area is covered and convex. They also consider restricting greedy routing to nodes whose Voronoi regions intersect the source–destination line.

Most literature uses the *unit-disk graph* model for sensing, which is similar to the unit-disk graph model used for communication. In this model, the sensor is able to monitor an event if and only if the distance from the sensor to the event is at most S, where S is its sensing radius. However, a closer look at the *physical layer* reveals that sensing ability decreases with distance. Instead of the unit-disk graph model, it is more realistic to use a model where the probability of sensing an event depends on the distance from the sensor to that event. Liu and Towsley [50] approached the coverage problem from a theoretical perspective and explored the fundamental limits of the coverage of a large-scale sensor network.

Zhang and Hou [51] studied the fundamental limits of sensor network lifetime that all algorithms can possibly achieve. If the lifetime of a sensor is T, they derived analytically and by simulation the minimum sensor density needed to achieve network lifetime kT. They observed that the increase in lifetime per unit of nodal density becomes marginal when the density exceeds a certain threshold.

11.4.1 Threshold-Based Protocols

Ye et al. [52] proposed a simple localized protocol (called PEAS) for dynamically selecting an area-dominating set in asynchronous sensor networks. Each sensor has the same probing radius P and the same maximum transmission radius R, which is also the monitoring radius. Any two active sensors must be at a distance of at least P, which is enforced by the scheme. Initially all sensors are in sleeping mode, with an exponentially distributed sleep-duration function. When sleeping time expires, the sensor sends a probing message using transmission radius P. Each active sensor that overhears this probing message should estimate whether or not its distance to the probing sensor is below P. Since they are able to detect signals from a greater distance, up to R, they should apply signal strength (which is considered an unreliable measurement due to fading effect) or time-delay measurements to make the judgment. If the distance is below P, then it sends (a/the) message to the probing sensor informing it about its activity. Upon receiving such a response, the probing sensor again selects a new sleeping duration and continues to sleep, waking up at a later predetermined time to reevaluate the decision. If the distance is above P, then no response is generated. If the sensor does not receive any response to its probing signal, it decides to wake up and monitor the area, up to radius R. Once a sensor wakes up, it continues to work until it dies. For this protocol, the probability of having full coverage of a monitored area is close to 1 if the threshold P is less than $1/(1 + \sqrt{5}) \approx 0.3$ of the sensing area's radius S, that is, $P < 0.3S$. The rationale is that otherwise activating the sensor has an insufficient contribution toward covering some new area, due to it being too close to an already active sensor. The method presented has a high degree of fault tolerance. However, this protocol is probabilistic and does not ensure full area coverage. Figure 11.4 illustrates this protocol, with the black nodes being active and the white nodes being in sleep mode, because each of them is contained within the threshold distance (smaller circles) to one of the active nodes. The larger circles indicate the communication radius for active nodes.

In ref. [53], three sensor-area covering schemes are proposed. In the probabilistic-based scheme, each node decides whether or not to remain active with a fixed probability, whose optimal value is derived based on the expected percentage of the sensing area coverage, which in turn depends on the number of neighbors

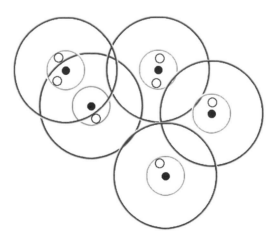

Figure 11.4 Threshold-based area coverage.

within transmission radius that announced active status and expected distance to them. In the nearest neighbor–based scheme, the decision is based on the expected distance (whose value is derived as a function of the ratio of the transmitting and sensing radii and the number of active neighbors) to the nearest of the active neighbors (this is similar to the PEAS scheme [52]). In the neighbor number–based scheme, the decision is based on a counterthreshold compared to the number of active nodes. All mentioned threshold values are determined numerically and experimentally, for use in the schemes, and do not guarantee area coverage.

11.4.2 Some Covering and Connectivity Properties

In refs. [48] and [54], it is proved that if the transmission range is at least twice the sensing range, and the area to be covered is convex, then the area coverage also implies connectivity among the covering sensors. This follows from observing that the distance between the centers of two intersecting circles of the same radius cannot exceed twice the radius, therefore two sensors whose sensing radii intersect are also communication neighbors. The distance between two nodes whose sensing ranges S intersect is $<2S$, which is within the transmission range R for $R > 2S$. Di Tian [55] generalized this proof by eliminating the need for the convexity condition.

When the sensing and transmission radii are equal, the coverage property can be tested by verifying whether or not the perimeter of the sensing circle is covered by other circles. This is illustrated in Figure 11.5. The number of uncovered arcs of a circle can be at most two.

When the communication range exceeds the sensing range, this simple test cannot be used. Finding the exact regions of intersection, or their size, is computationally sophisticated and time-consuming. However, one can apply the following well-known geometric theorem [48,54] to efficiently confirm that a sensing circle is fully covered by other sensing circles: It is shown that if there are at least two covering circles and any intersection point of two covering circles inside the sensing area is covered by a third covering circle, then the sensing area is fully covered.

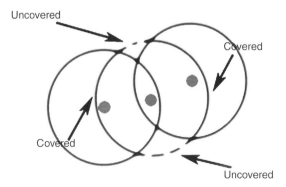

Figure 11.5   Testing the coverage property when sensing and communication radii are equal.

This preceding result is illustrated in Figure 11.6. Sensors within a communication edge of each other are joined by an edge. The central darker sensing circle is covered by other circles, and all the intersection points of other circles, which are inside it, are covered by a third circle.

This result provides an efficient method for testing the full-coverage criterion. However, it does not provide direct information about the possible size of the uncovered region. One possible estimate is to generate a certain number of points at random, test each for coverage with existing circles, and take the percentage of covered points as the estimate (this method can be computationally expensive if satisfactory precision is required). Another alternative is to make an estimate based on the distances and positions of the active sensors. There exists a need for designing more accurate and fast-coverage size-estimation protocols, including an efficient test for confirming full coverage.

11.4.3 Hexagonal Area Coverage

Zhang and Hou [47] described an efficient algorithm for selecting covering sensors. Sensors are assumed to be time synchronized, and they periodically make new decisions about sensors that remain active to cover the area. In each round, a single sensor starts the decision process, which then propagates to the whole network. New sensors are selected so that the priority is given to sensors located near optimal hexagonal area coverage, obtained when the area is ideally divided into equal regular hexagons. The coverage is indeed very optimal, given the *distributed* nature of the decisions. However, the need for a single sensor to start the process may cause problems in applying it. Since time is synchronized, and rounds are well defined, perhaps it is better to allow all sensors to make localized decisions without waiting for any specific sensor to start the process (especially if

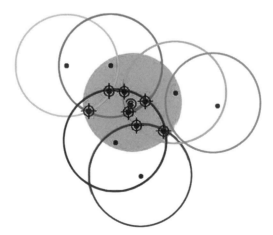

Figure 11.6 A circle is covered when all intersection points are covered.

the sensor somehow decided to start the process and failed to do so because of malfunctioning). The original sensing coverage may not be preserved (as shown by experimental results).

11.4.4 Area Coverage Based on Neighbor Cooperation

The algorithm presented in ref. [56] divides the area into small grids, and then covers each grid with a sensor. Each sensor that can cover a grid maintains a list of other sensors that can also cover it, in a priority order. All sensors covering the same grid can communicate with each other, since the communication range is at least twice the sensing range. When sensor density is significant, sensors need a lot of memory and processing time to maintain priority lists, plus the communication overhead for making covering decisions in cooperative manner is nontrivial.

Hsin and Liu [57] investigated random and coordinated area-coverage algorithms. Each sensor covers a circle of radius R. In their coordinated-coverage scheme, a sensor may decide to sleep after receiving "permission" from sponsoring neighbors, for the time such permission is given. A node that sponsors any other node must be active. The decisions are not synchronized, since each sensor can "negotiate" with its sponsors independently, and the scheme allows for several variants with (sophisticated) protocol details. The authors suggest that nodes collect information about residual energy from neighboring sensors. Sensors with high residual energy are more likely to enter the sleep state than sensors with low residual energy. Each sensor maintains its own delay counter, which is used for role alteration. Coordinated schemes performed better in their experiments. Although the Hsin and Liu's [57] coordinated scheme has some desirable properties, such as localized behavior, it may select too many sponsor nodes to be active, since there is no coordination between nodes for the selection of as many as possible common sponsor nodes.

11.4.5 Centralized Area-Coverage Protocols

Centralized (and distributed) schemes may be treated as localized schemes with extended communication range, where any node can reach any other node. In this scenario, obviously one node can make all sensing decisions for other nodes and communicate them.

In ref. [58], a centralized algorithm is given for finding a small-size connected sensor cover. A straightforward distributed version of the same algorithm is also given. The sensing circles are not necessarily of the same size. In their greedy algorithm, candidate sensors for inclusion are those sensors that partially (not fully) intersect with sensors previously included in the area coverage. For each such sensor, a shortest path from it to one of the sensors already selected is considered. Note that, if coverage circles were the same, the considered path consists of one hop only, since any two sensors whose coverage circles intersect must be neighbors. Circles of candidate sensors divide the area into subelements (each subelement is a small region belonging entirely to some circles and entirely outside the remaining circles). The length of each path is divided by the number of subelements. All sensors on the path with the maxim of such a ratio are added to the covering set (the sensor that starts the selected path is called "leader" here). In the distributed implementation, the sensor that was last added (the leader) initiates the search for a new sensor/path to add. It broadcasts the search message up to $2R$ hops, where R is the maximal hop distance between any two sensors whose circles intersect ($R = 1$, if all circles are equal). Sensors that receive such a message and have partial coverage perform the described iteration, to select a new path and a new sensor "leader." This process repeats until the entire query region is covered.

We observe that the algorithm presented in ref. [58] may not converge with full coverage of the region. For example, the corner of a region may just be fully covered by the last leader, and all sensors within distance $2R$ may be fully covered as well. On the other hand, uncovered regions may exist in other corners of the region. This problem, however, can be resolved by some additional protocols, such as time-out at sensors that activate if the region is partially covered, but no news is received within the given time-out. More detail regarding parallel actions by several such sensors needs to be added, and the quality of the final result may not differ significantly from the one obtained by centralized implementation.

For simplicity of analysis, consider the case of equal sensing and transmission radii ($R = 1$). Candidate search broadcast involves transmission from the leader, and retransmissions by several of the neighbors to reach all nodes at distance two (an MPR-like broadcasting method can be used), responses from each candidate sensor, and another broadcast to communicate the decision. In dense sensor networks, many sensors are candidate sensors, thus too much traffic for selecting each next sensor is easily generated. Let R be the transmission radius. Initially, all sensors at the distance in the interval $(0,2R)$ from the first leader are candidate sensors for the next leader. There is, unfortunately, no limit on their number inside this circle. It is also difficult to schedule so many transmissions at the MAC layer.

11.4.6 Localized Sensor Area Coverage

Tian and Georganas [46] proposed a solution for sensor area coverage in synchronous networks where sensing range is equal to the transmission range. It requires that every node know all its neighbors' positions before making its monitoring decision. At the beginning of each round, each node selects a time-out interval. At the end of the interval, if a node sees that neighbors that have not yet sent any messages together cover its monitoring area, the node transmits a "withdrawal" message to all its neighbors and goes into the sleep mode. Otherwise, the node remains active, but does not transmit any message. The process repeats periodically to allow for changes in monitoring status. There are several problems in this protocol. Neighboring active sensors may fail without notice, and neighboring sensors may not activate, believing that the sensor is "alive" and monitoring. This problem can be resolved if neighboring information is exchanged at the beginning of each round. However, this then involves significant communication overhead once sensors start to die between activity periods. The other problem is that covering sensors may not be connected; thus, reporting to a monitoring station may not succeed. The authors also discuss the case of different sensing radii at each sensor.

Jiang and Dou [59] describe several improvements to the algorithm in ref. [46]. They apply the criterion that a circle is covered completely if perimeters of other circles covering it are fully covered by other covering circles (note that it can be further simplified, as discussed later, to consider only intersection points). In the algorithm presented in ref. [59], at the beginning of each round, each node sends a hello message to inform about its position. The algorithm from ref. [46] is then applied (which relies on node withdrawals with negative acknowledgments) for all ratios of sensing and transmission radii, using criteria described here. Experimental data in ref. [59] show that this algorithm outperforms PEAS [52] with respect to the number of nodes needed in the coverage, while completely preserving sensing coverage of the original network.

Carle and Simplot-Ryl [28] described a localized algorithm for area coverage for the case of equal sensing and transmission radii. This approach has been generalized by Carle, Gallais, Simplot-Ryl, and Stojmenovic [60] for an arbitrary ratio of sensing and transmission radii. The approach, in addition to being fully localized, has a very small communication overhead. There are two variants in the approach. One requires each sensor to send exactly one message, while the other requires that only nodes that will remain active for covering the area send exactly one message. The two approaches have a trade-off, since one message sent by each sensor that will move to a sleep mode is expected to leave less active sensors for the area coverage.

The basic principle of the algorithm presented in ref. [60] is that each node selects a time-out and listens to messages sent by other nodes before the time-out expires. The time-out can be selected at random, or may depend on the sensor's remaining battery energy. Each received message provides information that a portion of the sensing range is covered. This information is derived from the position of the transmitting sensor, which is reported in the message. The reduction of the required area coverage for monitoring implies an extension of time-out. Nodes with smaller uncovered areas should receive a longer time-out, hoping that a message by a node that is able to cover more area will cover that small portion as well. At the end of time-out, the node verifies whether or not its sensing area is fully covered. If so, it goes to sleep mode in the current round. The two variants differ in whether or not the node then sends the message informing neighbors about the sleep status decision. These messages are called negative acknowledgments. If such a negative acknowledgment is sent by a node that will enter a sleep state, it still informs neighbors about a certain area that has already been covered by sensors that will remain active. The benefits of the negative acknowledgment message are illustrated in Figure 11.7 (where sensing and communication radii are equal). Assume that nodes 1–4 announced their active status. Although node 5 then decides to sleep, its withdrawal message reduces the area to be covered by node 6 by the shaded area in Figure 11.7 (more precisely, node 6 may now conclude that its sensing area is fully covered, which enables it to select sleep mode). That shaded area is covered by active nodes 3 and 4, which are not communication neighbors of 6. Sensor nodes whose sensing area is not fully covered (or fully covered but with a disconnected set of active sensors) when

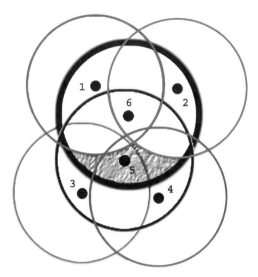

Figure 11.7 Negative acknowledgment by node 5 reduces the area to be covered by node 6.

the deadline expires decide to remain active for the considered round, and send a message to all its neighbors informing them about the decision. Such a message is called positive acknowledgment [60]. The process repeats in each round in synchronous fashion.

The details of the protocol given in ref. [60] include how the time-out is decided, and how the area coverage and connectivity tests are performed. First consider the case of equal sensing and transmission radii. One important property of the protocol is that no prior knowledge about neighbor existence and location is required. That is, there is no communication overhead coming from the preprocessing step to collect neighborhood information. The test for connectivity of covering circles must be performed whenever $2S > R$, where S and R are sensing and transmission radii.

The network can reselect covering nodes periodically to spread the sensing cost dynamically over all nodes in a fair manner. This method significantly extends the network's life. If the density is more than 30 nodes per unit area, the area-dominating graph is sparse, with nodes having on average three neighbors (this is valid when sensing and communication ranges are equal). In addition, the distance between its two neighboring nodes is typically two-thirds of the transmission radius. Hence, active nodes form a very simple network with a structure similar to regular hexagonal tiling.

11.4.7 Multiple Sensor Area Coverage

In ref. [53], the problem of covering each point in an area with at least k sensors (k-coverage) is reduced to the simpler problem of determining the similar coverage of all the intersection points of the sensing circles. A sensor is ineligible for turning active if all the intersection points inside its sensing circle are at least k-covered. To find all the intersection points inside its sensing circle, a sensor v needs to consider all the sensors in its sensing neighbor set, $SN(v)$. Set $SN(v)$ includes all the active nodes that are within a distance of twice the sensing range to v. The algorithm is then combined with SPAN activity-scheduling protocol [61], which is an inefficient version of Wu's dominating set definitions [25] published long before SPAN (see ref. [25] for details).

Abrams, Goel, and Plotkin [62] studied the problem of partitioning the sensors into covers so that the number of covers that include an area, summed over all k areas, is maximized. Three approximation algorithms, assuming k is fixed, are described. Randomized algorithm assigns to each sensor one of k covers at random. In the distributed greedy algorithm, each sensor sets a time-out and listens to decisions made by neighbors, increasing the counter in the appropriate set for each message announcing the decision by a neighbor (the communication radius is assumed to be twice the sensing radius). When time-out expires, each node selects a set for which the corresponding counter is minimal. The centralized greedy algorithm adds some weight, but otherwise runs a similar procedure. This article [62] does not discuss what the best value is for k, that is, how many layers of coverage could be reasonably achieved.

An adaptive localized multiple sensor area-coverage algorithm is proposed in ref. [63]. The algorithm [63] adjusts k dynamically to reflect the sensor density. Each sensor node selects a time-out, which depends on the portion of the area not covered by other sensors, and has some random number or other parameter in the formula to avoid simultaneous transmissions by neighbors. Suppose that node A received a message from a neighbor that informed about i, the cover-layer number selected by that neighbor, and its geographic coordinates. Node A adjusts the uncovered portion of layer i at the node, and extends appropriately its deadline. When the time-out expires, there are a few options for making a decision (which is then transmitted):

- Assign the layer j, which is a minimal number so that the area in layer j is not yet fully covered;
- Among layers covered partially by some neighbors, and not yet fully covered, choose one that maximizes the uncovered area;
- If all layers covered by some neighbors are fully covered, the sensor chooses a new layer, and informs its neighbors about covering it.

This algorithm can be extended to provide layers for activity scheduling in static ad hoc and sensor networks. Existing algorithms only select sensors for the next round, one round at a time. The difference is that the required area coverage is replaced by neighbor coverage. Each node again sets a time-out. The message received from a neighbor gives selected layer i covered by that neighbor. Time-out is extended, and covered neighbors at layer i are updated. At the end of the time-out, the node may select either minimal j, so that its neighbors are not covered at layer j, or layer j with a maximal number of uncovered neighbors. If all neighbors are covered for all known layers, the node announces participating in the next layer number.

11.4.8 Coverage Using the Physical-Layer Model of Sensing

Xing, Lu, Pless, and O'Sullivan [64] consider a probabilistic model of sensor coverage. A point A is covered in a sensor network if the probability at which a target located at A is detected by active sensors is above threshold β and the system false-alarm rate is below threshold α. The probability of correct detection by a sensor depends on the distance of the sensor. The authors describe a centralized algorithm for deciding which sensors should remain active. Using the active sensors' locations and local false-alarm rate, the location with minimal detection probability is found. If that probability is below β, then the closest sensor to the considered location is selected to become active. The authors also describe a distributed algorithm that divides the network into grids, selects one sensor in each grid to be coordinator, and then each coordinator follows the centralized algorithm to decide which sensors from its grid need to be active. Neighboring coordinators collaborate to improve the decisions in border areas.

A localized algorithm along these lines can be described as follows [65]. First, we need an approximate function for sensing probability with respect to distance. Then sensors select random time-outs and wait to hear from nearby sensors about their active status. For each received message, the sensor adjusts (normally prolongs) its time-out based on the measured coverage in its local area (e.g., the percentage of its local area having "satisfactory" coverage) and the measured benefit if that sensor is to become active. At the end of the time-out, if the sensor sees that its local area already has satisfactory coverage by other active sensors, it decides to sleep. Otherwise it decides to become active and informs its neighbors about it. The local area to be considered may be a small circle around the sensor that has high values for sensing. Some particular sample of points from the area can be taken to reduce computation time.

11.4.9 Variations of the Sensor Area-Coverage Problem

Gui and Mohapatra [66] observe that it is not necessary to achieve a perfect sensing coverage of a moving object. They found that the expected length of a straight-line path and object should move before hitting the boundary of any covered area, for a random sensor placement in the area, can be approximated by $|X|/(4nr)$, where $|X|$ is the area of a given field, n is number of sensors, and r is their sensing radius.

Cardei and Du [67] considered the point coverage problem. A certain number of points needs to be covered by sensors within sensing range of them. Each target point needs to be monitored by at least one sensor. The authors divide the sensors into disjoint sets, each covering target points, with sets being activated in turn. They prove that the problem is NP-complete and propose a centralized solution based on the heuristics of the disjoint set cover.

Shakkottai, Srikant, and Shroff [68] showed that the necessary and sufficient conditions for the random grid network of n nodes, arranged in a grid over a square region of unit area, to cover the unit square region as well as ensure that the active nodes that are connected are of the form $pr^2 = \theta(\log(n)/n)$, where r is the transmission radius of each node, and p is the probability that a node is active.

11.4.10 Mobile Sensors for Improved Area Coverage

Zou and Chakrabarty [69] proposed a virtual force algorithm as a sensor deployment strategy to enhance the coverage after an initial random placement of sensors. It is assumed that sensors can move by "virtual force" with the force's strength determined by node distance.

Cao, Wang, La Porta, and Zhang [70] considered the problem of moving some sensors from their initial random placement in order to cover some areas that were not covered by either the nature of randomness or some other effects such as wind. It is also assumed that sensors can move after gathering some information from neighbors. The algorithm proceeds in rounds. In each round, sensors communicate to local neighbors in order to construct Voronoi diagrams. Each sensor then subtracts its sensing area from its Voronoi polygon, and moves in the direction of the largest uncovered piece of area. The process repeats until no further improvement is possible. The approach appears suitable when robots, equipped with sensors, are monitoring an area, which can also be monitored by some static sensors. Voronoi diagram construction, however, may not always be locally constructed, and it may be better to use localized versions such as the partial Delauny triangulation [71]. The Gabriel graph can also be used. An alternative approach may be to use face routing [72] to estimate the size of a hole, find its centroid, estimate the number of sensors that should move toward the centroid, and provide the best possible information to sensors for their move.

Wang, Cao, and La Porta [73] propose a proxy-based sensor deployment protocol. Instead of moving iteratively, sensors calculate their target locations based on a distributive iterative protocol. Current proxy sensors advertise the service of mobile sensors to their neighborhoods (up to certain parameter distance), searching for a better coverage location. They collect bidding messages and choose the highest bid. Then they delegate the bidder as the new proxy. The iterative moves are logical, not physical. Actual movement only occurs when sensors determine their final locations. If the bidding process is local, the sensor movement and the area-coverage gains may be restricted. If the bidding process includes neighbors at several hops distance, the communication overhead for bidding becomes significant. Bidding decisions are based on price (number of logical movements made so far) and distance that the moving sensors are physically supposed to move altogether. A procedure to prevent multiple healing is described, which includes some message overhead. The bidding criterion does not include lost area coverage for moving out of the current position. It is not certain whether the described procedure is always loop-free and always converging. The difference between sensing and transmission radii (the ratio is not discussed in ref. [73]) has a direct impact on message complexity.

The iterative nature of logical moves in ref. [73] may still easily lead each mobile sensor to a local minimum. It may be better to apply an expanding ring strategy [74] in search of the best proxy, by using the increasing sizes of distances from each mobile sensor, and asking sensors within the ring to respond with their biddings (these responses may be suspended on the way to the mobile sensor by intermediate nodes that learn about better bids). The bidding price also includes the traveling distance for the mobile sensor, which chooses the best bid. Some mechanisms for avoiding multiple healing of the same hole need to be added in the protocol, such as reporting to only one mobile sensor (note that this is not sufficient, since few sensors can be located around the same hole). Instead of moving, mobile sensors send the best bid so far to the next ring, asking sensors from that ring to report their bids only if they have a better bid to offer. This protocol should reduce the number of reports, because substantially more free area needs to be made available to justify longer movement.

Wu and Yang [75] propose a scan-based movement-assisted sensor deployment method (SMART) that uses scan and dimension exchange to achieve a balanced state. In SMART, a given rectangular sensor field is first partitioned into a 2-D mesh through clustering. Each cluster corresponds to a square region and has a clusterhead which is in charge of bookkeeping and communication with adjacent clusterheads. Clustering is a widely used approach in sensor networks for its support for design simplification. In fact, it is shown in ref. [76] that clustering is the most efficient for sensor network where data are continuously transmitted. A hybrid approach is used for load balancing, where the 2-D mesh is partitioned into 1-D arrays by row and by column. Two scans are used in sequence: one for all rows, followed by the other for all columns. Within each row and column, the scan operation is used to calculate the average load and then to determine the amount of overload and underload in clusters. The load is shifted from overloaded clusters to underloaded clusters in an optimal way to achieve a balanced state. By optimal, we mean the minimum number of moves and minimum total moving distance and minimum number of moves. By a balanced state, we refer to a state where the maximum cluster size (the number of sensors in a cluster) and the minimum cluster size are different by at most 1. Using this 2-D scan without global information, each sensor moves at most twice, although it may not be globally optimal in terms of total moving distance in 2-D meshes. SMART addresses a unique problem called *empty cells* in sensor networks and provides a local solution to it.

Mobile and static sensors can use the perimeter created by a Gabriel graph to make moving decisions after only one iteration, as elaborated in ref. [72]. First, static sensors will locally communicate to ensure that their biddings are made for nonintersecting coverage areas. They then send their bidding by the GFG routing protocol [72] (which guarantees delivery in connected unit-disk graphs) in an arbitrary direction. Such routing will end up creating a loop along the perimeter. The node that detects the loop will store the bid. Mobile sensors also will search for the best bid by routing in arbitrary directions, ending on a perimeter. A similar idea has been described in ref. [77] for the purpose of providing location service. If the network of static sensors is disconnected, then mobile sensors will send one message to each connected component and search several perimeters. Mobile sensors will set a criterion for selecting the bid, which will include the cost for moving to a new location, and gain made for changing the coverage area (the difference between current and new coverage). After making a full traversal along the perimeter, the message sent by the mobile sensor will select the best bid and return it to the node responsible for the bid, which in turn will eliminate the bid to prevent other mobile sensors from taking it. The message is then routed back to the mobile sensor, which performs the indicated move. Note that the proposed protocol has only one iteration, flooding type of message circulation is avoided, and the message cost is made quite uniform.

11.5 RELATED SURVEY ARTICLES

Because of space limitations, this chapter did not cover all relevant aspects of the considered problems. For a more complete coverage, the reader is referred to several complementary book chapters [11,78–81]. In particular, ref. [78] contains comprehensive coverage of the topology aspects of these problems, ref. [11] discusses broadcasting with directional antennas and reliable broadcasting, refs. [79] and [81] give comprehensive coverage for broadcasting with adjustable transmission powers with omnidirectional and directional antennas (that is, the minimum energy broadcasting problem). Further, ref. [79] contains comprehensive coverage of broadcasting reliability issues, deciding transmission radii, and resource-aware broadcasting. Probabilistic broadcasting protocols are covered in refs. [11] and [79] (see also the recent article, ref. [82]). Finally, ref. [80] describes routing and broadcasting schemes for hybrid ad hoc and sensor networks.

ACKNOWLEDGMENTS

This research was supported by grants from NSERC, INRIA, and NSF. The authors are grateful to Roger Wattenhoffer for a fruitful e-mail discussion that contributed to clarification of major aspects of dominating sets and subsequently to an improved presentation of this chapter. We are also grateful to Stefano Basagni for a careful reading of the manuscript and several useful suggestions.

EXERCISES

11.1. Prove that intergateway and gateway nodes in Wu's concept [14,25] create dominating sets [26].

11.2. Write a procedure for deciding whether or not a node is an intermediate, intergateway or gateway node in Wu's concept [14,25,26].

11.3. Give a formal definition of an enhanced dominating set, generalizing the case of coverage by one neighbor presented in this chapter. Describe the appropriate efficient algorithm, and prove that the new set is indeed a CDS [30].

11.4. To increase reliability, double (and in general *t*-coverage) dominating sets can be considered. In this approach, every neighbor needs to be covered by two (in, general, *t*) neighbors, instead of only one. Describe some backbone construction methods based on double domination, and some broadcasting schemes that would require each node to receive the message at least twice [30].

11.5. Assume that each node knows its geographic location, but has no knowledge about the existence or position of its neighbors. Describe a beaconless broadcasting scheme that will work with such assumptions and will minimize the number of retransmissions [81,83].

11.6. Suppose that broadcast messages need to be acknowledged. Describe a protocol that will minimize the number of acknowledgment packets for reliable broadcasting [81].

11.7. Generalize the sensor area-coverage scheme [60], described in this chapter, for the case of unequal sensing radii at sensor nodes.

11.8. Give an example showing that GAF [21] can disconnect the network [17]. (*Hint*: Consider scenarios with nodes near corners of grids and near some empty grids.)

11.9. Suppose that sensor nodes are placed at vertices of a regular hexagonal tiling with side length *r* corresponding to the transmission radius. Prove that the side length that minimizes the total transmission power used when all nodes retransmit the packet is $r = (2c/(\alpha - 2))^{1/\alpha}$ [43].

REFERENCES

1. F. Kuhn and R. Wattenhoffer. Constant-time distributed dominating set approximation. In *Proceedings of the 22nd ACM Symposium on the Principles of Distributed Computing (PODC)*, Boston, Massachusetts, July 2003.

2. L. Barriere, P Fraigniaud, L. Narajanan, and J. Opatrny. Robust position-based routing in wireless ad hoc networks with irregular transmission ranges. *Wireless Communications and Mobile Computing*, 3(2):141–153, 2003.

3. I. Stojmenovic, A. Nayak, J. Kuruvila, F. Ovalle-Martinez, and E. Villanueva-Pena. Physical layer impact on the design and performance of routing and broadcasting protocols in ad hoc and sensor networks. *Computer Communications*, **28**(10), 1138–1151, June 2005.

4. M. Agarwal, J. H. Cho, L. X. Gao, and J. Wu Energy Efficient Broadcast in Wireless Ad Hoc Networks with Hitch-hiking, to appear in *ACM/Kluwer MONET*.

5. H. Chan and A. Perrig. ACE: An emergent algorithm for highly uniform cluster formation. In *Proceedings of the European Workshop on Wireless Sensor Networks (EWSN 2004)*, pages 154–171, Berlin, Germany, January 2004.

6. D. A. Fisher and H. F. Lipson. Emergent algorithms: A new method for enhancing survivability in unbounded systems. In *Proceedings of the 32nd Annual Hawaii International Conference on System Sciences (HICSS-32)*, Maui, Hawaii, January 1999.

7. D. Niculescu. Positioning in ad hoc sensor networks. *IEEE Networks*, **18**:24–29, July 4, 2004.

8. L. M. Feeney. Energy efficient communication in ad hoc wireless networks. In *Mobile Ad Hoc Networking*, S. Basagni, M. Conti, and S. Giordano, and I. Stojmenovic (eds.), pages 301–327, IEEE/Wiley, 2004.

9. V. Raghunathan, C. Schurger, S. Park, and M. B. Srivastava. Energy-aware wireless microsensor networks. *IEEE Signal Processing Magazine*, **19**:40–50, 2002.

10. M. L. Sichitiu. Cross-layer scheduling for power efficiency in wireless sensor networks. In *Proceedings of IEEE INFOCOM 2004*, Hong Kong, China, March 2004.

11. I. Stojmenovic and J. Wu. Broadcasting and activity scheduling in ad hoc networks. In *Mobile Ad Hoc Networking*, pages 205–229, S. Basagni, M. Conti, S. Giordano, and I. Stojmenovic (eds.), IEEE Press, 2004.

12. S. Basagni, M. Mastrogiovanni, and C. Petrioli. A performance comparison of protocols for clustering and backbone formation in large scale ad hoc networks. *In Proceedings of them 1st IEEE International Conference on Mobile Ad Hoc and Sensor Systems (MASS 2004)*, Fort Landerdale, Florida, October 2004.

13. C. R. Lin and M. Gerla. Adaptive clustering for mobile wireless networks. *IEEE Journal of Selected Areas in Communications*, **15**(7):1265–1275, 1997.

14. I. Stojmenovic, M. Seddigh, and J. Zunic. Dominating sets and neighbor elimination based broadcasting algorithms in wireless networks. *IEEE Transactions on Parallel and Distributed Systems*, **13**(1): 14–25, January 2002.

15. S. Basagni. Distributed clustering for ad hoc networks. In *Proceedings of the 1999 International Symposium on Parallel Architectures Algorithms, and Networks (ISPAN 99)*, pages 310–315, Freemantle, Australia, June 1999.

16. P.-J. Wan, K. M. Alzoubi, and O. Frieder. Distributed construction of connected dominating sets in wireless ad hoc networks. *Mobile Networks and Applications (MONET)*, **9**(2):141–149, April 2004.

17. S. Basagni, A. Carosi, and C. Petrioli. Sensor-DMAC: Dynamic topology control for wireless sensor networks. *In Proceedings of the IEEE VTC*, Los Angeles, California, September 2004.

18. J. Wu and F. Dai. A distributed formation of a virtual backbone in MANETS using adjustable transmission ranges. In *Proceedings of the 24th IEEE International Conference on Distributed Computing Systems (ICDCS)*, pages 372–379, Tokyo, Japan, March 2004.

19. F. Kuhn, T. Moscibroda, and R. Wattenhofer. Initializing newly deployed ad hoc and sensor networks. In *Proceedings of the 10th Annual International Conference on Mobile Computing and Networking (MobiCom)*, pages 260–274, Philadelphia, Pennsylvania, September 2004.

20. J. Gao, L. Guibas, J. Hershberger, L. Zhang, and A. Zhu. Discrete mobile centers. In *Proceedings of the 17th Annual Symposium on Computational Geometry (SCG)*, pages 188–196, Boston, Massachusetts, June 2001.

21. Y. Xu, J. Heidemann, and D. Estrin. Geography-informed energy conservation for ad hoc networks. In *Proceedings of the 7th Annual International Conference on Mobile Computing and Networking (MobiCom)*, pages 70–84, Rome, Italy, July 2001.

22. L. Lovasz. On the ratio of optimal integral and fractional covers. *Discrete Mathematics*, **13**:383–390, 1975.

23. C. Adjih, P. Jacquet, and L. Viennot. Computing Connected Dominating Sets with Multipoint Relays. Research Report #4597, INRIA, October 2002. To appear in *Ad Hoc & Sensor Wireless Networks.* Vol. 1, No. 1–2, 27–40.

24. J. Wu. An enhanced approach to determine a small forward node set based on multipoint relay. In *Proceedings of the IEEE VTC,* Orlando, Florida, October 2003.

25. J. Wu and H. Li. On calculating connected dominating set for efficient routing in ad hoc wireless networks. In *Proceedings of the 3rd International Workshop on Discrete Algorithms and Methods for Mobile Computing and Communications (DIAL-M),* pages 7–14, Seattle, Washington, August 1999.

26. I. Stojmenovic. Comments and corrections to "Dominating Sets and Neighbor Elimination-Based Broadcasting Algorithms in Wireless Networks." *IEEE Transactions on Parallel and Distributed Systems,* **15**(11), 1054–1055, November 2004.

27. F. Dai, and J. Wu. An extended localized algorithm for connected dominating set formation in ad hoc wireless networks. *IEEE Transactions on Parallel and Distributed Systems,* **15**(10), 2004.

28. J. Carle and D. Simplot-Ryl. Energy efficient area monitoring for sensor networks. *IEEE Computer,* **37**(2):40–46, February 2004.

29. F. Ingelrest, D. Simplot-Ryl, and I. Stojmenovic. Smaller connected dominating sets in ad hoc and sensor networks based on coverage by two-hop neighbors. Submitted for publication.

30. I. Stojmenovic. Data gathering and activity scheduling in ad hoc and sensor networks, Proc. International Workshop on Theoretical Aspects of Wireless Ad Hoc, Sensor, and Peer-to-Peer Networks, Chicago, USA, June 11–12, 2004.

31. J. Shaikh, J. Solano, I. Stojmenovic, and J. Wu. New metrics for dominating set based energy efficient activity scheduling in ad hoc networks. In *Proceedings of the IEEE Conference on Local Computer Networks,* pages 726–735, Bonn, Germany, October 2003.

32. S. Y. Ni, Y. C. Tseng, Y. S. Chen, and J. P. Sheu. The broadcast storm problem in a mobile ad hoc network. In *Proceedings of the 5th Annual ACM/IEEE International Conference on Mobile Computing and Networking (MobiCom),* pages 152–162, Seattle, Washington, August 1999.

33. H. Lim and C. Kim. Multicast tree construction and flooding in wireless ad hoc networks. In *Proceedings of the 3rd ACM International Workshop on Modeling, Analysis and Simulation of Wireless and Mobile Systems (MSWiM '00),* pages 61–68, Boston, Massachusetts, August 2000; see also *Computer Communications,* **24**:3–4, 353–363, February 2001.

34. W. Peng and X.-C. Lu. On the reduction of broadcast redundancy in mobile ad hoc networks. In *Proceedings of the 1st ACM International Symposium on Mobile and Ad Hoc Networking and Computing,* pages 129–130 Boston, Massachusetts, August 2000.

35. I. Stojmenovic and M. Seddigh. Broadcasting algorithms in wireless networks. In *Proceedings of the International Conference on Advances in Infrastructure for Electronic Business, Science, and Education on the Internet SSGRR,* L'Aquila, Italy, July–August 2000.

36. J. Wu and F. Dai. A generic distributed broadcast scheme in ad hoc wireless networks. In *Proceedings of the 23rd IEEE International Conference on Distributed Computing Systems (ICDCS),* pages 460–467, May 2003.

37. G. Calinescu, I. Mandoiu, P. J. Wan, and A. Zelikovsky. Selecting forwarding neighbors in wireless ad hoc networks. In *Proceedings of the 5th International Workshop on*

Discrete Algorithms and Methods for Mobile Computing and Communication (DIAL M 2001), pages 34–43, Rome, Italy, 2001.

38. A. Qayyum, L. Viennot, and A. Laouiti. Multipoint relaying: An efficient technique for flooding in mobile wireless networks. In *Proceedings of the 35th Annual Hawaii International Conference System Sciences (HICSS '02)*, Big Island, Hawaii, January 2002.

39. M. T. Sun and T. H. Lai. Location aided broadcast in wireless ad hoc network systems. In *Proceedings of the IEEE Symposium on Ad Hoc Wireless Networks, at GLOBECOM*, November 2001.

40. W. Peng and X. Lu. AHBP. An efficient broadcast protocol for mobile ad hoc networks. *Journal of Science and Technology* (Beijing, China), 2002.

41. W. Peng and X. Lu. Efficient broadcast in mobile ad hoc networks using connected dominating sets. In *Proceedings of the 7th International Conference on Parallel and Distributed Systems (ICPADS 2000)*, Iwate, Japan, July 2000.

42. J. Wu and W. Lou. Extended multipoint relays to determine connected dominating sets in MANETs. In *Proceedings of the 1st IEEE Communications Society Conference on Sensor and Ad Hoc Communications and Networks (SECON 2004)*, Santa Clara, California, October 2004.

43. F. Ingelrest, D. Simplot-Ryl, and I. Stojmenovic. Target transmission radius over LMST for energy-efficient broadcast protocol in ad hoc networks. In *Proceedings of the IEEE International Conference on Communications (ICC)*, Paris, June 2004.

44. M. Cardei and J. Wu. Energy-efficient coverage problems in wireless ad hoc sensor networks. *Computer Communications*, forthcoming.

45. T. Clouqueur, V. Phipatanasuphorn, P. Ramanathan, and K. K. Saluja. Sensor deployment strategy for target detection. In *Proceedings of the 1st ACM International Workshop on Wireless Sensor Networks and Applications (WSNA)*, pages 42–48, Atlanta, Georgia, September 2002.

46. D. Tian and N. D. Georganas. A coverage-preserving node scheduling scheme for large wireless sensor networks. *Wireless Communications and Mobile Computing*, **3**:271–290, 2003.

47. J. Wu and S. Yang. Coverage and connectivity in sensor networks with adjustable ranges. In *Proceedings of the 2004 International Workshop on Mobile and Wireless Networks (MWN)*, Montreal, Canada, August 2004.

48. H. Zhang and J. C. Hou. Maintaining sensing coverage and connectivity in large sensor networks. *Ad Hoc and Sensor Wireless Networks, an International Journal*, **1**(1–2): 89–124, 2005.

49. G. Xing, C. Lu, R. Pless, and Q. Huang. Greedy geographic routing is good enough in sensing covered networks. INFOCOM 2004.

50. B. Liu and D. Towsley. A study of the coverage of large-scale sensor networks. In *Proceedings of the 1st IEEE International Conference on Mobile Ad Hoc and Sensor Systems (MASS 2004)*, pages 475–483, Fort Lauderdale, Florida, October 2004.

51. H. Zhang and J. Hou. On deriving the upper bound of alpha-lifetime for large sensor networks. In *Proceedings of the 5th ACM International Symposium on Mobile Ad Hoc Networking and Computing (MobiHoc)*, pages 121–132, Tokyo, Japan, May 2004.

52. F. Ye, G. Zhong, J. Cheng, S. Lu, and L. Zhang. PEAS: A robust energy conserving protocol for long-lived sensor networks. In *Proceedings of the 23rd International*

Conference on Distributed Computing Systems (ICDCS), page 28, Providence, Rhode Island, May 2003.

53. Di Tian and N. D. Georganas. Location and calculation-free node scheduling schemes in large wireless sensor networks. *Ad Hoc Networks*, **2**: 65–85, 2004.

54. X. Wang, G. Xing, Y. Zhang, C. Lu, R. Pless, and C. D. Gill. Integrated coverage and connectivity configuration in wireless sensor networks. In *Proceedings of the 1st ACM Conference on Embedded Networked Sensor Systems (SenSys '03)*, Los Angeles, California, November 2003.

55. Di Tian. Node Activity Scheduling Schemes in Large-Scale Wireless Sensor Networks. Ph.D. thesis, SITE, University of Ottawa, 2004.

56. T. Yan, T. He, and J. A. Stankovic. Differentiated surveillance for sensor networks. In *Proceeding of the 1st International Conference on Embeded Networked Sensor Systems (SenSys '03)*, pages 51–62, Los Angeles, California, November 2003.

57. C. F. Hsin and M. Liu. Network coverage using low duty-cycled sensors: Random and coordinated sleep algorithms. In *Proceedings of the 3rd International Symposium on Information Processing in Sensor Networks*, Berkeley, California, April, 2004.

58. H. Gupta, S. R. Das, and Q. Gu. Connected sensor cover: Self-organization of sensor networks for efficient query execution. In *Proceedings of the 4th ACM International Symposium on Mobile Ad Hoc Networking and Computing (MobiHoc)*, Annapolis, Maryland, June 2003.

59. J. Jiang and W. Dou. A coverage preserving density control algorithm for wireless sensor networks. In *Proceedings of the 3rd International Conference on Ad-Hoc Networks and Wireless (ADHOC-NOW '04)*, (LNCS 3158), pages 42–45, Vancouver, July 2004.

60. J. Carle, A. Gallais, D. Simplot-Ryl, and I. Stojmenovic. Localized Sensor Area Coverage with Small Communication Overhead. *5th Scandinavian Workshop on Wireless Ad-Hoc Networks (ADHOC '05)*, Stockholm, May 3–4, 2005.

61. B. Chen, K. Jamieson, H. Balakrishnan, and R. Morris. Span: An energy-efficient coordination algorithm for topology maintenance in ad hoc wireless networks. *In Proceedings of the 7th Annual International Conference on Mobile Computing and Networking (MobiCom)*, Rome, Italy, July 2001.

62. Z. Abrams, A. Goel, and S. Plotkin. Set k-cover algorithms for energy efficient monitoring in wireless sensor networks. In *Proceedings of the 3rd International Symposium on Information Processing in Sensor Networks (ISPN)*, pages 424–432, Berkeley, California, April 2004.

63. I. Stojmenovic. An Adaptive Localized Algorithm for Multiple Sensor Area Coverage. In preparation.

64. G. Xing, C. Lu, R. Pless, and J. O'Sullivan. Co-Grid: An efficient coverage maintenance protocol for distributed sensor networks. In *Proceedings of the 3rd International Symposium on Information Processing in Sensor Network (IPSN)*, Berkeley, California, April 2004.

65. I. Stojmenovic. A Localized Algorithm for Sensor Area Coverage Based on a Realistic Physical Layer. In preparation.

66. C. Gui and P. Mohapatra. Target Tracking and Surveillance Using Sensor Networks. Paper presented at the International Workshop on Theoretical and Algorithmic Aspects of Sensor, Ad Hoc Wireless and Peer-to-Peer Networks, Fort Lauderdale, Florida, February 2004.

67. M. Cardei and D. Z. Du. Improving wireless sensor network lifetime through power aware organization. *Wireless Networks*, **11**(3): 333–340, May 2005.

68. S. Shakkottai, R. Srikant, and N. Shroff. Unreliable sensor grids: Coverage, connectivity, and diameter. In *Proceedings of IEEE INFOCOM 2003*, San Francisco, California, June 2003; See also *Ad Hoc Networks*, forthcoming.

69. Y. Zou and K. Chakrabarty. Sensor deployment and target localization based on virtual forces. In *Proceedings of IEEE INFOCOM 2003*, pages 1293–1303, San Francisco, California, March 2003.

70. G. Cao, G. Wang, T. La Porta, and W. Zhang. Distributed Algorithms for Deploying Mobile Sensors. Paper presented at the International Workshop on Theoretical and Algorithmic Aspects of Sensor, Ad Hoc Wireless and Peer-to-Peer Networks, Fort Lauderdale, Florida, February 2004.

71. Xiang-Yang Li, Ivan Stojmenovic, and Yu Wang. PartialDelaunay triangulation and degree limited localized Bluetooth multihop scatternet formation. *IEEE Transactions on Parallel and Distributed Systems*, Vol. 15, No. 4, April 2004, 350–361.

72. P. Bose, P. Morin, I. Stojmenovic, and J. Urrutia. Routing with guaranteed delivery in ad hoc wireless networks. *Wireless Networks*, 7(6):609–616, November 2001.

73. G. Wang, G. Cao, and T. La Porta. Proxy-based sensor deployment for mobile sensor networks. In *Proceedings of the 1st IEEE International Conference on Mobile Ad Hoc and Sensor Systems (MASS 2004)*, Fort Lauderdale, Florida, October 2004.

74. I. Stojmenovic. Deploying Mobile Sensors for Improved Area Coverage. In preparation.

75. J. Wu and S. Yang. SMART: A Scan-based Movement-Assisted sensoR deploymenT Method in Wireless Sensor Networks. Paper presented at INFOCOM 2005, Miami, Florida, March 2005.

76. W. Heinzelman. Application-Specific Protocol Architectures for Wireless Networks. Ph.D. thesis, Massachusetts Institute of Technology, 2000.

77. I. Stojmenovic. *A Scalable Quorum Based Location Update Scheme for Routing In Ad Hoc Wireless Networks*. Technical Report TR-99-09, SITE, University of Ottawa, September 1999.

78. X.Y. Li and I. Stojmenovic. Broadcasting and topology control in wireless ad hoc networks. In *Handbook of Algorithms for Mobile and Wireless Networking and Computing*, A. Boukerche and I. Chlamtac (eds.), CRC Press, forthcoming.

79. F. Ingelrest, D. Simplot-Ryl, and I. Stojmenovic. Energy-efficient broadcasting in wireless mobile ad hoc networks. In *Resource Management in Wireless Networking*, Mihaela Cardei, Ionut Cardei, and Ding-Zhu Du (eds.), pages 543–582, Kluwer, 2005.

80. François Ingelrest, David Simplot-Ryl, and Ivan Stojmenovic. Routing and broadcasting in hybrid ad hoc and sensor networks. In *Theoretical and Algorithmic Aspects of Sensor, Ad Hoc Wireless and Peer-to-Peer Networks*, Jie Wu (ed.), CRC Press.

81. Justin Lipman, Johnson Kuruvila, and Ivan Stojmenovic. Localized broadcasting in ad hoc networks. In *Wireless Ad Hoc and Sensor Networks*, Ahmed Safwat (ed.), Kluwer, forthcoming.

82. L. Orecchia, A. Panconesi, C. Petrioli, and A. Vitaletti. Localized techniques for broadcasting in wireless sensor networks. In *Proceedings of the DIALM-PODC Joint Workshop, on Foundations of Mobile Computing*, pages 41–51, Philadelphia, Pennsylvania, October 2004.

83. I. Stojmenovic. Beaconless Area Based Broadcasting. In preparation.

FURTHER READING

Z. Hu and B. Li. On the fundamental capacity and lifetime limits of energy-constrained wireless sensor networks. In *Proceedings of the 10th IEEE Real-Time and Embedded Technology and Applications Symposium (RTAS '04)*, pages 38–47, Toronto, Canada, 2004.

CHAPTER 12

Geographic and Energy-Aware Routing in Sensor Networks

HANNES FREY

University of Trier, Trier, Germany

IVAN STOJMENOVIĆ

University of Ottawa, Ontario, Canada

Typical communication patterns within a sensor network are data delivery from sensor nodes to one of selected information sinks, and information sinks requesting a certain physical phenomenon or requesting sensor nodes lying within a sensed area. In general, addressing is achieved by utilizing sensor locations. Geographic routing algorithms allow routers to be nearly stateless since packet forwarding is achieved by utilizing location information about candidate nodes in vicinity and the location of the final destination only. By their localized nature, geographic routing algorithms are highly scalable solutions which do not require any additional control overhead when network topology changes due to mobility or energy conserving sleep cycles. Recent work investigated that location information may be utilized to define new link metrics aiming on energy and physical layer optimized routing paths instead of only minimizing the number of hops needed to reach the desired destination. This chapter reviews geographic and energy aware routing algorithms for sensor networks. It includes simple heuristic greedy forwarding strategies, strategies which obtain guaranteed delivery by memorizing information about all ongoing routing tasks, memoryless recovery strategies, energy aware routing strategies aiming on increased network lifetime, and routing without information about their neighbor nodes. The majority of geographic routing protocols assume a simplified network model which does not take into account random variations in correct

Handbook of Sensor Networks: Algorithms and Architectures, Edited by Ivan Stojmenović
Copyright © 2005 John Wiley & Sons, Inc.

message receipt. This chapter also discusses physical layer impact on both greedy geographical routing strategies and their recovery strategies.

12.1   INTRODUCTION

Sensor networks are typically composed of hundreds to thousands of small collaborating wireless sensor nodes that have limited computation and communication capabilities. Communication patterns within a sensor network are data delivery from sensor nodes to one or a subset of selected information sinks, and information sinks requesting a certain physical phenomenon or requesting sensor nodes lying within a sensed area. In general, addressing is achieved by utilizing sensor locations or querying all sensor nodes matching a certain criterion instead of utilizing individual node addresses. Ease of deployment and the fact that sensor nodes are small and closely located at the measured phenomenon makes an external power supply, recharging batteries, or replacing depleted batteries impractical or even impossible. Consequently, the lifetime of a sensor node is directly related to its on-board power supply, and thus energy is the most sensitive resource with respect to the whole network lifetime.

Communication between sensor nodes and information sinks can be achieved by setting an appropriate transmission power (if possible) and sending data or control messages directly to the desired recipient. However, this simple communication form may degrade the bandwidth of the limited shared wireless communication media, and moreover will drastically increase energy consumption at sender nodes, since signal attenuation increases significantly with the distance to the message recipient. If no fixed networked infrastructure is additionally available, a resource-saving communication may only be achieved by multihop ad hoc routing techniques, where communication between any two network nodes requires collaborating with intermediate next-hop forwarding nodes.

Since location information is often available due to the very nature of sensor networks, the special class of geographic routing algorithms may be a good choice in order to build a scalable resource-saving communication infrastructure. Geographic routing algorithms allow routers to be nearly stateless, since packet forwarding is achieved by utilizing location information about candidate nodes in the vicinity and the location of the final destination only. By their localized nature, geographic routing algorithms are highly scalable solutions that do not require any additional control overhead when network topology changes due to mobility or energy-conserving sleep cycles. In particular, due to the addressing scheme of sensor networks there is no need for an additional location service (producing an additional network load), which is used in other ad hoc network scenarios in order to acquire location information about individual network nodes before communication can take place. Finally, location information about all neighbor nodes can be used in order to estimate the signal strength needed to reach a certain neighbor node. Recent work investigated that location information can be utilized to define new link metrics aiming at energy and physical-layer optimized routing paths instead of only minimizing the number of hops needed to reach the desired destination.

12.2 GREEDY ROUTING ALGORITHMS

Greedy routing algorithms limit forwarding decisions on information about the position of all nodes in the vicinity and forward a message to the "best" neighbor regarding the position of the final destination and the metric being optimized. Each forwarding node applies this greedy principle until the final destination (if possible) is finally reached. The current required location information about neighbor nodes is maintained by proactively exchanging short beacon messages (containing node ID and location) transmitted with maximum signal strength.

12.2.1 Progress, Distance, and Direction

The first geographic routing algorithm was described by Takagi and Kleinrock in the mid-1980s [1]. They introduced the notion of *progress* in order to define the *most forward within radius (MFR)* greedy routing scheme. The distance between the current node S and the projection A' of a neighbor node A onto the line SD connecting S and final-destination node D is termed progress (see node A in Fig. 12.1). MFR selects the neighbor node that maximizes progress, while nodes with negative progress are ignored (e.g., MFR selects node A in Fig. 12.1).[1] Alternatively, *distance-based* greedy forwarding considers *Euclidean distance* instead of progress. Finn [3] proposed the first distance-based greedy routing scheme, which selects a node closer and minimizes the distance d to the final destination (e.g., node B in Fig. 12.1). This scheme is the most widely applied greedy strategy in the literature, and it will subsequently be referred as *GREEDY*. In recent years *direction-based* (*DIR*) greedy routing, which considers the angle between the next hop, current, and destination nodes, was investigated as a third alternative of greedy forwarding. The DIR method, described by Kranakis et al. [4], selects the next hop forwarding node, minimizing the deviation from the line connecting the current and the destination node (e.g., node C in Fig. 12.1).

There are several variants of nodes that will be considered, along with the stopping criterion in progress or distance-based routing schemes. In originally proposed articles, greedy forwarding based on progress or distance considers nodes in the forward direction (respectively closer to destination) only (e.g., nodes A, B, C, and E in Fig. 12.1), since choosing a node in the backward direction (e.g., nodes F and G in Fig. 12.1) might lead to a packet loop. Ref. [5] considers all nodes, but routing stops at a node whose best choice is to return the packet to a neighbor that sent the packet to it (the loop-free property has been proved for this variant).

---

[1]More precisely, in their original work MFR also considered nodes with negative progress. However, in later studies (e.g., ref. [2]) MFR was often described to consider nodes with nonnegative progress only. This chapter will use this variant when speaking of MFR.

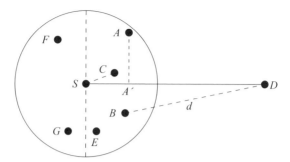

Figure 12.1   Node A is maximizing progress, node B has the least distance to D, and node C lies closest in direction to D.

A node where packet forwarding is stopped due to lack of a neighbor in the forward direction, or by applying the described stoppage criterion, is termed a *concave node*.

12.2.2   Sensing Coverage and Greedy Routing

Many applications of sensor networks (e.g., tracking of moving targets) require a specific class of sensor networks that provide *sensing coverage*, that is, every point of a geographic area must be within the sensing range of at least one sensor node. A simplified formal model assumes that every node N has the same sensing range, which is a circle with radius R_s centered at N. Thus, for a sensor network covering an area A, the union of the circular sensing ranges of all network nodes must at least contain the area A. In a similar way, a simplified formal communication model can be defined by using a unique communication range R_c. The communication network, which is often referred to as the *unit-disk graph*, has a bidirectional communication link between any two sensor nodes X and Y, if and only if the Euclidean distance between X and Y is less than R_c.

Xing et al. [6] investigated properties of greedy routing algorithms in sensing covered networks having the *double-range property*, that is, $R_c/R_s \geq 2$. Focus on this special class is motivated by the geometric analysis from Wang et al. [7], which showed that a sensing covered network is always connected if it has the double-range property. The geometric analysis and simulation results from ref. [6] demonstrate that greedy geographic routing is a viable and effective routing scheme in sensing covered networks, and it turns out that the *range ratio R_c/R_s* has a significant impact on the quality of greedy routing in sensing covered networks.

The qualitative properties of greedy routing in sensing covered networks may be expressed in terms of *network* and *Euclidean dilation*. A subgraph H of G is termed a *network t-spanner* of G if the length (measured in hops) of the shortest path between any two nodes U and V in H is at most t times longer than the shortest path produced in G. The value t is termed the *network stretch factor* of the spanner H. Network dilation represents the stretch factor of a graph G relative to an ideal network producing a minimum number of hops of about $|SD|/R_c$ between source S and destination D. *Euclidean stretch factor* and *dilation* can be defined in a similar way by utilizing Euclidean distance instead of hop count.

Xing et al. [6] studied the dilation properties of sensing covered networks by utilizing a known upper bound of the Euclidean stretch factor of *Delaunay triangulations (DTs)*. This well-studied graph structure can be defined as the twin of the *Voronoi diagrams*. For a set of n nodes in two-dimensional (2D) space, the Voronoi diagram partitions the plane into n *Voronoi regions Vor(U)*, while each Voronoi region contains all points in the plane that are closest to U (see the face surrounding node F in Fig. 12.2, for instance). The DT can be obtained by connecting each node pair (U, V) that shares a common boundary in the Voronoi diagram (see the dotted lines in Fig. 12.2).

It is proved in ref. [6] that the DT is a subgraph of a sensing covered network when the double-range property holds. This result and the known upper bound of the Euclidean stretch factor of DTs is finally used to derive a constant upper bound for the network dilation in a sensing covered network that has the double-range property. Additionally, it is observed that any DT edge is shorter than $2R_s$. Thus, when range ratio R_c/R_s increases, the shortest path found in the DT gets significantly longer than the shortest possible path in the complete network, since all edges longer than $2R_s$ are ignored by the DT.

Besides the results based on DT, properties of greedy routing are investigated in ref. [6] as well. It is proved that greedy routing will always find a routing path between any two nodes if the sensing covered network has a convex network boundary and the double-range property holds. Additionally, the progress made in each routing step is at least $R_c - 2R_s$ closer to the destination than the current forwarding node. Furthermore, the latter result is used in order to estimate the quality of the routing path produced by greedy routing. It is observed that in a sensing covered network with the double-range property the path (from source S to destination D) found by greedy routing is always no longer than about $|SD|/(R_c - 2R_s)$.

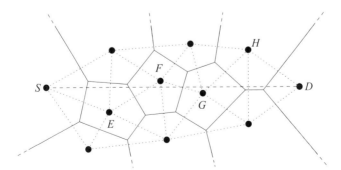

Figure 12.2 The bounded Voronoi greedy forwarding (BVGF) routing algorithm will select the node closest to D, but considers only nodes having a Voronoi region intersecting the straight line SD.

In summary, the result of DT and greedy routing motivate the *bounded Voronoi greedy forwarding* (*BVGF*) algorithm [6] as follows. Greedy routing applied in sensing covered networks will produce satisfactory path lengths if the range ratio R_c/R_s is significantly greater than 2, while, on the other hand, the upper bound of the path lengths produced $|SD|/(R_c - 2R_s)$ tends to infinity when the range ratio is close to 2. The shortest path found in a DT is always upper bound by a constant, while the upper bound becomes very conservative when the range ratio is increased. Thus, the BVGF algorithm is a combination of both methods, greedy routing, and routing along the edges of a DT. A node holding a packet addressed from the source node S to the final destination D will consider only those neighbor nodes U, where the line segment connecting S and D intersects $Vor(U)$ or coincides with one of the boundaries of $Vor(U)$. From these subsets of one-hop neighbors, the node closest to D will be selected as the next hop node. For instance, in Figure 12.2 only nodes E, F, G, and H can be visited by BVGF. Note that BVGF is not constrained to the edges of the DT. For instance, if source S is able to reach node F, it will send the message to node F directly, since node F is closer to D than E.

A theoretical analysis shows that BVGF will always find a path in sensing covered networks with the double-range property. An additional result shows that each node visited by the path produced by BVGF has a distance of at most R_s from the line connecting source node S and final destination D. Finally, this result is used to prove for the sensing covered network that the network dilation of the paths produced by BVGF is always upper bound by a constant value, provided that the range ratio is at least 2. In addition to this theoretical analysis, the average dilation of BVGF has been investigated in a simulation environment. The simulation results show that network dilations produced by BVGF are comparable to greedy forwarding, while Euclidean dilations are always better for all range ratios within 2 and 10.

12.2.3 Real-Time Communication in Sensor Networks

Sensor network applications such as surveillance systems may require sensor nodes to meet certain "soft" real-time communication constraints. Only a few results that adequately address such real-time requirements exist for sensor networks. The *SPEED* protocol by Lu et al. [8] is the first greedy-based protocol addressing real-time guarantees for sensor networks. SPEED utilizes the notion of *relay speed* in order to select one "best" next hop node in a greedy manner. Relay speed toward a next hop node A is calculated by dividing the advance in distance by the estimated send delay toward A. The single-hop delay toward a neighbor A can be estimated by continuously measuring the round-trip delay between current unicast data transmissions and the receipt of related acknowledgments. The estimated single-hop delay is calculated by means of an exponential weighted moving average over the previous average with the current single-hop delay. The latter is obtained by subtracting the receiver-side processing time from the round-trip delay experienced by the sender.

Before selecting the next hop node, a set of candidate nodes is calculated by selecting all nodes closer to the final destination than the current node and removing all nodes having a relay time smaller than a certain threshold s, which is a system-dependent parameter. The next hop node is selected according to a discrete exponential distribution, while the node with the fastest relay speed is selected with the highest probability. Selecting only nodes with a relay speed greater than a certain threshold assures that this routing scheme, if successful, will guarantee delivery of a packet within time d/s, with d being the distance between source and final destination. Furthermore, the randomized selection scheme provides traffic balance, and thus reduced congestion, since packets are dispersed into a large relay area. The authors of ref. [8] also propose neighborhood feedback-loop and back-pressure rerouting mechanisms.

Huang, Dai, and Wu [9] considered a quality of service (QoS) routing scheme, using progress instead of distance metric to advance toward the destination. The selected neighbor is one that maximizes the ratio of progress and delay in sending to a neighbor, where progress from node S when forwarding to neighbor A and with destination D can be measured as $SD \cdot SA$ (the dot product of vector SD and SA), and delay can be replaced by any other additive QoS metric. The authors also proposed several ticket-based multipath schemes to search for QoS paths. They also proposed a backward checking method that corresponds to the iterative improvement method described here for power and cost aware routing protocols.

12.3 GUARANTEED DELIVERY BASED ON MEMORIZATION

Stojmenović and Lin proposed neighbor flooding as a recovery mechanism at concave nodes, while every intermediate node handles received messages using the basic routing algorithm (named *f-GEDIR*, *f-MFR*, and *f-DIR*, for instance) [5]. Each concave node memorizes message IDs and rejects further copies of the same message (more precisely, neighbors learn about their concave status from the packet and do not select them as forwarding nodes). In original *f-GEDIR* or *f-MFR*, each neighbor of a concave node initiates a separate routing task toward destination D. Lin et al. proposed *component routing* [10], a more elaborate recovery strategy where concave nodes determine connected components in the subgraph of its neighbors and forward the message to only one "best" node in each component. The number of routing tasks initiated due to concave nodes is thus reduced significantly, since there are at most four connected components of neighbors of any concave node in the unit graph model [10].

The *geographical routing algorithm* (*GRA*) by Jain et al. [11] maintains a routing table that maps locations on next-hop forwarding nodes. A node receiving a message addressed to destination D, will look up its routing table and find the position p that is closest to the final destination D. The message will then be forwarded to the neighbor node that is assigned with position p. Initially, the routing table contains position information about neighbor nodes only, thus, operation of GRA is the same as greedy forwarding. Message forwarding is deferred and a route discovery is invoked if the routing table contains no position closer to the destination than the position of the current forwarding node S itself. The route discovery will find an acyclic path from the current node to the final destination and update the routing tables of all nodes lying on that path. A new entry is added to the routing table that maps the location of destination D on the next hop along the discovered path. Jain et al. propose *breadth-first search* (*BFS*) and *depth-first search* (*DFS*) as two possible route discovery mechanisms. BFS is the same as flooding, that is, a node receiving a discovery message appends its address on the path discovered so far and rebroadcasts the packet. Additional broadcast packets received subsequently are ignored. DFS yields only a single acyclic path from node S to destination D. Similar to BFS, a node receiving a route-discovery packet puts its address into the packet, but forwards it to a single neighbor Y that has not been visited so far and that minimizes the sum of the distance between S and Y and Y and D. If all neighbors have been visited, the current forwarding node removes its address from the path discovered so far and returns the packet to the node from which it was first received. Once the final destination receives a route-discovery packet, it is able to send an acknowledgment to the originator of the route discovery by utilizing the reverse of the path stored in the discovery packet. All nodes along that path will receive the destination acknowledgment and will update their routing table accordingly. On receipt of the destination acknowledgment the originator S will continue to forward the original message toward destination D.

Independently a localized DFS-based routing algorithm was proposed by Stojmenović et al. [12]. In contrast to the GRA algorithm, nodes do not store any routing-table entries and the list of visited nodes is not stored in the DFS packet. In order to enable DFS in a distributed manner, each node remembers if it has already been visited by the DFS traversal. Additionally, each node memorizes the node from where the packet was received for the first time. Packet forwarding is performed by sorting all neighbor nodes with respect to their distance from the final destination D and selecting the node that is closest to D. Already-visited neighbor nodes have already transmitted a forward packet, therefore neighboring nodes can overhear it and can learn their status and do not select them for another forwarding. A returned message will be sent to the next choice in the sorted list of all next-hop nodes. If all neighbors already have been visited or have returned the packet, then the message will be returned to the neighbor node that sent the message for the first time. In addition to the basic algorithm, Stojmenovic et al. discussed a possible improvement with respect to QoS support. By utilizing information about its own physical location and periodically updated position information about all neighbor nodes, a node A can estimate the current speed and send direction of itself and its neighbor B and can thus estimate how long the link between A and B will remain stable. This link measure can be used in order to construct a path that provides a specific connection-time requirement. Each node visited by the search message will simply ignore all adjacent links that do not match this QoS requirement. In addition, a minimum bandwidth requirement and maximum delay may be considered as well during DFS traversal. In a simplified model, total delay is decomposed into the number of hops × propagation delay per hop (which is directly related to the bandwidth requirement per hop). In order to find a path with the required maximum propagation delay, DFS traversal is limited to a maximum path length and will consider only edges that have at least the minimum required bandwidth. A node will return the search message immediately if the maximum number of hops is exceeded or no outgoing edge matches the minimum bandwidth requirement. Nodes located along the path found will memorize the uplink and downlink edges of the path, which finally enables communication between source S and destination D within the established QoS requirements. The DFS-based QoS routing protocol can also be designed by using an advance (distance- or progress-based) per delay metric over links with sufficient bandwidth and connection times. The search for such a path proceeds until the destination is found and overall delay is acceptable.

12.4 MEMORYLESS GUARANTEED DELIVERY

Bose et al. described *FACE*, the first memoryless single-path recovery mechanism with guaranteed delivery [13] in a unit-disk graph model of communication (assuming ideal medium-access control (MAC) layer and connectivity). The FACE algorithm is an improvement of the routing algorithm due to Kranakis et al. [4], which guaranteed delivery in connected geometric planar graphs. A geometric planar graph partitions the plane into faces bounded by the polygons made up of the edges of the graph, and the nodes are described by geographic positions. A geometric graph is said to be planar if there is no intersection between any two edges of the graph (see the graph depicted in Fig. 12.3, for example).

The main idea of the FACE algorithm is to route a packet along the interiors of the faces intersected by the straight line connecting the source node S and destination D (see Fig. 12.3). Each face interior is traversed by applying the *right-hand rule*

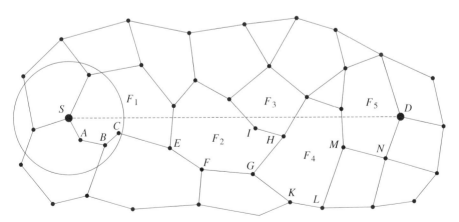

Figure 12.3 Face routing of a packet sent from source S to destination D leads to the path *SABCEFGHIHGKLMND* if the right-hand rule is applied.

or the *left-hand rule*, that is, a packet is forwarded along the next edge clockwise or counterclockwise from the edge where it arrived. When the packet arrives at an edge intersecting the line connecting S and D, the next face intersected by this line is handled in the same way. For example, in Figure 12.3 a packet routed from source S to destination D visits the faces $F_1, \ldots, F_5$. The algorithm proceeds until the destination node is eventually reached or if the first edge of current face traversal is traversed twice in the same direction. In the latter case, the destination node is not reachable. Face routing is shown to be loop-free and to guarantee delivery in static connected planar geometric graphs [13]. There are two main variants of FACE routing: the *before crossing* and *after crossing* protocols. They differ in the selection of the next edge after the current node detects that the face for traversing needs to be changed. The example in Figure 12.3 shows the before-crossing variant. The after (before) crossing variant selects (does not select, respectively) the edge that is intersected by the straight line SD. Note that nodes cannot be certain locally whether they are following the right-hand or left-hand rule, because an open face has the opposite orientation to the closed faces, and nodes are not aware locally whether or not they are on the open face.

Ad hoc and sensor wireless networks can be modeled as unit-disk graphs. However, the unit-disk graph is not planar in general. Thus, before the FACE recovery procedure can be performed, a planar subgraph has to be extracted from the complete network graph. In the description of FACE, Bose et al. [13] proposed a distributed algorithm for extracting a planar subgraph from a unit-disk graph, which is based on *Gabriel graphs (GG)* [14], a well-known geometric planar graph construction. A GG for a finite-point set S is constructed by connecting any two nodes X and Y of S if and only if the circle with diameter (X, Y) contains no other node of S. This test can be performed by each node without any message exchange with neighbors, other than "hello" messages to learn their position. It is proved in ref. [13] that the minimal spanning tree belongs to the intersection of the GG and the unit-disk graph, therefore the network connectivity is preserved.

When the average density (average number of neighbors) increases, edges in GG become smaller, therefore the routes in FACE routing become longer. The other problem is that the routes may be long if an external face is encountered on the route. On the other hand, the path produced by successful greedy routing is comparable to the one produced by Dijkstra's single-source shortest path algorithm. Thus, Bose et al. [13] proposed a combination of the FACE algorithm with distance-based greedy routing, called *GFG* (*greedy-face-greedy*). A packet arriving at a concave node is switched into recovery mode and routed along the faces until reaching a node closer to the destination than the position of the concave node where recovery mode was entered. At this node, routing is again performed in greedy mode. The integration of GFG algorithm [13] with IEEE 802.11 was later implemented in the *greedy perimeter stateless routing* (*GPSR*) protocol by Karp and Kung [15]. Their GPSR protocol is the same as GFG. More precisely, they use the before-crossing instead of after-crossing variant, and also discuss the relative neighborhood graph (RNG) as an alternative to the GG. These modifications do not improve the performance of the routing protocol.

12.4.1 Connected Dominating Sets and Shortcuts

Face routing has an increased hop count compared to Dijkstra's single-source short-est path algorithm, since planar graph construction based on GGs favors short edges over long ones. Datta et al. [16] improved the performance of GFG by the concept of *connected dominating sets* (*CDS*), *shortcut-based routing*, and a combination of both. Localized dominating-set construction in unit graphs is only possible with one-hop neighbor information, while shortcut-based routing also requires infor-mation about 2-hop neighbors.

A subset *S* of all network nodes *G* is called a *dominating set* if each node of *G* is either an element of *S* or has at least one neighbor in *S*. If the dominating set is connected, FACE routing constrained on CDS will produce shorter paths, since the corresponding GG edges will be longer on average. If a concave node is not in CDS, then it forwards the message to one of its adjacent nodes from CDS. Face routing (in recovery mode of GFG) then proceeds using only nodes from CDS. In greedy mode, the GFG algorithm works somewhat better on the whole set than on CDS, since there are more neighbor choices and longer edges can be used. The construction of CDS for unit-disk graphs is discussed in the chapter on backbones (Chapter 11) in this book.

In addition to the next forwarding node, there might be more neighbor nodes on the same path produced by FACE routing. For example, in Figure 12.3 the nodes *A*, *B*, and *C* on the path produced by traversal of face F_1 are all within transmission range of node *S* (the circle around *S*). When information about 2-hop neighbors is available, the concept of shortcut-based routing can be applied at each node. A forwarding node locally constructs the part of the planar graph seen by all its neigh-bors. Based on this information a node can make a shortcut by sending the message to the last known hop directly instead of forwarding it to the next hop along the path. For example, in Figure 12.3 node *S* could send the packet to node *C* directly.

12.4.2 Asymptotic Optimality of Face Routing

In order to analyze asymptotic behavior of combined greedy and face routing algo-rithms, Kuhn et al. [17] constructed a family of networks where each localized memoryless algorithm will produce a routing path that has quadratic cost compared to the cost of the shortest weighted path. The cost of a path is calculated by summing the cost produced by each path edge, while the theoretic results from ref. [17] are valid for all cost metrics that are polynomial in the Euclidean distance. Due to the lower-bound argument given in ref. [17], a localized memoryless algorithm produ-cing at most quadratic path costs (compared to the shortest weighted path) for any network configuration in the worst case can be denoted as asymptotic optimal.

It can be observed [18] that asymptotic optimality is sacrificed if face traversal is switching back to greedy mode when the line connecting concave node *S* and final destination *D* is intersected for the first time (i.e., GFG and its previously described variants are not asymptotic optimal). On the other hand, the combination of greedy and face routing becomes asymptotic optimal when packets in face mode traverse the complete face and change back to greedy mode at the face edge that is closest to the destination D. However, successful greedy routing is more efficient than face routing in the average case, that is, even when not optimal in the worst case, switching back to greedy mode as soon as possible may be the better strategy in practice.

Kuhn et al. [19] described *greedy other adaptive face routing plus (GOAFR+)*, a greedy routing algorithm that overcomes the trade-off between asymptotic optimality and average case efficiency of combined greedy and face routing. The efficient operation of face routing depends on the decision in the starting node of whether a face is being traversed in the clockwise or counterclockwise direction. For example, in Figure 12.4 applying the left-hand rule to traverse the outer face F_1 leads to the path $MLK \ldots UTSR$ until arriving at the edge (R, Q) intersecting the line connecting source S and destination D. In contrast, if the face traversal was started in the opposite direction, the packet is forwarded along the significantly shorter path $MNOPQ$ before switching to face F_2.

In order to cope with that suboptimality, Kuhn et al. proposed an extension of the GFG algorithm limiting the searchable area during face traversal. The GOAFR+ algorithm uses a circle C centered at the destination node D in order to restrict face traversal to the searchable area C. The radius of C is initially set to $\rho_0|SD|$ with $\rho_0 \geq 1$ so that source node S is also included within C (see the dashed circular arc centered at D in Fig. 12.4). The greedy mode is applied as long as there is a next-hop node closer to the destination D, and whenever possible the radius r_C of C is exponentially decreased ($r_C = r_C/\rho$ with $\rho > \rho_0$) as long as the currently visited node stays within C. Whenever the greedy mode encounters a local minimum at a node U, the algorithm continues with a modified version of face routing. When the face is traversed completely without hitting the current circle C, the packet will be sent to the node visited so far that is closer to D than U (and handled in greedy mode again). However, if no visited node is closer to D than U, the algorithm will terminate and report that no path from S to D exists. When C is hit for the first time, face traversal is reversed and face exploration is applied in the opposite

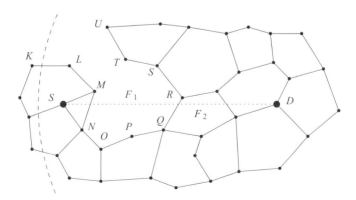

Figure 12.4 The GOAFR+ algorithm limits exploration to a circle centered at D and containing at least the node where the recovery procedure was invoked.

direction. If C is hit for the second time and none of the visited nodes is closer to D than U, face exploration is continued as if started at node U, but the radius of circle C is exponentially increased ($r_C = \rho r_C$). In order to avoid a complete face exploration, the algorithm applies an elaborate "early fallback" technique to return to greedy routing as soon as possible. However, it is proved in ref. [18] that algorithms will lose their asymptotic optimality when resuming greedy routing as soon as they arrive at the first node closer to the destination D than the concave node U. GOAFR+ maintains two counters to keep track of the number of nodes closer to and the number of nodes not closer to the destination than the starting node U of the current face traversal. If face exploration has visited up to a constant factor σ more nodes closer to D, GOAFR+ will interrupt face traversal, advance to the node seen so far that is closest to the destination D, and the packet will be handled in greedy mode again. Thus, GOAFR+ does not explore the complete face in general, but on the other hand, greedy routing is not resumed at the first node closer to destination D than concave node U. The latter property of GOAFR+ is finally used in ref. [19] in order to prove its asymptotic optimality. Finally, from simulation results it turned out that $\rho_0 = 1.4$, $\rho = \sqrt{2}$, and $\sigma = 1/100$ are good choices for practical purposes.

A simplified example of GOAFR+ is depicted in Figure 12.4, where source node S will forward in greedy mode to node M, which has no neighbor closer to destination D. Thus, the recovery strategy of GOAFR+ will begin exploration of face F_1 in the clockwise direction. At node L, face traversal hits the circle centered at node D, and the algorithm switches to exploration in the opposite direction. Each routing step updates the number p of nodes closer and the number q of nodes not closer to the destination D. When arriving at node P, a certain threshold condition $p > 1/3q$ holds ($p = 2$, $q = 3$, and $\sigma = 1/3$) and the message will be handled in greedy mode again.

For arbitrary unit-disk graphs (i.e., no restrictions regarding minimum node distance and maximum node degree), cost metrics divide into two classes, *linearly bounded* and *superlinear* cost functions. The first ones are lowerbound by a linear function, while for the latter there exists no such function. A theoretical result from ref. [19] reveals that for any localized memoryless routing algorithm A, there exists a node configuration where the cost of the path produced by A is unbounded with respect to the path produced by the shortest weighted-path algorithm if a superlinear cost function is considered. Thus, discussion of asymptotic optimality is reasonable only if restricted minimum node distance, maximum node degree, or linearly bounded cost metrics are considered. Standard cost metrics like hop count or Euclidean distance are linearly bound from below, while energy metrics defined as a polynomial d^α, with $\alpha > 1$ and d being the distance between sender and receiver, fall into the class of superlinear functions. However, as discussed in Section 12.5, "Routing with Energy-Aware Cost Metrics," from a practical point of view even energy-aware metrics are often considered to be of the form $d^\alpha + c$, with $c > 0$, and are thus linearly bound in practice.

In order to prove asymptotic optimality for linearly bound cost metrics and arbitrary unit-disk graphs, Kuhn et al. described an improved version of GOAFR+, which utilizes a *routing backbone* instead of using all possible edges from the unit disk graph. Before routing takes place GOAFR+ precomputes a subgraph of the unit disk graph that forms a connected dominating set of bounded degree. The distributed construction of such a graph structure is not described in ref. [19], but it is referred to in existing methods described in refs. [20–22]. Similar to GFG-I, the execution of GOAFR+ is restricted on the routing backbone, that is, a message is first sent to a dominating set member (if necessary), and from there on routing takes place along the routing backbone only until reaching a dominating set member that has the final destination D in its neighborhood.

12.4.3 Routing Along Geographical Clusters

Typical sensor network scenarios assume that sensor nodes are densely deployed in the monitored area. Greedy routing algorithms applied on uniformly distributed and densely deployed network nodes perform close to the shortest-path algorithm and are thus the first choice for such a scenario. However, even for densely deployed network nodes a recovery strategy may still be necessary, since greedy forwarding might get stuck at convex network boundaries or at network voids resulting from an inhomogeneous node distribution. Such an inhomogenity may be due to the physical properties of the monitored surface (e.g., a lake inside the monitored area where sensor nodes cannot be deployed).

Frey and Görgen [23] observed that such an inhomogeneous node distribution can have a significant impact on the recovery strategy being applied. Simulation experiments show that performance of face routing and the internal nodes concept may even degrade when node density is increased in a network scenario with an inhomogeneous node distribution. On the other hand, face recovery in combination with the shortcut procedure is almost unaffected by such an inhomogenity. However, the shortcut procedure has an increased message complexity compared to face routing or the internal nodes concept, since information about all 2-hop neighbors is required.

Frey and Görgen [23] described the *geographic cluster routing* (*GCR*) algorithm, which is based on the concept of GFG and assumes that the network is modeled as a unit-disk graph in the 2D Euclidean space as well. Routing in GCR is not performed on a per-node basis, but packets are forwarded along the edges of adjacent *geographical clusters*. In order to define geographical clusters, the plane is partitioned by an infinite mesh of regular hexagons (see Fig. 12.5), while each hexagon defines one cluster. Two geographical clusters C_1 and C_2 are denoted as adjacent, if there are at least two connected nodes with one located in C_1 and the other located in C_2. The graph resulting from adjacent clusters is not necessarily planar, thus, before face routing can be applied, a planar subgraph has to be extracted in advance. This is obtained by a variant of the localized planar graph construction applied by GFG. However, the method used by GCR may produce a disconnected subgraph even when the original graph is connected. Thus, in contrast with GFG and its variants, GCR cannot guarantee delivery, even if there is a path from source to destination. However, simulation results reveal that the delivery rate quickly tends to 100% when network degree is increased. In particular, in densely deployed networks

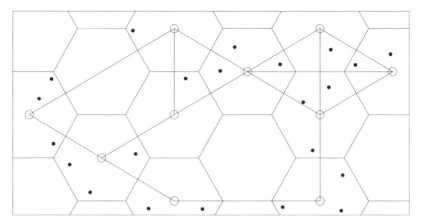

Figure 12.5 Geographic cluster routing explores the faces resulting from a planar graph extracted from the graph defined by all connected clusters.

GCR achieves a comparable performance to that of GFG combined with the shortcut procedure. However, message complexity is significantly decreased to an exchange of one-hop neighbor information only. Thus, GCR is a good choice to apply as a recovery strategy in densely deployed sensor networks.

12.4.4 Multicast Routing

A sensor network request may simultaneously address several different network nodes or network regions. This can be achieved by sending a unicast message to each individual entity. However, a resource-saving multicast strategy may be the better choice in order to reduce the bandwidth requirement when the same packet has to be delivered to multiple destinations. The majority of multicast protocols addressed to wireless networks require a distribution structure for the delivery of multicast messages. Mauve et al. [24] described a quasi-stateless protocol that achieves multicast addressing based on destination positions and neither requires construction and maintenance of a distribution structure nor resorts to some sort of flooding. The proposed *position-based multicast* (*PBM*) algorithm is a generalization of the GFG algorithm, with rules for splitting multicast greedy packets and a repair strategy for concave nodes that includes one or more addressed destinations.

Minimizing the path length for individual nodes and reducing the total number of message transmissions are two desirable and potentially conflicting properties of multicast forwarding strategies. The greedy routing part of PBM utilizes a localized criterion aimed at optimizing both objectives. In order to achieve short path lengths, greedy routing may select for each destination D the neighbor node that is closest to D. Applied as the sole optimization criterion, this strategy would lead to splitting message forwarding as soon as there is no single node that is optimal with respect to progress toward all destination nodes. Thus, while this criterion is a good choice to minimize the path length for individual nodes, the total number of message transmissions remains suboptimal. On the other hand, reducing bandwidth usage can be obtained by sending along a single path as long as possible, that is, the message will be duplicated only if no neighbor node exists that is closer to all the destination nodes considered. However, it can be observed that splitting a packet too late may again increase the total number of hops.

Mauve et al. derived an optimization criterion for greedy multicast forwarding that combines both objectives into one expression. The function described in ref. [24] depends on a parameter λ within [0, 1] that can be used to bias the expression between both extremes. A value close to 0 will result in splitting a message as soon as possible, while the total number of single-hop transmissions is likely to decrease when λ is increased up to a value $s < 1$. Simulation results show that there exists an optimal value for λ within [0, 1] regarding the total network load produced.

Greedy multicast forwarding may arrive at nodes with no node closer to some of the addressed destination nodes. Thus, similar to unicast greedy routing, a recovery strategy must be employed in order to guarantee delivery to all destination nodes. Mauve et al. generalized the face routing algorithm to support message forwarding with multiple destinations. If a node has no neighbor with forward progress with regard to one or more destination nodes, face recovery will be invoked for all these destinations, while all other destinations are handled in a greedy mode further on. Face recovery is started by sending the recovery packet to the next edge in a counterclockwise direction to the line connecting the current node and a virtual position averaged over all affected destination nodes. A node receiving a recovery packet checks to see if it is closer to some of the destinations addressed by the packet. For all destinations where the receiving node is closer than the node where face routing was invoked, the packet will revert to the greedy mode again. For all remaining destinations face recovery is continued by transmitting to the next edge in a counterclockwise direction from the last edge at which the packet arrived.

Splitting a packet into a greedy and a recovery copy may lead to redundant message transmissions, since the greedy packet and the recovery packet may travel the same edges for some hops. In order to reduce the load due to such redundant transmissions, PBM combines the greedy and recovery packet in one transmission as long as possible, that is, the greedy packet will follow the path of the recovery packet as long the next hop node selected by face exploration also provides progress toward all destinations addressed by the greedy packet.

12.4.5 Routing Toward a Single Information Sink

All known memoryless routing strategies for arbitrary sender/receiver pairs resort to some variant of face traversal in order to provide guaranteed delivery in a connected network. However, for a typical sensor network scenario where each sensor node is aware of the location of a single information sink D, a quasi-stateless alternative (some nodes do need to memorize some information to facilitate routing) to face routing has recently been proposed that enables a reliable traffic flow from each

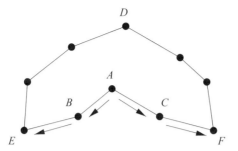

Figure 12.6 The algorithm PAGER establishes an acyclic graph leading toward one information sink.

sensor node toward D. The *partial-partition avoiding geographic routing (PAGER)* algorithm by Zou et al. [25] is a two-phase distributed and stateless construction of an acyclic graph leading toward the information sink D.

In the first phase the algorithm subsequently finds all *shadow nodes* where greedy routing toward the information sink would fail. Concave nodes are declared to be shadow nodes, and recursively other nodes are declared shadow nodes as well. A node becomes a shadow node when all its neighbors, closer to destination D, already became shadow nodes. For instance, in Figure 12.6, nodes A, B, and C are shadow nodes as far information sink D is concerned.

Greedy routing started at nonshadow nodes is always successful when shadow nodes are ignored. In order to enable successful traffic flow from all sensor nodes, shadow nodes establish exit pointers as follows. Each shadow node that has a non-shadow neighbor, or a neighbor with an already established pointer, will point to that neighbor. Packets originated in shadow nodes will follow the exit pointers until the first nonshadow node is reached. Routing then follows the greedy strategy until the final destination D is finally reached. An example is given in Figure 12.6. Nodes B and C will establish an exit pointer to nodes E and F, respectively. Afterwards, node A will establish an exit pointer to both nodes B and C. Traffic originated in A will follow exit pointers until reaching node E for instance. From there on, the packet will be delivered successfully in greedy mode.

12.5 ROUTING WITH ENERGY-AWARE COST METRICS

Sensor nodes are typically equipped with small low-power batteries, and it is impossible to recharge them in most sensor network scenarios. Thus, the lifetime of a sensor network is directly related to the energy consumption produced by the routing mechanism applied. If sensor nodes are able to adjust their signal strength, routing algorithms could attempt to reduce power consumption by selecting next-hop nodes within optimal transmission range. Geographic information can be incorporated in order to enable a localized computation of the best next-hop node by means of a power metric, which is a function that depends on the distance to the receiving node. However, a power-metric considers the optimal transmission range only, thus, single nodes might be selected by many routing tasks, which will result in their premature failure. Using a cost-metric or a combination of both power- and cost-metric might cause the nodes remaining battery power to increase the total network lifetime by spreading the energy consumption evenly among all network nodes. Such energy-aware metrics have been used to define novel energy-aware routing algorithms, replace traditional link-metrics in existing routing algorithms, and finally have also implicitly been applied to existing routing protocols by restricting the selection of next-hop forwarding nodes on an energy-optimized subgraph that has been constructed in advance.

12.5.1   Making Existing Protocols Energy Aware

The total energy needed for communication between two devices S and D might be reduced if the communication were relayed over an intermediate node R, while R and S transmit with the minimal power needed to reach nodes S and D, respectively. However, possible energy reduction depends on the position of the relay node R and the additional energy dissipation at the receiving device. This observation was used by Rodoplu and Meng [26] in order to define the *minimum energy communication network* (*MECN*) algorithm[2] that constructs power-optimized paths between a set of source nodes to one master node (i.e., the information sink in a sensor network scenario). It is implicitly assumed that each node is able to reach each other node in the network by transmitting with appropriate signal strength. In order to find all power-efficient routes to the master node, the algorithm first extracts a subnetwork (termed *enclosure graph*) containing at least all shortest-path edges (with respect to the power metric being optimized) leading from source nodes to the master node. This is achieved by a localized algorithm utilizing position information about all neighbor nodes and eliminating all nodes A for which it takes less power to send messages over a relay node instead of sending it directly to A. As a result each node obtains a reduced set of immediate neighbors, and thus in a second phase optimal routes can be constructed in a more power-efficient way, since communicating with neighbors in the enclosure graph requires less power than communicating with all neighbors from the original network. Optimal routes are found in ref. [26] by applying the distributed Bellman–Ford shortest-path algorithm. Each node calculates the minimum cost it can attain given the cost values of all its neighbors from the enclosure graph and the power needed to transmit a message to that neighbor. When the cost value of any neighbor is reduced, the current minimum cost value is recalculated, and if it was reduced, the new value is announced again to all immediate neighbors from the enclosure graph. The initial route setup from all sources to the sink can be obtained by broadcasting from the sink using only the edge of the enclosure graph, until all sources are reached.

---

[2]The algorithm was not termed MECN in the original work. However, this chapter will follow subsequent publications (e.g., ref. [27]), which referred this algorithm as MECN.

12.5.2 Localized Power- and Cost-Aware Routing

The majority of energy-aware geographic routing schemes described in the literature utilizes the distance to neighbors in the vicinity and apply some sort of distributed shortest weighted path algorithm to that information in order to construct a path from the source to the final destination. Stojmenovic and Lin [28] were the first investigating localized energy-aware greedy routing algorithms, that is, according to the greedy routing principle a received message will be forwarded to the best node regarding the energy metric being optimized.

According to refs. [26] and [29], a general power metric can, depending on node distances, be derived based to the most common channel model used for radio frequency systems. The received signal power for radio frequency communication decreases by a factor $1/d^\alpha$ (referred to as *path loss model*), with $\alpha \geq 2$ and d denoting the distance between the sending and receiving device. The correct choice of α depends on the system being used and can be determined from field measurements. A value of $\alpha = 2$ is often used to model radio propagation at short distances (referred as the free-space propagation model), while $\alpha = 4$ is used for radio transmission at longer distances (referred to as the two-ray ground reflection model). Additionally, the expression may be normalized by t, which denotes the predetection threshold at the receiver. Altogether this leads to an expression td^α, which denotes the minimum power the sender has to radiate in order to enable a signal detection at distance d. Besides power consumption at the sender there is additional power consumption at the receiver that is independent of the distance d and can thus be described as a constant $c > 0$. Summing the power expenditure for one signal transmission altogether amounts to $td^\alpha + c$. (*Note*: The constant c may also incorporate additional power expenditure due to computer processing and encoding/decoding on the sending and receiving devices.)

Assuming that additional nodes can be placed arbitrarily between source S and destination D, the polynomial power consumption $u(d) = td^\alpha + c$ in case of direct transmission can be converted to a linear function in d, producing minimal power consumption. More precisely, there is an optimal number $n = dc_1$ of equally spaced intermediate nodes producing a minimal total power consumption of $v(d) = dc_2$, where the constant values c_1 and c_2 are calculated from the constant power metric parameters t, c, and α [28]. In reality, it is not possible to insert equally spaced intermediate nodes. However, assuming that the power consumption for the rest of the path is equal to the optimal one, this result can be used to define the power-aware greedy routing algorithm *POWER*, where each intermediate node S selects the best next-hop neighbor E closer to the final destination D, which minimizes the sum $u(s) + v(t)$, with $s = |SA|$ and $t = |AD|$. For example, in Figure 12.7, node S will select node C as the next forwarding node, since the power $u(s)$ needed to transmit a message directly to C and the minimal power $v(t)$ needed to forward the packet over the remaining distance between C and D is minimal compared to all other neighbors.

The theoretical result of the power optimal number of equally spaced intermediate nodes is directly related to the polynomial power-consumption formula $td^\alpha + c$.

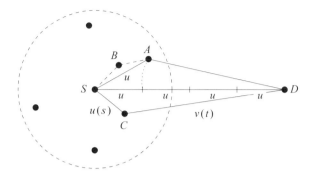

Figure 12.7   *S* selects nodes *C*, *A*, and *B* in the POWER, PowerProgress, and IPowerProgress methods, respectively.

If other power metrics were applied (e.g., a metric changing at a reference distance between free-space propagation and the two-ray ground model) a new theoretical analysis will be necessary in order to calculate these optimal values again before the POWER routing algorithm can be applied. In particular, if the power metric is given by empirical values only, an approximation by an appropriate choice of t and α may be necessary. Kuruvila et al. [30] proposed a novel power-aware greedy routing scheme, *PowerProgress*, which does not suffer from this fact. Let $u(s)$ be the power needed to transmit a message from node S to neighbor node A at distance s. If all subsequent forwarding nodes make the same progress toward the destination, the minimum number of forwarding steps amounts to $d/(d-t)$, with d being the distance between S and D and t being the remaining distance between A and D. When each forwarding step consumes the same amount $u(s)$ of power, the total power consumption will be at least $u(s)d/(d-t)$. Thus, a forwarding node applying the PowerProgress routing strategy will select the neighbor node A that minimizes $u(s)/(d-t)$ (with $t < d$), that is, the power spent per unit of progress made in terms of getting closer to the destination D (see Fig. 12.7).

Additionally, the *IPowerProgress* algorithm, which is an iterative refinement of the optimal node found by the PowerProgress method, was described by Kuruvila et al. In the first iteration step a node S applies the PowerProgress selection criterion in order to find the optimal next-hop node E regarding the distance between S and final destination D. However, sending the packet to node E may still be optimized locally, that is, it might still be more power efficient to send the packet over a relay node instead of sending it directly to E. Thus, the next iteration step selects (if possible) a neighbor F of both S and E, which has a distance to D less than the distance between S and D and which minimizes the sum R of power needed to send from S to F and finally from F to E. However, the relay node F is selected only if the power r needed to relay the packet is less than sending it directly to E. If such a node F is found, the original next hop node E is replaced by F and the iterative refinement method is applied again. The procedure repeats until no better node can be found and the packet is sent to the last optimal relay node found. Note that the node E, which was found initially, is not necessarily visited by the selected routing path. An example of the algorithm is depicted in Figure 12.7. First, node S will select A according to the PowerProgress method. However, there is an optimal relay node B producing less power consumption than would be spent by sending directly to node A. The algorithm will terminate at node B, since there is no additional node U that would further improve the power consumption when sending to B over relay node U.

Kuruvila et al. [30] also defined the *ProjProgress* and *IProjProgress* algorithms, which differ from PowerProgress and IPowerProgress in terms of measuring the progress made in each routing step. Instead of calculating distances, the progress made by the projection of neighbor node A onto the line SD connecting source node S with destination node D is considered. A node S, applying the ProjProgress, will forward a message to the neighbor node A, minimizing the expression $u(s)/(SD \cdot SA)$, where $SD \cdot SA$ denotes the dot product of vectors SD and SA (cf. difference between MFR and GREEDY). The IProjProgress method is very similar to IPowerProgress, but differs in the first iteration step, which selects the best node by applying the ProjProgress method instead.

Singh et al. [31] have observed that minimizing hop count, delay, or the power consumption of the paths produced by routing algorithms may be misguided in the long term. A longer path passing through nodes that have plenty of energy may be a better solution in terms of total network lifetime. In order to avoid energy-critical nodes and to maximize the number of successful routing tasks, a cost metric $f(A)$ expressing a node's reluctance to forward a packet is defined in ref. [31]. It is an expression proportional to the inverse of the node's remaining battery power, thus, a node's reluctance increases significantly when its battery power approaches 0. Stojmenovic and Lin [28] proposed a localized algorithm, *COST*, which is based on that cost metric. The cost to route a packet addressed to D via a neighboring node A is the sum of the cost $f(A)$ and the estimated cost produced to send along the remaining distance between node A and final destination D. The cost of the remaining path is assumed to be proportional to the number of hops between A and D, which in turn can be estimated as td/R, with d being the distance between A and D, R expressing a node's sending radius, and time t set to an appropriate value (empirical results showed $t = f(A)$ to be a good performing definition). A node holding a packet addressed to D will select a next-hop node A closer to the destination, which minimizes the expression $c(A) = f(A) + td/R$ (with $d = |AD|$). In their recent paper, Kuruvila et al. [30] also investigated the principle of proportional progress in combination with the cost metric defined in ref. [31] and defined the CostProgress routing scheme, which selects the forwarding neighbor closer to the destination, which minimizes $f(A)/(d - x)$. An iterative improvement like IPowerProgress cannot be defined for CostProgress, since the overall cost increases by adding intermediate nodes on a path.

Stojmenovic and Lin [28] also investigated combinations of power and cost metrics in one expression in order to achieve both objectives, reducing energy dissipation of the current message forwarding and increasing total network lifetime for many routing tasks. A multiplicative expression termed power/cost metric can be defined as $powercost(S, A) = f(A)u(r)$ (with $r = |AS|$ and $u(r) = r^\alpha + c$). Based on that metric, a forwarding node running the PowerCost routing algorithm will select the neighbor node A, minimizing the expression $powercost(S, A) + v(d)f'(A)$ with $r = |SA|$, $d = |AD|$, and $f'(A)$ being the average reluctance of A and its neighbors. A simpler algorithm is proposed in ref. [30], by selecting a neighbor that minimizes $powercost(S, A)/(|SD| - |AD|)$ (the algorithm is named *Power-Cost Progress*). The *ProjPowerCostProgress* proposed in ref. [30] applies the same metric, but a node S will forward a message to the neighboring node A closer to the destination, which is minimizing the expression $powercost(S, A)/(SD \cdot SA)$. Finally, similar iterative versions of to Power-Cost Progress and ProjPowerCostProgress protocols are also described in ref. [30].

12.5.3   Energy-Aware Guaranteed Delivery

The localized energy-aware greedy routing algorithms described so far do not guarantee delivery even if there is a path from source to destination. Stojmenovic and Datta [32] investigated a combination of face routing with power, cost, and power/cost greedy routing algorithms (called *PFP*, *CFC*, and *PcFPc*, respectively), which guarantee delivery in connected unit-disk graphs. More precisely, routing will start with a power-, cost-, or power/cost routing scheme, respectively. As with GFG, a message that encounters a concave node E also will be handled by the face routing mechanism until the final destination D is reached or a node having a neighbor that is closer to D than the distance between E and D is found. In the latter case, the message is sent to the best of these neighbors and is again handled by the corresponding PFP, CFC, or PcFPc routing method. The choice of such nodes enables it to be proved that the combined routing mechanism remains loop-free and guarantees delivery.

Energy savings of PFP, CFC, and PcFPc result from the energy efficiency of the greedy methods being applied when not in recovery mode. An additional performance gain can be achieved by providing energy awareness in recovery mode, too. Stojmenovic and Lin [32] investigated the impact of CDS construction on the energy-efficiency of PFP, CFC, and PcFPc. A static selection of CDS results in a shorter lifetime of nodes from CDS, which ultimately leads to a shorter lifetime of the whole network. Thus, with the same argument applied to cost-routing, a cost metric might be applied to the construction of the dominating set, taking the node's remaining battery power into consideration. This kind of energy-aware dominating-set construction has been proposed by Wu et al. [33]. Roughly, the algorithm is an extension of the basic distributed dominating-set construction from ref. [34], where the energy level of each node serves as the primary key when comparing two identifiers for a decision about including a CDS (the details are in Chapter 11 in this book).

An additional improvement has been achieved in ref. [32] by applying the shortcut procedure during the recovery mode of PFP, CFC, and PcFPc. In contrast to the original shortcut procedure, the forwarding node considers an energy metric instead of a hop-count metric. To apply this shortcut procedure, 2-hop neighbor information is required.

12.6 BEACONLESS ROUTING

The greedy forwarding mechanisms described need periodic hello messages (beaconing) transmitted with maximum signal strength by each node in order to provide current position information to all one-hop neighbors. This proactive component of greedy routing leads to additional energy consumption, which occurs independently of current data traffic.

Heissenbüttel and Braun [35] proposed the *beaconless routing (BLR)* algorithm. The *contention-based forwarding (CBF)* by Füssler et al. [36] and *implicit geographic forwarding (IGF)* by Blum et al. [37] implement the same idea, focusing on the integration of BLR with the IEEE 802.11 MAC layer. Since no beacons are transmitted, a node, currently holding the packet with the known destination, is generally not aware of any of its neighboring nodes and simply broadcasts a data packet. The main idea of BLR is that each neighboring node receiving the packet calculates a small transmission time-out before forwarding the packet, depending on its position relative to the last node and the destination. The node located at the "best" position introduces the fewest delays and retransmits the packet first. The remaining nodes then cancel the scheduled packet.

For example, in Figure 12.8, node B is closest to the destination D, but its transmission is not heard by node F, also closer to destination than S. To ensure that all potential forwarding nodes detect this transmission, only nodes within a certain forwarding area may be allowed as candidate nodes for the next forwarding step. The forwarding area has the property that each node is able to overhear the transmission of every other node within that area. Heissenbüttel and Braun show that the circle with a diameter equal to the transmission radius, centered at the line SD with S as one endpoint (the dotted circle in Fig. 12.8) is a good forwarding area with regard to progress and successful hops before greedy routing fails. Several delay functions are investigated, resulting in different forwarding behavior.

The authors of ref. [36] also propose a technique called the *active selection method.* A forwarding node sends a control packet instead of the full message to all its neighbors. Neighbors that provide forward progress respond after a time-out

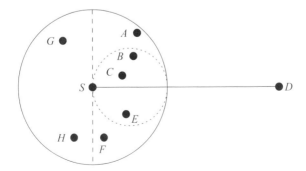

Figure 12.8 A possible forwarding region for BLR.

that depends on their distance to the destination. If a neighbor hears another neighbor's response, it does not respond itself (it is suppressed). The forwarding node then sends the full message, indicating which of its neighbors will forward the message. In a similar way, Zorzi [38] proposed to avoid duplicate forwarding in a BLR scheme by applying the request-to-send/clear-to-send (RTS/CTS) MAC scheme known from IEEE 802.11. The current node sends an RTS signal instead of the message. Afterwards, the node waits for a node to respond with a CTS signal. If several responses are received, the node selects the one that appears to be the best for forwarding and then sends the packet to that neighbor directly.

The principle of sending a control message before selecting the appropriate next-hop node can also be applied in order to describe a beaconless GFG (or alternative protocol, for example, beaconless GOAFR+) scheme [39]. If no CTS signal is received, the node assumes that no neighbor closer to the destination exists and sends another RTS packet to enter the recovery mode. The following procedure is repeated at each intermediate node S during the recovery phase of the beaconless GFG protocol. Each receiving neighbor of S sets a time-out based on the distance to S, so that closer neighbors have a smaller time-out, following the preference of the localized planar graph extraction method. All neighbors participate (including those closer to the destination) in competing for the forwarding neighbor for recovery mode. When the time-out at neighbor A expires, A makes a decision whether or not to report to S. If A heard a transmission from any node B such that B is located inside the circle with diameter AS, then A cancels reporting to S. Otherwise, A reports. Note that A reports even if it learns in the process that it will not be selected as forwarding neighbor, because its report may prevent other nodes, not in GG, from falsely reporting to S, which may contribute to the wrong choice of forwarding neighbor at S. After receiving all replies from GG neighbors, node S selects, among all neighbors from the GG, the one that creates the smallest angle in relation to the incoming packet direction, in the direction decided (clockwise or counterclockwise), following the GFG (or alternative, e.g., GOAFR+) scheme.

12.7 PHYSICAL-LAYER IMPACT ON ROUTING

Almost all existing literature on geographic routing uses the unit-graph communication model, which does not take into account random variations in received signal strengths. It was demonstrated by Schmitz et al. [40] that signal-strength fluctuations have a significant impact on ad hoc network performance metrics, sometimes "outperforming" the impact of node mobility. Thus, nondeterministic radio fluctuations cannot be ignored when designing robust ad hoc network protocols based on ad hoc network simulation and analysis. For instance, in order to find the positions of neighbors, nodes need to resort to a hello message exchange. This is a simple procedure in the unit-graph model, accomplished by each node sending one hello packet, which is then received by all neighbors located within transmission radius R. However, with a realistic physical layer, hello message operation requires a closer look [41].

Independent of the physical-layer model being used (e.g., the combined Friis and two-ray ground model used in ref. [45] or the log normal shadowing model used in ref. [41], protocols dealing with physical-layer impact require nodes to estimate the probability of receiving a bit or a packet based on either signal strength, distance between nodes, or merely by deriving statistics from a number of bits or packets recently sent between two nodes. The basic property of each of realistic physical modeling is a rapidly decreasing packet reception probability. For example, in the shadowing model used in ref. [41], the packet reception probability $p(x)$ depends on the probability of receiving a bit successfully, the length of the packet, and the distance x between two nodes. Suppose R can be determined in that way, so that the packet error rate at distance R is 0.5. Then the function $p(x)$ may have approximately the following values: $p(0) = 1$, $p(0.1R) \approx 1$, $p(0.5R) \approx 0.9$, $p(R) = 0.5$, $p(1.5R) \approx 0.25$, and $p(2R) = 0$. The given values are only an illustration, but give a sufficient intuition on how to design physical layer–aware routing schemes.

Kuruvila et al. [42,43] described geographic routing schemes that are amenable to any realistic physical-layer model (which follows the basic properties of the wireless medium) and consider two basic medium access-layer approaches, with fixed and variable packet lengths, while cases both with and without acknowledgments are being considered. The described methods assume that all nodes use the same transmission power for sending messages, and, in most cases, optimize the expected (packet or bit) hop count on a route. In the case of routing without acknowledgments, the goal is to find the route with the maximal probability of delivering a packet at the destination.

In order to apply position-based routing, the first step is to find a reasonably accurate approximation for the bit and packet reception probabilities for the given physical-layer model. In refs. [42] and [43], Kuruvila et al. considered the lognormal shadowing model and used the following function $P(q, x)$ as an approximation within 5% accuracy of the actual one. The functions $P(q, x) = 1 - (x/R)^{q\beta}/2$ for $x < R$ and $P(q, x) = (2 - x/R)^{q\beta}/2$ otherwise, where β is the power attenuation factor (between 2 and 6). The constant R is determined so that the value of the considered probability at distance R is $P(q, R) = 0.5$. The value q depends on the length of the considered entity. Bit reception probability is $b(x) = P(1, x)$, while, for instance, for packets 120 bits long the packet reception probability is $p(x) = P(2, x)$. The reason for using the approximation rather than the actual function is to reduce computation time at each node (if the protocol is used in practice) and in order to simplify the analyses and simulation of the protocol.

First, consider the case of routing with fixed-size packets and acknowledgments using the same packet size. If the acknowledgments are of a different packet size, the algorithms described are still applicable by changing only the corresponding formulas involving acknowledgments. Let C be the node currently holding the message, D be destination node, A the forwarding neighbor considered, $c = |CD|$, $a = |AD|$, and $x = |CA|$ (see Fig. 12.9). Several localized position-based algorithms are described in ref. [42]. The following discussion describes only the best performing ones, which also apply a general design principle. The progress made by

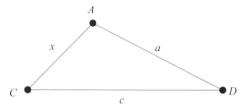

Figure 12.9 Several physical-layer optimized localized routing schemes can be defined by considering the probability of a successful transmission $p(x)$ and the progress $c - a$.

forwarding from C to A is $c - a$, and this progress is probabilistic. In the *aEPR* (expected progress routing) algorithm [42], the node C currently holding the packet will forward it to a neighbor A (closer to the destination than itself), which maximizes the expected progress, which is the product of the probability of successful delivery and acknowledgment of the packet from C to A (which is $p^2(x)$) and the progress made ($c - a$) by forwarding to A. Thus in aEPR, the neighbor A that maximizes $p^2(x)(c - a)$ is selected.

The progress that can be made by sending a packet to A can also be considered with respect to the cost measure for making this progress. The cost measure considered is the expected hop count. The expected hop count depends on the distance and the selected number u of acknowledgments. The progress made could be measured in different ways. In the aEPR-1 algorithm [42], the node C currently holding the packet will forward it to a neighbor A (closer to the destination than itself), which maximizes the ratio of expected progress and the cost of the progress made. Since the considered cost, the expected hop count, is $1/p(x)^2 + 1/p(x)$, aEPR-1 will select the neighbor A, which maximizes $(c - a)/(1/p(x)^2 + 1/p(x))$. This derivation is based on a single acknowledgment for each packet, which is best only if packet reception probability is over 0.5. The optimal number of acknowledgment retransmissions u is approximated as $u \approx 1/p(x)$. The expected hop count is then $f(u, x) = 2/(p(x)(1 - (1 - p(x))^u))$. This variant, called *aEPR-u*, selects the neighbor that maximizes $(c - a)/f(u, x)$.

The *iterative EPR (IaEPR)* algorithm is an improved variant of aEPR-u. The algorithm can be described as follows. As in aEPR-u, the node C currently holding a message will first find a neighbor A that maximizes $(c - a)/f(u, x)$. Then, an intermediate common neighbor node B (closer to the destination than C, if it exists, $b = |BD|$) is found, which minimizes $f(u_1, |CB|) + f(u_2, |BA|)$, where $u_1 \approx 1/p(|CB|)$ and $u_2 \approx 1/p(BA|)$. If $f(u_1, |CB|) + f(u_2, |BA|) < f(u, x)$, then B becomes the new forwarding neighbor, taking the role of A. This process is iteratively repeated until no improvement is possible. Node C will forward the message to the selected neighbor A, which then again applies the same scheme for its own forwarding.

Consider now the model that does not have hop-by-hop acknowledgments. Localized protocols for this model are described in ref. [43]. It was proved in ref. [43] that the packet delivery rate approaches 1 if a large number of intermediate nodes is placed between the source and the destination nodes at distances between consecutive hops approaching 0. Following this observation, a localized algorithm can be described as in ref. [43]: The node C currently holding a message will forward it to its nearest neighbor A, which is closer to the destination than itself. The process continues until the destination is reached or a node is reached that has no neighbor closer to the destination.

A somewhat better performance is obtained by the following alternative scheme [43]. The progress made by forwarding from C to A is $c - a$. This progress is probabilistic. In the *non-acknowledged EPR* (*nEPR*) algorithm [43], the node C currently holding a message will forward it to a neighbor A (closer to the destination than itself), which maximizes the expected progress, which is the product of the probability of successful delivery $p(x)$ of the message from C to A and the progress $(c - a)$ made by forwarding to A. Therefore, the neighbor A that maximizes $p(x)(c - a)$ is chosen to forward the message.

The *iterative EPR* (*InEPR*) algorithm [43] is an improved variant of nEPR. The algorithm can be described as follows. As in nEPR, the node C currently holding a message will first find a neighbor A that maximizes $p(|CA|)(|CD| - |AD|)$. Then, if it exists, an intermediate node B (closer to the destination than C and a neighbor to both C and A) is found that satisfies $p(|CB|)p(|BA|) > p(|CA|)$ and has the maximum $p(|CB|)p(|BA|)$ measure. If found, then B becomes the new forwarding neighbor, taking the role of A. This process is iteratively repeated until no improvement is possible. Node C will forward the message to the selected neighbor A, which then again applies the same scheme for its own forwarding.

Now consider the case of variable packet lengths on each hop, and routing with hop-by-hop acknowledgments [44]. The localized algorithms described remain the same, with the following differences. Instead of the expected hop count in terms of packets, the schemes measure the expected number of transmitted bits. The expected hop count $f(u, x)$ in aEPR-u and IaEPR is replaced by the expected bit count $g(b, k)$ for routing with acknowledgments. If the aEPR variant is considered, then the criterion maximizing $p^2(x)(c - a)$ is replaced by the criterion maximizing $g(b, k)$ $(c - a)$. Observe here that k, the packet length corresponding to the optimal expected bit count $g(b, k)$ (determined in ref. [45]), is not a constant, since each neighbor, being at a different distance, has its own optimal value for k. The case of variable packet length and routing without hop-by-hop acknowledgments was also considered in ref. [44].

The algorithms described so far are physical layer–based solutions for greedy position–based routing. Routing with guaranteed delivery for the unit-graph model and an ideal MAC layer, as described in ref. [13], applies greedy routing as long as possible, and when a node has no neighbor closer to the destination than itself, it resorts to face-recovery mode until a node closer to it is found. The recovery procedure is based on a planar graph locally defined. This procedure can be adapted to the physical layer in a straightforward manner. The edges of the planar graph are normally short ones, and therefore have relatively high reception probabilities. They are therefore good choices for edge selection. Thus, the recovery mode for the physical-layer impact routing may proceed in the same way as in the unit-graph model. Only greedy mode needs to be changed.

Finally, beaconless routing can be adapted to the physical layer by modifying the criterion for selecting the best forwarding neighbor and the appropriate time-out. The time-out can be based on the formulas already described here for selecting the best forwarding neighbor. If a given node announces the request for forwarding the packet several times, the best forwarding neighbors will receive it, and the best will respond a few times to make sure the response was received and it was selected.

EXERCISES

12.1   Show that any greedy routing algorithm that selects only nodes closer to the destination is always loop free. Find a representation of MFR in terms of the dot product and show in a similar way that MFR is also a loop-free routing scheme. Finally, construct an example where DIR will end in a packet loop [5].

12.2   Suppose a node configuration $S = (0, 0)$, $A = (1, 1)$, $B = (1, -1)$, and $D = (3, 0)$ ((x, y) represents the node position), while nodes S, A, B can mutually reach each other and node D is disconnected from all other nodes. Show that MFR forwarding from node S to D will end in a loop and that such a loop can also be constructed even when there is a path from source S to destination D. Explain why this does not contradict the proof of the loop-free property of MFR [5] and show how this loop can easily be repaired in a practical implementation.

12.3   The GEDIR [5] algorithm is an improvement of GREEDY that considers all (i.e., even those in the reverse direction) neighbor nodes and selects the node closest to the final destination. A message is dropped if it would be sent back to the node where it was previously sent from. Show that GEDIR is a loop-free routing algorithm and construct an example where GEDIR is successful while GREEDY is not.

12.4   The *nearest with forward progress* (*NFP*) algorithm [2] is a progress-based routing strategy which selects the neighbor with least forward progress as the next hop node. Investigate whether the method provides loop-free operation or if there is a node configuration where this routing strategy will produce a packet loop.

12.5   Derive an expression that estimates the expected forward progress of MFR applied on uniformly spatial distributed network nodes [46]. Use this expression to estimate a lower bound on the average number of hops produced by MFR when routing a packet over a distance d.

12.6   Greedy routing is often used as a single-path strategy, that is, at any time there is only one instance of the message in the network. In contrast, localized multipath strategies perform routing along a few recognizable paths simultaneously. Extend the GREEDY method presented to a multipath strategy that forwards the packet along c recognizable paths (however, paths can have edges in common). Define the different rules for nodes receiving the same greedy packet more than once [47].

12.7 Restricted directional flooding allows each intermediate node to forward a packet to possibly more than one neighbor lying in the general direction of the final destination [48,49]. Suppose the destination node D is located somewhere inside a circle C centered at the last known position of D. Apply the concept of restricted directional flooding and generalize distance-, progress-, and direction-based greedy routing methods in order to address all possible destination locations within the circle C [50].

12.8 Show that in a sensing covered network with the double-range property any node U lying within the Voronoi region $Vor(V)$ is also covered by node V. Use this result in order to prove that the Delaunay triangulation is always a subgraph of a sensing covered network with double-range property.

12.9 Show that the combination of GREEDY with the recovery strategy FACE is a loop-free routing strategy, while GEDIR and FACE may produce a packet loop.

12.10 Construct an example to show that face routing may not be successful in a connected unit-disk graph if it is applied in a subgraph that is not planar.

12.11 Suppose a weakened planar graph construction where the edges do not intersect in one intersection point but where collinear edges may intersect. Is face traversal always successful in such a graph construction?

12.12 Assume the following simple implementation of face routing. Face traversal is always performed in the clockwise direction. When face exploration encounters an edge intersected by the straight line connecting the last intersection point and final destination D, the next face is determined by simply skipping the intersected edge and continuing face exploration with the next edge clockwise from the intersected edge. Give an example where this simple implementation of face routing will lead to a packet loop. What additional condition must be checked in order to provide a loop-free operation of this algorithm?

12.13 It can be observed that face routing can produce a forwarding loop when the network topology changes due to node mobility. Show that both adding a new edge and removing an edge during face traversal may lead to a packet loop. Find a solution utilizing the creation time of both message and edges that will guarantee loop-free operation in the case where new edges are added during face traversal.

12.14 Assume a GFG implementation with the following simplified recovery strategy. The current face, which is traversed due to a packet recovery started at a node A, will never be changed. The recovery strategy will fall back to greedy mode when arriving at a node B, which lies closer to the destination D than the distance between A and D. Construct an example planar graph where this algorithm will end in a loop.

12.15   Construct a family of unit-disk graphs in order to show that any memoryless geographic routing algorithm with guaranteed delivery can produce a path of length $O(c^2)$, where c is the length of the shortest path [17].

12.16   Addressing all nodes lying within a certain geographical area is termed geocasting. Define an algorithm based on planar graph traversal that achieves geocasting with guaranteed delivery when the area is a circle centered around a given center position [51].

12.17   The concept of *Gabriel graphs* [14] can be used to define a localized planar graph construction for unit-disk graphs. By using information about all neighbor nodes, a node U preserves an edge to its neighbor V if and only if the circle with diameter $|UV|$ passing the nodes U and V does not contain any other neighbor. Show that for a connected unit-disk graph the resulting subgraph is planar and connected [13].

12.18   The localized planar graph construction from the previous exercise will not work correctly if the unit-graph property is missing. Give an example where two neighbor nodes U and V will produce an inconsistent view, that is, node U will preserve edge UV, while node V will remove that edge.

12.19   Suppose a generalization of the unit-graph concept where any node can have a sending range that varies within a maximum r and minimum $r/\sqrt{2}$ transmission range [52,53]. An edge exists between two nodes if they are mutually included in their sending ranges (i.e., only bidirectional connections are considered). Show that by additional message exchange a localized planar graph construction is also possible for this generalized network class. Investigate whether localized planar graph construction is also possible for variations in transmission range ratios that are larger than $\sqrt{2}$.

12.20   Face traversal needs exact location information about neighbor nodes and the final destination. Construct two examples where planar graph routing will end in a routing failure due to imprecise location information about the destination and the neighbor nodes, respectively [54].

12.21   Design and analyze a routing algorithm that will consider two types of errors: transmission failures (the receiver node is believed to be within the transmission radius, but it is not), and backward progress (the receiver node is believed to be closer to destination node than the sender node, but it is not) [55].

12.22   Planar graph routing applied on the network defined by connected geographical clusters may suffer from the fact that there are connected node configurations where any extracted planar graph will be disconnected. Give an example node configuration that proves this claim [23].

12.23 Assume a simplified power metric d^α with $\alpha \geq 2$ and d is the distance between the sender and the receiver. Show that it is always better to relay traffic along an intermediate collinear node. Does this proposition also hold if the power metric is extended by an additive constant $c > 0$?

REFERENCES

1. Hideaki Takagi and Leonard Kleinrock. Optimal transmission ranges for randomly distributed packet radio terminals. *IEEE Transactions on Communications*, **32**(3):246–257, March 1984.

2. Ting-Chao Hou and Victor O. K. Li. Transmission range control in multihop packet radio networks. *IEEE Transactions on Communications*, **34**(1):38–44, January 1986.

3. Gregory G. Finn. *Routing and Addressing Problems in Large Metropolitan-Scale Internetworks*. Technical Report ISI/RR-87-180, Information Sciences Institute (ISI), March 1987.

4. Evangelos Kranakis, Harvinder Singh, and Jorge Urrutia. Compass routing on geometric networks. In *Proceedings of the 11th Canadian Conference on Computational Geometry (CCCG '99)*, pages 51–54, Vancouver, Canada, August 1999.

5. Ivan Stojmenovic and Xu Lin. Loop-free hybrid single-path/flooding routing algorithms with guaranteed delivery for wireless networks. *IEEE Transactions on Parallel and Distributed Systems*, **12**(10):1023–1032, October 2001.

6. Guoliang Xing, Chenyang Lu, Robert Pless, and Qingfeng Huang. On greedy geographic routing algorithms in sensing-covered networks. In *Proceedings of the 5th ACM International Symposium on Mobile Ad Hoc Networking and Computing (MobiHoc '04)*, pages 31–42, Tokyo, Japan, May 2004.

7. Xiaorui Wang, Guoliang Xing, Yuanfang Zhang, Chenyang Lu, Robert Pless, and Christopher Gill. Integrated coverage and connectivity configuration in wireless sensor networks. In *Proceedings of the 1st International Conference on Embedded Networked Sensor Systems (Sensys '03)*, Los Angeles, California, November 2003.

8. Chenyang Lu, John A. Stankovic, Tarek Abdelzaher, and Tian He. SPEED: A stateless protocol for real-time communication in sensor networks. In *Proccedings of the 23rd International Conference on Distributed Computing Systems (ICDCS 2003)*, Providence, Rhode Island, May 2003.

9. C. Huang, F. Dai, and J. Wu. On-demand location-aided QoS routing in ad hoc networks. In *Proceedings of the 2004 International Conference on Parallel Processing (ICPP)*, pages 502–509, Montreal, Canada, 2004.

10. Xu Lin, Mouhsine Lakshdisi, and Ivan Stojmenovic. Location based localized alternate, disjoint, multi-path and component routing schemes for wireless networks. In *Proceedings of the 2001 ACM Symposium on Mobile Ad Hoc Networking and Computing (MobiHoc 2001)*, pages 287–290, Long Beach, California, October 2001.

11. Rahil Jain, Anuj Puri, and Raja Sengupta. Geographical routing using partial information for wireless ad hoc networks. *IEEE Personal Communication*, pages 48–57, February 2001.

12. Ivan Stojmenovic, Mark Russell, and Bosko Vukojevic. Depth first search and location based localized routing and QoS routing in wireless networks. *Computers and Informatics*, **21**(2):149–165, 2002.

13. Prosenjit Bose, Pat Morin, Ivan Stojmenovic, and Jorge Urrutia. Routing with guaranteed delivery in ad hoc wireless networks. In *Proceedings of the 3rd ACM International Workshop on Discrete Algorithms and Methods for Mobile Computing and Communications (DIALM '99)*, pages 48–55, Seattle, Washington, August 1999.

14. K. R. Gabriel and R. R. Sokal. A new statistical approach to geographic variation analysis. *Systematic Zoology*, **18**:259–278, 1969.

15. Brad Karp and H. T. Kung. GPSR: Greedy perimeter stateless routing for wireless networks. In *Proceedings of the 6th ACM/IEEE Annual International Conference on Mobile Computing and Networking (MobiCom-00)*, pages 243–254, New York, August 2000.

16. Susanta Datta, Ivan Stojmenovic, and Jie Wu. Internal node and shortcut based routing with guaranteed delivery in wireless networks. In *Proceedings of the IEEE International Conference on Distributed Computing and Systems (Wireless Networks and Mobile Computing Workshop [WNMC])*, pages 461–466, Phoenix, Arizona, April 2001.

17. Fabian Kuhn, Roger Wattenhofer, and Aaron Zollinger. Asymptotically optimal geometric mobile ad-hoc routing. In *Proceedings of the 6th International Workshop on Discrete Algorithms and Methods for Mobile Computing and Communications (DIALM-02)*, pages 24–33, New York, September 2002.

18. Fabian Kuhn, Roger Wattenhofer, and Aaron Zollinger. Worst-case optimal and average-case efficient geometric ad-hoc routing. In *Proceedings of the 4th ACM International Symposium on Mobile Computing and Networking (MobiHoc 2003)*, pages 267–278, Annapolis, Maryland, 2003.

19. Fabian Kuhn, Roger Wattenhofer, Yan Zhang, and Aaron Zollinger. Geometric ad-hoc routing: Of theory and practice. In *Proceedings of the 22nd ACM International Symposium on the Principles of Distributed Computing (PODC)*, pages 63–72, Boston, Massachusetts, July 2003.

20. Khaled M. Alzoubi, Peng-Jun Wan, and Ophir Frieder. Message-optimal connected dominating sets in mobile ad hoc networks. In *Proceedings of the 3rd ACM International Symposium on Mobile Ad Hoc Networking and Computing (MobiHoc)*, pages 157–164, Lausanne, Switzerland, 2002.

21. Jie Gao, Leonidas J. Guibas, John Hershberger, Li Zhang, and An Zhu. Discrete mobile centers. In *Proceedings of the 17th Annual Symposium on Computational Geometry (SCG)*, pages 188–196, 2001.

22. Yu Wang and Xiang-Yang Li. Geometric spanners for wireless ad hoc networks. In *Proceedings of the 22nd International Conference on Distributed Computing Systems (ICDCS '02)*, pages 171–180, July 2002.

23. Hannes Frey and Daniel Görgen. Planar graph routing on geographical clusters. *Ad hoc Networks*, forthcoming.

24. Martin Mauve, Holger Füßler, Jörg Widmer, and Thomas Lang. *Position-Based Multicast Routing for Mobile Ad-Hoc Networks*. Technical Report TR-03-004, Department of Computer Science, University of Mannheim, Germany, 2003.

25. Le Zou, Mi Lu, and Zixiang Xiong. Pager: A distributed algorithm for the dead-end problem of location-based routing in sensor networks. In *Proceedings of the 13th International Conference on Computer Communications and Networks (ICCCN '04)*, pages 509–514, Chicago, Illinois, October 2004.

26. Volkan Rodoplu and Teresa H. Meng. Minimum energy mobile wireless networks. *IEEE Journal on Selected Areas in Communications*, **17**(8):1333–1344, August, 1999.

27. Li Li and Joseph Y. Halpern. Minimum-energy mobile wireless networks revisited. In *Proceedings of the 2001 IEEE International Conference on Communications (ICC 2001)*, Volume 1, pages 278–283, June 2001.

28. Ivan Stojmenovic and Xu Lin. Power-aware localized routing in wireless networks. *IEEE Transactions on Parallel and Distributed Systems*, **12**(11):1122–1133, November 2001.

29. Wendi Rabiner Heinzelman, Anantha Chandrakasan, and Hari Balakrishnan. Energy-efficient communication protocol for wireless microsensor networks. In *Proceedings of the 33rd Hawaii International Conference on System Sciences (HICSS-33)*, Volume 8, page 8020, Maui, Hawaii, January 2000.

30. Johnson Kuruvila, Amiya Nayak, and Ivan Stojmenovic. Progress based localized power and cost aware routing algorithms for ad hoc networks. In *Proceedings of the 3rd International Conference on AD-HOC Networks & Wireless (ADHOC-NOW '04)*, pages 294–299, Vancouver, British Columbia, July 2004.

31. Suresh Singh, Mike Woo, and C. S. Raghavendra. Power-aware routing in mobile ad hoc networks. In *Proceedings of the 4th Annual ACM/IEEE International Conference on Mobile Computing and Networking (MobiCom-98)*, pages 181–190, New York, October 1998.

32. Ivan Stojmenovic and Susanta Datta. Power and cost aware localized routing with guaranteed delivery in unit graph based ad hoc networks. *Wireless Communications and Mobile Computing*, **4**:175–188, 2004.

33. Jie Wu, Fei Dai, Ming Gao, and Ivan Stojmenovic. On calculating power-aware connected dominating sets for efficient routing in ad hoc wireless networks. *Journal of Communications and Networks*, **4**(1), March 2002.

34. Jie Wu and Hailan Li. On calculating connected dominating set for efficient routing in ad hoc wireless networks. In *Proceedings of the 3rd International Workshop on Discrete Algorithms and Methods for Mobile Computing and Communications (DIALM '99)*, pages 7–14, Seattle, Washington, August 1999.

35. Marc Heissenbüttel and Torsten Braun. BLR: Beacon-less routing algorithm for mobile ad-hoc networks. *Computer Communications (Elsevier)*, **27**(11):1076–1086, July 2004.

36. Holger Füßler, Jörg Widmer, Michael Käsemann, Martin Mauve, and Hannes Hartenstein. Contention-based forwarding for mobile ad-hoc networks. *Ad Hoc Networks*, **1**(4):351–369, November 2003.

37. Brian M. Blum, Tian He, Sang Son, and John A. Stankovic. *IGF: A State-Free Robust Communication Protocol for Wireless Sensor Networks*. Technical Report CS-2003-11, Department of Computer Science, University of Virginia, April 21, 2003.

38. M. Zorzi. A new contention-based MAC protocol for geographic forwarding in ad hoc and sensor networks. In *Proceedings of the IEEE International Conference on Communications (ICC 2004)*, Volume 16, pages 3481–3485, Paris, 2004.

39. Ivan Stojmenovic. Beaconless Position Based Power Aware Routing and Routing with Guaranteed Delivery. In preparation.

40. R. Schmitz, M. Torrent-Moreno, H. Hartenstein, and W. Effelsberg. The impact of wireless radio fluctuations on ad hoc network performance. In *Proceedings of the 4th International IEEE Workshop on Wireless Local Networks (WLN 2004)*, Tampa, Florida, November 2004.

41. I. Stojmenovic, A. Nayak, J. Kuruvila, F. Ovalle-Martinez, and E. Villanueva-Pena. Physical layer impact on the design and performance of routing and broadcasting

protocols in ad hoc and sensor networks. *Computer Communications* (Elsevier), forthcoming.

42. Johnson Kuruvila, Amiya Nayak, and Ivan Stojmenovic. Hop count optimal position based packet routing algorithms for ad hoc wireless networks with a realistic physical layer. In *Proceedings of the 1st IEEE International Conference on Mobile Ad-hoc and Sensor Systems (MASS)*, Fort Lauderdale, Florida, October 2004.

43. Johnson Kuruvila, Amiya Nayak, and Ivan Stojmenovic. Greedy localized routing for maximizing probability of delivery in wireless ad hoc networks with a realistic physical layer. In *CD Proceedings of the 1st International Workshop on AlgorithmS for Wireless And mobile Networks (A-SWAN), Personal, Sensor, Ad-hoc, and Cellular Workshop* (at MobiQuitous), Boston, Massachusetts, August 2004.

44. Ivan Stojmenovic, Amiya Nayak, and Johnson Kuruvila. Design guidelines for routing protocols in ad hoc and sensor networks with a realistic physical layer. *IEEE Communications Magazine* (Ad Hoc and Sensor Networks Series), **43**(3):101–106, March 2005.

45. T. Nadeem and A. Agrawala. IEEE 802.11 fragmentation-aware energy-efficient ad-hoc routing protocols. In *Proceedings of the 1st IEEE International Conference on Mobile Ad-hoc and Sensor Systems (MASS)*, pages 90–103, Fort Lauderdale, Florida, October 2004.

46. Pedro Acevedo Contla and Milos Stojmenovic. Estimating hop counts in position based routing schemes for ad hoc networks. *Telecommunication Systems*, **22**:109–118, 2003.

47. Xu Lin and Ivan Stojmenovic. Location-based localized alternate, disjoint and multipath routing algorithms for wireless networks. *Journal of Parallel and Distributed Computing*, **63**:22–32, 2003.

48. Stefano Basagni, Imrich Chlamtac, Violet R. Syrotiuk, and Barry A. Woodward. A distance routing effect algorithm for mobility (DREAM). In *Proceedings of the 4th Annual ACM/IEEE International Conference on Mobile Computing and Networking (MobiCom-98)*, pages 76–84, Dallas, Texas, October 1998.

49. Young-Bae Ko and Nitin H. Vaidya. Location-aided routing (LAR) in mobile ad hoc networks. In *Proceedings of the 4th Annual ACM/IEEE International Conference on Mobile Computing and Networking (MobiCom-98)*, pages 66–75, Dallas, Texas, October 1998.

50. Ivan Stojmenovic, Anand Prakash Ruhil, and D. K. Lobiyal. Voronoi diagram and convex hull based geocasting and routing in wireless networks. In *Proceedings of the 8th IEEE Symposium on Computers and Communications (ISCC)*, pages 51–56, Kemer-Antalya, Turkey, July 2003.

51. Ivan Stojmenovic. Geocasting with Guaranteed Delivery in Sensor Networks. Paper presented at the International Workshop on Theoretical and Algorithmic Aspects of Sensor, Ad Hoc Wireless and Peer-to-Peer Networks, Fort Lauderdale, Florida, February 2004.

52. Lali Barriere, Pierre Fraigniaud, Lata Narajanan, and Jaroslav Opatrny. Robust position-based routing in wireless ad hoc networks with unstable transmission ranges. In *Proceedings of the 5th ACM International Workshop on Discrete Algorithms and Methods for Mobile Computing and Communications (DIALM '01)*, pages 19–27, Rome, Italy, July 2001.

53. Fabian Kuhn, Roger Wattenhofer, and Aaron Zollinger. Ad-hoc networks beyond unit disk graphs. In *Proceedings of the 2003 Joint Workshop of Foundations of Mobile Computing (DIALM-POMC)*, pages 69–78, San Diego, California, September 2003.

54. Karim Seada, Ahmed Helmy, and Ramesh Govindan. *On the Effect of Localization Errors on Geographic Face Routing in Sensor Networks*. Technical Report 03-797, University of Southern California USC, 2003.

55. S. Kwon and N. B. Shroff. Geographic Routing in the Presence of Location Errors. Submitted for publication.

CHAPTER 13

Data-Centric Protocols for Wireless Sensor Networks

IVAN STOJMENOVIĆ

University of Ottawa, Ontario, Canada

STEPHAN OLARIU

Old Dominion University, Norfolk, Virginia

This chapter reviews a number of emerging topics pertaining to a data-centric view of wireless sensor networks. These topics include data-driven routing, tracking mobile objects, constructing and maintaining reporting trees, dynamic evolution of a monitoring region for moving targets (mobicast), disseminating monitoring tasks, data gathering, receiving reports from a particular area of interest, and sending information and task assignment from a sink to all the sensors inside a geographic region (geocasting). The chapter also discusses various other issues, including sensor training options, data aggregation, data storage, as well as design guidelines for data aggregation and clustering, and rate-based data propagation in wireless sensor networks.

13.1 INTRODUCTION

13.1.1 Sensors and Sensor Networks

Recent technological advances have enabled the development of low-cost, low-power, and multifunctional sensor devices. These nodes are autonomous devices with integrated sensing, processing, and communication capabilities. A sensor is an electronic device that is capable of detecting environmental conditions such as temperature, sound, or the presence of certain objects. Sensors are generally equipped with data-processing and communication capabilities. The sensing circuitry measures parameters from the environment surrounding the sensor and

Handbook of Sensor Networks: Algorithms and Architectures, Edited by Ivan Stojmenović
Copyright © 2005 John Wiley & Sons, Inc.

transforms them into electric signals. Processing such signals reveals some properties about objects located and/or events happening in the vicinity of the sensor. The sensor sends such sensed data, usually via radio transmitter, to a command center either directly or through a data collection station (a base station or sink). To conserve the power, reports to the sink are normally sent via other sensors in a multihop fashion. Retransmitting sensors and the base station can perform fusion of the sensed data in order to filter out erroneous data and anomalies and to draw conclusions from the reported data over a period of time. For example, in a reconnaissance-oriented network, sensor data indicates detection of a target, while fusion of multiple sensor reports can be used for tracking and identifying the detected target.

The block diagram of a typical sensor is depicted in Figure 13.1. The functionality of the sensing circuitry depends on the sensor capabilities. In general, the sensing circuitry generates analog signals whose properties reflect the surrounding environments. These signals are sampled using the analog/digital (A/D) converter and stored in the on-board memory as a sequence of digital values. The sensed data can be further processed using a data processor (microprocessor or digital signal processor (DSP)) prior to sending them over to the base station using the radio transceiver. The capabilities of the data processor are subject to a trade-off. A powerful DSP can be advantageous, since it will allow the sensor to transmit only important findings rather than excessive raw readings. Reducing the sensor's traffic generation rate can save the energy consumed by the radio transmitter and can decrease radio signal interference and collisions among the deployed sensors. On the other hand, sophisticated data processing can consume significant energy and can be a cost and a design burden by increasing the complexity of the sensor design. In all cases, the sensor has to include some control logic to coordinate the interactions among the different functional blocks. Such a control function also can be performed by the data processor, if included. Individual sensors have severely limited bandwidth and battery power. State-of-the-art sensors use one-to-all communication provided by omnidirectional antennas and communication on a single common

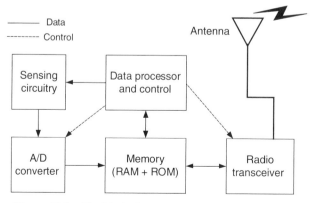

Figure 13.1   The block diagram design of a typical sensor.

channel (sensors using several frequencies, frequency hopping, or several transceivers and receivers are also being considered). Variants of IEEE 802.11 (designed to operate efficiently at low poser consumption) are candidate medium-access control protocols for sensor networks, while Bluetooth appears to be an energy-expensive solution (Chapter 8 in this book is devoted to medium access). Sensor memory and processing capabilities are limited. Routing tables, if used at all, must be small. Data-compression and error-control schemes for sensor networks must be carefully selected. Secure operation is difficult to provide. There exists a great risk when using sensors. Sensor nodes can be defective, lost, damaged, compromised, or expired. Sensors in the active state spend considerably more energy than sensors in the sleep state, as discussed in several chapters in this book.

13.1.2 Applications and Physical Properties of Wireless Sensor Networks

Once deployed, the sensors are expected to self-configure into a *wireless network*. Sensor networks consist of a large number of sensor nodes that collaborate together using wireless communication and asymmetric many-to-one data. Indeed, sensor nodes usually send their data to a specific node called the *sink node* or *monitoring station*, which collects the requested information. The limited energy budget at the individual sensor level implies that in order to ensure longevity, the transmission range of individual sensors is restricted, perhaps of the order of a few meters. In turn, this implies that wireless sensor networks should be multihop. An important difference between wireless sensor networks and conventional networks is that sensor nodes do not need node addresses (e.g., medium-access control (MAC) address and Internet protocol (IP) address). In conventional networks (e.g., Internet), the node address is used to identify every single node in the network. Various communication protocols and algorithms are based on this low-level naming. However, wireless sensor networks are information-retrieval networks, not point-to-point communication networks. That is, wireless sensor network applications focus on collecting data, rather than on providing communication services between network nodes. Node address is not essential for sensor network applications.

Wireless sensor networks are a special case of ad hoc networks. However, there are several major differences between wireless sensor networks and ad hoc networks. To begin, the nodes of a wireless sensor network are generally densely deployed (e.g., hundreds or thousands of sensors may be placed, mostly at random, either very close or inside the phenomenon to be studied). Also, the number of nodes is typically not the same: while there are hundreds or thousands of sensors, the number of nodes (laptops, personal digital assistants (PDAs), palmtops, etc.) in an ad hoc network normally ranges from tens to hundreds. The sensors have a larger failure rate and feature lower data reliability, and are subject to stringent limitations in the energy budget, computing capacity, and memory. The nodes of an ad hoc network are normally distinguished by their IP addresses or other identifiers, while sensors are usually *anonymous*, lacking fabrication-time identifiers. Consequently, they are being addressed and named using various strategies that either endow sensors with temporary IDs or else rely on data or position-driven naming. While ad hoc networks normally rely on topological information in their operation (e.g., knowledge of one-hop and often times 2-hop neighbors), such information may not be available in wireless sensor networks simply because of the lack of IDs at the individual sensor level. In some cases, however, the sensors benefit from a sense of relative geographic position with respect to the monitored environment and/or with respect to a sink. Thus, positional information (covered in Chapter 9 in this book) may be essential in some applications of sensor networks, although it may not be essential for ad hoc networks.

Depending on the application, different architectures and design goals/ constraints have been considered for wireless sensor networks. We attempt to capture architectural design issues and highlight their implications on the network infrastructure and operation models proposed in the literature. We use the routing protocol as a vehicle for discussion in order to highlight how the infrastructure has been set to fit the network operational model and to deal with the specific architectural issue.

There are three main components in a sensor network. These are the sensor nodes, the sink, and the monitored events. Aside from the few architectures that utilize mobile sensors, most of the network architectures assume that sensor nodes are stationary. On the other hand, supporting the mobility of sinks, clusterheads (CHs), or gateways is sometimes deemed necessary. Routing messages from or to moving nodes is more challenging, since route stability becomes an important optimization factor, in addition to energy, bandwidth, and the like. The sensed event can be either dynamic or static depending on the application. For instance, in a target detection/ tracking application, the event (phenomenon) is dynamic, whereas forest monitoring for early fire prevention is an example of static events. Monitoring static events allows the network to work in a reactive mode, simply generating traffic when reporting. Dynamic events in most applications require periodic reporting, and consequently generate significant traffic to be routed to the sink.

An important design consideration is the topological deployment of nodes. This is usually application-dependent and affects the performance of the communication protocol. The deployment is either deterministic or self-organizing. In deterministic situations, the sensors are manually placed and data are routed through predetermined paths. In addition, collision among the transmissions of the different nodes can be minimized through the prescheduling of medium access. However, in self-organizing systems, the sensor nodes are scattered randomly, creating an infrastructure in an ad hoc manner. In that infrastructure, the position of the sink or the CH is also crucial in terms of energy efficiency and performance. When the distribution of nodes is not uniform, optimal clustering becomes a pressing issue to enable energy-efficient network operation. During the creation of an infrastructure, the process of setting up the network topology is greatly influenced by energy considerations. Since the transmission power of a wireless radio is proportional to distance squared or even higher order in the presence of obstacles, multihop routing will consume less energy than direct communication. However, multihop routing introduces significant overhead for topology management and MAC. Direct routing would perform well enough if all the nodes were very close to the sink. Most of the time sensors are scattered randomly over an area of interest, and multihop routing becomes unavoidable. Arbitrating medium access in this case becomes cumbersome.

Depending on the application of the wireless sensor network, the data-delivery model to the sink can be continuous, event-driven, query-driven, and hybrid. In the continuous-delivery model, each sensor sends data periodically. In event-driven and query-driven models, the transmission of data is triggered when an event occurs or when a query is generated by the sink. Some networks apply a hybrid model using a combination of continuous, event-driven, and query-driven data delivery. The routing and MAC protocols are highly influenced by the data-delivery model, especially with regard to the minimization of energy consumption and route stability. For instance, it has been concluded in that for a habitat monitoring application where data are continuously transmitted to the sink, a hierarchical routing protocol is the most efficient alternative. This is due to the fact that such an application generates significant redundant data that can be aggregated en route to the sink, thus reducing traffic and saving energy. In addition, in the continuous data-delivery model time-based medium access can achieve significant energy saving, the since it will enable turning off sensors' radio receivers. Carrier sense multiple access (CSMA) medium-access arbitration is a good fit for event-based data-delivery models, since the data are generated sporadically.

In a wireless sensor network, different functionalities can be associated with the sensor nodes. In the early work on sensor networks, all sensor nodes are assumed to be homogenous, having equal capacity in terms of computation, communication, and power. However, depending on the application a node can be dedicated to a particular special function, such as relaying, sensing, and aggregation, since engaging the three functionalities at the same time on a node might quickly drain the energy of that node. Some of the hierarchical infrastructures proposed in the literature designate a CH different from the normal sensors. While some networks have selected CHs from the deployed sensors in other applications a CH is more powerful than the sensor nodes in terms of energy, bandwidth, and memory. In such cases, the burden of transmission to the sink and aggregation is handled by the CH.

13.1.3 Transport Layer Issues in Wireless Sensor Networks

The transport layer in wireless sensor networks is different from its counterpart in ad hoc and other types of wireless networks. There are several reasons for this. First, the reduced amount of traffic in sensor networks implies fewer congestion problems. Second, traditional end-to-end reliability does not usually apply in wireless sensor networks. Additionally, acknowledgments consume significant amounts of energy and are consequently avoided; similarly, the small on-board memory makes data-significant buffering at the individual sensor nodes level infeasible. The reliability of *individual* sensor measurement is low, and the goal is to provide good reliability of global sensor *network* measurement. Finally, quality of service (QoS) issues are of a different type in sensor networks: here it is more important to provide the reliability of a small amount of information rather than providing bandwidth or delay guarantees. Therefore, the transport control protocols designed for wired networks or for other kinds of wireless networks cannot be used for wireless sensor networks.

When an event occurs, there is usually a multiple correlated data flow from the event to sink. A spatial correlation exists among the data reported. Several reports may arrive at the sink, or several reports can be combined at intermediate nodes to reduce communication (data fusion). The sink makes a decision on the event based on these reports, which has a certain degree of *collective reliability*. The transport-layer problem in wireless sensor networks can be defined concisely as follows: to configure the reporting rate to achieve the required event detection reliability at the sink with minimum resource utilization.

13.1.4 Query Processing

In other types of networks, queries are normally *address-centric* in the sense that they are sent to an individual node using, for example, IP-based routing. By contrast, the anonymity of sensors suggests that in wireless sensor networks queries be either *location-centric* or *data-centric*. Queries are addressed to a geographic region rather than to individual sensors. Since, as we discussed, the sensors do not have unique IDs, routes are created based on the nature and value of data collected by sensors. An example of data-driven routing is the response to a query that is asking to report all sensor readings with temperature over 40°C.

Queries can be distinguished along several orthogonal axes. Spatially, queries may be *global* and be sent to the *entire* deployment area, or *area-specific*, in which case they are addressed to a *geocasting* region (where only sensors inside a geographic region are asked to report), or to *multigeocasting* regions (where all sensors located inside several geographic regions are asked to report). In terms of the reporting mechanism there are several possible types of queries. We only mention the following three: event-driven, on-demand, and persistent. In an *event-driven query*, the sensor itself decides when it has something to report (for instance, when it measures high temperature, which may indicate incipient fire). In an *on-demand query*, the request comes from the end user via the sink. In a *persistent query*, the end user expresses a long-term interest in an event or a disjunction of events. The various sensors tasked with answering the persistent query report whenever a trigger event occurs during the lifetime of the interest.

13.1.5 Data Aggregation in Wireless Sensor Networks

When data are measured or arrive from a neighbor, the sensor needs to decide whether or not they are important enough to forward them. The coding techniques used need to minimize the number of forwarded bits. The new data may also be combined with other received data, in order to minimize the number of bits to forward. Such *data aggregation* (also referred to as *fusion*) from multiple sensors is important, because of severe energy and bandwidth limitations as well as for numerous other reasons, including reliability. The reliability of individual measurements depends on the sensing distance and other factors. For instance, some sensors may be malfunctioning (there are also some security issues, see Chapter 7 in this book). The process of data aggregation from multiple sensors is also referred to as *collaborative signal processing*. Some sensors may aggregate data by doing some computation, such as computing the average of received values, computing the sum total of received values, and computing the largest/smallest of the received values. In order to maximize efficiency, wireless sensor networks may espouse division or work and functional specialization of sensors. For example, based on their relative position and remaining energy level, some sensors may forgo sensing, limiting their activities to data aggregation and data forwarding, while some other sensors may engage in a larger spectrum of activities or even in all the activities for which they qualify. An interesting aspect of the division of work is that it is done dynamically, balancing the load of the various sensors in order to extend as much as possible the useful life of the network.

13.1.6 Deployment Strategies, Time Synchronization and Position Awareness

There are several strategies for deploying wireless sensor networks. The sensors can be *embedded* in the ambient environment, be embedded in the asphalt covering streets and highways, in the walls of building, in trees, and so on. They can be placed *deterministically* by humans or robots, or incorporated in the paint coating walls, or deployed in a purely random fashion. Most research is devoted to *random placement*, where the sensors are dispersed randomly by plane, artillery, humans, or robots. Further, the initial deployment may be followed by later redeployment, as necessary.

Wireless sensor network self-organization includes a time component. One aspect of the problem is the time at which each sensor starts to operate. In many protocols, there exists an implicit assumption that all sensors start to operate at the same time, which could be preprogrammed, or may be externally decided and communicated. The later option is avoided because sensors need to be in the idle state to receive any instruction, which is much more energy-consuming compared to the sleep state (when receivers are turned off). Sensor network operation may require *time synchronization* (covered in Chapter 7 in this book), whether or not all sensors follow the same time or at least have synchronized time slots. Time synchronization can be provided by a global positioning system (GPS), by collaborative efforts, or can be achieved by some other means.

Some applications benefit or even require that the sensory data collected by sensors be supplemented with location information, which encourages the development of communication protocols that are location aware and perhaps location dependent. The practical deployment of many sensor networks will result in sensors initially being *unaware* of their location: they must acquire this information postdeployment. In fact, in most of the existing literature, the sensors are assumed to have learned their geographic position. The *location-awareness* problem is for individual sensors to acquire location information either in absolute form (e.g., geographic coordinates) or relative to a reference point. The *localization* problem is for individual sensors to determine, as precisely as possible, their geographic coordinates in the area of deployment. One simple solution to the localization problem is to use a GPS, where sensors receive signals from several satellites and decide their position directly. However, for tiny sensors such direct position learning may not be possible or may not be sufficiently accurate enough (if a GPS signal is not provided with sufficient accuracy) for the assigned task. However, due to limitations in form factor, cost per unit, and energy budget, individual sensors are not expected to be GPS-enabled. Moreover, in many occluded environments, including those inside buildings, hangars, or warehouses, satellite access is drastically limited.

Since direct reliance of GPS is specifically proscribed, in order to obtain location awareness individual sensors exchange messages to *collaboratively* determine their own geographic position (absolute or relative) in the network. The vast majority of collaborative solutions to the localization problem are based on multilateration or multiangulation. These solutions assume the existence of several *anchors* that are aware of their geographic position (e.g., sinks or specialized sensors that can engage in satellite communication). By exchanging messages with their neighbors, individual sensors can conceivably measure signal strengths and/or time delays in communication. Some approaches are based on hop-count distances to reference points. Sensors receiving location messages from at least three sources can approximate their own locations. For a good survey on localization protocols for wireless sensor networks, we refer the interested reader to the relevant Chapter 9 in this book.

In some other applications, exact geographic location is not necessary: all that individual sensors need is *coarse-grain* location awareness. There is an obvious trade-off; coarse-grain location awareness is lightweight, but the resulting accuracy is only a rough approximation of the exact geographic coordinates. One can obtain this coarse-grain location awareness by a *training* protocol that imposes a coordinate system onto the sensor network. Olariu et al. [1] have shown that an interesting by-product of such a training protocol is that it provides partitioning into clusters and a structured topology with natural communication paths. The resulting topology will make it simple to avoid collisions between transmissions of nodes in different clusters, between different paths, and also between nodes on the same path. This is in contrast with the majority of papers that assume routing along spanning trees with frequent collisions. In the training protocol of Olariu et al. [1] the deployment area is endowed with a virtual infrastructure (for details see a dedicated Chapter 4 in this book). To make the presentation self-contained, however, we now outline the idea. Referring to Figure 13.2, the coordinate system divides the sensor network area into equiangular wedges. In turn, these wedges are divided into sectors by means of concentric circles or coronas centered at the sink. The task of training the wireless sensor network involves establishing:

- *Coronas:* The deployment area is covered by coronas determined by concentric circles centered at the sink
- *Wedges:* The deployment area is ruled into a number of angular wedges centered at the sink.

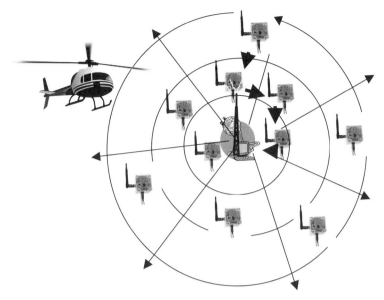

Figure 13.2 Training a wireless sensor network.

Individual sensors can acquire the desired coarse-grain location awareness by learning the identity of the corona and the wedge to which they belong. As it turns out, the training protocol is lightweight and does not require sensors to have IDs; moreover, sensors are not aware of their neighbors within the same sector. It is worth noting that location awareness is modulo the sector to which the sensor belongs. Since accurate position information is unreliable because of shadowing, scattering, multi-paths, and time synchronization problems, training provides a viable alternative.

13.1.6 Topology Control and Area Coverage

In addition to gaining a sense of their location, sensors also need to gain some sense of their neighborhood. This can be achieved with various degrees of self-organization. For example, if sensors have IDs, they can discover neighbors by exchanging "hello" messages, and decide which neighbors and links are needed for their best operation, or what transmission range to select, to provide a certain density for reliable reporting and route construction. The communication may be critical for sparse networks, while for dense networks collisions, congestions and excessive energy expenditures may occur.

Since sensor batteries cannot be recharged under present-day technology, energy consumption is considered the most important parameter contributing to the longevity of the network. The best energy-conservation method is to have as many sensors as possible in *sleep* mode, where energy consumption is minimal. The network must be connected to remain functional, so that the monitoring station can receive the message sent by any of the active sensors. An intelligent strategy for selecting and updating a set of active sensors that are connected is needed in order to extend the network lifetime. After learning about neighboring sensors, sensors decide whether to remain active or to go to sleep if their sensing areas are already covered. This problem is known as the connected *area coverage* problem, which aims to dynamically activate and deactivate sensors while maintaining the full coverage of the monitoring area. Efficient solutions to the connected area coverage problem are discussed in Chapter 11 in this book. When this coverage step is performed first, the large sensor network becomes reasonably sparse, but remains connected.

In the case of training [1], the optimal solution might be to keep a few active sensors in each sector, which can be decided by a simple leader election process. For example, each sensor may choose a time-out based on its remaining energy, and send a packet containing its sector information and remaining lifetime, so that other sensors in the same sector can hear that message, cancel their own transmission, and decide how long they could sleep.

Topology in sensor networks may change more frequently, because of failures, changes in sleep/active periods, and perhaps mobility. Designing efficient protocols for many operations requires a backbone, which is a subset of sensors, so that each sensor is either in a backbone or near it. Backbone examples include clustering and connected dominating sets. Active sensors can organize themselves into *clusters.* In a clustering process, some sensors may be selected as clusterheads (CHs), and every other sensor is assigned to one of the clusters. The alternative organization is to create backbones via *connected dominating sets* (each node is in such a set or is a neighbor of a node from the set). Backbone creation and sensor area coverage (which decides activity schedules) are covered in Chapter 11 in this book.

13.1.7 Localized versus Centralized Protocols

Estrin, Govindan, Heidemann, and Kumar [2] promoted the design of *localized* rather than *centralized* protocols in wireless sensor networks. Due to a number of factors, the topology of wireless sensor networks changes frequently and self-organization must be adaptive to local changes. *Centralized* protocols require global network information at each sensor (sink only, respectively, with sink making decisions) for making sensor decisions. This includes the use of topological structures, such as minimal spanning tree (MST), whose local links cannot be locally determined. There are a number of combinatorial optimization formulations of sensor network design problems with linear programming solutions. These protocols can perform well only when sensor networks are small. We do not discuss centralized approaches further, since we believe in and assume large-scale wireless sensor networks where centralized protocols do not work well.

Localized protocols only require local knowledge for making decisions, and a limited (usually constant) amount of additional information (e.g., the position of the sink). Some localized protocols may require preprocessing, such as constructing a suitable topology for further operation. One typical example is setting up a cluster structure. In addition to localized protocol operation, it is also important to consider the maintenance cost of such topology. For instance, if the cluster structure is adopted, what happens when CHs move or fail? Does the update procedure remain local, and, if so, what is the quality of the maintained structure over time? Some maintenance procedures may not remain local. This happens when local change triggers message propagation throughout the network. Of course, *localized maintenance* is preferred, meaning that local topology changes should be performed by a procedure that always remains local, involving only the neighborhood of the affected sensors.

A number of protocols in the literature are localized, but use an excessive number of messages between neighboring sensors. For instance, some topology control and position determination protocols require over a dozen (sometimes even thousands of) messages to be exchanged between neighbors. Because of the severely limited bandwidth and energy budget and medium-access problems caused by excessive messaging, messages between neighbors to construct/maintain topology, determine position, or perform any other operation should be minimized, possibly avoided entirely (e.g., some backbone construction methods do not require any message after hello messages to learn that neighbors have been exchanged).

13.1.8 Roadmap of the Chapter

This chapter concentrates on localized protocols, featuring localized maintenance, and a limited number of messages between neighboring sensors. We begin by discussing *data gathering*—the most fundamental problems in wireless sensor networks. Data gathering has an implicit routing component, with or without involving data aggregation. Protocols for reporting an event (upon detecting it) by a single sensor are described in Sections 13.2 and 13.4. These protocols can be considered as responses to an event-driven query. The event may be detected by a group of sensors, but a single sensor reports it after data are aggregated first. Section 13.2 is devoted to protocols where a report is sent to the sink based on its position or merely distance to it (the later suffices in the case of direct transmission with omnidirectional antennas), without using any local information inferred by a dissemination originating from the sink. Section 13.3 discusses various ways for disseminating monitoring tasks from sink to sensor nodes. This is mainly done by applying broadcasting and geocasting protocols. Section 13.4 is devoted to data-gathering methods that are based on broadcast trees, which are constructed during the task-dissemination process from the sink, with sensors memorizing certain information that is later used for reporting. Section 13.5 and 13.6 discuss the case when all sensors in an area are requested to report as a reply to an on-demand query. Section 13.5 focuses on the case where data aggregation is not applied, whereas Section 13.6 looks at data aggregation as well. Data aggregation can be applied to all active sensors, or only to the active sensors within a region or a cluster. Section 13.7 discusses the case of mobile sinks or sensors. In Section 13.8, we discuss the problem of sending enough reports about an object to the sink so that the sink can accurately determine the position of the object. It also discusses tracking mobile objects using tree reconfiguration and mobicast protocols. Section 13.9 discusses the problem of rate-based data propagation in sensor networks. Section 13.10 discusses an important corollary of the data-centric view of wireless sensor networks, namely, anonymity. Conclusions, exercises, and references complete this chapter.

13.2 DATA GATHERING WITHOUT MEMORIZING LINKS TOWARD THE SINK

13.2.1 Direct Reporting by Individual Sensors

The simplest way of reporting an event is to simply send a packet with sufficient power to reach the sink. If communication is omnidirectional, the exact position of the sink is not needed, since the approximate signal strength needed to reach the sink will suffice. Since sinks can always operate with more power than sensors, they can send a packet to all sensors announcing its presence, or assigning a task, or perhaps informing about its location. Sensors can also apply the power-increasing method to reach the sink, for example, to double the power applied for transmission until the sink acknowledges the receipt of the report. Since such direct communication may be over a long distance, it will drain the power quickly, and will drain it from all sensors. Therefore this method is presented here only for completeness, and has not been seriously considered as a viable option except for some small-size networks such as the home environment.

13.2.2 Direct Reporting by Cluster Heads

Heinzelman, Chandrakasan, and Balakrishnan [3] described the low-energy adaptive clustering hierarchy (LEACH) protocol for reporting data to the sink. Each node randomly decides whether or not to become a CH. The parameter used in decision making is the percentage of desired CHs. Sensors that decide to become CHs send a packet with their decision. Each node reports to the CH with the highest signal strength, and therefore clusters correspond to Voronoi diagrams of CHs. The CHs assign to each sensor from their cluster a time slot for reporting, aggregate data received from individual sensors, and send aggregated data directly to the sink. The selection of CHs is repeated periodically, to balance energy consumption. The optimal number of clusters is not investigated. LEACH is illustrated in Figure 13.3. The major problem with LEACH is that the sink may be very far from many CHs, therefore direct reporting may be extremely energy-consuming or even impossible. This basic method has variants that depend on how clusters are created. In some scenarios (e.g., military applications, with sensors attached to soldiers), there may exist natural cluster organization, especially if different types of sensors are being used. Different methods for forming reporting clusters are investigated in ref. [4]. Each sensor chooses a time-out interval. If no message is heard during that interval, the sensor decides to form a cluster and to report; otherwise, it becomes the follower of the sensor that sent the message.

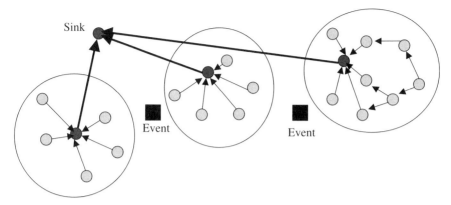

Figure 13.3   Data aggregation at CHs and direct reporting to sink.

13.2.3   Design Guidelines for Clustering and Aggregation in Sensor Networks

Mhatre and Rosenberg [5] considered the organization of sensors into clusters. Sensors could use either a single- or multihop mode of communication to send their data to their respective CHs. The CHs send their data directly to base stations. The energy needed for communication between two nodes at distance r is proportional to $r^{\alpha} + c$, where α is a power attenuation factor (between 2 and 6), and c is a constant that accounts for minimum reception energy and energy to run circuitry. The goal is to minimize and balance energy consumption. The authors analyze two modes of communicating between sensors and base stations, and derive conditions under which single-hop transmission by all nodes is best. One of the conclusions made is that, for $\alpha = 2$, there is no benefit from multihop communication. When multihop communication is better, each CH is assumed to be at the center of a circle divided into rings of equal width (equal to the used transmission radius). Therefore, they assume that each multiple hop is of approximately equal length and they find the optimal forwarding distance for each hop. The authors [5] do not prove that it is indeed optimal to use each hop of equal length (i.e., that the rings indeed all need to have equal width for optimality). Their result is based on minimizing the energy in a ring that is found to be critical. However, other rings may not be critical at that time. Our analysis [6], presented below, shows that, in fact, the rings are not of the same size for the optimal case (i.e., the sensor uses different transmission radii for maximizing network lifetime). We also note that the communication from CHs to base stations can also be multihop, via other CHs or even other sensors, instead of being single-hop. Finally, overall analysis is based on each sensor having an equal amount of data to report, which may not hold in a real application. Mhatre and Rosenberg [5] also studied the problem of determining and selecting the optimal number of CHs and required battery energy. Their derivations, however, are very complex.

Suppose that N sensors are randomly placed in a circle with fixed radius P. The task is to subdivide the circle into n rings, and determine their widths $R_1, R_2, \ldots, R_n$, so that the network lifetime is optimized, where $R_1 + R_2 + \cdots + R_n = P$ (that is, the sum is fixed). Optimization variables are therefore n, $R_1, R_2, \ldots, R_n$. The sink is in the center of the first circle, of radius R_1. It is assumed that the energy required for communication between two sensors (or sensor and sink) is proportional to $d^\alpha + c$, for some constants α ($2 \leq \alpha \leq 6$) and c. For instance, two particular models considered in ref. [7] are $\alpha = 2$, $c = 10^4$, and $\alpha = 4$, $c = 10^8$. For simplicity, the energy is charged fully to the transmitting node. Suppose that the sensor distribution is uniform, and therefore the number of sensors N_i in the ith ring is proportional to the areas of the rings. That is, $N_i/N_j = R_i^2/R_j^2$, and the sum $N_1 + N_2 + \cdots + N_n = N$ is fixed, equal to the total number of sensors. It is assumed that each sensor helps a proportional number of sensors from the rings farther from the sink in retransmitting. To send a message from sensor in the ith ring to a sensor in the $(i-1)$th ring, we assume that the energy needed is proportional to $R_i^\alpha + c$, as an average amount with respect to ring size, or worst-case amount for the first ring (transmitting to sink, which is the zeroth ring). We also assume that the maximum transmission radius is limited, equal to T.

The sensors in the first ring spend (in the worst case) energy proportional to $R_1^\alpha + c$ to send their own message directly to the sink, and to retransmit each message. In addition, each of them retransmits a proportional number of messages from sensors in all other rings. The number of sensors it helps is therefore $(N_2 + \cdots + N_n)/N_1 = (N - N_1)/N_1$, and the total number of messages it sends is $1 + (N - N_1)/N_1 = N/N_1 = P^2/R_1^2$. Thus the energy needed for these transmissions is $(R_1^\alpha + c)P^2/R_1^2$. This is a function of one variable, which has the minimum at $r_1 = (2c/(\alpha - 2))^{1/\alpha}$. For $\alpha = 2$, the energy is minimized for the maximal transmission range. The (unrealistic) case $c = 0$ is easy to discuss. We therefore continue the discussion only for the case $c > 0$ and $\alpha > 2$. Interestingly, the minimal energy for the first ring is obtained for the target radius that does not depend on P, N, and even n! The target radius for the first ring is adopted if $r_1 < T$, otherwise, it must be changed to $r_1 = T$. If $P \leq r_1$, then the optimal number of rings is $n = 1$. Moreover, in this case $r_1 = P$. We assume that $P > r_1$ in the remaining analysis.

Now consider sensors in the last ring. They send only one message to sensors in the previous ring, which requires energy proportional to $R_n^\alpha + c$. The sensor network will maximize its lifetime when all sensors die at approximately the same time. Otherwise, the sensor network will not be able to either monitor or report the event. Therefore the optimal value of R_n is obtained when energies in the first and last rings are equal, that is, when $R_n^\alpha + c = (r_1^\alpha + c)P^2/r_1^2$ (here the optimal value for the first ring is already assumed). This equation has a straightforward solution for r_n (see ref. [6] for the formula) as the optimal ring size. Interestingly, the solution again does not depend on n and N, but it does depend on P, the overall circle size. If this optimal ring size is $> T$, it should be reduced to T. Note that $r_n \neq r_1$ (unless they are both "trimmed" to the same value T), which can be easily verified from the equation.

If $r_1 + r_n \geq P$, then the optimal value for n is $n = 2$. In the case of strict inequality, further energy savings cannot be achieved, because the limit on the first ring width does not depend on P. If $r_1 + r_n < P$, then more rings are needed, and the process can continue iteratively, from the last rings toward the first ring. Sensors in ring R_i will forward messages from a proportional number of sensors from rings R_j for $j > i$. The number of such sensors is $(N_n + \cdots + N_{i+1})/N_i = (R_n^2 + \cdots + R_{i+1}^2)/R_i^2$. Therefore, in the worst case, the sensors in the ith ring are expected to spend energy $(1 + (R_n^2 + \cdots + R_{i+1}^2)/R_i^2)(R_i^k + c)$. Assume that the optimal values for rings $n, \ldots, i+1$ are already determined. The equation to be solved is then $(1 + (r_n^2 + \cdots + r_{i+1}^2)/R_i^2)(R_i^k + c) = r_n^\alpha + c$. This is the equation of one variable, and the function has its minimum, obtained by standard calculus methods (finding a derivative) [6]. Let the optimal solution be $R_i = r_i$. If this solution is larger than T, then it should be changed to T. If $r_1 + r_n + r_{n-1} + \cdots + r_i \leq P$, then $i = 2$, which determines the final value of n. Otherwise, it continues with the next value of i, effectively increasing n by 1. Note that, in the analysis presented, all sensors are assumed to be active.

Assume now that the transmission radii of all sensors are the same and fixed to T. Energy consumption can then be balanced by applying nonuniform sensor distribution. This problem was studied in ref. [8], with solution techniques involving sleep periods and energy consumption for routing tasks. We will extend the preceding solution to the case of nonuniform densities, following ref. [6], keeping all sensors active. Suppose that N sensors are randomly placed in a circle with fixed radius P. The task is to subdivide the circle into n rings of the same fixed widths $T = R_1 = R_2 = , \ldots, = R_n$, and corresponding sensor densities $\rho_1, \rho_2, \ldots, \rho_n$ in these rings so that the network lifetime is optimized. The number of rings n is therefore $n = P/T$, since $R_1 + R_2 + \cdots + R_n = P$. Optimization variables are therefore $\rho_1, \rho_2, \ldots, \rho_n$. The sink is in the center of the first circle, of radius T and density ρ_1. Since ring areas are the same, the number of sensors N_i in the ith ring is proportional to their densities. That is, $N_i/N_j = \rho_i/\rho_j$, and the sum $N_1 + N_2 + \cdots + N_n = N$ is fixed, equal to the total number of sensors. If all densities were the same, then balancing energy consumptions would not be possible, because sensors in rings closer to the sink are getting an increasing number of forwarding tasks, and the transmission energy per task is fixed. Therefore, for balanced energy consumption we have $\rho_1 \geq \rho_2 \geq \cdots \geq \rho_n$. Suppose that $\rho_n = 1$, since other values would simply result in multiplying other densities by the same factor. The energy consumption is proportional to the numbers of messages sent. Sensors in the last ring send one message per considered time unit (which depends on reporting rate), as a result of their monitoring. Sensors in the first ring, being more densely spread, require a lower reporting rate for their own monitoring. They therefore send $1/\rho_1$ reports in the same time frame. Similarly, sensors in the ith ring generate $1/\rho_i$ reports. Sensors in the first ring retransmit a proportional number of messages from sensors in all other rings. The number of sensors each of them helps is therefore $(N_2/\rho_2 + \cdots + N_n/\rho_n)/N_1 = (n-1)/\rho_1$, and the total number of messages it sends is $1/\rho_1 + (n-1)/\rho_1 = n/\rho_1$. Thus, $n/\rho_1 = 1$ or $\rho_1 = n$. Continuing this discussion, we conclude that $\rho_i = n + 1 - i$. With this solution, each sensor sends on average one message per unit time, independently on the ring it is contained in.

13.2.4 Data Aggregation with Consensus

A data-aggregation and -consensus algorithm for object location and tracking by a sensor network is described by Kumar, Schwiebert, and Brockmeyer [9]. The first node that detects an event will first generate consensus by obtaining a quorum from nodes having similar interests and area of coverage. If more than half of the sensors close to the event confirm the same observation by acknowledging to the initiating node, the node will report the event.

13.2.5 Multihop Reporting among Nodes or Clusters

The direct communication from CHs to sinks may be impossible because of distance or can be extremely energy-consuming. Further, even communication from any sensor to its own CH can have such problems. If multihop reporting is applied, packets can be forwarded among CHs only until they reach the sink (if transmission power is adjusted to reach a neighboring CH), or the route could include bridge (or gateway) sensors between adjacent clusters. Note that this multihop forwarding using other CHs can be applied with or without further data aggregation, beyond the initial one within each cluster.

Alternatively, cluster organization may not be necessary. Sensors may react to an event by first finding consensus among other nearby sensors that detect the same event (e.g., the consensus method [9]), and then the lead sensor applies multihop reporting, that is, routing via other active sensors in the network toward the sink.

13.2.6 Reporting with Energy-Efficient Routing

Multihop reporting can be performed with a routing algorithm that aims at minimizing hop count. Alternatively, the algorithm may attempt to minimize the energy expenditure needed for a given routing task, or to maximize the network lifetime by considering the remaining energy when selecting forwarding neighbors. In Chapter 12 in this book, routing protocols with mentioned optimization criteria are surveyed. For the sake of completeness, we summarize here a few relevant protocols.

Schurgers and Srivastava [10] propose that nodes collect several packets intended for the same neighbor into a single packet. They claim that compression can be achieved in this way, leading to more energy efficiency. They also propose stochastic schemes where the best neighbor is chosen at random, an energy-based scheme where the best neighbor is selected based on its energy, and a stream-based scheme where busy nodes inform their neighbor by asking them to select other forwarding nodes instead.

Chatzigiannakis and Nikoletseas [11] describe a *routing protocol for sensors* that have the sense of direction, but do not know their coordinates. The monitoring center is a wall known to sensors, and wider than the width of the sensor network. The task of reporting from a sensor to the wall proceeds by a greedy algorithm, which follows the direction orthogonal to the wall. At each step, the node currently holding the message broadcasts a search message looking for another sensor within an angular range with respect to the wall direction and at a certain minimum distance (and maximum distance, which is the transmission radius). Thus, each sensor has the ability to estimate the distance to neighboring sensors. Each awake sensor located in the desired cyclic sector will report back to current node A, but only the first such node B will receive the full message from the current node. Current node A will wait to hear the forwarded message from B to one of its neighbors. If successful, A will go to sleep. If there is no closer node to the wall from B, a failure message is generated and the message is backtracked to A. Note that no two consecutive backtrack steps are possible, so this simple greedy routing may fail. Note also that the *greedy-face-greedy* (*GFG*) algorithm [12] can be used to guarantee delivery.

13.2.7 Sector Routing

In the case of *sector training* [1], messages are not directed toward any particular sensor. Instead, they are directed toward a sector. All active sensors inside the sector receive the message. One of them decides to retransmit, and others in the same sector (if more than one in a given sector is active) can overhear this transmission, which prevents them from their own retransmissions. This assumes that inside a sector the sensors are within the communication range of each other. This may or may not be true in general. In any case, forwarding toward the sink then follows a sector-routing principle: a route is created from sector to sector, with an arbitrary sensor from each sector participating (see Fig. 13.2, showing a route from the top sector toward the sink). In case of empty sectors, a variant of routing with guaranteed delivery [12] can be applied, since sectors are creating a planar graph. In recovery mode, face routing can be employed using direction orthogonal to the wall (that is, with the destination being imagined at infinity).

13.2.8 Data-Centric Storage

In some particular scenarios, wireless sensor networks can operate, at least temporarily, without a sink. In this case, reports by sensors need to be stored in the sensor network itself. Ratnasamy, Estrin, Govindan, Karp, Shenker, Yin, and Yu [13] described a data-centric storage system for application in wireless sensor networks. Sensors have a tiny memory, and therefore limited storage capacity. Therefore, they need to distribute storage among themselves. The algorithm in ref. [13] is to apply a hash function to a keyword assigned to a file, datum, information, or an object, which will map it to a point with geographic coordinates. The hash function needs to be carefully selected so that the obtained point is inside the geographic region containing the sensors. A planar graph over sensor network can be obtained by applying a Gabriel graph (GG) structure (described in Chapter 10). The information is stored in all sensors on the face containing the mapped point. In order to retrieve the information, GFG routing that guarantees delivery [12] can be applied. Since the mapped node is generally not in the sensor network, routing will create a loop along the face containing it. Sensors on that face have already stored the information and can provide it to the requester.

13.3   DATA DISSEMINATION FROM THE SINK

13.3.1   Broadcasting Short Packets from the Sink

This subsection discusses various ways the sink assigns monitoring tasks. The tasks assigned to sensors need to be propagated to all active sensors in the network (broadcasting), or to all the active sensors located inside a region of interest (geocasting), consisting of currently active sensors (e.g., sensors selected for area coverage). One simple solution, if it is assumed to have sufficient transmission power to cover the entire deployment area, is that the sink sends one message that reaches all the sensor nodes in the network. If the sink does not have sufficient power, the sensors themselves need to retransmit such messages. If the packet containing the task and the location of the sink is relatively short then the data dissemination can be fulfilled by any of the broadcasting protocols, covered in Chapter 11. Most of these protocols assume that the sensors know the position of their local neighbors. Otherwise, *blind flooding* can be applied, meaning that each sensor receiving the packet for the first time will rebroadcast it. This method is the most popular in the existing literature (for instance, it was applied in protocols given in refs. [14] and [15]). Intelligent flooding (broadcasting) schemes are surveyed in Chapter 11. Some sensors do not need to retransmit the task packet, and the task can still be distributed to all the sensors (assuming an ideal medium-access protocol). One such method is beaconless area-based broadcasting [16], where a sensor whose communication area is completely covered by transmissions from other sensors does not need to retransmit. Note that this method does not require prior knowledge of neighbors. When sensors need to report using a broadcast tree, they can link themselves to one of the nodes from which the packet was received.

Acknowledgments for the receipt of monitoring tasks may or may not be sent. If requested, it can be provided, for instance, as follows. Lipman, Boustead, and Chicharo [17] proposed to send acknowledgments only to neighbors along *local minimum spanning tree* (*LMST*) edges. Each sensor then on average sends only at most two acknowledgments, because of the sparse LMST structure. To construct a LMST [18], each node first constructs a MST of its local neighbors (knowing their geographic positions), and keeps edges that are included in such local MSTs by both end points (see more details in Chapter 10). On average, each node will send one acknowledgment only, since the average degree (average number of neighbors) of a LMST is only slightly larger than two.

13.3.2   Broadcasting Long Packets from Sink

If the message containing a detailed assignment, or other type of message that needs to be disseminated, is relatively long, then an alternative is to send two types of messages instead: a short message is sent first that offers a long message to sensors, followed by a long message sent only to those sensors that require it. In the *sensor protocol for information via negotiation* (*SPIN*) [19], each node that receives the datum (full message) that is being broadcast will forward the corresponding *metadatum* (short message) that has a considerably shorter bit length (e.g., 16 bytes instead of 500) to all its neighbors. The metadatum is thus flooded. Neighboring nodes that did not yet receive the full message will reply to the short message with a request to receive it. The sensor will then respond by sending the full message to all nodes that requested it. If an omnidirectional antenna is used, sensors may retransmit the full message upon receipt of the first request for it. Note, however, that, short request messages may be sent back to the transmitting node only if it is a neighbor in a selected sparse, connected structure, as observed in ref. [20], greatly reducing the amount of short messages needed. For example, if a LMST [18] is used as the sparse structure, the reduction is about $2/d$ times, where d is the average number of neighbors in the network. This reduction is possible if nodes have 2-hop topological or one-hop positional information about their neighbors.

13.3.3 Geocasting in Wireless Sensor Networks

Data dissemination, or task allocation, from the sink does not need to be propagated to all active sensors. If only sensors that are close to a monitoring event (e.g., a factory that pollutes the environment) need to be alerted, then only sensors located inside a geographic region need to receive the task. This problem is known as *geocasting*. A survey of existing geocasting schemes is given in ref. [21]. It was shown that most existing geocasting schemes do not guarantee delivery to all nodes inside a region, the main reasons being either the partitioning of the network inside the region, or applying greedy routing instead of one that guarantees delivery.

Yu, Govindan, and Estrin [22] considered a geocasting variant of the data-gathering problem. They describe the *geographic and energy-aware routing* (*GEAR*) algorithm, which uses energy-aware neighbor selection to route the packet toward the target region, and recursive geographic forwarding, or restricted blind flooding algorithm, to disseminate the packet inside the destination region. Recursive forwarding applies GEAR to send messages to four subregions in the geocast region, which repeats until the region has a single node inside it. Blind flooding does not guarantee delivery to all sensors inside the region, because of possible partition inside the region (but connectivity outside it), and can be replaced by a more intelligent scheme (see Chapter 11 in this book). The GEAR algorithm selects a forwarding neighbor (among those that are closer to the destination), which minimizes a linear combination of their distance to the destination and the energy they already spent. This is almost equivalent to the cost-aware localized scheme by Stojmenović and Lin [7], originally proposed in 1998 (described in detail in Chapter 12). Yu, Govindan, and Estrin [22] also claim that GEAR can avoid holes by applying a learning A^* algorithm-based approach, without presenting details. To avoid holes, one can use, for example, the depth-first search (DFS) approach [23]. This approach requires memorizing past traffic at nodes. Unfortunately, it does not guarantee delivery to all sensors in the geocasting region, because of possible partitioning inside the geocasting region.

We observe that, to guarantee delivery to all sensors in a geocasting region, and also to avoid memorization, GFG [12] can be applied first, while some optimizations (described in ref. [22]) can follow later on recursively. Note also that ref. [24] further elaborated on the use of GEAR for various forms of data dissemination, without giving its description.

Three existing geocasting algorithms that guarantee delivery to all nodes inside the geocasting region (subject to the ideal medium-access layer and connectivity of these nodes to the source) are described in refs. [21] and [25]. One algorithm [21,25] is based on multicasting to entrance zones, and flooding from entrance zones to nodes inside the geocasting region. Bose, Morin, Stojmenović, and Urrutia [12] observed that a geocasting algorithm will guarantee delivery if all faces of a planar graph that are inside or intersect the geocasting region are traversed. The geocasting algorithm [12] is based on a DFS of the face tree, constructed from a node inside the geocasting region.

Seada and Helmy [26] observed that it is sufficient to traverse only faces that intersect the boundary of a given geocasting region, and proposed the following algorithm. Source node first uses *GFG* algorithm [12] to forward the packet toward the region. Each node that is inside region will retransmit the packet when receiving it for the first time ("regional flooding"). If the node also has neighbors outside geocast region, it will instruct them to perform face traversals using "right-hand" rule (see chapter on routing in this book for details). The first node inside the region to receive the face traversal packet floods it inside the region or ignores it if that packet was already received and flooded before [26]. Figure 8 in ref. [26] shows that face traversal was not assumed in cases when an outer node brings the packet inside the region (the receiving node then only floods the region, but does not instruct the sender node outside the region to then also perform face traversal). Therefore, as elaborated in ref. [21], the algorithm [26] does not guarantee delivery, despite the claim. A protocol that does guarantee delivery (with proof of it) was described in refs. [21] and [26].

Algorithm Geocast_traversal_intersecting_faces

- The source node *S* sends the message toward the geocasting region, using the GFG algorithm [12];
- Each node inside the region retransmits the message when receiving it for the first time, and ignores it when receiving it again;
- Each internal border node (node inside a region having neighbor(s) on planar graph outside the region) will instruct (together with retransmission) all its perimeter neighbors outside the region to perform right-hand-based face traversals;
- Each external border node (node outside the region having neighbor(s) on the planar graph inside the region) will initiate right-hand-based face traversal(s) with respect to all edges leading to internal-perimeter neighbors, after receiving the first copy of the message, and will ignore further received copies unless a packet is received from an external neighbor following a different "external" face (in which case it forwards it along that face, as requested). Each traversal is performed until another node that is inside the region is found.

13.3.4 Multicasting in Wireless Sensor Networks

A monitoring task can also be disseminated to all the sensors located in several geo-casting regions. Assuming that these regions are relatively small, a position-based multicasting protocol [27] can be applied. Mauve, Fusler, Widmer, and Lang [27] proposed two multicasting schemes, with some optimizations. In the *optimal-paths* method, each node receiving a multicasting message for a group of nodes will forward it to each neighbor that is closest to one of the group members. More precisely, each group member is assigned to the neighbor that is closest to it (provided that neighbor is closer to it than the current node). In the *aggregate-paths* method, for each neighbor A, the number of destinations for which A is the closest node is determined. Then a covering algorithm is applied. Basically, a neighbor is chosen that covers the maximum number of destinations, these destinations (and other nodes for which a selected node makes some progress) are eliminated from the list, then another neighbor is chosen that covers the maximal number of remaining destinations, and so on. The forwarding list of multicast group is similarly changed as in the previous algorithm [27]. In both schemes, if no neighbor is closer to one or more destinations, then the recovery mode in the GFG algorithm [12] is applied. The virtual destination used for the recovery mode is calculated as the position representing the average of the positions of the affected destination nodes. When a node receives a multicast packet in recovery mode, it checks for each destination, if it is closer to that destination than the node where the packet entered recovery mode. For all destinations where this is the case, greedy multicast forwarding can be resumed, as described in the corresponding scheme. For all other destinations, recovery mode is continued, with an updated average of positions of affected nodes (those not recovered yet). Both optimal- and aggregate-path methods can be modified by considering metrics different from hop count, such as power, cost, or delay. Greedy routing can be replaced by power and/or cost-aware routing (see Chapter 12), and forwarding neighbors will be judged based on the metric in question, combined with their coverage ability, for their selection.

13.4 DATA GATHERING BASED ON MEMORIZED BROADCASTING TREES

Sensors (or their CHs, when they are clustered), may report back to the sink using a tree structure that is constructed together with the task allocation from the sink. This tree is referred to here as the *broadcasting tree*, since it is usually set during the *broadcasting* operation. It sends task allocation and other (e.g., sink position) information from the sink to all the sensors in the network. The use of broadcasting tree also implies the need to memorize some information made available during broadcasting process, which is then used in the reporting phase. The broadcasting tree consists of links along which the sensors learned about the position of the sink(s). Therefore, the sink monitoring implicitly informs the sensors that link to be used for replying. Sensors then create links for reporting along the *reverse broadcast tree*. This term is used, since reporting is normally applied in the direction that is opposite to the direction of request propagation from the sink. Most of the literature considers this type of sensor training for reporting. Note that the broadcasting operation is applied when all active sensors are alerted to report possible events. The broadcasting tree can also be set during a *geocasting* operation, where the monitoring station requests reports only from sensors located inside a geographic region. This can be further generalized to *multigeocasting* operations, which disseminates a request to all sensors located inside several geocasting regions. Because of the volatility of individual sensors (failures or changes between sleep and active periods), the use of broadcasting trees for reporting has certain risks, since a particular node or link may not be available although demanded by the memorized response path.

Carle and Simplot-Ryl [28] proposed the following framework for wireless sensor network operation. Sensor-area monitoring consists of three phases or subproblems. The first one is to select the sensors that are needed for *connected-area coverage*, placing other sensors in sleep mode. The second phase is to construct a *broadcasting tree* from the sink to all *active* sensors. They consider two types of trees, minimum energy broadcasting or dominating set based. The last phase is to *report* events using the reverse broadcast tree.

13.4.1 Directed Diffusion

Directed diffusion [14] is a often cited scheme for data gathering by using a data-centric routing scheme. The data sink identifies a set of attributes and propagates an *interest* message throughout the network. The interest is flooded throughout the network (apparently blind flooding was used). Each receiving node records the interests and establishes the so called *gradient*, the state indicating the next hop direction for other nodes to report data of interest. When an interest arrives at a data producer, data are being forwarded to the sink along established gradients. Note that the algorithm is similar to the well-known ad hoc on-demand distance vector (*AODV*) routing scheme [29], considered as a possible routing standard. Flooding the interest with attribute-based addressing corresponds to the route discovery with IP or ID addressing. Instead of comparing their *address* with the destination address as in AODV, sensors in directed diffusion compare the *interest* from the packet with the data they measure and their location if the interest is location specific. Therefore various AODV optimizations that exist in the literature are applicable in the context of directed diffusion. Although it is an on-demand localized scheme that does not require prior "hello" messages, the scalability is questionable. If the interest is location specific, then obviously it is much more efficient to route the request (using, e.g., a protocol that guarantees delivery [12]) toward the location of interest instead of flooding it to the whole network. The protocol described in ref. [14] uses path memorization for reporting the sensor measurements back toward sink.

13.4.2 Reporting via Neighbor with Smaller Hop Count

Ding, Sivalingam, Kashyapa, and Chuan [15] considered the problem of finding a route from a sensor to the single sink in a wireless sensor network. Following a reactive route discovery strategy, the sink floods the network and sets the routes. The difference is that each sensor does not memorize the whole route, or a single pointer to the previous sensor on the route, but instead memorizes its hop-count distance to sink. When a packet is sent toward the sink, any neighbor at one less hop distance can forward it, instead of reporting back to the first node that sent the task assignment packet to it. For instance, a report can be sent to the neighbor with the highest energy and smaller hop count, or any neighbor that sent the packet with a smaller hop count from the sink [15]. The node can memorize few such alternatives during the setup phase and try them one by one. Alternatively, a neighbor at one less hop distance can simply retransmit, and the node can block further retransmissions by a separate blocking packet.

Fujiwara, Iida, and Watanabe [30] proposed a mechanism that allows nodes to maintain their routes to the base station via multihopping, if needed. If a direct link between any node and its base station is broken, the node starts monitoring communications in its neighborhood to find a node that is still connected to the base station, either directly or by multihopping. When the node finds a connected neighbor, which should be one hop nearer, it marks it as its router and sends to it the packets that must be sent to the access point. This allows nodes to always be able to connect to their base station. The authors consider only the case of a single access point.

Zou, Nikolaidis, and Harms [31] described several localized schemes for constructing reporting trees for sensors to the sink. The tree construction starts at a sink node, which floods a message in the network. Upon receipt of several copies of the message, a given node may decide which of the nodes that sent the message is best to use for reporting data back to the sink. Authors described several possible localized criteria for selecting the best neighbor: minimum distance to the next hop, maximum distance to the next hop, random next hop, maximum degree of the next hop neighbor, and the maximum size of the 2-hop neighbor set of the next hop neighbors.

Wireless sensor networks with multiple sinks are special cases of hybrid wireless networks considered in ref. [32]. The hop distances to the closest sink, and therefore the routes, can be similarly determined as in the case of the single sink [15,30], as described in ref. [32]. Each access point sends messages toward sensors to establish reporting links. Each sensor may receive such messages from multiple sinks, but will forward them only from the closest sink. This will reduce the amount of traffic, without affecting the choice of the closest sink. If all sinks start the process synchronously, at the same time, then only one message is forwarded by each sensor. Otherwise, sensors will forward a new message only if it comes from a sink that is closer than the previously closest sink from which such a message was already received. This algorithm constructs the reporting trees from each sink to all sensors for which that sink is the closest one.

13.4.3 Reporting via Alternate Paths in a Broadcast Tree

Most of the current methods for reporting sensor data first construct a broadcast tree from the source, then use this tree for reporting in reverse order. Nodes in the tree

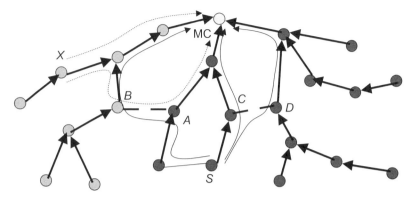

Figure 13.4   Reporting from source *S* to *MC* via three alternate paths.

may first be selected so that they make minimal connected-area coverage. The problem is that if a node in the tree fails or is malfunctioning, then the reports can be lost or compromised. To enhance the security of reporting, secondary and ternary paths for reporting are proposed in refs. [33] and [34]. A primary broadcast tree is constructed from the monitoring center (MC). All nodes are labeled based on the first hop out of the MC. Thus the network is effectively partitioned based on which neighbor of the MC delivers the report to it. Now consider an edge *AB* in the network, where *A* and *B* have two different labels. Such edges can serve as bridges for the second copy of messages. Suppose that node *S* wants to send a report to the MC (see Fig. 13.4). Node *S* can send one copy of the report directly to the MC using the reverse broadcast tree. The second copy can be sent instead to a node *A* with the same label as *S*, which has a neighbor *B* with a different label. Then the report can be sent from *S* to *A*, and from *A* to *B*, and from *B* using the reverse broadcast tree to the MC. The two paths should be disjoint, if possible. Similarly, *S* may find even a ternary path through another bridge, on the other side of the tree, for even higher security, or to provide the majority consensus to the MC, which can choose two out of three reports. The construction of the secondary and ternary paths may proceed as follows. Both end points *A* and *B* of each bridge edge *AB* can initiate the construction of their private trees within the neighborhood with the same label, in the same way the MC constructs its own broadcast tree. These trees are labeled with the label of the other end point. That is, the private tree of *A* within the neighborhood of *A* with the same primary label as *A* is labeled with the label of *B*, and vice versa. Each node *Q* receiving such a message for the first time will join the corresponding broadcast tree, and forward the message so that the tree is enlarged. If the message is already received from another bridge node, with the same "bridge" label, *Q* ignores it to avoid too much traffic and constructing additional trees that may not significantly enhance security. If a message is received with a different secondary label, *Q* accepts the participation in yet another tree and enlarges it by rebroadcasting the message. Routing the report then follows the constructed trees. The source node S sends the primary report on the primary tree, and uses the secondary and perhaps the ternary tree to which it belongs to send additional copies of the same report. Security can be jeopardized with the new method if several nodes close to the sink, one or two hops away, located on several branches of the tree, are all compromised. To prevent that, the sink must be particularly responsible for the authentication of all nodes that are one hop and perhaps two or three hops away from it in the broadcast tree.

13.5 PERIODIC REPORTS BY ALL SENSORS

13.5.1 PEGASIS: Chain of Reporting Nodes

Lindsey, Raghavendra, and Sivalingam [35] proposed a framework for energy-efficient data gathering algorithms in wireless sensor networks. Their power-efficient gathering in sensor information system (PEGASIS) protocol [35] first organizes sensors into a chain, by a centralized algorithm (e.g., the sink can decide about ordering of reports). Thus, sensors are initialized as $c_0, c_1, \ldots, c_{n-1}$. Data gathering is performed in rounds. In round k, first find $i = k \bmod n$. Each round consists of n iterations. In each iteration, only one sensor is sending a message, containing data gathered by that sensor. Iterations are performed as follows: c_0 sends to c_1, c_1 to $c_2, \ldots, c_{i-1}$ to c_i. Then c_{n-1} sends to c_{n-2}, c_{n-2} to $c_{n-3}, \ldots, c_{i+1}$ to c_i. Finally, $c_{i\cdot}$ sends the gathered data to the MC. The distance to the MC is assumed to be larger than the distances between the sensors. Chains can be difficult to construct in multi-hop sensor networks. For single-hop networks, initialization algorithm needs to run, or the MC needs to assign reporting indices to individual sensors. Once constructed, when sensors change activity status (from active to passive) or stop functioning, the order scheme needs to run again, or a maintenance procedure is needed. The scheme is also not sensitive to the energy levels of the sensors, as different sensors consume different amount of energy, depending mainly on their distances to the MC.

13.5.2 LMST- and Geocasting-Based Data Gathering

Several localized solutions are proposed in refs. [33] and [34]. One is a localized algorithm that first constructs LMST (or other sparse structure such as the relative neighborhood graph (*RNG*)). Instead of creating a chain, a token is circulated in the network. The node currently having the token will send it to one of its LMST neighbors. This can be done in different ways. Nodes can forward with equal probability of sending to one of its neighbors (not returning to the neighbor it came from). Since the average degree of LMST is about 2.04 [18], there is on average one such neighbor to forward the token. The forwarding probability may depend on node densities. Neighbors with more LMST neighbors should have a smaller probability of getting the token (since they may get tokens from more neighbors in the process). Next, neighbors with more energy left may have a higher probability of getting the token. Finally, in the case of monitoring an event that can be geographically located, sensors nearby need to preserve more energy, and thus they may decide to postpone reporting to the MC. Thus, instead of reporting every nth time to the MC, the frequency may also be decided probabilistically, depending on the energy level of the node. This scheme does not offer an immediate alternative for the data aggregation. LMST may be converted to MST [36], or a spanning tree may be constructed by broadcasting a message from the node holding the token. The constructed tree may be used for data gathering from other sensors, before the node holding the token sends the report to the MC. MST can also be used for data aggregation, since reports can be sent toward the node holding the token, and aggregated on the way.

Another solution proposed in refs. [33] and [34] is to apply the geocasting algorithm [12,37], which follows a single path from the source while visiting all nodes in the region. The algorithm guarantees to see all nodes, and on average it does so twice during a single geocasting process, which can be repeated periodically. If the sink is fixed, preprocessing can de done to decide the entry edges and reduce communication time, as described in refs. [21] and [25]. The advantage over the solution just described is to guarantee the participation of each node on a fairly regular basis.

13.6   DATA GATHERING WITH DATA AGGREGATION

Data collection, known as data gathering or data dissemination, can be considered as a reverse multicasting task, with all nodes from the multicasting group reporting their data to the MC. There are several cases that need to be distinguished in these tasks. The data collected may or may not be aggregated at intermediate sensor nodes. Data aggregation is applied when sensor measurements are correlated, which is reasonable to assume when they measure the same event in nearby positions. In this case, it is not necessary that each individual report (which may not be sufficiently reliable) reaches the monitoring center. Intermediate nodes may combine (aggregate) data received from several neighbors, and possibly one measured by itself, into a single report, and forward it toward the MC. In the case of data aggregation, a distinction can be based on whether or not forwarding sensors have their own data measured. Obviously, all sensors that want to report data need to be included in the reporting tree. If data aggregation is not applied, then clearly each reporting sensor may apply one of the routing algorithms with guaranteed delivery (e.g., ref. [12]) for sending its report to the MC. If data aggregation is applied, a distinction can be made between protocols applied within the geocasting region, and outside of it. Outside the geocasting region, not all sensors need to participate in reporting. The problem appears, then, to be the inverse of the multicasting problem, that is, the multicast tree that is set while sending the request to the sensors can be used to report data back from the sensors (assuming an on-demand query was issued to all sensors).

13.6.1   Power-Efficient Data-Gathering and Aggregation Protocol

Tan and Korpeoglu [38] proposed a power-efficient data-gathering and aggregation protocol (PEDAP) that assumes that locations of all nodes are known by the sink a priori. The sink constructs a MST which is then used for data gathering and aggregation. In the power-aware version of the same protocol, the MST is constructed with weights on each edge calculated as the product of power consumption on the edge and the reluctance of a neighbor to receive the packet (reluctance is the inverse of the remaining energy at the node).

13.6.2   LMST-Based Data-Aggregation

Inside the geocast region, two cases for data aggregation may occur: all nodes in the region sense the event, or some nodes are there only to forward traffic. If all sensors within the region are reporting, the optimal tree to use is apparently the MST (as observed in ref. [39]). The existing distributed algorithm for constructing a MST require $O(n \log n)$ messages, with a high constant involved. The algorithm presented in ref. [36] is based on breaking all cycles created in the LMST. Each of the LMST's links is broken by identifying the longest edge in it and removing it [36]. The removal of one such edge may lead to a longer cycle, which is broken in the next iteration. The LMST can also be broken differently, in the considered context. The new solution [33,34] is to also start with the *LMST* structure. The (MC) will create a tree out of the LMST by forwarding its request, within the geocasting region, only along the LMST's edges. When a node receives a message from the MC, it will forward it on only on its remaining LMST edges, if any. However, if a neighbour already received a message from the MC, that link is not used; an LMST cycle is broken that way. The obtained tree is not necessarily MST, but its approximation, which is expected to be very close to it in performance. Figure 13.5 illustrates the construction of a LMST-Based data aggregation tree. The edges of the LMST are drawn with thick edges, and all but one (drawn with a dashed line) is included in the broadcasting tree. Edge *WF* is the only one that is in the LMST, but not in the MST.

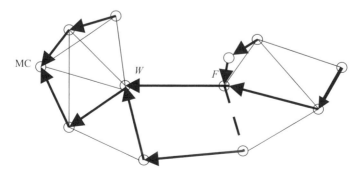

Figure 13.5   LMST-based energy-efficient data-aggregation tree.

13.7 MOBILE SINKS OR SENSORS

13.7.1 Two-Tier Data Dissemination

Data dissemination for large-scale wireless sensor networks was considered in ref. [40], for the case when multiple mobile sources send information constantly to multiple mobile destinations. The authors [40] proposed a two-tier data-dissemination approach, source to destination and destination to source, with grid subdivision of the area and greedy forwarding. The initial routes are set by flooding (if the destination locations are not known) or greedy forwarding (otherwise). Sources create grids that contain sensors that are closest to the grid intersections. These sensors act as a backbone for routing. When sources or sinks move to a new cell, they flood new cells to find a new backbone sensor, so that the existing paths can be extended. This article [40] does not give any concrete scheme for path optimization to avoid using long paths.

The grid backbone assumes large sensor density, with all sensors being active. However, if a sensor-area coverage scheme is applied to select active sensors, the grid backbone can be replaced by a more efficient and more natural backbone of covering sensors (discussed in a separate chapter in this book). Grid division is unnecessary overhead, and greedy forwarding may fail. GFG [12] can replace greedy forwarding. If either sources or sinks are fixed, then mobile components will initiate route maintenance toward stationary components. Let S be a fixed source, and D be the initial position of a destination. Path extension from the new destination position D' toward the old position D can be applied up to a certain traveled distance, then D' can initiate routing toward the source until it reaches it. In fact, the maintenance can stop when a node that already knows the path to S is found. Alternatively, the new path search may terminate after reaching a node A' that is a neighbor of a node A on the original path SAD. Instead of $SADD'$ the new path, is then $SAA'D'$ [33,34].

Further optimizations can be achieved my merging some reporting streams toward same sink or the same source. Suppose that messages from sensors A and B, sent toward the same sink or source, are heard by sensor C, their common neighbor. Sensor C can then offer to merge these streams, reporting its position. Nodes A and B evaluate the gain obtained by each of such candidates C, in the case of several such offers, and select the best one [33,34]. Note that A and B may or may not be neighbors themselves, which results in two different protocols. The problem exists when both sources and sinks can be mobile, since then the updates do not have precise destinations. The procedure then is an alternate the application of the described procedure from both ends until the packets meet somewhere in the network.

13.7.2 Mobile Collectors

Tirta, Li, Lu, and Bagchi [41] proposed using a mobile collector, such as an airplane or a vehicle, to collect sensor data from remote fields. The sensor network is clustered and only CHs report data. They present three different schedules for the collector.

In the *round-robin* scheme, each CH is visited in a predefined order, regularly for same amount of time. In the *rate-based* scheme, the frequency of visits depends on the amount of data reported. In the *min-movement* scheme, CHs are visited in specific order, but the time spent with each of them depends on the amount of data to report (more precisely, the collector stays with each CH until all data are collected).

13.7.3 Mobile Sensors

Taherian and O'Keefe [42] proposed an energy-aware event-dissemination protocol for mobile sensor networks. In this protocol, each sink proactively constructs a redundant tree in the network. This redundant tree is combined with probabilistic forwarding. The main idea is to limit the number of parent and sibling nodes as the redundant tree grows. The proposed redundant tree is not guaranteed to be connected, that is, be useful for event dissemination. It is also not guaranteed to provide full coverage of the sensor network area. Routing follows tree pointers if they exist; otherwise, it applies a probabilistic forwarding scheme (which is similar to existing beaconless routing schemes reviewed in Chapter 12).

13.8 TRACKING OBJECTS IN SENSOR NETWORKS

13.8.1 Tracking Objects Without Data Aggregation

We now discuss problems associated with tracking an object, possibly moving. This section considers problem aspects when data aggregation is not involved. Each report therefore needs to be sent directly to the base station. The problem is to send a sufficient number of reports so that the position of the object can be reliably determined by the base station, while minimizing the number of sensors that send the report. We assume that each sensor knows its own location, and location of its neighboring sensors. Although in reality the reliability of sensor observation depends on the distance to the object, we consider a simplified model, assuming that all sensors located within the sensing radius from the object can reliably detect it. After detecting the object, sensors can measure either the distance to it, or the direction (angle) toward it. The case of distance is similar to the position determination problem, discussed in Chapter 9. We therefore now study the case of measuring an angle toward an object, without knowing the distance to it. All reports will be assumed accurate, although in reality some reports may be false, and security and report reliability issues need to be studied as well. Estrin, Govindan, Heidemann, and Kumar [2] were the first to investigate this problem. In their solution, the sensor network is clustered first. CHs for which all neighboring CHs lie on the same side of a line drawn toward the object elect themselves to participate in object location. The goal is to elect sensors that form the longest baseline for triangulation. There are several problems with this approach. In general, there are two such nodes, which are tangent nodes from the object to the convex hull of CHs. If one of them for any reason does not see the object (because of obstacles),

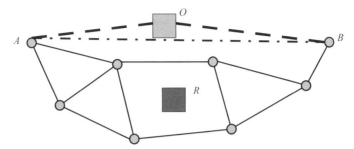

Figure 13.6 Longest baseline may not accurately determine the location of the object O; the object R is inside the convex hull of the observing sensors.

the object cannot be accurately determined. Next, the longest baseline is not always the best choice for accuracy (see Fig. 13.6 where the object R in drawn as a square). Small or large angles in the triangle containing the baseline may cause large errors in object positioning. Short and long baselines, with respect to other network measurements, also cause large errors. What is a better criterion? It appears that it is better to maximize the minimal angle in the triangle used to decide object location [43].

The next problem occurs when the object is inside the convex hull of participating CHs, as object R in Figure 13.6. In this case, no sensor reports the direction of the object in ref. [2]. Therefore, there is a need for designing new protocols. In fact, the problem appears to be quite challenging. Some proposals were given in ref. [43] as follows. Each node can decide whether or not it is locally a northernmost (N), easternmost (E), westernmost (W), or southernmost (S) node, by verifying whether or not any sensor exists on the corresponding horizontal or vertical line passing through the sensor. Similarly, NE, NW, SE and SW sensors can be found, by considering directions $\pi/4$ and $3\pi/4$. Every sensor that can see the object could report it to its locally northernmost node, by sending/forwarding a message to its northernmost neighbor, using the greedy routing or routing scheme [12]. The exact protocol details and analysis are still under investigation [43]. The best approach appears to be to use the multilateration technique, which is applied, for instance, in ref. [4] for position determination based on distances. It is also interesting to consider the different ways of selecting which sensors will participate. In addition, clustering organization could be replaced by another backbone, for example, connected dominating sets. Any type of backbone could be applied on an area covering the set of sensors.

13.8.2 Tree Reconfiguration for Tracking Mobile Objects

Zhang and Cao [44] discussed how to monitor an object by sensors located inside a monitoring region, such as a circle. These sensors are organized into a tree, with one of the sensors serving as the root. The root collects all reports, aggregates them, and routes them to one or more sinks (base stations). In this method, the root keeps monitoring its distance to the target. When the distance becomes larger than a certain threshold d, it will be replaced by the node that is closest to the center of the current monitoring region.

13.8.3 Mobicast Protocol for Tracking Mobile Objects

One particular application of geocasting is tracking mobile objects. Mobile objects create geocast regions that are time dependent, and data collection is performed by the sensors in the vicinity of a moving object. The sink may collect reports from the sensors in the vicinity of the object, and may send periodic signals to the sensors adjusting the geocasting region, following the trajectory along with the object advances. In mobicast application [45], however, the sensors themselves adjust the geocasting region.

   Huang, Lu, and Roman [45,46] proposed a mobicast protocol where the nodes that belong to the forward, time-dependent region, or belong or are about to enter the geocast region, retransmit the message. Their algorithm presented in ref. [45] is an improved version of the one in ref. [46]. In their problem formulation, the MC is not used to decide and inform about the geocast region. Instead, the sensors themselves cooperate, follow the movement of an object, and inform the sensors, which are approached by the object, to start monitoring. This is achieved by considering the planar graph of covering sensors, and forwarding messages to the faces of the planar graph in the direction of object movement, with proper timing corresponding to the arrival time of the object at the considered face. This is illustrated in Figure 13.7. Suppose that the rectangle object, shown with dashed lines, moved from the far left and is continuing toward the far right. All faces that the object intersected are traversed by messages from sensors in these faces. They are marked in Figure 13.7 by clockwise arrows along the face edges (following the

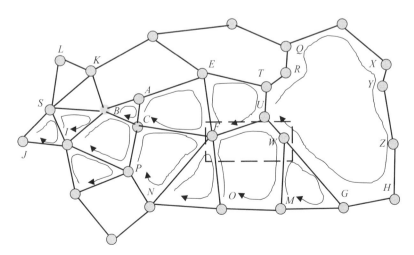

Figure 13.7   Reliable mobicast.

left-hand rule for face traversal). Node F, after receiving the signal from neighbors C, E, or N, starts monitoring. It then estimates when the rectangle will reach two other faces containing F, and will send a signal at the appropriate time, so that nodes in these faces are alerted in time. In turn, node W decides when the rectangle will reach that face, and sends the alert signal at the best possible time, so that all the nodes on the long face are alerted just before the arrival of the rectangle. In that way, alert messages are able to cross over obstacles, or areas without sensors. For instance, node Z will be alerted although there is no direct neighbor of it that monitored the same rectangle before the rectangle reached Z. In the algorithmic details given by the authors [45], each node needs to learn all nodes that are on the same face as the given node, which may not always be available from local knowledge. However, we believe that this may be avoided by slightly changing the description. Proper timing for reaching node Z is also an issue, since the face size may be significantly larger than the average one, and the message may even follow the outer boundary of the network. Face routing may not be necessary for alerting sensors ahead of the arriving object. Greedy routing may be used, or, more precisely, GFG [12]; it is its combination with face routing that guarantees delivery. If the area is convex, completely covered by sensors, and the communication radius is over twice as large as the sensing area, ref. [47] showed that greedy routing guarantees delivery. The problem can even be considered as a variant of geocasting, with sensors themselves issuing geocast messages toward the future location of the object, alerting sensors in a certain area. If several such messages arrive at a given node, it will forward only one of them, to control the overhead.

13.9   RATE-BASED DATA PROPAGATION IN SENSOR NETWORKS

In sensor networks, data sources sample data and propagate them to potential consumers. A consumer may subscribe to a data item at a certain rate, and the subscription rate may vary for different consumers. The problem is to construct a data propagation tree to efficiently disseminate data from a source node at the required rate to each of its consumers.

Singh, Pujar, and Das [48] proposed a breadth-first search-based protocol for a one-to-one network model, assuming that the connections between sources and consumers are with a wired network. We shall consider a sensor network, that is, a wireless network, for connection between sources and consumers. In this scenario, communication is one-to-all. The main difference from Singh et al.'s model [48] is that the rates are marked at each node, instead of at each edge.

The proposed solution [49] is a generalized multicasting position-based protocol (following a solution proposed in ref. [27]), which applies the *multipoint relay* (*MPR*) strategy (see Chapter 11) in determining forwarding neighbors and their rates.

In a preprocessing step, each consumer reports to the source (assume that there is only one source, for simplicity). Thus the source is informed about customers and their preferred rates. The source then creates a list of customers and their rates and makes forwarding decisions, that is, which neighbors will retransmit and at what rates. Other nodes, receiving decisions to retransmit, will follow a very similar protocol to determine their forwarding nodes.

Let S be the current node, and let $C_1, C_2, \ldots, C_n$ be consumers that S needs to serve. Let $R_1, R_2, \ldots, R_n$ be their preferred rates. Let $A_1, A_2, \ldots, A_m$ be neighbors of S. For each consumer, only neighbors that are closer to it than S can be considered for serving and covering. Each possible covering of consumer C_i by neighbor $A_j (|A_j C_i| < |SC_i|)$ is associated with the cost per progress, defined as $R_i/(|SC_i| - |A_j C_i|)$. More precisely, since A_i, once selected, will cover all consumers it is closer, the cost of selecting A_i is associated with the progress it makes toward all consumers it could serve (it is closer than S to them). To avoid notational difficulties, let $B_1, B_2, \ldots, B_k$ be those consumers among $C_1, C_2, \ldots, C_n$ that are considered for serving by A_i (they are all closer to A_i then to S, but not all such nodes need to be selected). Let $P_1, P_2, \ldots, P_k$ be their corresponding preferred rates. Then the cost per progress for selecting A_i is $\max(P_1, P_2, \ldots, P_k)/(|SB_1| - |A_i B_1| + |SB_2| - |A_i B_2| + \cdots + |SB_k| - |A_i B_k|)$. An alternative measure is to consider each progress individually: $P_1/(|SB_1| - |A_i B_1|) + P_2/(|SB_2| - |A_i B_2|) + \cdots + P_k/(|SB_k| - |A_i B_k|)$. However, this is not likely to be a better criterion, since one small progress can easily undermine a number of good progresses made. The selection of covering neighbors and their rates then can proceed in the following manner:

- If there is any consumer served by a single neighbor, then that neighbor is selected; moreover, the selected neighbor will also cover other consumers that are closer to it than to S.
- Select one of the remaining consumers with maximal preferred rate, and consider the cost of each neighbor serving it, and the additional benefit such a choice makes overall. Select the node that then minimizes its own $\max(P_1, P_2, \ldots, P_k)/(|SB_1| - |A_i B_1| + |SB_2| - |A_i B_2| + \cdots + |SB_k| - |A_i B_k|)$.
- Repeat previous step until all consumers are covered.

13.10 ANONYMITY ISSUES IN WIRELESS SENSOR NETWORKS

In many applications, safeguarding *output data assets*, that is, data produced by the wireless sensor network and consumed by the end user (application), against loss or corruption is a major security concern. In these application domains, a wireless sensor network is typically deployed in a hostile target environment for a relatively long period of time. The network self-organizes and works to generate output data that is of import to the application. For example, a wireless sensor network may be deployed across a vast expanse of enemy territory ahead of a planned attack; the network system monitors the environment and produces reconnaissance data that are *absolutely essential* to a mission planning application. Periodically, during the network lifetime, a mobile gateway, mounted on a person, land or airborne vehicle, or a satellite, collects the output data assets from the network system, to maintain an up-to-date state. This means the network system must store the output data assets from the time it is produced until it is collected. Therefore, securing the output data assets in the network is an important problem in this class of applications.

We view an attack on the output data assets in the sensor network as a type of denial of service attacks. This view is based on the abstraction that output data are stored in a logical *repository*, and that *access* to this output data repository constitutes, in effect, a "service" provided by the network system to the application; corruption or loss of output data denies the application access to that service.

13.10.1   What Is Anonymity?

Anonymity protects the identity of the sender or receiver and guarantees that both parties involved in a communication remain anonymous to each other. Recent years have seen a flurry of activity, and many anonymous communication systems have been developed for the Internet. Most of the work on anonymity is concerned with *sender* anonymity, *receiver* anonymity, and *mutual* anonymity. Quite recently, *traffic* anonymity has also received well-deserved attention in the literature. Recently, the problem of securing ad hoc networks has received a great deal of well-deserved attention. To the best of our knowledge, the anonymity problem has not been adequately addressed in wireless sensor networks [50].

The threat model assumed by Wadaa et al. [50] comes from a data-centric view of wireless sensor networks. The model is predicated on the assumption that *the end-goal of anonymity attacks on the wireless sensor network is to identify and eliminate the minimum number of sensors to inflict maximum loss of data assets*; eliminating a sensor means disabling it so that it is permanently nonoperational. For any operation cycle, if a sink suffers a permanent failure before transferring the contents of its data repository to the gateway, then a portion of the data assets corresponding to the cycle is irrevocably lost. The goal of the adversary is to eliminate all sinks. This can be accomplished in two ways.

13.10.2   The Anonymity Threat Model

13.10.2.1   Brute-Force (Sink Nodes Not Identified)   This can take the form of randomly eliminating nodes in the network on the assumption that, statistically, some sinks will be eliminated in the process. Coarse sink granularity and sink redundancy mitigate the risk of loss of data assets as a result of this type of attack. A straightforward special case is the massive elimination of all sensors in the network.

13.10.2.2   Smart (Sink Nodes Identified)   The adversary analyzes network traffic to deduce information about topology, traffic flow patterns, and other system attributes. The goal is to discover sink nodes and to eliminate them. In this chapter we assume the adversary engages in smart elimination attacks. The specifics of the architecture and the implementation of the adversary system are assumed to be unknown.

13.10.3 Sender and Path Anonymity

Sender anonymity is most commonly achieved by transmitting a message to its destination through one or more intermediate nodes in order to hide the true identity of the sender. The message is thus effectively *rerouted* along what is called a *rerouting path*. It is important to study rerouting-based anonymous communication systems in terms of their ability to protect sender anonymity. The selection of rerouting paths is critical for this kind of system. Olariu et al. [51] investigated how different path selection strategies affect the ability to protect sender anonymity. For a given anonymous communication system, they measure this ability by determining how much uncertainty this system can provide in order to hide the true identity of a sender. They call this measure the *anonymity degree*. In ref. [51] the authors assume a *passive* adversary model: the adversary can compromise *one or more* nodes in the system. An adversary agent at such a compromised node can gather information about messages that traverse the node. If the compromised node is involved in the message rerouting, it can discover and report the immediate predecessor and successor nodes for each message traversing the compromised node. We assume that the adversary collects all the information from its agents at the compromised nodes and attempts to derive the identity of the sender of a message.

Common sense indicates that the degree of anonymity increases with increasing number of intermediate nodes between the sender and the receiver. Olariu et al. [51] call this number of intermediate nodes the *path length* of the rerouting path. There is a point, however, beyond which increasing the path length actually *decreases* the degree of anonymity. The authors give a quantitative analysis of how path length affects the degree of anonymity. Rerouting schemes give rise either to paths with *fixed length* (where messages are forwarded to the receiver after traversing a fixed number of intermediate nodes) or *variable length* (for example, where every intermediate node randomly decides whether to forward the message to the receiver directly or to another intermediate node). The authors show that variable path-length strategies perform better than fixed path-length strategies in terms of degree of anonymity. However, when the expected path length is sufficiently long, the difference of anonymity degree is relatively small between different variable and fixed path-length strategies. As a result of this study, Olariu et al. [51] argue that several well-known anonymous communication systems are not using the best path selection strategies. They go on to propose an optimal method to select path lengths, by showing that the path selection problem can be cast as an optimization problem, whose solution yields an optimal path-length distribution that maximizes the degree of anonymity.

13.11 CONCLUSIONS

We considered some relevant aspects of the process of issuing requests and collecting data, with sensor ad hoc networks as the primary application of the presented methods. Protocol efficiency was the primary goal, with efficiency defined by some metrics or design characteristics (such as localized behavior of protocols).

Ad hoc and sensor networks have recently attracted exponentially increasing interest, including the creation of new conferences, new journals, and publication of a number of books. We envision that this trend will continue in the short term, and we envision that data-centric operation problems, discussed in this chapter, will continue to be intensively studied. We hope that the research efforts will lead toward real applications of ad hoc networks, especially sensor networks.

Sensor networks pose a number of research challenges. In addition to the problems discussed in this and other chapters in this book, we mention two more problem areas. One is about the design of sensor network protocols for heterogeneous sensor networks, the other is the investigation of various scenarios and protocols for wireless sensor and actor networks. Actor nodes are active nodes, with higher energy and computation capabilities, that are able to perform some actions and are able to move around.

EXERCISES

13.1. Describe a routing algorithm based on sector training [1] that will guarantee delivery in when there are empty sectors.

13.2. Derive a formula for the error involved when the position of an object is determined based on the angles measured from three given sensors [43]. Show that the error is minimized when the minimal angle in the triangle created by three measuring sensors is maximized.

13.3. Prove that the *geocast_traversal_intersecting_faces* algorithm guarantees delivery to all nodes inside the geocasting region, which is connected to the source [21,25].

13.4. Design an algorithm for finding optimal ring sizes (for extending network lifetime) for reporting to a CH with data aggregation for the case of n rings [49].

13.5. Design some protocols for moving the sink to a new position near the old position so that the overall energy consumption for reporting from last-hop sensors is reduced. Design another procedure for moving the sink to reduce the number of incoming reports that violate delay constraints [52].

13.6. Design an energy-efficient data-aggregation protocol for the following scenario. There are two types of sensors in a geocasting region, plus sensors outside the region. Some sensors inside the geocasting region are sensing and can perform data aggregation, while some other sensors are not sensing, but can only perform data aggregation, if needed. Sensors outside the geocasting region can only perform data aggregation, if needed, or can simply forward the traffic.

13.7. In a heterogeneous sensor network, there are two kinds of sensors. Some supersensors have high-energy resources and can communicate with each other and with the sink with much smaller delays than communication between regular sensors. Suppose that each node knows the distance to and label of the nearest supersensor, and that this information is communicated to neighboring sensors. Describe a broadcasting protocol in this environment [32].

13.8. Describe a localized protocol for general multigeocasting problems, where a monitoring task is to be disseminated from the sink to all the sensors located inside several geographic regions that are of arbitrary sizes, shapes, and locations, known to the sink.

ACKNOWLEDGMENT

This research is partially funded by NSERC Discovery grant.

REFERENCES

1. S. Olariu, A. Wadaa, L. Wilson, and M. Eltoweissy. Wireless sensor networks: Leveraging the virtual infrastructure. *IEEE Network*, pages 51–56, July/August 2004.

2. D. Estrin, R. Govindan, J. Heidemann, and S. Kumar. Next century challenges: Scalable coordination in sensor networks. In *Proceedings of the 5th Annual ACM/IEEE International Conference on Mobile Computing and Networking (MobiCom'99)*, pages 263–270, Seattle, Washington, August 1999.

3. W. R. Heinzelman, A. Chandrakasan, and H. Balakrishnan. Energy-efficient communication protocol for wireless microsensor networks. In *Proceedings of the 33rd Hawaii International Conference on System Sciences (HICSS)*, Volume 8, page 8020, Maui, Hawaii, January 2000.

4. R. Nagpal, H. Shrobe, and J. Bachrach. Organizing a global coordinate system from local information on an ad hoc sensor network. In *Proceedings of the 2nd International Workshop in Information Processing in Sensor Networks (IPSN '03)*, pages 333–348, Palo Alto, California, April 2003.

5. V. Mhatre and C. Rosenberg. Design guidelines for wireless sensor networks: Communication, clustering and aggregation. *Ad Hoc Networks*, **2**(1): 45–63, 2004.

6. S. Olariu and I. Stojmenovic. Design Guidelines for Clustering and Aggregation in Sensor Networks. In preparation.

7. I. Stojmenović and X. Lin. Power aware localized routing in wireless networks. *IEEE Transactions on Parallel and Distributed Systems*, **12**(11):1122–1133, November 2001.

8. J. Lian, K. Naik, and G. B. Agnew. Data capacity improvement of wireless sensor networks using non-uniform sensor distribution. *International Journal of Distributed Sensor Networks*, forthcoming.

9. M. Kumar, L. Schwiebert, and M. Brockmeyer. Efficient data aggregation middleware for wireless sensor networks. Paper presented at the 1st IEEE International Conference on Mobile Ad-hoc and Sensor Systems (MASS 2004), Fort Lauderdale, Florida, October 2004.

10. C. Schurgers and M. Srivastava. Energy efficient routing in wireless sensor networks. In *Proceedings of MILCOM 2001*, pages 357–361, Vienna, Virginia, October 2001.

11. I. Chatzigiannakis and S. Nikoletseas. A sleep-awake protocol for information propagation in smart dust networks. In *Proceedings of the 17th International Parallel and Distributed Processing Symposium (IPDPS 2003)*, page 225, Nice, France, April 2003.

12. P. Bose, P. Morin, I. Stojmenović, and J. Urrutia. Routing with guaranteed delivery in ad hoc wireless networks. In *Proceedings of the 3rd ACM Inernational Workshop on Discrete Algorithms and Methods for Mobile Computing and Communications (D/ALM'99)*, pages 48–55, Seattle, Washington, August 1999. See also in *Wireless Networks*, **7**(6):609–616, 2001.

13. S. Ratnasamy, D. Estrin, R. Govindan, B. Karp, S. Shenker, L. Yin, and F. Yu. Data-centric storage in Sensornets with GHT, a geographic hash table. *Mobile Networks and Applications (MONET)*, **8**:427–442, August 2003.

14. C. Intanagonwiawat, R. Govindan, and D. Estrin. Directed diffusion: A scalable and robust communication paradigm for sensor networks. In *Proceedings of the 6th Annual ACM/IEEE International Conference on Mobile Computing and Networking (MobiCom '00)*, pages 56–67, Boston, Massachusetts, August 2000. See also *IEEE/ACM Transactions on Networking*, **11**(1):2–16, February 2003.

15. J. Ding, K. M. Sivalingam, R. Kashyapa, and L. J. Chuan. A multi-layered architecture and protocols for large-scale wireless sensor networks. In *Proceedings of the IEEE Vehicular Technology Conference* (VCT2003), Orlando, Florida, October 2003.

16. I. Stojmenovic. Beaconless Area Based Broadcasting. In preparation.

17. J. Lipman, P. Boustead, and J. Chicharo. Reliable minimum spanning tree flooding in ad hoc networks. *IEEE Transactions on Vehicular Technology*, forthcoming.

18. N. Li, J. C. Hou, and L. Sha. Design and analysis of an MST-based topology control algorithm. In *Proceedings of IEEE INFOCOM*, Volume 3, pages 1702–1712, San Francisco, California, April 2003.

19. W. R. Heinzelman, J. Kulik, and H. Balakrishnan. Adaptive protocols for information dissemination in wireless sensor networks. In *Proceedings of the 5th Annual ACM/IEEE International Conference on Mobile Computing and Networking (MobiCom '99)*, pages 174–185, Seattle, Washington, August 1999.

20. M. Seddigh, J. Solano Gonzalez, and I. Stojmenovic. RNG and internal node based broadcasting algorithms for wireless one-to-one networks. *Mobile Computing and Communications Review*, **5**(2):37–44, 2001.

21. I. Stojmenovic. *Geocasting in Ad Hoc and Sensor Networks.* Technical Report TR-2004-02, Computer Science, SITE, University of Ottawa, March 2004. See also in *Theoretical and Algorithmic Aspects of Sensor, Ad Hoc Wireless and Peer-to-Peer Networks*, Jie Wu (ed.), CRC Press, pages 79–97, 2005.

22. Y. Yu, R. Govindan, and D. Estrin. *Geographic and Energy Aware Routing: A Recursive Data Dissemination Protocol for Wireless Sensor Networks.* Technical Report TR-01-0023, Computer Science, University of California, Los Angeles, August 2001.

23. I. Stojmenović, M. Russell, and B. Vukojevic. Depth first search and location based localized routing and QoS routing in wireless networks. *Computers and Informatics*, **21**(2):149–165, 2002.

24. J. Heidemann, F. Silva, and D. Estrin. Matching data dissemination algorithms to application requirements. In *Proceedings of the 1st International Conference on Embedded Networked Sensor System (SenSys)*, pages 218–229, Los Angeles, California, November 2003.

25. I. Stojmenovic. Geocasting with guaranteed delivery in sensor networks. *IEEE Wireless Communications Magazine*, **11**(6):29–37, December 2004.

26. K. Saeda and A. Helmy. Efficient geocasting with perfect delivery in wireless networks. In *Proceedings of the IEEE Wireless Communications and Networking Conference (WCNC 2004)*, Volume 5, pages 2555–2560, Atlanta, Georgia, March 2004.

27. M. Mauve, H. Fusler, J. Widmer, and T. Lang. *Position-Based Multicast Routing for Mobile Ad Hoc Networks*, Technical Report TR-03-004, Department of Computer Science, University of Mannheim, March 2003. See also Poster: Position-based multicast routing for mobile ad-hoc networks, In *Proceedings of the 4th ACM International Symposium on Mobile Ad Hoc Networking and Computing (MobiHoc '03)* (electronic edition), Annapolis, Maryland, June 2003.

28. J. Carle and D. Simplot-Ryl. Energy efficient area monitoring by sensor networks. *Computer (IEEE)*, **37**(2):40–47, February 2004.

29. C. Perkins and E. M. Royer. Ad hoc on-demand distance vector (AODV) routing. In *Proceedings of the IEEE Workshop on Mobile Computing Systems and Applications (WMCSA '99)*, pages 90–100, New Orleans, Louisiana, February 1999.

30. T. Fujiwara, N. Iida, and T. Watanabe. An ad hoc routing protocol in hybrid wireless networks for emergency communications. In *Proceedings of the 24th International Conference on Distributed Computing Systems Workshops—W6: WWAN (ICDCSW '04)*, pages 748–754, Tokyo, Japan, March 2004.

31. S. Zou, I. Nikolaidis, and J. J. Harms. Efficient data collection trees in sensor networks with redundancy removal. In *Proceedings of the 3rd International Conference on AD-HOC Networks and Wireless (ADHOC-NOW 2004)*, pages 252–265, Vancouver, British Columbia, July 2004.

32. F. Ingelrest, D. Simplot-Ryl, and I. Stojmenovic. Routing and broadcasting in hybrid ad hoc and sensor networks. In *Theoretical and Algorithmic Aspects of Sensor, Ad Hoc Wireless and Peer-to-Peer Networks*, Jie Wu (ed.), CRC Press, pages 415–426, 2005.

33. I. Stojmenović. *Geocasting, Data Gathering and Activity Scheduling in Ad Hoc and Sensor Networks*. Technical Report TR-2003-05, Computer Science, SITE, University of Ottawa, August 2003.

34. I. Stojmenović. Data Gathering and Activity Scheduling in Ad Hoc and Sensor Networks. Paper presented at the International Workshop on Theoretical Aspects of Wireless Ad Hoc, Sensor, and Peer-to-Peer Networks, Chicago, Illinois, June 2004.

35. S. Lindsey, C. Raghavendra, and K. Sivalingam. Data gathering algorithms in sensor networks using energy metrics. *IEEE Transactions on Parallel and Distributed Systems*, **13**(9):924–935, September 2002.

36. F. J. Ovalle-Martinez, I. Stojmenovic, F. Garcia-Nocetti, and J. Solano-Gonzalez. Finding minimum transmission radii and constructing minimal spanning trees in ad hoc and sensor networks. *Journal of Parallel and Distributed Computing*, **65**(2):132–141, February 2005.

37. P. Morin. Online Routing in Geometric Graphs. Ph.D. thesis, School of Computer Science, Carleton University, January 2001.

38. H. O. Tan and I. Korpeoglu. Power efficient data gathering and aggregation in wireless sensor networks. *ACM SIGMOD Record*, **32**(4):66–71, December 2003.

39. M. Khan, G. Pandurangan, and B. Bhargava. *Energy-Efficient Routing Schemes for Sensor Networks*, Technical Report CSD TR 03-013, Purdue University, July 2003.

40. F. Ye, H. Luo, J. Cheng, S. Lu, and L. Zhang. A two-tier data dissemination model for large-scale wireless sensor networks. In *Proceedings of the 8th International ACM Conference on Mobile Computing and Networking (MobiCom)*, pages 148–159, Atlanta, Georgia, September 2002.

41. Y. Tirta, Z. Li, Y. H. Lu, and S. Bagchi. Efficient collection of sensor data in remote fields using mobile collectors. In *Proceedings of the 13th International Conference on Computer Communications and Networks (ICCCN 2004)*, pages 515–520, Chicago, Illinois, October 2004.

42. S. Taherian and D. O'Keefe. Event dissemination in mobile wireless sensor networks. Paper presented at the 1st IEEE International Conference on Mobile Ad Hoc and Sensor Systems (MASS 2004), Fort Lauderdale, Florida, October 2004.

43. I. Stojmenovic. Object Location in Sensor Networks. In preparation.

44. W. Zhang and G. Cao. Optimizing tree reconfiguration to track mobile targets in sensor networks. *Mobile Computing and Communications Review*, 7(3):39–40, July 2003.

45. Q. Huang, C. Lu, and G. C. Roman. Reliable mobicast via face-aware routing. In *Proceedings of IEEE INFOCOM*, Hong Kong, China, March 2004.

46. Q. Huang, C. Lu, and G. C. Roman. *Mobicast: Just-in-Time Multicast for Sensor Networks under Spatiotemporal Constraints*, Technical Report TR WUCS-02-42, Washington University, St. Louis, Missouri, December 2002.

47. G. Xing, C. Lu, R. Pless, and Q. Huang. Greedy geographic routing is good enough in sensing covered networks. *IEEE INFOCOM* 2004.

48. G. Singh, S. Pujar, and S. Das. Rate-Based Data Propagation in Sensor Networks. Paper presented at the IEEE Wireless Communications and Networking Conference (WCNC 2004), Atlanta, Georgia, March 2004.

49. I. Stojmenovic. Rate Based Data Propagation in Sensor Networks. In preparation.

50. A. Wadaa, S. Olariu, L. Wilson, M. Eltoweissy, and K. Jones. On providing anonymity in wireless sensor networks. In *Proceedings of the 10th International Conference on Parallel and Distributed Systems (ICPADS-2004)*, pages 411–418, Newport Beach, California, July 2004.

51. S. Olariu, A. Wadaa, L. Wilson, K. Jones, and M. Eltoweissy. Enforcing Anonymity in Wireless Sensor Networks. In preparation.

52. K. Akkaya, M. Younis and M. Bangad. Sink repositioning for enhanced performance in wireless sensor networks. *Computer Networks*, forthcoming.

CHAPTER 14

Path Exposure, Target Location, Classification, and Tracking in Sensor Networks

KOUSHA MOAVENI-NEJAD and XIANG-YANG LI

Illinois Institute of Technology, Chicago, Illinois

Algorithms designed for sensor networks should be self-organizing (should not depend on global infrastructure), robust (be tolerant to node failures and range errors), and energy efficient (i.e., require little computation and, especially, communication). This chapter briefly discusses what information is available to the nodes whose location is unknown and the methods that this information can be used to derive the location of the object. The localization and tracking problems are introduced and methods to solve it are discussed. Some experimental location and tracking systems are then reviewed.

14.1 INTRODUCTION

Wireless sensor networks are large-scale distributed embedded systems composed of small devices that integrate sensors, actuators, wireless communication, and microprocessors. With advances in hardware, it will soon be feasible to deploy dense collections of sensors to perform distributed microsensing of physical environments. Sensor networks will serve as a key infrastructure for a broad range of applications, including precision agriculture, intelligent highway systems, emergent disaster recovery, and surveillance.

Sensor networks are an emerging technology that promises the unprecedented ability to monitor and instrument the physical world. Sensor networks consist of a large number of inexpensive wireless devices (nodes) densely distributed over

Handbook of Sensor Networks: Algorithms and Architectures, Edited by Ivan Stojmenović
Copyright © 2005 John Wiley & Sons, Inc.

the region of interest and have wireless connectivity. They are typically battery powered with limited communication and computation abilities. Also each node in a wireless sensor network is equipped with a variety of sensing modalities, such as acoustic, seismic, and infrared.

Having location information can be very useful and has so many applications, it can answer questions like: Are we almost to the campsite? What lab bench was I standing by when I prepared these tissue samples? How should our search-and-rescue team move to quickly locate all the avalanche victims? Can I automatically display this stock devaluation chart on the large screen I am standing next to? Where is the nearest cardiac defibrillation unit? and so on.

Service providers can also use location information to provide some novel location-aware services. The navigation system based on a global positioning system (GPS) is an example. A user can tell the system his destination and the system will guide him there. Phone systems in an enterprise can exploit locations of people to provide follow-me services.

Researchers are working to meet these and similar needs by developing systems and technologies that automatically locate people, equipment, and other tangibles. Indeed, many systems over the years have addressed the problem of automatic location-sensing. Because each approach solves a slightly different problem or supports different applications, they vary in many parameters, such as the physical phenomena used for location determination, the form factor of the sensing apparatus, power requirements, infrastructure versus portable elements, and resolution in time and space.

For outdoor environments, the most well-known positioning system is the GPS [1], which uses 24 satellites set up by the U.S. Department of Defense to enable global three-dimensional positioning services, and it provides an accuracy of around 20 to 50 m. In addition to the GPS system, positioning can also be done using some wireless networking infrastructures. Taking the personal communications service (PCS) cellular networks as an example, the $E911$ emergency service requires determining the location of a phone call via the base stations of the cellular system.

In GPS, triangulation uses ranges to at least four known satellites to find the coordinates of the receiver, and the clock bias of the receiver. For our node-location purposes, we are using a simplified version of the GPS triangulation, as we only deal with distances, so there is no need for clock synchronization. Because of the following reasons GPS is not suitable for wireless sensor networks and much work has been dedicated recently to positioning and location tracking in the area of wireless sensor networks. The reasons are:

- It is not available in an indoor environment because satellite signals cannot penetrate buildings.
- For more fine-grained applications, higher accuracy is usually necessary in the positioning result.
- Sensor networks have their own battery constraint, which requires special design.

Many applications of sensor networks require knowledge of physical sensor positions. For example, target detection and tracking are usually associated with location information [2]. Further, knowledge of sensor location can be used to facilitate network functions such as packet routing [3–5], and collaborative signal processing [6]. Sensor position can also serve as a unique node identifier, making it unnecessary for each sensor to have a unique ID assigned prior to its deployment.

In sensor networks the capabilities of individual nodes are very limited and nodes are often powered by batteries only. To conserve energy, collaboration between nodes is required and communication between nodes should be minimized. To achieve these goals nodes in wireless sensor networks (WSNs) need to determine a device's context, and since each node has limited power, we want to determine the location of individual sensor nodes without relying on external infrastructure (base stations, satellites, etc.).

Location information not only can be used to minimize the communication but also can be used to improve the performance of wireless networks and provide new types of services. For example, it can facilitate routing in a wireless ad hoc network to reduce routing overhead. This is known as geographic routing [7,8]. Through location-aware network protocols, the number of control packets can be reduced. Other types of location-based services include geocast [9], by which a user can request to send a message to a specific area, and temporal geocast, by which a user can request to send a message to a specific area at a specific time. In contrast to traditional multicast, such messages are not targeted at a fixed group of members, but rather at members located in a specific physical area.

However, location discovery in wireless sensor networks is very challenging. First, the positioning algorithm must be distributed and localized in order to scale well for large sensor networks. Second, the localization protocol must minimize communication and computation overhead for each sensor, since nodes have very limited resources (power, CPU, memory, etc.). Third, the positioning functionality should not increase the cost and complexity of the sensor, since an application may require thousands of sensors. Fourth, a location-detection scheme should be robust. It should work with accuracy and precision in various environments, and should not depend on sensor-to-sensor connectivity in the network.

The localization problem has received considerable attention in the past, as many applications need to know where objects or persons are, and hence various location services have been created. Undoubtedly, the GPS is the most well-known location service in use today. The approach taken by GPS, however, is unsuitable for low-cost, ad hoc sensor networks, since GPS is based on extensive infrastructure (i.e., satellites). Likewise, solutions developed in the area of robotic [10–12] and ubiquitous computing [13] are generally not applicable for sensor networks, as they require too much processing power and energy. Recently, a number of localization systems have been proposed specifically for sensor networks [14–16]. We are interested in truly distributed algorithms that can be employed on large-scale ad hoc sensor networks (100+ nodes). Such algorithms should be:

- Self-organizing (i.e., do not depend on global infrastructure)
- Robust (i.e., be tolerant of node failures and range errors)
- Energy efficient (i.e., require little computation and, especially, communication)

These requirements immediately rule out some of the proposed localization algorithms for sensor networks. The rest of this chapter is organized as follows. In Section 14.2, we briefly discuss what information is available to the nodes whose location is unknown, and then we discuss the methods that use this information to find the location of the object. In Section 14.3 the localization problem is introduced and methods to solve it are discussed. Similarly in Section 14.4 the target-tracking problem is introduced and methods to solve it are discussed. Section 14.5 reviews some experimental location and tracking systems, and finally Section 14.6 concludes the chapter.

14.2 NAVIGATION TECHNIQUES TO DERIVE LOCATION

There are several methods, such as *time of arrival* (*ToA*), *time difference of arrival* (*TDoA*), and *signal strength*, with which an object can estimate its distance or its relative location to a reference point. In this section several approaches, which use the methods previously cited to calculate an object's location, will be discussed.

14.2.1 Lateration

Lateration computes the position of an object by measuring its distance from multiple reference positions. Calculating an object's position in two dimensions requires distance measurements from three noncollinear points, as shown in Figure 14.1(a), and in three dimensions, distance measurements from four noncoplanar points are

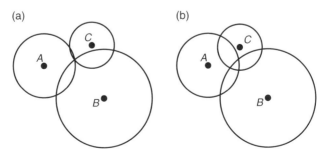

Figure 14.1 Trilateration method: (a) ideal situation and (b) real situation with error.

required. Domain-specific knowledge can reduce the number of required distance measurements. For example, if all reference points are above the object, then distance measurements from only three reference points are required. We now describe two different lateration techniques.

14.2.1.1 *Trilateration*

Trilateration is a well-known technique in which the positioning system has a number of *beacons* at known locations. These beacons can transmit signals so that other devices can determine their distances to these beacons based on received signals. If a device can hear at least three beacons, its location can be estimated. Figure 14.1(a) shows how trilateration works: A, B, and C are beacons with known locations. From A's signal, one can determine the distance to A, and thus that the object should be located at the circle centered at A with a radius equal to the estimated distance. Similarly, from B's and C's signals, it can be determined that the object should be located at some circles centered at B and C, respectively. Thus, the intersection of the three circles is the estimated location of the device. The preceding discussion has assumed an ideal situation; however, as mentioned earlier, distance estimation always contains errors that will, in turn, lead to location errors. Figure 14.1(b) illustrates an example in practice. The three circles do not intersect at a common point. In this case, the maximum-likelihood method can be used to estimate the device's location. Let the three beacons A, B, and C be located at $(x_A, y_A), (x_B, y_B)$, and (x_C, y_C), respectively. For any point (x, y) on the plane, a difference function is computed:

$$\sigma_{x,y} = \left| \sqrt{(x - x_A)^2 + (y - y_A)^2} - r_A \right| + \left| \sqrt{(x - x_B)^2 + (y - y_B)^2} - r_B \right|$$
$$+ \left| \sqrt{(x - x_C)^2 + (y - y_C)^2} - r_C \right|$$

where r_A, r_B, and r_C are the estimated distances to A, B, and C, respectively. The location of the object can then be predicted as the point (x, y) among all points such that $\sigma_{x,y}$ is minimized. In addition to using the ToA approach for positioning, the angle of arrival (AoA) approach can be used. For example, in Figure 14.2, the unknown node D measures the angle of $\angle ADB$, $\angle BDC$, and $\angle ADC$ by the received signals from beacons A, B, and C. From this information, D's location can be derived [16].

14.2.1.2 *Multilateration*

The trilateration method has its limitation in that at least three beacons are needed to determine a device's location. In a sensor network, in which nodes are randomly deployed, this may not be true. Several multilateration methods are proposed to relieve this limitation. The ad hoc localization system (AHLoS) [17] enables nodes to discover their locations by using a set of distributed iterative algorithms. Figure 14.3 shows an example in which, initially, beacon nodes contain only nodes marked as reference points. Device nodes A, B, and C are at

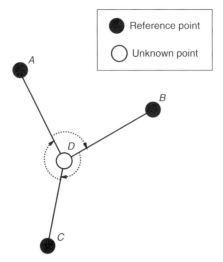

Figure 14.2 Angle measurement from three beacons A, B, and C.

unknown locations. In the first iteration, as Figure 14.3(a) shows, the locations of nodes A and C will be determined.

Once the location of a device is estimated, its role is changed to a beacon node so as to help determine other devices' locations. This is repeated until all host locations are determined (if possible). As Figure 14.3(b) shows, in the second iteration, the location of node B can be determined with the help of nodes A and B, which are now serving as beacons.

If the distance or regular estimation is precise, we can show that the order in which each node determines its location and then serves as a beacon node will

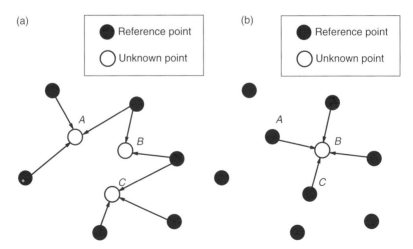

Figure 14.3 (a) Atomic multilateration and (b) iterative multilateration.

not affect the number of nodes whose positions can be computed. However, when the information is not precise, it does affect this and further, the precision of the system.

14.2.2   Pattern Matching Using Database

Pattern matching (also known as *fingerprinting*) tries to compare the received signal pattern against the training patterns in the database and to determine the likelihood that the device is currently located in a position. A typical solution has two phases:

1. *Off-Line Phase.* The purpose of this phase is to collect signals from all base stations at each training location, thus the received signal strengths are recorded in the database. For higher accuracy, one can establish multiple entries in the database for the same training location. From the database, some positioning rules, which form the positioning model, will then be established.
2. *Real-Time Phase.* With a well-trained positioning model, one can estimate a device's location, given the signal strengths collected by the device from all possible base stations. The positioning model can determine a number of locations, each associated with a probability.

There are several similarity searching methods in the matching process in the literature, such as *nearest-neighbor algorithms* (*NNSS*) [18,19] and *probability-based algorithms* [20].

14.2.3   Network-Based Tracking

At the network level, location tracking may be done via the cooperation of sensors. Tseng and colleagues [21] addressed these issues using an agent-based paradigm. Once a new object is detected by the network, a mobile agent will be initiated to track the roaming path of the object. Then the agent invites some nearby slave sensors to cooperatively position the object and inhibit other irrelevant (i.e., farther) sensors from tracking the object. More precisely, only three agents will be used for the tracking purpose at any time, and they will move as the object moves. The trilateration method is used for positioning.

Figure 14.4 shows an example. The sensor network is deployed in a regular manner, and it is assumed that each sensor's sensing distance equals the distance between two neighboring sensors. Initially, each sensor is in the idle state, searching for new objects. Once it detects a target, a sensor will transit to the election state, trying to serve as the master agent. The nearest sensor will win. The master agent will then dispatch two neighboring sensors as the slave agents; master and slave agents will cooperate to position the object. In the figure, the object is first tracked by sensors $\{S_0, S_1, S_2\}$ when resident in A_0, then by $\{S_0, S_2, S_3\}$ when in A_1, by

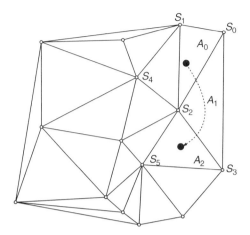

Figure 14.4   Roaming path of an object (dashed line).

$\{S_2, S_3, S_5\}$ when in A_2, etc. The master agent is responsible for collecting all sensing data and performing the trilateration algorithm. It also conducts data fusion by keeping the tracking results while it moves around. At the proper time, the master agent will forward the tracking result to the data center.

14.3   LOCALIZATION IN WIRELESS SENSOR NETWORKS

In this section, we address the issue of localization in ad hoc sensor networks. That is, we want to determine the location of individual sensor nodes without relying on external infrastructure (base stations, satellites, etc.). Undoubtedly, the GPS is the most well-known location service in use today. The approach taken by GPS, however, is unsuitable for low-cost, ad hoc sensor networks, since GPS is based on extensive infrastructure (i.e., satellites). Location service has many applications, for example, it enables routing in sufficiently isotropic large networks, without the use of large routing tables.

We assume that only a limited fraction of nodes, which are called *anchor* nodes, have a self-location capability. Note that, in wireless ad hoc sensor networks, there exist no fine control over the placement of the sensor nodes when the network is installed (e.g., when nodes are dropped from an airplane). Consequently, we assume that nodes are randomly distributed across the environment. For simplicity and ease of presentation, we limit the environment to two dimensions, but all algorithms are capable of operating in three dimensions. Figure 14.5 shows an example network with 25 nodes; pairs of nodes that can communicate directly are connected by an edge. The connectivity of the nodes in the network (i.e., the average number of neighbors) is an important parameter that has a strong impact on the accuracy of most localization algorithms.

Figure 14.5 Example network topology.

In some application scenarios, nodes may be mobile. Here, however, we focus on static networks, where nodes do not move for a reasonably short period of time, since this is already a challenging condition for distributed localization. Note that anchor nodes have the same capabilities (processing, communication, energy consumption, etc.) as all other sensor nodes with unknown positions. Ideally, the fraction of anchor nodes should be as low as possible to minimize the installation costs.

14.3.1 Ad Hoc Positioning System

The ad hoc positioning system (APS) [16] is a distributed, hop-by-hop positioning algorithm that provides the approximate location for all nodes in a network where only a limited fraction of nodes have the self-location capability. Also, APS is appropriate for indoor location-aware applications.

14.3.1.1 Algorithm It is not desirable to have the landmarks emit maximum power to cover the entire network for several reasons: collisions in local communication, high power usage, coverage problems when moving. Also, it is not acceptable to assume some fixed positions for the landmarks, as the applications envisioned by APS systems are either in deployments from the air over inaccessible areas, or possibly involving the movement and reconfiguration of the network. In this case, one option is to use the hop-by-hop propagation capability of the network to forward distances to landmarks. In general, they aim for the same principle as GPS, the difference being that the landmarks are contacted in a hop-by-hop fashion, rather than directly, as ephemerides are. Once an arbitrary node has estimated a number

(≥ 3) of landmarks, it can compute its own position in the plane, using a procedure similar to the one used in GPS position calculation described in the previous section. The estimate we start with is the centroid of the landmarks collected by a node. In what follows we refer to one landmark only, as the algorithm behaves identically and independently for all the landmarks in the network. It is clear that the immediate neighbors of the landmark can estimate the distance to the landmark by direct signal-strength measurement. Using some propagation method, the second-hop neighbors then are able to infer their distance to the landmark, and the rest of the network follows, in a controlled flood manner, initiated at the landmark. Complexity of signaling is therefore driven by the total number of landmarks, and by the average degree of each node.

14.3.1.2 *Distance to Anchors*   The APS system uses three methods of hop-to-hop distance propagation and examines advantages and drawbacks for each of them. Each propagation method is appropriate for a certain class of problems, as it influences the amount of signaling, power consumption, and position accuracy achieved.

Nodes that can communicate with anchor nodes directly are able to find their distance from anchor nodes, but this information is not available to all nodes. Nodes share information to collectively determine the distances between individual nodes and the anchors, so that an (initial) position can be calculated. None of the alternatives engage in complicated calculations, so finding the distance to anchors is communication bounded. Most of the distributed localization algorithms share a common communication pattern: information is flooded into the network, starting at the anchor nodes. A networkwide flood by some anchor *A* is expensive, since each node must forward *A*'s information to its (potentially) unaware neighbors. This implies a scaling problem: flooding information from all anchors to all nodes will become too expensive for large networks, even with low anchor fractions. Fortunately a good position can still be derived with knowledge (position and distance) from a limited number of anchors. Therefore, nodes can simply stop forwarding information when enough anchors have been *located*. This simple optimization (have a *flood limit*) has been proved to be highly effective in controlling the amount of communication. We now list three methods for determining the location.

- *Sum-dist*. This method is also known as *DV-distance* [16]. The most simple solution for determining the distance to the anchors is simply adding the ranges encountered at each hop during the network flood. Sum-dist starts at the anchors, which send a message including their identity, position, and a path length set to 0. Each receiving node adds the measured range to the path length and forwards (broadcasts) the message if the flood limit allows it to do so. Another constraint is that when the node has already received information about the particular anchor, it is only allowed to forward the message if the current path length is less than the previous one. The end result is that each node will have stored the position and minimum path length to at least flood-limit anchors.

- *DV-hop*. A drawback of Sum-dist is that range errors accumulate when distance information is propagated over multiple hops. This cumulative error becomes significant for large networks with few anchors (long paths) and/or poor ranging hardware. A robust alternative is to use topological information by counting the number of hops instead of summing the (erroneous) ranges. This approach was named DV-hop by Niculescu and Nath [16], and Hop-TERRAIN by Savarese et al. [22]. The DV-Hop propagation method is the most basic scheme, and it first employs a classic distance vector exchange so that all nodes in the network discover distances, in hops, to the landmarks. Essentially, DV-hop consists of two flood waves. After the first wave, which is similar to Sum-dist, nodes have obtained the position and minimum hop count to at least flood-limit anchors. The second calibration wave is needed to convert hop counts into distances such that nodes can compute a position. This conversion consists of multiplying the hop count by an average hop distance. Whenever an anchor A_1 infers the position of another anchor A_2 during the first wave, it computes the distance between them, and divides that by the number of hops to derive the average hop distance between A_1 and A_2. When calibrating, an anchor takes all remote anchors into account that it is aware of. Nodes forward (broadcast) calibration messages only from the first anchor that calibrates them, which reduces the total number of messages in the network.

- *Euclidean*. A drawback of DV-hop is that it does not work for highly irregular network topologies, where the variance in actual hop distances is very large. Niculescu and Nath [16] have proposed another method, named Euclidean, which is based on the local geometry of the nodes around an anchor. Again anchors initiate a flood, but forwarding the distance is more complicated than in the previous cases. When a node has received messages from two neighbors that know their distance to the anchor, and to each other, it can calculate the distance to the anchor. Figure 14.6 shows a node X that has two neighbors n_1 and n_2 with distance estimates (a and b) to an anchor. Together with the known ranges c, d, and e, there are two possible values (r_1 and r_2) for the distance of the node to the anchor. Niculescu describes two methods to determine which, if any, distance to use. The neighbor vote method can be applied if there exists a third neighbor n_3 that has a distance estimate to the anchor and that is connected to either n_1 or n_2. Replacing n_2 (or n_1) by n_3 will again yield a pair of distance estimates. The correct distance is part of both pairs, and is selected by a simple vote. Of course, more neighbors can be included to make the selection more accurate.

14.3.1.3 *Node Position*

Now nodes can determine their position using lateration, min–max (presented by Savvides et al. [23]), or other methods based on the distance estimates to a number of anchors provided by one of the three alternatives (Sum-dist, DV-hop, or Euclidean). The determination of the node positions does not involve additional communication.

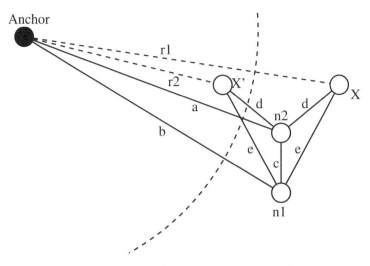

Figure 14.6   Determining distance using Euclidean.

14.3.2 Time-Based Positioning Scheme

The time-based positioning scheme (TPS) is meant for use with outdoor wireless sensor networks. Many applications of outdoor sensor networks require knowledge of physical sensor positions. For example, target detection and tracking is usually associated with location information. Further, knowledge of sensor location can be used to facilitate network functions such as packet routing and collaborative signal processing. Sensor position can also serve as a unique node identifier, making it unnecessary for each sensor to have a unique ID assigned prior to its deployment.

TPS relies on an RF signal, which performs well compared to ultrasound, infrared, and so on, in outdoor environments. They measure the difference in arrival times (TDoA) of beacon signals. TPS does not need the specialized antennae generally required by an AoA positioning system. This time-based location-detection scheme avoids the drawbacks of many existing systems for outdoor sensor location detection. Compared to existing schemes proposed in the context of outdoor sensor networks, the TPS scheme has the following characteristics and advantages:

- Time synchronization of all base stations and nodes is not required in TPS.
- There are no requirements for an ultrasound receiver, second radio, or specialized antennae at base stations or sensors.
- The TPS algorithm is not iterative and does not require a complicated refinement step.
- TPS has low computation cost.
- Sensors listen passively and are not required to make radio transmissions.

14.3.2.1 Network Model Assume that the sensors are deployed randomly over a two-dimensional monitored area (on the ground). Each sensor has limited resources (battery, CPU, etc.), and is equipped with an omnidirectional antenna. Three base stations A, B, C, with known coordinates (x_a, y_a), (x_b, y_b), and (x_c, y_c), respectively, are placed beyond the boundary of the monitored area, as shown in Figure 14.7. Let us assume A is the master base station. Assume the monitored area is enclosed within the angle $\angle BAC$. Let the unknown coordinates of a sensor be (x, y), which will be determined by TPS. Each base station can reach all sensors in the monitored area. One restriction on the placement of these base stations is that they must be noncollinear, as otherwise, the sensor locations will be indistinguishable. If the monitored area is so large that three base stations cannot cover the whole area completely, we can always divide the area into smaller subareas and place more base stations.

14.3.2.2 Positioning Scheme The TPS time-based location-detection scheme consists of two steps. The first step detects the time difference of signal arrival times from three base stations. These time differences are transformed into range differences from the sensor to the base stations. In the second step, we perform trilateration to transform these range estimates into coordinates.

Given the locations (x_a, y_a), (x_b, y_b), and (x_c, y_c) of base stations A, B, and C, respectively, the TPS system determines the location (x, y) of sensor S, as shown in Figure 14.6.

- *Range Detection.* Let A be the master base station, which will initiate a beacon signal every T seconds. Each beacon interval begins when A transmits a beacon signal. Sensor S, base stations B and C will all receive A's beacon signal, respectively. Base station B will reply to A with a beacon signal conveying the difference between the time the signal from A was received and the time

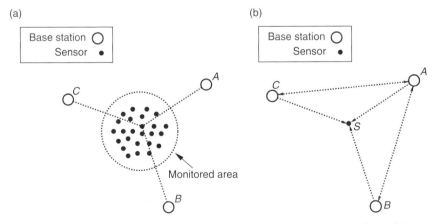

Figure 14.7 TPS example: (a) sensor networks; (b) node S determines its position.

the replay was sent. This signal will reach S. After receiving beacon signals from both A and B, C will reply to A with a beacon signal conveying the difference between the time the signal from A was received by C and the time the replay was sent. This signal will also reach S, based on triangle inequality.

- *Location Computation*. Node S know the time the signal was sent from A and the time it was received by B, C, and also by itself. Node S also has the same information about the signal sent by B and C. Now node S can calculate its position using trilateration.

14.3.3   GPS-less Low-Cost Outdoor Localization for Very Small Devices

GPS solves the problem of localization in outdoor environments for PC-class nodes. However, for large networks of very small, inexpensive, and low-power devices, practical considerations such as size, form factor, cost, and power constraints of the nodes preclude the use of GPS on all nodes. The GPS-less system [14] addresses the problem of localization for such devices, with the following design goals.

- *RF-Based*. They focus on small nodes that have some kind of short-range RF transceiver. The primary goal is to leverage this radio for localization, thereby eliminating the cost, power, and size requirements of a GPS receiver.
- *Receiver-Based*. In order to scale well to a large distributed networks, the responsibility for localization must lie with the receiver node that needs to be localized, and not with the reference points.
- *Ad hoc*. In order to ease deployment, a solution that does not require preplanning or extensive infrastructure is desired.
- *Responsiveness*. We need to be able to localize within a fairly low response time.
- *Low Energy*. Small, untethered nodes have modest processing capabilities, and limited energy resources. If a device uses all of its energy localizing itself, it will have none left to perform its task. Therefore, we want to minimize computation and message costs to reduce power consumption.
- *Adaptive Fidelity*. In addition, we want the accuracy of our localization algorithms to be adaptive to the granularity of available reference points.

This scheme uses an idealized radio model and proposes a simple connectivity-based localization method for such devices in unconstrained outdoor environments. It leverages the inherent RF communications capabilities of these devices. A fixed number of nodes in the network with overlapping regions of coverage serve as *reference* points and transmit periodic beacon signals. Nodes use a simple connectivity metric to infer proximity to a given subset of these reference points and then localize themselves to the centroid of the selected (proximate) reference points.

14.3.3.1   Localization Algorithm   We considered two approaches to engineer an RF-based localization system, based on measurements of received signal strength and connectivity, respectively. The first approach for RF-based localization is to use the measured signal strength of received beacon signals to estimate distance, as in the RADAR system [19], with an outdoor radio signal-propagation model. We discarded this approach for several reasons relating to our short-range (10-m) radios. First, signal strength at short ranges is subject to unpredictable variation due to fading, multipaths, and interferences. It does not therefore correlate directly with distance. Moreover, short range does not allow much gain in density of reference points when considering signal strength. We have found an idealized radio model useful for predicting bounds on the quality of connectivity-based localization. We chose this model because it was simple and easy to reason about mathematically. This subsection presents this idealized model. To our surprise, this model compares quite well to outdoor radio propagation in uncluttered environments, as we explore in the next subsection. We make two assumptions in our idealized model:

1. Perfect spherical radio propagation
2. Identical transmission range (power) for all radios

Multiple nodes in the network with overlapping regions of coverage serve as reference points (labeled R_1 to R_n). They are situated at known positions, containing their respective positions (X_1, Y_1) to (X_n, Y_n), that form a regular mesh and transmit periodic beacon signals every T seconds. We assume that neighboring reference points can be synchronized so that their beacon signal transmissions do not overlap in time. Furthermore, in any time interval, each of the reference points would have transmitted exactly one beacon signal.

Each mobile node listens for a fixed time period t and collects all the beacon signals that it receives from various reference points. We characterize the information per reference point R_i by a *connectivity metric*, defined as:

$$CM_i = \frac{N\,recv(i,t)}{N\,sent(i,t)} \times 100$$

where $N\,recv(i,t)$ is the number of beacons sent by R_i that have been received in time t, and $N\,sent(i,t)$ is the number of beacons that have been sent by R_i. In order to improve the reliability of our connectivity metric in the presence of various radio propagation vagaries, we would like to base our metric on a sample of at least S packets, where S is the sample size, a tunable parameter of our method (i.e., $N\,sent(i,t) = S$). Since we know T to be the time period between two successive beacon signal transmissions, we can set t, the receiver's sampling time as,

$$t = (s + 1 - \varepsilon)T \qquad (0 < \varepsilon \ll 1)$$

From the beacon signals that it receives, the receiver node infers proximity to a collection of reference points for which the respective connectivity metrics exceed a certain threshold. We denote the collection of reference points by $R_{i1}, R_{i2}, \ldots, R_{ik}$.

The receiver localizes itself to the region that coincides with the intersection of the connectivity regions of this set of reference points, which is defined by the centroid of these reference points:

$$(X, Y) = \left(\frac{X_{i1} + X_{i2} + \cdots + X_{ik}}{k}, \ \frac{Y_{i1} + Y_{i2} + \cdots + Y_{ik}}{k} \right)$$

14.3.4   Computational Complexity of Sensor Network Localization

The localization problem for sensor networks is to reconstruct the positions of all of the sensors in a network, given the distances between all pairs of sensors that are within some radius r of each other. In the past few years, many algorithms for solving the localization problem were proposed, without knowing the computational complexity of the problem. Aspnes et al. [24] showed that no *polynomial-time* algorithm can solve this problem in the worst case, even for sets of distance pairs for which a unique solution exists, unless $RP = NP$.

Although the designs of the previous schemes have demonstrated clever engineering ingenuity, and their effectiveness is evaluated through extensive simulations, the focus of these schemes is on algorithmic design, without knowing the fundamental computational complexity of the localization process. In sensor network localization, since only nodes that are within communication range can measure their relative distances, the graphs formed by connecting each pair of nodes that can measure each other's distance are better modeled as unit disk graphs. Such constraints could have the potential of allowing computationally efficient localization algorithms to be designed.

The localization problem considered here is to reconstruct the positions of a set of sensors given the distances between any pair of sensors that are within some unit-disk radius r of each other. Some of the sensors may be beacons, sensors with known positions, but our impossibility results are not affected much by whether beacons are available. To avoid precision issues involving irrational distances, it is assumed that the input to the problem is presented with the distances squared. If we make the further assumption that all sensors have integer coordinates, all distances will be integers as well.

For the main result, we consider a decision version of the localization problem, which we call UNIT-DISK GRAPH RECONSTRUCTION. This problem essentially asks if a particular graph with given edge lengths can be physically realized as a unit-disk graph with a given disk radius in two dimensions. The input is a graph G where each edge uv of G is labeled with an integer l_uv^2, the square of its length, together with an integer r^2 that is the square of the radius of a unit disk. The output is *yes* or *no*, depending on whether there exists a set of points in R^2 such that the distance between u and v is l_uv whenever uv is an edge in G and exceeds r whenever uv is not an edge in G.

The main result, is that UNIT-DISK GRAPH RECONSTRUCTION is *NP-hard*, based on a reduction from CIRCUIT SATISFIABILITY. The constructed graph for a circuit with m wires has $O(m^2)$ vertices and $O(m^2)$ edges, and the number of solutions to the resulting localization problem is equal to the number of satisfying assignments for the circuit. In each solution to the localization problem, the points can be placed at integer coordinates, and the entire graph fits in an $O(m)$-by-$O(m)$ rectangle, where the constants hidden by the asymptotic notation are small. The construction also permits a constant fraction of the nodes to be placed at known locations. Formally:

Theorem 14.1 There is a polynomial-time reduction from CIRCUIT SATIS-FIABILITY to UNIT-DISK GRAPH RECONSTRUCTION, in which there is a one-to-one correspondence between satisfying assignments to the circuit and solutions to the resulting localization problem.

A consequence of this result is:

Corollary 14.1 There is no efficient algorithm that solves the localization problem for sparse sensor networks in the worst case unless $P = NP$.

It might appear that this result depends on the possibility of ambiguous reconstructions, where the position of some points is not fully determined by the known distances. However, if we allow randomized reconstruction algorithms, a similar result holds even for graphs that have unique reconstructions.

Corollary 14.2 There is no efficient randomized algorithm that solves the localization problem for sparse sensor networks that have unique reconstructions unless $RP = NP$.

Finally, because the graph constructed in the proof of Theorem 14.1 uses only points with integer coordinates, even an approximate solution that positions each point to within a distance $\varepsilon < \frac{1}{2}$ of its correct location can be used to find the exact locations of all points by rounding each coordinate to the nearest integer. Since the construction uses a fixed value for the unit disk radius r (the natural scale factor for the problem), we have the following corollary.

Corollary 14.3 The results of Corollary 14.1 and Corollary 14.2 continue to hold even for algorithms that return an approximate location for each point, provided the approximate location is within εr of the correct location, where ε is a fixed constant.

What we do not know at present is whether these results continue to hold for solutions that have large positional errors, but that give edge lengths close to those in the input. Our suspicion is that edge-length errors accumulate at most polynomially across the graph, but we have not yet carried out the error analysis necessary to prove this. If our suspicion is correct, we would have:

Conjecture 14.1 The results of Corollary 14.1 and Corollary 14.2 continue to hold even for algorithms that return an approximate location for each point, provided the relative error in edge length for each edge is bounded by ε/n^c for some fixed constant c.

14.4 TARGET TRACKING AND CLASSIFICATION

One of the most important areas where the advantages of sensor networks can be exploited is for tracking mobile targets. Scenarios where this network may be deployed can be both military (tracking enemy vehicles, detecting illegal border crossings) and civilian (tracking the movement of wild animals in wildlife preserves). Typically, for accuracy, two or more sensors are simultaneously required for tracking a single target, leading to coordination issues. Additionally, given the requirements to minimize the power consumption due to communication or other factors, we want to select the minimum number of sensors dedicated for the task, while all other sensors should preferably be in the sleep or off state. In order to simultaneously satisfy requirements like power saving and improving overall efficiency, we need large-scale coordination and other management operations. These tasks become even more challenging when one considers the random mobility of the targets and the resulting need to coordinate the assignment of the sensors best suited for tracking the target as a function of time. In this section we discuss managing and coordinating a sensor network for tracking moving targets.

The power limitation due to the small size of the sensors, the large numbers of sensors that need to be deployed and coordinated, and the ability to deploy sensors in an ad hoc manner give rise to a number of challenges in sensor networks. Each of these needs to be addressed by any proposed architecture in order for it to be realistic and practical.

- *Scalable Coordination.* A typical deployment scenario for a sensor network comprises a large number of nodes reaching in the thousands to tens of thousands. At such large scales, it is not possible to attend to each node individually due to a number of factors. Sensor nodes may not be physically accessible, nodes may fail, and new nodes may join the network. In such dynamic and unpredictable scenarios, scalable coordination and management functions are necessary that can ensure the robust operation of the network. In the light of target tracking, the coordination function should scale with the size of the network, the number of targets to be tracked, number of active queries, and so on.
- *Tracking Accuracy.* To be effective, the tracking system should be accurate and the likelihood of missing a target should be low. Additionally, the dynamic range of the system should be high while keeping the response latency, sensitivity to external noise, and number of false alarms low. The overall architecture should also be robust against node failures.
- *Ad hoc Deployability.* A powerful paradigm associated with sensor networks is their ability to be deployed in an ad hoc manner. Sensors can be thrown in an area affected by a natural or man-made disaster or air dropped to cover a geographical region. Thus sensor nodes should be capable of organizing themselves into a network and achieving the desired objective in the absence of any human intervention or fixed patterns in the deployment.

- *Computation and Communication Costs.* Any protocol being developed for sensor networks should keep in mind the costs associated with computations and communication. With current technology, the cost of computation locally is lower than that of communication in a power-constrained scenario. As a consequence, emphasis should be put on minimizing the communication requirements.
- *Power Constraints.* The available power in each sensor is limited by the battery lifetime due to the difficulty or impossibility of recharging the nodes. As a consequence, protocols that tend to minimize the energy consumption or power-aware protocols that adapt to the existing power levels are highly desirable. Additionally, efforts should be made to turn off the nodes themselves, if possible, in the absence of sensing or coordination operations.

14.4.1 Collaborative Signal Processing

Power consumption is a critical consideration in a wireless sensor network. The limited amount of energy stored at each node must support multiple functions, including sensor operations, on-board signal processing, and communication with neighboring nodes. Thus, one must consider power-efficient sensing modalities, low sampling rates, low-power signal-processing algorithms, and efficient communication protocols to exchange information among nodes. To facilitate monitoring of a sensor field, including detection, classification, identification, and tracking of targets, global information in both space and time must be collected and analyzed over a specified space–time region. However, individual nodes only provide spatially local information. Furthermore, due to power limitation, temporal processing is feasible only over limited time periods. This necessitates *collaborative signal processing* (*CSP*) (i.e., collaboration between nodes to process the space–time signal). A CSP algorithm can benefit from the following desirable features:

- *Distributive Processing.* Raw signals are sampled and processed at individual nodes, but are not directly communicated over the wireless channel. Instead, each node extracts relevant summary statistics from the raw signal, which are typically of smaller size. The summary statistics are stored locally in individual nodes and may be transmitted to other nodes upon request.
- *Goal-Oriented, On-Demand Processing.* To conserve energy, each node only performs signal-processing tasks that are relevant to the current query. In the absence of a query, each node retreats into a standby mode to minimize energy consumption. Similarly, a sensor node does not automatically publish extracted information (i.e., it forwards such information only when needed).
- *Information Fusion.* To infer global information over a certain space–time region from local observations, CSP must facilitate efficient, hierarchical information fusion and progressively lower bandwidth information must be shared between nodes over progressively large regions. For example, (high bandwidth) time series data may be exchanged between neighboring nodes for classifi-

cation purposes. However, lower bandwidth *closest point of approach* (*CPA*) data may be exchanged between more distant nodes for tracking purposes.

- *Multiresolution Processing.* Depending on the nature of the query, some CSP tasks may require higher spatial resolution involving a finer sampling of sensor nodes, or higher temporal resolution involving higher sampling rates. For example, reliable detection may be achievable with a relatively coarse space–time resolution, whereas classification typically requires processing at a higher resolution.

14.4.2   Target Tracking Using Space–Time Cells

14.4.2.1   Introduction   Each object in a geographical region generates a time-varying space–time signature field that may be sensed in different modalities, such as acoustic, seismic, or thermal. The nodes sample the signature field spatially, and the density of nodes should be commensurate with the rate of spatial variation in the field. Similarly, the time series from each sensor should be sampled at a rate commensurate with the required bandwidth. Thus, the rate of change of the space–time signature field and the nature of the query determines the required space–time sampling rate. A moving object in a region corresponds to a peak in the spatial signal field that moves with time. Tracking an object corresponds to tracking the location of the spatial peak over time.

14.4.2.2   Using Space–Time Cells   To enable tracking in a sensor network, the entire space–time region must be divided into *space–time cells* to facilitate local processing. The size of a space–time cell depends on the velocity of the moving target and the decay exponent of the sensing modality. It should approximately correspond to a region over which the space–time signature field remains nearly constant. In principle, the size of space–time cells can be dynamically adjusted as new space–time regions are created based on predicted locations of targets. Space–time signal averaging can be done over nodes in each cell to improve the signal-to-noise ratio. We note that the assumption of a constant signature field over a space–time cell is at best an approximation in practice due to several factors, including variations in terrain, foliage, temperature gradients, and the nonisotropic nature of the source signal. However, such an approximation can be judiciously applied in some scenarios for the purpose of reducing intrasensor communication, as well to improve algorithm performance against noise.

14.4.2.3   Single-Target Tracking   One of the key premises behind the networking algorithms being developed at Wisconsin [2] is that routing of information in a sensor network should be geographic-centric rather than node-centric. In other words, from the viewpoint of information routing, the geographic locations of the nodes are the critical quantities rather than their arbitrary identities. In the spirit of space–time cells, the geographic region of interest is divided into smaller regions (spatial cells) that facilitate communication over the sensor network. Some

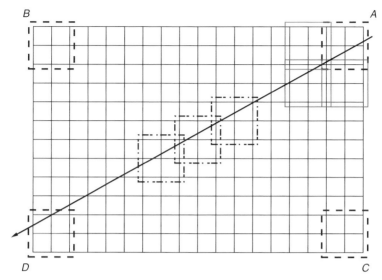

Figure 14.8 A schematic illustrating detection and tracking of a single target.

of the nodes in each cell are designated as *manager nodes* for coordinating signal processing and communication in that cell.

Figure 14.8 illustrates the basic idea of region-based CSP for detection and tracking of a single target. Under the assumption that a potential target may enter the monitored area via one of the four corners, four cells, *A*, *B*, *C*, and *D*, are created by the UW-API protocols [2]. Nodes in each of the four cells are activated to detect potential targets.

Each activated node runs an energy-detection algorithm whose output is sampled at an a priori fixed rate, depending on the characteristics of expected targets. Suppose a target enters cell *A*. Tracking of the target consists of the following five steps:

Step 1. Some and perhaps all of the nodes in cell *A* detect the target. These nodes are the active nodes and cell *A* is the active cell. The active nodes also yield CPA time information. The active nodes report their energy detector outputs to the manager nodes at *N* successive time instants.

Step 2. At each time instant, the manager nodes determine the location of the target from the energy detector outputs of the active nodes. The simplest estimate of target location at an instant is the location of the node with the strongest signal at that instant. However, more sophisticated algorithms for target localization can be used. Such localization algorithms justify their higher complexity only if the accuracy of their location determination is finer than the node spacing.

Step 3. The manager nodes use locations of the target at the *N* successive time instants to predict the location of the target at $M(<N)$ future time instants.

Step 4. The predicted positions of the target are used by the UW-API protocols [2] to create new cells that the target is likely to enter. This is illustrated in Figure 14.8, where the three dotted cells represent the regions that the target is likely to enter after the current active cell (cell *A* in Fig. 14.8). A subset of these cells is activated by the UW-API protocols for subsequent detection and tracking of the target.

Step 5. Once the target is detected in one of the new cells, it is designated as the new active cell and the nodes in the original active cell (cell *A* in Fig. 14.8) can be put in the standby state to conserve energy.

Steps 1–5 are repeated for the new active cell, and this forms the basis of detecting and tracking a single target. For each detected target, an information field containing tracking information, such as the location of the target at certain past times, is usually passed from one active cell to the next one. This is particularly important in the case of multiple targets.

14.4.2.4 *Multiple-Target Tracking*

Figure 14.8 illustrates detection and tracking of a single target. If multiple targets are sufficiently separated in space or time, that is, they occupy distinct space–time cells, essentially the same procedure as described in Section 14.4.2.3 may be used: a different track is initiated and maintained for each target. Sufficient separation in time means that the energy detector output of a particular sensor exhibits distinguishable peaks corresponding to the CPAs of the two targets. Similarly, sufficient separation in space means that at a given instant the spatial target signatures exhibit distinguishable peaks corresponding to nodes that are closest to the targets at that instant. The assumption of sufficient separation in space and/or time may be too restrictive in general. In such cases, classification algorithms are needed that operate on spatiotemporal target signatures to classify them. This necessarily requires a priori statistical knowledge of typical signatures for different target classes.

14.4.2.5 *Target Classification*

Here, we focus on single-node (no collaboration between nodes) classification based on temporal target signatures: a time series segment is generated for each detected event at a node and processed for classification. Some form of temporal processing, such as a fast Fourier transform (FFT), is performed and the transformed vector is fed to a bank of classifiers corresponding to different target classes. The outputs of the classifiers that detect the target, active classifiers, are reported to the manager nodes as opposed to the energy detector outputs. Steps (1) to (5) in Subsection 14.4.2.3 are repeated for all the active classifier outputs to generate and maintain tracks for different classified targets. In some cases, both energy-based CPA information and classifier outputs may be needed.

Now we briefly describe the three classifiers explored in this chapter. Given a set of N-dimensional feature vectors $\{x; x \in R^N\}$, we assume that each of them is assigned a class label, $\omega_c \in \Omega = \{\omega_1, \omega_2, \ldots, \omega_m\}$, that belongs to a set of m elements. We denote by $p(\omega_c)$ the prior probability that a feature vector belongs to class ω_c. Similarly, $p(\omega_c|x)$ is the posterior probability for class ω_c given that x is observed.

A *minimum error classifier* maps each vector x to an element in. such that the probability of misclassification (i.e., the probability that the classifier label is different from the true label) is minimized. To achieve this minimum error rate, the optimal classifier decides x has label ω_i if $p(\omega_i|x)$ for all $j \neq i$, $\omega_i, \omega_j \in \Omega$. In practice, it is very difficult to evaluate the posterior probability in closed form. Instead, one can use an appropriate discriminant function $g_i(x)$ that satisfies $g_i(x) > g_j(x)$ if $p(\omega_i|x) > p(\omega_j|x)$ for $j \neq i$, for all x. Then minimum error classification can be achieved as: decide x has label ω_i if $g_i(x) > g_j(x)$ for $j \neq i$. The minimum probability of misclassification is also known as the *Bayes* error, and a minimum error classifier is also known as a Bayes classifier or a maximum posterior probability (MAP) classifier. Below, we briefly discuss three classifiers that approximate the optimal Bayes classifier.

k-Nearest-Neighbor Classifier   The k-nearest-neighbor (kNN) classifier uses all the training features as the set of prototypes $\{p_k\}$. During the testing phase, the distance between each test vector and every prototype is calculated, and the k prototype vectors that are closest to the test vector are identified. The class labels of these k-nearest prototype vectors are then combined using majority vote or some other method to decide the class label of the test vector. When $k = 1$, the kNN classifier is called the nearest-neighbor classifier. It is well-known [6] that asymptotically (in the number of training vectors), the probability of misclassification of a nearest-neighbor classifier approaches twice the (optimal) Bayes error. Hence, the performance of a nearest-neighbor classifier can be used as a baseline to gauge the performance of other classifiers. However, as the number of prototypes increases, a kNN classifier is not very suitable for actual implementation, since it requires too much memory storage and processing power for testing.

Maximum-Likelihood Classifier   Using the Gaussian mixture density model in this classifier, the distribution of training vectors from the same class is modeled as a mixture of Gaussian density functions. That is, the likelihood function is modeled as:

$$p(x|\omega_i) \propto G_i(x|\theta_i) = \sum_k |\Lambda_{ik}|^{-N/2} \exp\left(-\frac{1}{2}(x - m_{ik})^T \Lambda_{ik}^{-1}(x - m_{ik})\right) \quad (14.1)$$

where $\theta_i = [m_{i1}, m_{i2}, \ldots, m_{ip}, \Lambda_{i1}, \Lambda_{i2}, \ldots, \Lambda_{ip}]$ are the mean and covariance matrix parameters of the P mixture densities corresponding to class ω_i. These model parameters can be identified by applying an appropriate clustering algorithm, such as the k-means algorithm, or the expectation-maximization algorithm to the training vectors of each class. The discriminant function is computed as $g_i(x) = G_i(x|\theta_i)$ $p(\omega_i)$, where the prior probability $p(\omega_i)$ is approximated by the relative number of training vectors in class i. In the numerical examples, the can also be modeled data as Gaussian rather than a Gaussian mixture ($P = 1$). Furthermore, you can use the maximum-likelihood (ML) classifier (uniform prior probabilities).

Support Vector Machine Classifier   A support vector machine (SVM) is essentially a linear classifier operating in a higher dimensional space. Consider a binary classification problem without loss of generality. Let $\{\varphi(x)\}_{i=1}^{M}$ be a set of nonlinear transformations mapping the N-dimensional input vector to an M-dimensional feature space ($M > N$). A linear classifier, characterized by the weights $w_1, w_2, \ldots, w_M$, operates in this higher dimensional feature space $g(x) = \sum_{j=1}^{M} w_j \varphi_j(x) + b$, where b is the bias parameter of the classifier. The optimal weight vectors for this classifier can be represented in terms of a subset of training vectors, termed the support vectors $w_j = \sum_{i=1}^{Q} \alpha_i \varphi_j(x_i), j = 1, 2, \ldots, M$. Using the preceding representation for the weight vectors, the linear classifier can be expressed as $g(x) = \sum_{i=1}^{Q} \alpha_i K(x, x_i) + b$, where $K(x, x_i) = \sum_{j=1}^{M} \varphi_j(x)\varphi_j(x_i)$ is the symmetric kernel representing the SVM. In the numerical examples presented in this chapter, we use a third-degree polynomial kernel: $K(x\,mmax_i) = (x^T x_i + 1)^3$. In practice, the SVM discriminant function $g(x)$ is computed using the kernel representation, bypassing the nonlinear transformation into the higher dimensional space [13]. The classifier design then corresponds to the choice of the kernel and the support vectors. By appropriately choosing the kernel, an SVM can realize a neural network classifier as well. Similar to neural networks, the training phase can take a long time. However, once the classifier is trained, its application is relatively easy. In general, a different SVM is trained for each class. The output of each SVM can then be regarded as an estimate of the posterior probability for that class and the MAP decision rule can be directly applied.

14.4.3   Target Tracking Based on Cooperative Binary Detection

14.4.3.1   Introduction   Unlike other sensor network–based methods, which depend on determining distance to the target or the angle-of-arrival of the signal, the cooperative tracking approach requires only that a sensor be able to determine if an object is somewhere within the maximum detection range of the sensor. Cooperative tracking is proposed as a method for tracking moving objects and extrapolating their paths in the short term. By combining data from neighboring sensors, this approach enables tracking with a resolution higher than that of the individual sensors being used. In cooperative tracking, statistical estimation and approximation techniques can be employed to further increase the tracking precision, and enables the system to exploit the trade-off between accuracy and timeliness of the results. This work focuses on acoustic tracking; however, the presented methodology is applicable to any sensing modality where the sensing range is relatively uniform.

Cooperative tracking is a solution for tracking objects using sensor networks, and may achieve a high degree of precision while meeting the constraints of sensor network systems. The approach uses distributed sensing to identify an object and determine its approximate position, and local coordination and processing of sensor data to further refine the position estimate. The salient characteristics of the cooperative tracking approach are that it achieves a resolution that is finer than that of the individual sensors being used and that it provides early estimates of the object's position and velocity. Thus cooperative tracking is useful for short-term extrapolation of the object's path. Here an acoustic tracking system for wireless sensor networks is considered as a practical application of the cooperative tracking methodology. Acoustic tracking relies on a network of microphone-equipped sensor nodes to track an object by its characteristic *acoustic signature*.

14.4.3.2 Model In the real world, objects can move arbitrarily, that is, possibly changing speed and direction at any time. The representation of such arbitrary paths may be cumbersome and unnecessarily complex for the purpose of tracking the object's path with a reasonable degree of precision. Instead, an approximation of the path can be considered. Cooperative tracking uses *piecewise-linear approximation* to represent the path of the tracked object. Although the object itself may move arbitrarily, its path is considered as a sequence of line segments along which the object moves with a constant speed. The degree to which the actual path diverges from its representation depends on several factors, including the speed and turning radius of the object itself. For vehicles such as cars driving along highways the difference is quite small, whereas for a person walking a curved route with tight turns, it may be significant. In either case, accuracy can be improved by increasing the resolution of the sensor network, either through increasing sensor density or by other means.

In cooperative tracking, it is assumed that each node is equipped with a sensor (in the case of acoustic tracking, a microphone) and a radio for communication with nearby nodes. Since these embedded systems are designed to be small and inexpensive, the sensors they are equipped with are unlikely to be very sophisticated. Traditionally, tracking relies on sensors that are long range and can detect the direction of an object and the distance to it. This is not the case with sensor networks: the microphones used for acoustic tracking are likely to be short range, nondirectional, and poorly suited for detecting the distance to the sound source. The method presented here assumes that only binary (on–off) detection can be used. It is possible to generalize this analysis if multilevel detection is feasible. Moreover, without proper calibration the detection range may be neither uniform nor exact. Figure 14.9 shows the model of a sensor considered in this chapter. Given a sensor with a nominal (noncalibrated) range R, the object will always be detected if it is distance $R - e$ or less away from the sensor, detected some of the time between $R - e$ and $R + e$, and never detected beyond that range. We found that setting $e = 0.1R$ comes fairly close to the actual behavior of the sensors used in our experiments.

To track an object, it must be identified and its presence detected. For acoustic tracking, objects are identified based on their acoustic signature, which is a characteristic sound pattern or a set of frequencies unique to that object. For simplicity, it is assumed that the object emits the sound of a frequency not present in the environment, so there are no false positives. However, the results are fairly robust with respect to intermittent detection (false negatives) during the period of observations.

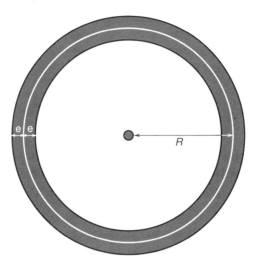

Figure 14.9   Model of a sensor. For nominal sensing range R, the object is always detected when it is $R - e$ away or closer, never detected beyond $R + e$, and has a nonnegative chance of detection between $R - e$ and $R + e$.

It is worth noting that the sensor model is generic enough to encompass other sensing modalities beyond acoustic. All that is required is a sensor with a relatively uniform range, as defined earlier, that is capable of differentiating the target from the environment. The magnetometer, a device that detects changes in magnetic fields, is one such sensor.

14.4.3.3   Algorithm   The simplest distributed tracking algorithm entails simply recording the times when each sensor detects the object, and then performing line fitting on the resulting set of points. While simple, this approach is not very precise: it can only track the object with a resolution of the sensor range R. Moreover, if a sensor detects the object more than once as it moves through the sensor's detection range, that information is lost.

The position of a stationary object, or a moving object for that matter, which is determined using this method is not very precise and depends heavily on the number, the detection range, and precision of sensors that detect the sound. Instead of looking at a single position measurement, we are interested in the path of a moving object, which is a sequence of positions over a period of time. Combining a large number of somewhat imprecise position estimates distributed over space and time may yield surprisingly accurate results. Cooperative tracking addresses the problem of high-resolution tracking using sensor networks. It improves accuracy by combining information from neighboring sensors. The only requirement for cooperative tracking to be used is that the density of sensor nodes must be high enough for the sensing ranges of several sensors to overlap. When the object of interest enters the region where multiple sensors can detect it, its position can be pinned down with a higher degree of accuracy, since the intersection area is smaller than the detection area of a single node. Below is the outline of a generic cooperative tracking algorithm.

- Each node records the duration for which the object is in its range.
- Neighboring nodes exchange these times and their locations.
- For each point in time, the object's estimated position is computed as a weighted average of the detecting nodes' locations.
- A line-fitting algorithm is run on the resulting set of points.

Several of these steps require careful consideration. First, the algorithm implicitly assumes that the nodes' clocks are synchronized, and that the nodes know their locations. Second, we obtain a position reading by a weighted average of the locations of the nodes that detected the sound at a given instant, but the exact weighting scheme is not specified. This is an important issue, as selecting an appropriate scheme will improve accuracy, while a poor choice might be detrimental to it.

The simplest choice is to assign equal weights to all sensors' readings. This effectively puts the estimate of the object's position at the centroid of the polygon, with sensors acting as vertices. This is a safe choice, and intuitively it should be more accurate than noncooperative tracking. However, it is possible to do even better. Consider Figure 14.10: sensors that are closer to the path of the target will stay in sensor range for a longer duration. Thus to increase accuracy, the weight of a sensor's reading should be proportional to some function of the duration for which the target has been in range of that sensor.

Once the individual position estimates are computed, the final step of the line-fitting algorithm can begin. Least-square regression can be used to find the equation of the line. It is interesting to note that the duration-based weighting scheme for pos-

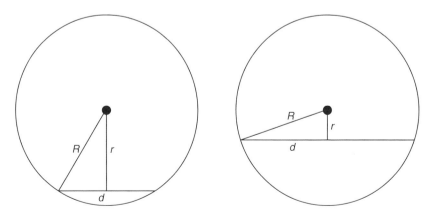

Figure 14.10 If the object's speed is constant, detection time is directly proportional to the path segment d and inversely proportional to the distance r from the sensor to the object's trajectory.

ition estimates moves the points closer to the actual path, thus reducing variance in the least-square computation. Also important is the fact that the multistep approach enables early estimates of the path to be computed, so that continuous refinement is possible, as more data points become available. The resulting equation of the line extrapolates the path of the object until it changes course sharply. This information may be used by the system, for example, for asynchronous wake-up of nodes likely to be in its path.

14.4.3.4   Data Aggregation   The final step of the algorithm involves performing a line-fitting computation on the set of all the position estimates (or some subset of them). Unlike position estimates, which can be performed in a distributed manner with only local communication, this necessitates collecting sensor readings from many sensor nodes at a centralized location for processing. This process is called *data aggregation*, and it is present in one form or another in virtually all sensor network applications. The main concerns for data aggregation are timeliness and resource usage. Timeliness, with respect to sensor data, is critical to real-time monitoring and control applications where stale data are useless or even detrimental. Resources, in particular network bandwidth and message buffers, are quite scarce in networked-embedded systems. Low bandwidth of small wireless transmitters and the potential for contention with other messages drastically limit the amount of data that can pass through the network.

We assume that some nodes in the sensor network are gateway nodes connected to outside networks such as the Internet. To process the data from the sensor network, they need to be sent through one of these gateway nodes to the more powerful computers connected to the outside network. To do this efficiently, a tree rooted at each gateway is constructed and spanning the entire network. Each sensor node in the tree collects data from its children and sends them up the tree to either the closest or the least busy gateway. This scheme addresses the conflicting requirements of low bandwidth usage and timeliness of data: a near-shortest path is always taken, unless its links are overloaded. The method just discussed is practical only if the outside network is low latency and high bandwidth, so it does not matter to which gateway the data are sent.

14.4.4   Distributed Prediction Tracking

The distributed prediction tracking (DPT) algorithm is specifically aimed at addressing the various challenges outlined in Section 14.4 while accurately tracking moving targets. As the name suggests, this algorithm does not require any central control point, eliminating the possibility of a single point of failure and making it robust against random node failures. The tracking task is carried out distributively by sequentially involving the sensors located along the track of the moving target. DPT assumes a cluster-based architecture for the sensor network, and the choice was motivated by the need to ensure the sensor network's scalability and energy efficiency. Any suitable clustering mechanism from those proposed in the literature may be used, also note that DPT does not impose any specific requirements or restrictions on the choice of clustering algorithm.

14.4.4.1 *Assumptions of the DPT Algorithm*

While no assumption is made on the choice of the clustering algorithm, we assume that the *clusterhead* (*CH*) has the following information about all sensors belonging to its cluster: (1) sensor identity, (2) location, and (3) energy level. When tracking a moving target and deciding which sensors to use for tracking, the CH's decision-making procedure will be based on this information. The assumptions about the sensors are enumerated below. These assumptions are realistic and targeted at reducing the energy cost and prolonging the whole network's lifetime as well.

1. All sensors have the same characteristics.
2. Sensors are randomly distributed across the whole sensing area with uniform density.
3. Each sensor has two sensing radii, normal beam r and high beam R. The default operation uses the low beam, and the high beam is turned on only when necessary. The following relationship holds between the energy consumed by the low and high beams:

$$\frac{E_{low\ beam}}{E_{high\ beam}} = \frac{r^2}{R^2} \qquad (14.2)$$

4. A sensor's communication and sensing channels stay in the hibernation mode most of the time where they consume minimal energy. The communication channel will wake up routinely to receive possible messages from its CH. The sensor will perform sensing according to its CH's requirements.

In order to produce information that is accurate enough to locate the moving target, DPT requires that at any given time there should be at least three sensors to sense the target jointly. The number 3 is chosen as a compromise between increasing accuracy and minimizing the consumed energy (note that this is not a hard assumption, and depending on the sensor node specifications, the number may vary).

No specific assumptions are made about the movement pattern of the targets. However, DPT assumes that the targets originate outside the sensing region and then move inside. Also, it is assumed that the movement of each tracked target needs to be forwarded to a central location, which we term the sink. In reality, the sink could be either a special node or a terminal associated with a human.

14.4.4.2 *DPT Algorithm*

The DPT algorithm comes into play after sensors are deployed and clusters are formed. DPT distinguishes between the *border sensors*, sensors located within a given distance of the border, and *nonborder sensors* in terms of their operation. While border sensors are required to keep sensing all times in order to detect all targets that enter the sensing region, the nonborder sensor's sensing channel hibernates unless it is specifically asked to sense by its cluster head. Since the target is assumed to move from outside into the sensing area, it will be detected by the border sensors when trespassing the border. As soon as a target is detected, a sequence of tasks in the order of *sense–predict– communicate–sense* are carried out distributively by a series of sensors that are located along the target's track. This forms the essential idea behind the DPT algorithm.

Let $CH_1, CH_2, \ldots, CH_N$ denote the sequence of CHs that become involved with tracking the target as it proceeds from its very first location to the last. The information gathered by each CH is sent all the way back to the sink (either sent intact or after being aggregated) for further processing as well as to the downstream clusterhead CH_{i+1}. The *target identity* is created when the target is first detected. This identity is unique and all CHs that cotrack this target use it to identify the target. In order to facilitate the smooth tracking of the target, CH_i predicts the future location of the moving target, and informs the downstream clusterhead CH_{i+1} ahead of time about this target. The accuracy of the *prediction* is very important if downstream CHs are to be identified accurately and the overall tracking mechanism is to be effective. Many prediction mechanisms are possible, the simplest one is a linear predictor, which uses only the previous two locations to linearly predict the third location. Higher-order prediction can also be adopted, it predicts the nth location information based on previous $n - 1$ actual locations. Higher-order prediction results in more accurate results, though, at the cost of greater energy consumption.

Sensor Selection Algorithm   After clusterhead CH_i predicts the location of the target, the downstream clusterhead CH_{i+1}, toward which the target is headed receives a message from CH_i indicating this predicted location. With information of all the sensors belonging to CH_{i+1} available in its database, the search algorithm running at CH_{i+1} is able to locally decide the sensor-triplet to sense the target. The selection rule chooses three sensors (if possible) such that their distances to the predicted location are not only less than the sensor's normal beam r, but also the smallest. After the sensor-triplet is chosen, CH_{i+1} sends them a *wake-up* message so that they are ready to sense the target. If the prediction and selections process succeeds, after sensing, each sensor will send a location message to CH_{i+1}. If CH_{i+1} is unable to find enough sensors eligible for this sensing task with the normal sensing beam, it will try to search for eligible sensors within a distance R, the higher sensing beam, from the predicted location. The selected sensors, whose distance from the predicted location is greater than r and lower than R, will now be contacted and instructed to sense with their high beam, while the rest of the sensors in the triplet use their normal beam. If CH_{i+1} is unable to find enough sensors even with high sensing beams, it asks its neighboring CHs for help.

Failure Recovery   Let us first identify two possible failure scenarios. As described in the previous subsections, each upstream CH sends a message to the expected downstream CH. If the upstream CH does not get any confirmation from the downstream CH after a given period of time, then it assumes that the downstream CH is no longer available and the target has been lost. Another failure scenario occurs when the target changes it direction or speed so abruptly that it moves significantly away from the predicted location and falls out of the detectable region of the sensor-triplet selected for the sensing task. In both of these failure scenarios a straightforward solution is to wake up all sensors within a given area, which is calculated based on the target's previous actual location. The *recapture* radius σ is an important parameter in this process and is decided by the target's moving speed and time elapsed since it was last sensed.

14.5 EXPERIMENTAL LOCATION AND TRACKING SYSTEMS

In this section, several location systems are introduced. Although they may not be specially designed for wireless sensor networks, these design concepts and experiences will benefit future implementations of positioning systems in wireless sensor networks.

14.5.1 Active Badge and Bat

14.5.1.1 Introduction Efficient location and coordination of staff in any large organization is a difficult and recurring problem. Hospitals, for example, may require up-to-date information about the location of staff and patients, particularly when medical emergencies arise. See ref. [25] for detail information.

14.5.1.2 An Active Badge Design A solution to the problem of automatically determining the location of an individual has been to design a tag in the form of an "Active Badge" that emits a unique code for approximately one-tenth of a second every 15 seconds (a beacon). These periodic signals are picked up by a network of sensors placed around the host building. A master station, also connected to the network, polls the sensors for badge "sightings," processes the data, and then makes it available to clients that may display it in a useful visual form. The badge was designed in a package that is roughly $55 \times 55 \times 7$ mm and weighs a comfortable 40 g. Pulsewidth-modulated infrared (IR) signals are used for signaling between the badge and sensor [26], mainly because IR solid-state emitters and detectors can be made very small and very inexpensively (unlike ultrasonic transducers); they can be made to operate with a 6-m range; and the signals are reflected by partitions and therefore are not directional when used inside a small room. Moreover, the signals will not travel through walls, unlike radio signals that can penetrate the partitions found in office buildings. An active signaling unit consumes power; therefore, the signaling rate is an important design issue. First, by only emitting a signal every 15 seconds, the mean current consumption can be very small, with the result that "badge-sized" batteries will last for about one year. Second, it is a requirement that several people in the same locality be detectable by the system. Because the signals have a duration of only one-tenth of a second, there is approximately a 2/150 chance that two signals will collide when two badges are placed in the same location. For a small number of people, there is a good probability they will all be detected. The Active Badge also incorporates a light-dependent component that, when dark, turns the badge off to conserve battery life. Reduced lighting also increases the period of the beacon signal to a time greater than 15 seconds. In ambient lighting conditions in a room, this effect only slightly modifies the period, but it is another factor that ensures that synchronized badges will not stay synchronized very long. If the badge is placed in a drawer out of office hours, at weekends, and during vacation, the effective lifetime of the batteries is increased by a factor of 4. A disadvantage of an infrequent signal from the badge is that the location of a badge is only known, at best, to a 15-s time window. However, because in general a person tends to move relatively slowly in an office building, the information the Active Badge system provides is very accurate. An Active Badge signal is transmitted to a sensor through an optical path. This path may be found indirectly through a surface reflection, for example, from a wall. The Active Badge location system was developed at Olivetti Research Laboratory and now AT&T at Cambridge.

14.5.1.3 Bat System   A successor of the Active Badge system is the Bat system [27], which consists of a collection of wireless transmitters, a matrix of receiver elements, and a central RF base station. The wireless transmitters, called bats, can be carried by a tagged object and/or attached to equipment. The sensor system measures the time of flight of the ultrasonic pulses emitted from a bat to receivers installed in known and fixed positions and it uses the time difference to estimate the position of each bat by trilateration. The RF base station coordinates the activity of bats by periodically broadcasting messages to them. The location of the bat can be determined within 3 cm of error in a three-dimensional space at 95% accuracy. This accuracy is quite enough for most location-aware services; however, the deployment cost is high.

14.5.2 Cricket

Cricket is a location-support system for in-building, mobile, location-dependent applications. It allows applications running on mobile and static nodes to learn their physical location by using listeners that hear and analyze information from beacons spread throughout the building. Cricket is the result of several design goals, including user privacy, decentralized administration, network heterogeneity, and low cost. Rather than explicitly tracking user location, Cricket helps devices learn where they are and lets them decide to whom to advertise this information; it does not rely on any centralized management or control and there is no explicit coordination between beacons; it provides information to devices regardless of their type of network connectivity; and each Cricket device is made from off-the-shelf components and costs less than $10. See ref. [28] for more information about Cricket.

By not tracking users and services, user privacy concerns are adequately met. We emphasize that Cricket is a *location-support* system, rather than a conventional *location-tracking* system that tracks and stores location information for services and users in a centrally maintained database.

Cricket uses a combination of RF and ultrasound to provide a location-support service to users and applications. Wall- and ceiling-mounted beacons are spread through the building, publishing location information on an RF signal. The beacon transmits a concurrent ultrasonic pulse with each RF advertisement. The listeners receive these RF and ultrasonic signals, correlate them to each other, and infer the space they are currently in. The beacons use a decentralized randomized transmission algorithm to minimize collisions and interference among each other. The listeners implement a decoding algorithm to overcome the effects of ultrasound multipath and RF interference.

14.5.3 RADAR: An In-Building RF-Based User Location and Tracking System

RADAR [19] is an RF-based system for locating and tracking users inside buildings. RADAR operates by recording and processing signal strength information at multiple base stations positioned to provide overlapping coverage in the area of interest. It combines empirical measurements with signal propagation modeling to determine user location, and thereby enable location-aware services and applications.

RADAR complements the data networking capabilities of RF wireless local area networks (LANs) with accurate user location and tracking capabilities, thereby enhancing the value of such networks. RADAR uses signal-strength information gathered at multiple receiver locations to triangulate the user's coordinates. Triangulation is done using both empirically determined and theoretically computed signal-strength information.

14.6 CONCLUSION

In this chapter, some fundamental techniques in positioning and location tracking have been discussed and several experimental systems reviewed. Location information may enable new types of services. Accuracy and deployment costs are two factors that may contradict each other, but both are important factors for the success of location-based services.

REFERENCES

1. P. Eng and P. Mirsa. Special issue on global positioning system. *Proceedings of the IEEE*, **87**:3–15, January 1999.

2. D. Li, K. Wong, Y. Hu, and A. Sayeed. Detection, classification, and tracking of targets. *IEEE Signal Processing Magazine*, **19**(2):17–30, March 2002.

3. P. Bose, P. Morin, I. Stojmenovic, and J. Urrutia. Routing with guaranteed delivery in ad hoc wireless networks. *Wireless Networks*, **7**(6):609–616, 2001. See also In *Proceed-*

ings of the 3rd International Workshop on Discrete Algorithms and Methods for Mobile Computing and Communications, pages 48–55, Seattle, Washington, August 1999.

4. Brad Karp and H. T. Kung. GPSR: Greedy perimeter stateless routing for wireless networks. In *Proceedings of the 6th Annual ACM/IEEE International Conference on Mobile Computing and Networking (MobiCom 2000)*, pages 234–254, Boston, Massachusetts, August 2000.

5. J. Li, J. Jannotti, D. De Couto, D. Karger, and R. Morris. A scalable location service for geographic ad-hoc routing. In *Proceedings of the 6th Annual ACM International Conference on Mobile Computing and Networking (MobiCom 2000)*, pages 120–130, Boston, Massachusetts, August 2000.

6. J. Heidemann and N. Bulusu. Using geospatial information in sensor networks. In *Proceedings of the CSTB Workshop on Intersection of Geospatial Information and Information Technology*, Arlingtan, Virginia, October, 2001.

7. Y. Ko and N. Vaidya. *GeoTORA: A Protocol for Geocasting in Mobile Ad Hoc Networks*. Technical Report 00-010, Department of Computer Science, Texas A&M University, March 2000.

8. Julio C. Navas and Tomasz Imielinski. GeoCast—geographic addressing and routing. In *Proceedings of the 3rd Annual ACM/IEEE International Conference on Mobile Computing and Networking (MobiCom '97)*, pages 66–76, Budapest, Hungary, September 1997.

9. Y. Ko and N. Vaidya. *Geocasting in Mobile Ad Hoc Networks: Location-Based Multicast Algorithms*. Technical Report TR-98-018, Texas A&M University, September 1998.

10. S. Atiya and G. Hager. Real-time vision-based robot localization. *IEEE Transactions on Robotics and Automation*, **9**(6):785–800, 1993.

11. J. Leonard and H. Durrant-Whyte. Mobile robot localization by tracking geometric beacons. *IEEE Transactions on Robotics and Automation*, **7**(3):376–382, 1991.

12. R. Tins, L. Navarro-Serment, and C. Paredis. Fault tolerant localization for teams of distributed robots. In *Proceedings of the IEEE/RSJ International Conference on Intelligent Robots and Systems*, Volume 2, pages 1061–1066, Maui, Hawaii, October 2001.

13. Jeffrey Hightower and Gaetano Borriella. Location systems for ubiquitous computing. *Computer (IEEE)*, **34**(8):57–66, 2001.

14. N. Bulusu, J. Heidemann, and D. Estrin. *Gps-less Low Cost Outdoor Localization for Very Small Devices*. Technical Report 00-729, Computer Science Department, University of Southern California, April 2000.

15. Srdan Capkun, Maher Hamdi, and Jean-Pierre Hubaux. GPS-free positioning in mobile ad-hoc networks. In *Proceedings of the 34th Annual Hawaii International Conference on Systems Science (HICSS-34)*, Volume 9, page 9008, Maui, Hawaii, January 2001.

16. Dragos Niculescu and Badri Nath. Ad hoc positioning system (APS). In *Proceedings of IEEE GLOBECOM 2001*, pages 2926–2931, San Antonio, Texas, November 2001.

17. Andreas Savvides, Chih-Chieh Han, and Mani B. Srivastava. Dynamic fine-grained localization in ad-hoc networks of sensors. In *Proceedings of the 7th Annual International Conference on Mobile Computing and Networking (MobiCom 2001)*, pages 166–179, Rome, Italy, July 2001.

18. P. Bahl, A. Balachandran, and V. Padmanabhan. *Enhancements to the RADAR User Location and Tracking System*. Microsoft Research Technical Report MSR-TR2000-12, Microsoft Research, Redmond, Washington, February 2000.

19. Paramvir Bahl and Venkata N. Padmanabhan. RADAR: An in-building RF-based user location and tracking system. In *Proceedings of INFOCOM* 2000, Volume 2, pages 775–784, Tel-Aviv, Israel, March 2000.

20. T. Roos, P. Myllymaki, H. Tirri, P. Misikangas, and J. Sievanen. A probabilistic approach to WLAN user location estimation. *International Journal of Wireless Information Networks*, **9**(3):155–164, 2002.

21. Y.-C. Tseng, S.-P. Kuo, H.-W. Lee, and C.-F. Huang. Location tracking in a wireless sensor network by mobile agents and its data fusion strategies. In *Proceedings of the 2nd International Workshop on Information Processing in Sensor Networks (IPSN)*, pages 625–641, Palo Alto, California, April 2003.

22. C. Savarese, J. Rabay, and K. Langendoen. Robust positioning algorithms for distributed ad-hoc wireless sensor networks. In *Proceedings of the USENIX Technical Annual Conference*, pages 317–328, Monterey, California, June 2002.

23. A. Savvides, H. Park, and M. Srivastava. The bits and flops of the n-hop multilateration primitive for node localization problems. In *Proceedings of the 1st ACM International Workshop on Wireless Sensor Networks and Applications (WSNA 2002)*, pages 112–121, Atlanta, Georgia, September 2002.

24. J. Aspnes, D. Goldengerg, and Y. R. Yang. On the computational complexity of sensor network localization. In *Proceedings of the 1st International Workshop on Algorithmic Aspects of Wireless Sensor Networks (ALGOSENSORS 2004)*, pages 32–44, Turku, Finland, July 2004.

25. Roy Want, Andy Hopper, Veronica Falcão, and Jonathan Gibbons. *The Active Badge Location System*. Technical Report 92.1, Olivetti Research Ltd. (ORL), Cambridge, England, 1992.

26. Satellite, Cable, and TV IC Handbook. In *Plessey Semiconductors*, pages 64, 67, 124, 1988.

27. Mike Addlesee, Rupert Curwen, Steve Hodges, Joe Newman, Pete Steggles, Andy Ward, and Andy Hopper. Implementing a sentient computing system. *Computer (IEEE)*, **34**(8):50–56, 2001.

28. Nissanka B. Priyantha, Anit Chakraborty, and Hari Balakrishnan. The cricket location-support system. In *Proceedings of the 6th Annual ACM/IEEE International Conference on Mobile Computing and Networking (MobiCom 2000)*, pages 32–43, Boston, Massachusetts, August 2000.

CHAPTER 15

Data Gathering and Fusion in Sensor Networks

WEI-PENG CHEN

Fujitsu Laboratories of America, Sunnyvale, California

JENNIFER C. HOU

University of Illinois at Urbana Champaign, Urbana, Illinois

Data gathering is one of the most important tasks in sensor networks. In most data-gathering applications, sensors extract useful information from the environment and either respond to queries made by users or take the active role of disseminating the information to one or more sinks. The information is then exploited by subscribers/users for environment monitoring, target tracking, and/or decision making. In some sense, a sensor network can be envisioned as a distributed database that provides a layer of query processing for users. How information can be effectively gathered, aggregated, and disseminated to users and how queries made by users can be effectively directed to sensors that have the corresponding information is the focus of this chapter. Specifically, an overview is given of research activities on data gathering and fusion along three research thrusts: (1) query processing in sensor database systems, (2) data-gathering and -dissemination mechanisms, and (3) data-fusion mechanisms. In each of the research thrusts, first an overview of the various mechanisms is presented and their design objectives and criteria are discussed. This is then followed by a taxonomy of existing mechanisms based on their design objectives and criteria. Also presented are several utility-based data-gathering algorithms that maximize the amount of information extracted, subject to constraints on flow conservation, energy, latency, and/or channel bandwidth.

Handbook of Sensor Networks: Algorithms and Architectures, Edited by Ivan Stojmenović

Copyright © 2005 John Wiley & Sons, Inc.

15.1   INTRODUCTION

Recent technological advances have led to the emergence of small, low-power devices that integrate sensors and actuators with limited on-board processing and wireless communication capabilities. Pervasive networks of such sensors and actuators open new vistas for constructing complex monitoring and control systems, ranging from habitat monitoring [1], target tracking [2], home automation [3], ubiquitous sensing for smart environments [4], construction of safety monitoring, and inventory tracking. In most sensor network applications, sensors extract useful information from the environment, and either respond to queries made by users or take an active role to disseminate the information to one or more sinks. The information is then exploited by subscribers/users for their decision making. In other words, one can envision sensor networks as a distributed database for users to query the physical world [5]. How information can be effectively gathered, aggregated, and disseminated to users and how queries made by users can be effectively directed to sensors that have the corresponding information is the focus of this chapter. Figure 15.1 depicts the simplified relationship between users and sensor networks.

The process of data gathering in sensor networks is nevertheless significantly different from conventional warehousing database systems, where data are extracted from sensors and stored in a centralized server that is responsible for query processing. Aside from the fact that sensor networks operate in a distributed fashion, they encompass several distinct characteristics, and hence pose more challenges [6,7]: (1) the convention that sensors are usually deployed with high nodal density pose a scalability problem; (2) the fact that these sensors are usually left unattended once deployed makes autonomous operations necessary; (3) the fact that the computing and communication environment is unreliable due to the irregular terrain, environment dynamics, energy depletion, and potential hardware defects requires that the design be robust; and (4) the resource constraints in energy, bandwidth, storage, and computation capability require that resources be more efficiently used. In general, the design criteria for data-gathering applications in sensor networks are: (1) scalability, (2) autonomy, (3) robustness, and (4) energy-efficiency. In addition,

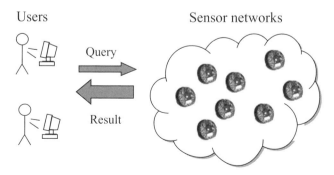

Figure 15.1   Query/result relationship between users and sensor networks.

there are several features that should be included in the design and implementation of data-gathering applications:

- *Devising Localized Algorithms.* In a localized algorithm, each node operates on the information locally collected. As compared to algorithms that rely on global topological knowledge, localized algorithms incur less communication overhead (and hence save power) in the case of topology changes (as a result of power depletion and/or environmental stimuli), and hence adapt better to these changes. Reduction in the communication overheads (and hence saving in the power) also contributes to system scalability.

- *Aggregating Data in the Process of Routing* [8]. Redundancy exists in sensor data in both the temporal and spatial domains. That is, readings collected by a single sensor at different times *or* among neighboring sensors may be highly correlated, and contain redundant information. Instead of transmitting all the highly correlated information to subscribers, it may be more effective for some intermediate sensor node(s) to digest the information received and come up with a concise digest, in order to reduce the amount of raw data to be transmitted (and hence the power incurred, and bandwidth consumed, in transmission). This technique is termed as *data fusion* (also called *data aggregation*). Data fusion can also be integrated with routing. Compared with traditional *address-centric* routing, which finds the shortest paths between pairs of end nodes, *data-fusion–centric* routing aims to locate routes that lead to the largest degree of data aggregation.

- *Being Adaptive to Topology Changes.* Due to environmental dynamics (such as channel fading due to weather effects) and node failure (as a result of power depletion and hardware failure), the network topology can change from time to time. In addition, the locations of the traffic source and destination, as well as the amount of traffic may vary. Adaptation to these changes is the key to making the system autonomous and efficient.

- *Increasing Node/Route Redundancy.* In second listed feature, we state that it is desirable to remove *data* redundancy in the time and spatial domains. On the other hand, deploying a sensor network with a high nodal density so as to increase node/route redundancy is likely to make the system more resilient and robust to all the aforementioned environment dynamics. Increasing node redundancy also extends network lifetime if subsets of nodes can be properly identified (each of which covers the entire monitoring area) and take turns carrying out the task of sensing the environment and monitoring.

In this chapter, we give a survey of research activities in the areas of data gathering, dissemination, and fusion. The survey is conducted along three research thrusts: (1) query processing in sensor database systems, (2) data-gathering and -dissemination mechanisms, and (3) data-fusion mechanisms. The categorization is made roughly based on the major focus of algorithms, although some algorithms consider both data dissemination and fusion jointly.

The rest of this chapter is organized as follows. In Section 15.2 we introduce sensor database systems and how queries are processed in such systems. In Section 15.3, we present an overview of data-gathering and dissemination mechanisms and two predominant factors that determine the system architecture. This is then followed by a taxonomy of data-gathering mechanisms based on storage locations, directions of diffusion, and structures of dissemination. In Section 15.4, we give an overview of data-fusion mechanisms, and then classify them based on functions of data fusion, system architectures, and trade-offs in the system design. Finally, in Section 15.5 we present several utility-based data-gathering algorithms that maximize the amount of information extracted.

15.2   SENSOR DATABASE

Sensor networks provide a new computing platform for users to readily access the data in the physical world [5]. They can be viewed as a large distributed database system. Consider an environment monitoring and alert system that is similar to the ALERT system (http://www.alertsystems.org). Several types of sensors, including rainfall sensors, water-level sensors, weather sensors, and chemical sensors, are used to record the precipitation and water level regularly, to report the current weather conditions, and to issue flood or chemical pollution warnings. In such a monitoring application, there are five types of queries that users typically make [5,9,10]:

1. *Historical Queries.* These queries are concerned with aggregate, historical information gathered over time and stored in a database system, for example, "What was the average level of rainfall of Champaign County in May 2000?"

2. *Snapshot Queries.* These queries are concerned with the information gathered from the network at a specific (current or future) time point, for example, "Retrieve the current readings of temperature sensors in Champaign County."

3. *Long-Running Queries.* These queries ask for information over a period of time, for example, "Retrieve every 30 minutes the highest temperature sensor reading in Champaign County from 6 P.M. to 10 P.M. tonight."

4. *Event-Triggered Queries* [9]. These queries prespecify the conditions that trigger queries, for example, "If the water level exceeds 10 meters in Champaign County, query the rain-fall sensors about the amount of precipitation during the past hour. If the amount of precipitation exceeds 100 mm, send an emergency message to the base station to issue a flood warning."

5. *Multidimensional Range Queries* [10]. These queries involve more than one attribute of sensor data and specify the desired search range as well, for example, "In Champaign County, list the positions of all sensors that detect water level between 5 to 8 meters and have temperatures between 50 and $60°F$."

A complete hierarchical architecture (four-tier) of sensor database systems for a monitoring application answering these five types of queries is depicted in

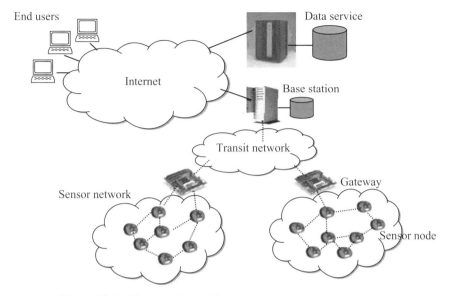

Figure 15.2 The complete architecture of a sensor database system.

Figure 15.2 [1]. The lowest level is a group of sensor nodes that perform sensing, computing, and in-network processing in a field. The data collected within the sensor network are first propagated to its gateway node (second level). Next the gateway node relays the data through a transit network to a remote base station (third level). Finally, the base station connects to a database replica across the Internet. Among the four tiers, the resource within the sensor networks is the most constrained. In most of the applications, the sensor network is composed of sensors and a gateway node (sink), as shown in Figure 15.3, although the number of sinks or sources might vary from application to application.

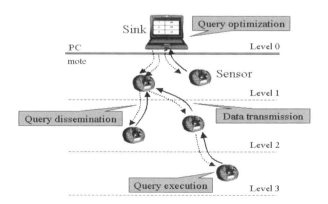

Figure 15.3 Procedures for query and data extraction in TinyDB [9,11].

15.2.1   Example Sensor Database System

The main purpose of a sensor database system is to facilitate the data-collection process. Users specify their interests via simple, declarative *structured query language – like (SQL)* queries. Upon receipt of a request, the sensor database system efficiently collects and processes data within the sensor network, and disseminates the result to users [11]. A query-processing layer between the application layer and the network layer provides an interface for users to interact with the sensor network. The layer should also be responsible for managing the resources (especially the available power).

Two of the most representative sensor database systems are TinyDB [9,12] and Cougar [11,13]. The former evolves from *tiny aggregation (TAG)*, and is built on top of the TinyOS operating system [14] (which operates on smart dusts, Motes, developed by University of California at Berkeley). The latter database system is developed by Cornell University. Both the TinyDB and Cougar architectures consist of a single base station (sink) and multiple sensors. The sink and sensors are connected in a routing tree, shown in Figure 15.3. A sensor chooses its parent node, which is one hop closer to the root (sink). The sink accepts queries from users outside the sensor network. Query processing can be performed in four steps: query optimization, query dissemination, query execution, and data dissemination.

Both TinyDB and Cougar provide a declarative SQL-like query interface for users to specify the data to be extracted. Similar to SQL, the acquisitional query language used in TinyDB, TinySQL, consists of a select-from-where clause that supports selection, join, projection, and aggregation. The data within sensor networks can be considered as virtually a table, each column of which corresponds to an attribute and each row of which corresponds to a sample measured at a specific location and time. An example in TinySQL is like:

SELECT region id, **AVG**(water level), **AVG**(precipitation)
FROM water level sensor (W), precipitation sensor (P)
WHERE W.location **IN** Champaign County **AND** P.location **IN**
 Champaign County
GROUP BY region
Having AVG(W.water level) > 10 meters
EPOCH DURATION 10 minutes
TRIGGER ACTION report an emergency warning

This query monitors the water level in all regions in Champaign County every 10 minutes. If the average water level of sensors in a region exceeds 10 meters, the system generates a flooding warning and sends the region ID and the value of the average water level and precipitation to the sink. The query language in the sensor database differs from SQL mainly in that its queries are continuous and periodic [11].

Upon reception of a query, the sink performs query optimization to reduce the energy incurred in the pending query process. Two query optimization techniques are commonly used in TinyDB: ordering of sampling operations and query aggregation. First, since the energy incurred in retrieving readings from different types of sensors is different, the sampling operations should be reduced for sensors that consume high energy. For instance, the energy consumed for sampling a magnetic reading is much higher than that for a light reading. The sampling energy can be saved if a proper ordering of sampling operations can be arranged in the evaluation of the **HAVING** clause. For another example, the query "**HAVING** light > 200 and mag > 100" consumes less energy than the query "**HAVING** mag > 100 and light > 200," because in the former case the sampling operation for magnetic readings can be skipped if the condition on the light reading fails. Second, by combining multiple queries for the same event into a single query, only one query needs to be sent.

After a query is optimized at the sink, it is broadcast by the sink and disseminated to the sensor network. When a sensor receives a query, it has to decide whether to process the query locally and/or rebroadcasts it to its children. A sensor only needs to forward the query to those child nodes that may have the matched result. To this end, a sensor has to maintain information on its children's attribute values. In TinyDB, a *semantic routing tree* (*SRT*) containing the range of the attributes of its children is constructed at each sensor. The attributes can be static information (e.g., location) or dynamic information (e.g., light readings). For attributes that are highly correlated among neighbors in the tree, SRT can reduce the number of disseminated queries.

One distinct characteristic of query execution in TinyDB is that sensors sleep during every epoch and are synchronized to wake up, receive, transmit, and process the data in the same time period.

15.3 DATA-GATHERING AND DISSEMINATION MECHANISMS

The wide variety of requirements and objectives for different applications in sensor networks impose various design criteria and lead to different solutions. Two major factors that determine the system architecture and design methodology are:

1. *The number of sources and sinks within the sensor network*: Sensor network applications can be classified into three categories: one-sink–multiple-sources, one-source–multiple-sinks, and multiple-sinks–multiple-sources. An environment monitoring application shown in Figure 15.3 falls in the one-sink–multiple-sources category, since the interaction between the sensor network and the subscribers is usually through a single gateway (sink) node. On the other hand, a traffic-reporting system that disseminates the traffic condition (e.g., an accident) at a certain location to many drivers (sinks) falls in the one-source–multiple-sinks category.

2. *The trade-offs between energy, bandwidth, latency and information accuracy*: An approach cannot usually optimize its performance in all aspects. Instead, based on the relative importance of its requirements, an application usually trades less important criteria for optimizing the performance with respect to the most important attribute. For instance, for mission-critical applications, the end-to-end latency is perhaps the most important attribute and needs to be kept below a certain threshold, even at the expense of additional energy consumption. We will treat this topic in Section 15.4.3.

In what follows, we categorize data-gathering and -dissemination mechanisms based on the following three factors: (1) storage location, (2) direction of diffusion, and (3) structure of devices.

15.3.1 Classification of Data-Gathering Mechanisms Based on the Storage Location

In order to process historical queries, data collected at different sensors have to be properly stored in a database system for future query processing. Figure 15.4 shows three scenarios of placing storage at different locations [15]:

1. *External Storage (ES)*: All the data collected at sensors in a sensor network are relayed to the sink and stored at its storage for further processing. For a sensor network with n sensor nodes, the cost of transmitting data to the external storage is $O(\sqrt{n})$. There is no cost for external queries, while the cost of a query within the network incurs a cost of $O(\sqrt{n})$.

2. *Local Storage (LS)*: Data are stored at each sensor's local storage and thus no communication cost for data storage is incurred. However, each sensor needs to process all queries and a query is flooded to all sensors. The cost of flooding a query is $O(n)$.

3. *Data-Centric Storage (DCS)*: DCS stores the data at a sensor (or a location) in the sensor network based on the content of the data. Data storage in a DCS system consists of two steps: first the sensor maps an event it detects to a label via a consensus hash function and then routes the data to a node according to the label. The label can be a location and the sensor can route the data via geographic routing. We will introduce two of the representative approaches relying on geographic information, GHT [15] and DIM [10] in the next subsection. Both data and query communication costs are $O(\sqrt{n})$.

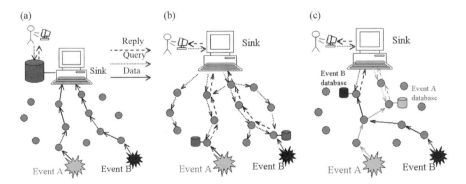

Figure 15.4 Three types of storage scenarios [15]. (a) External storage; (b) local storage; (c) data-centric storage.

15.3.1.1 *Database with Geographic Information* As just mentioned, one of the common hash functions in sensor database systems is to map the data to a location and then send the data via geographic routing to the sensor node that is closest to the mapped location for storage. If all of the sensors have the same hash function, a query with a specific content can be converted to a location where the data were stored for future retrieval. Geographic hash table (GHT) [15] and distributed index for multidimensional data (DIM) [10] are two of the representative databases with geographic information. Both of them adopt greedy perimeter stateless routing (GPSR) [16,17] as the underlying routing protocol, but differ slightly in the hash functions used.

In GHT, the input to the hash function is a reading of a single attribute or a specific type of event, and the hash result is a point in the two-dimensional space. If no sensor node is located at the precise coordinates of the hash result, the data are stored at the node closest to the hash result. With the use of the perimeter mode of GPSR, the data packet traverses the entire perimeter enclosing the location of the hash result, and the closest location can be identified.

DIM, on the other hand, is designed especially for multidimensional range queries. DIM maps a vector of readings with multiple attributes to a two-dimensional geographic zone. Two assumptions are made in DIM: first, sensors are aware of their own locations and field boundaries, and second, all the sensors are static. The entire field is divided recursively into zones, as shown in Figure 15.5. The sequence of divisions is vertical, horizontal, and so on. Each zone is encoded with a unique code based on the following rule: For a vertical division (the ith division where i is an odd number), the ith bit code of the zone is encoded as "1" if it is in the right region, and "0" otherwise. Similarly, the even bit of the code word is determined by whether the zone is above ("1") or below ("0") the divided line. For instance, the code word of the region in which node 6 resides in Figure 15.5 is "101." Due to the fact that sensors may not be uniformly deployed in an area, every zone just defined may not contain

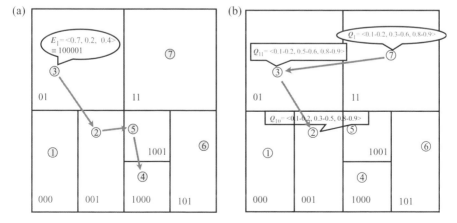

Figure 15.5 (a) Inserting an event; (b) issuing a multidimension range query [10].

a sensor. In other words, a sensor needs to determine the zone(s) it owns where no other sensors reside. This can be easily achieved when a node is aware of its neighbors' locations.

The encoding rule for mapping an event A with m normalized attributes $(A_1 \cdots A_m)$ $(0 \le A_i \le 1)$ to a zone with k divisions (k is a multiple of m) is based on the following rule:

> For $i = 1 \to m$, if $A_i < 0.5$, then the ith bit of the event $= 0$; otherwise, $= 1$.
>
> For $i = m + 1 \to 2m$, if $A_{i-m} < 0.25$ or $A_{i-m} = [0.5, \ 0.75)$, then the ith bit of the event $= 0$; otherwise, $= 1$.

Repeat the same procedure until all k bits are assigned.

With the encoding rules for both zones and events, the next task is to route the event to the node that owns the zone (code word) of the event. An example of inserting an event is illustrated in Figure 15.5(a). The event with two attributes $\langle 0.7, 0.2, 0.4 \rangle$ is routed to node 4, which owns the zone 1000. Similar encoding rules are applied to queries, except that when the range of a query is larger than the range of a zone, it has to be divided into several subqueries. An example of querying range event $\langle 0.1-0.2, \ 0.3-0.6, \ 0.8-0.9 \rangle$ is illustrated in Figure 15.5(b).

15.3.2 Classification of Data-Gathering Mechanisms Based on the Direction of Diffusion

The data-gathering process usually consists of two steps: query and reply. A sink (or user) sends a query to a sensor network and sensors that detect events matching the query send replies to the sink. Applications with different requirements opt for different communication paradigms. According to the direction of interest/data diffusion, there are three types of approaches [18]:

1. *Two-Phase Pull Diffusion.* The most representative approach in this category is *directed diffusion* [19]. Both the queries for events of interest and the replies are initially disseminated via flooding, and multiple routes may be established from a source to the sink. In the second pull phase, the sink *reinforces* the best route (usually with the lowest latency) by increasing its data rate (i.e., gradient). Data are then sent to the sink along this route. We present in detail the directed diffusion mechanism later. Two-phase pull diffusion is especially well suited for applications with many sources and only a few sinks.

2. *One-Phase Pull Diffusion* [18]. The overheads of flooding of both queries and replies are high in the cases that (1) there exist a large number of sinks or sources, and (2) the rate of queries for different events is high. One-phase pull diffusion skips the flooding process of data diffusion. Instead, replies are sent back to neighbors that first send the matching queries. In other words, the reverse path is the route with the least latency. One-phase pull

diffusion is well-suited for scenarios in which a large number of disparate events are being queried.

3. *Push Diffusion.* In the push-diffusion mechanism, a source actively floods the information collected when it detects an event and sinks subscribe to events of interest via positive enforcements. Push diffusion is well-suited for: (1) applications in which there exist many sinks and only a few sources, and sources generate data only occasionally, and (2) target tracking applications [2] in which data sources constantly change with time, and hence data routes cannot be established effectively via reinforcement. Sensor protocol for information via negotiation (SPIN) [20,21] can be classified as a protocol built upon the push-diffusion mechanism. We will present SPIN in detail below.

With the knowledge of geographic scooping of either sources or sinks, one can apply the energy- and location-aware routing protocols [22–24] to further reduce the flooding region, and hence save more energy.

15.3.2.1 *Directed Diffusion*

Directed diffusion [19] is a two-phase pull routing mechanism in which data consumers (sinks) search for the data sources matching their interests and the sources find the best routes to route their data back to the subscribers. Directed diffusion consists of three phases: interest propagation, data propagation, and reinforcement (Fig. 15.6). Sinks first broadcast interest packets to their neighbors. When a node receives an interest packet, the packet is cached and rebroadcast to other neighbors if it is new to this node. Propagation of interest packets also sets up the gradient in the network to facilitate data delivery to the sink. A gradient specifies both a data rate and a direction to relay data. The initial data rate of the gradient is set to be a small value and will be increased if the gradient along the path is enforced.

When a node matches an interest (e.g., it is in the vicinity of the event in the target-tracking application), it generates a data packet with the data rate specified in the gradient. The data packet is unicast individually to the neighbors from which the interest packet is received. When a node receives a data packet matching a query in its interest cache, the data packet is relayed to the next hop toward the sink. Both interest and data propagation are exploratory, but the initial data rate is low.

When a sink receives data packets from some neighbors, it *reinforces* one of the neighbors by increasing the data rate in the interest packet. Usually this neighbor is the one on the least-delay path. If a node receives an interest packet with a higher data rate, it also reinforces the path in the interest cache. Since the entries in the interest cache are kept as soft state, eventually only one path remains while other paths are torn down.

15.3.2.2 *SPIN*

SPIN [21,25] is a push-diffusion mechanism in which data sources initiate the data-sending activities. SPIN consists of three-stage handshaking operations (Fig. 15.7), including *ADV* (advertisement), *REQ* (request for data), and *DATA* (data message). Instead of directly flooding new data, the description of new

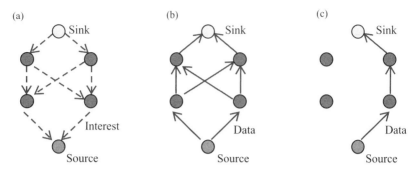

Figure 15.6   Three phases in directed diffusion [19]. (a) Interest propagation; (b) data propagation; (c) data delivery along reinforced path.

data, that is, metadata, is exchanged in the first two advertisement–subscription phases to reduce message overhead. If a node receives an advertisement with new information that is of interest to it, it replies with a request packet. The real data are then transmitted in the third phase upon receipt of such a request. Propagation of new information is executed hop-by-hop throughout the entire network.

15.3.3   Classification of Data-Gathering Mechanisms Based on the Structure of Dissemination

The number of sources and sinks in sensor network applications not only determines the direction of diffusion but also plays a crucial role in laying the structure of

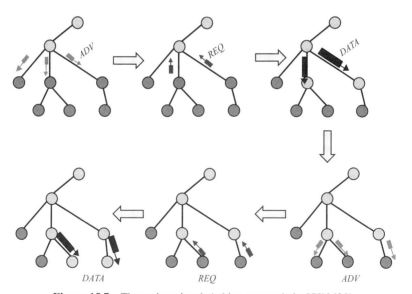

Figure 15.7   Three phase hand-shaking protocols in SPIN [25].

dissemination in the system, especially when it is considered in conjunction with data fusion. In what follows, we introduce four types of configurations, including tree, grid, cluster, and chain, and their representative approaches.

1. *Tree.* One of the most common dissemination structures used in sensor networks is a tree that is rooted at the sink and spans the set of sources from which the sink will receive information. It is usually constructed in the reverse multicast fashion. TAG [26] and TinyDB [9] are two examples that use sink trees for data dissemination. On the other extreme, in the scenario of a single source and multiple sinks, a tree is rooted at a source and constructed in the usual multicast fashion. The self-organizing multicast forwarding tree proposed by Mirkovic et al. [27] to disseminate reports from stimuli to multiple sinks falls in this category. The sinks broadcast their interest packets for certain events. Upon receipt of an interest packet, each sensor updates its distance to the sink and forwards the packet if it is new to the sensor. Each of the interest packets that record a minimum distance from some sink will be used by the source to construct the shortest path tree. The tree grows from the root and follows the reverse paths to reach sinks. A sensor node with a new stimulus joins the tree at the on-tree sensor that is closest to it, thus creating a new branch of the tree.

In the scalable energy-efficient asynchronous dissemination protocol (SEAD) [28], a dissemination tree is built to deliver data from a source (root) to multiple mobile sinks (leaves). The tree is built upon an underlying geographical routing protocol. When a mobile sink would like to receive data from a source, it connects to the dissemination tree through one of its neighboring sensors, called an *access node*. Similar to the home agent in Mobile IP (Internet protocol), the access node acts as an anchor node to relay data to the sink. When the sink moves out of the transmission range of its access node, it informs its access node of its new whereabouts by sending a *PathSetup* message. The latter will then forward all the data packets that are of interest to the node. When the distance to the original access node exceeds a predetermined threshold, the mobile sink joins a new access node. In order to reduce the number of messages transmitted over the tree, a source node duplicates its data at several replicas. The criterion for placing a replica on the tree is to minimize the extra cost of constructing a branch for a new join request.

2. *Grid.* Similar to SEAD, two-tier data dissemination (TTDD) [29] is designed for scenarios with a single source and multiple *mobile* sinks. Unlike SEAD, a grid structure is adopted as the dissemination structure in TTDD. An example grid structure originated from the source is shown in Figure 15.8. In the higher tier, a source that detects an event proactively constructs a grid structure where sensors close to the grid points are elected as dissemination nodes. In the lower tier, a mobile sink sends a query to, and receives data from, its nearest grid point on the local grid. When a sink moves to another grid, it can quickly connect to the grid structure and the information access delay thus incurred is reduced. One of the applications for which TTDD is particularly well suited is target tracking in the battlefield.

3. *Cluster.* When data-fusion is integrated with data dissemination, data generated by sensors are first processed locally to produce a concise digest, which is then

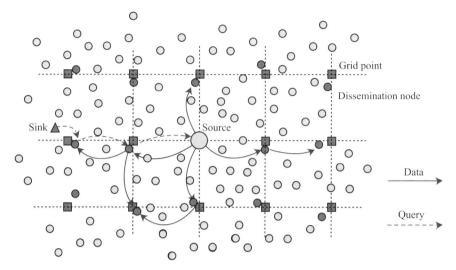

Figure 15.8 Two-tier data dissemination (TTDD) grid structure [29].

delivered to a sink. A hierarchical cluster structure [20,30,31] is better suited for this purpose. The low-energy adaptation clustering hierarchy (LEACH) [20] is a two-level clustering mechanism in which sensors are partitioned into clusters. Each sensor volunteers to become a clusterhead (CH) with a certain probability such that the task of being CHs is evenly distributed and rotated among all sensors. Once a sensor elects itself as the CH, it broadcasts a message to notify other nearby sensor nodes of the fact that it is willing to be a CH. The remaining sensors then select a minimum transmission power to join their closest CHs. Within the cluster, a CH uses time-division multiple access (TDMA) to allocate time slots to cluster members (so that the latter can relay their readings to the CH), compresses received data, and transmits a digested report directly to the base station (sink). Bandyopadhyay and Coyle [30] propose a multilevel hierarchical clustering algorithm. Similar to LEACH, this approach aims to realize the objective of balancing the load of sensors and achieving energy efficiency.

Chen et al. [32] devise and evaluate a fully decentralized, light-weight, dynamic clustering algorithm for target tracking. A cluster is dynamically formed and a CH becomes active when the acoustic signal strength detected by the CH exceeds a predetermined threshold. The active CH then broadcasts an information solicitation packet, asking sensors in its vicinity to join the cluster and provide their sensing information. With the use of a Voronoi diagram, they devise solution approaches to determine (I1) how CHs cooperate with one another to ensure that only one CH (preferably the CH that is closest to the target) is active with high probability; (I2) when the active CH solicits for sensor information, instead of having all the sensors in its vicinity reply, only a sufficient number of sensors respond with nonredundant, essential information to determine the target location; and (I3) both the packets that sensors send to their CHs and packets that CHs report to subscribers do not incur significant collision.

4. *Chain.* If the energy efficiency and bandwidth usage requirement is more important than the latency requirement, the chain structure that allows aggregation of data along a path ending at a sink is a competitive solution. The power-efficient gathering in sensor information system (PEGASIS) [33] is designed to aggregate data collected by all sensors in the entire network. Only one leader is elected each time, and the leadership is rotated among all the sensors. Under the assumption that the network topology is a complete graph, the leader is able to connect all the sensors with the chain structure. Starting from the sensor at one end of the chain, data are propagated and aggregated along the chain toward the leader. Then the data dissemination and aggregation processes continue from the other end. The aggregations from both ends arrive at the leader, which directly transmits the aggregation result to the sink.

15.4 DATA-FUSION MECHANISMS

As mentioned in Section 15.1, in most of the sensor network applications, sensors are deployed over a region to extract environmental data. Once data are gathered by multiple sources (e.g., sensors in the vicinity of the event of interest), they are forwarded perhaps through multiple hops to a single destination (sink). This, coupled with the facts that the information gathered by neighboring sensors is often redundant and highly correlated, and that the energy is much more constrained (because once deployed, most sensor networks operate in the unattended mode), necessitates the need for data fusion. Instead of transmitting all the data to a centralized node for processing, data are processed locally and a concise digest is forwarded (perhaps through multiple hops) to sinks. Data fusion reduces the number of packets to be transmitted among sensors, and thus the usage in bandwidth and energy. Its benefits become manifest, especially in a large-scale network. For a network with n sensors, the centralized approach takes $O(n^{3/2})$ bit-hops, while data fusion takes only $O(n)$ bit-hops to transmit data [34].

When data fusion is considered in conjunction with data gathering and dissemination, the conventional *address-centric* routing, which finds the shortest routes from sources to the sink, is no longer optimal. Instead, data-centric routing, which considers in-network aggregation along the routes from multiple sources to a sink, achieves better energy and bandwidth efficiency, especially when the number of sources is large, and/or when the sources are located closely to one another and far from the sink [8]. Figure 15.9 gives a simple illustration of data-centric routing versus address-centric routing. Source 1 chooses node A as the relaying node in address-centric routing, but node C as the relaying and data aggregation node in data-centric routing. As a result, a smaller number of packets are transmitted in data-centric routing.

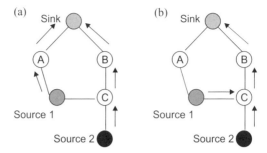

Figure 15.9 Address-centric routing vs. data-centric routing [8]. (a) Address-centric routing; (b) data-centric routing.

Existing research activities of data fusion can be categorized with respect to the following aspects:

- *Fusion Function.* Data-fusion is generally applied for:
  (a) *Basic Operations.* The most basic operations for data fusion include: COUNT, MIN, MAX, SUM, and AVERAGE [26].
  (b) *Redundancy Suppression.* Data-fusion, in this case, is equivalent to data compression [35,36].
  (c) *Estimation of a System Parameter.* Based on the observations from several pieces of sensor data, the data-fusion function aims to solve an optimization problem to minimize the estimation error of a system parameter [34].
- *System Architecture.* Besides the sources and sinks, a sensor network that considers data fusion has an additional component—the data aggregator. There exist a wide variety of ways to determine the location of the data aggregator.
- *Trade-Offs of Resources.* Depending on the resource constraints in a sensor network, there exist the following trade-offs: energy vs. estimation accuracy [34,37], energy vs. aggregation latency [38,39], and bandwidth vs. aggregation latency [36].

15.4.1 Classification of Data-Fusion Mechanisms Based on Functions

The major purpose of incorporating data fusion into the data-gathering and dissemination process is to reduce the number of packets to be transmitted, and hence the energy incurred in transmission. There are two types of data aggregation: "Snapshot aggregation" is data fusion for a single event, such as tracking a target, while "periodic aggregation" periodically executes the data-fusion function, such as monitoring an environment parameter periodically [37]. Depending on the application requirements, three types of data-fusion functions can be used: basic aggregation functions, redundancy suppression, and estimation of a system parameter.

15.4.1.1 Basic Aggregation Function Basic aggregation functions include five SQL-like operations: COUNT, MIN, MAX, SUM, and AVERAGE [26]. Here we use the structure of aggregates and the AVERAGE function used in TAG as an example to explain the procedure of data aggregation. An aggregation component consists of three functions: a merging function f, an initializer i, and an evaluator e. The aggregation process starts with one sensor specifying the initial states for initializer i, $\langle x, 1 \rangle$, where the first entry in the 2-tuple is the sensor value of the starting node, and the second entry represents the number of readings in the first entry. The aggregation packet including the initializer is propagated to the next hop, and the merging function f is executed there for data aggregation. The merging function f is one of the five functions just mentioned and in the case of AVERAGE its function is expressed below:

$$f(\langle S_1, C_1 \rangle, \langle S_2, C_2 \rangle) = \langle S_1 + S_2, \ C_1 + C_2 \rangle$$

which means that the first entry is the sum of sensor readings along the aggregation path and the second entry is the count of sensor readings. Finally, when the aggregation packet arrives at the sink (or the subscriber), the evaluator e calculates the final result $e(\langle S, \ C \rangle) = S/C$.

Although the basic functions share the same aggregation structure, the characteristics of different functions differ in three aspects (as summarized in Table 15.1) [26]:

1. *Duplicate Sensitive.* Duplicate sensitivity indicates whether the result of the aggregation evaluator is affected by a duplicated reading from a single sensor. In the case of duplicate sensitive aggregates such as COUNT or AVERAGE, sending aggregation packets over multiple paths will lead to incorrect results.

2. *Exemplary or Summary.* The result of exemplary aggregates might depend on any one value from the set of all sensor readings, while summary aggregates compute some property over all values. The results of exemplary aggregates are not predictable when one critical reading is lost.

3. *Monotonic.* An aggregate is monotonic if and only if for any two partial states, s_1 and s_2, and their aggregate state s', either $e(s') \geq$

TABLE 15.1 Classes of Aggregation Functions

	MAX, MIN	COUNT, SUM	AVERAGE	MEDIUM
Duplicate sensitive	No	Yes	Yes	Yes
Exemplary (E)/summary (S)	E	S	S	E
Monotonic	Yes	Yes	No	No
Partial state	Distributive	Distributive	Algebraic	Holistic

Source: Ref. [26].

$MAX(e(s_1), \ e(s_2))$ or $e(s') \leq MIN(e(s_1), \ e(s_2))$ holds. The monotonic property is an index of whether some predicates (e.g., **HAVING**) can be applied to partial states before the final aggregate state is acquired. If applicable, an early predicate evaluation reduces the search space by avoiding traversing an unlikely aggregation tree.

4. *Partial State.* The characteristic of partial state for aggregation functions represents the size of intermediate states. *Distributive* aggregates like MAX, COUNT, and SUM keep a single entity in their partial states. *Algebraic* aggregates such as AVERAGE update more than one entity in the partial states, but the number of entities kept is a constant. In *Holistic* aggregates, the size of entities in a partial state is proportional to the number of the aggregated states. That is, aggregation takes no effect and all the data have to be transmitted to a centralized node for further processing. Function MEDIUM falls into this category.

15.4.1.2 *Redundancy Suppression*   Due to the fact that correlation exists in sensor data both in the spatial and temporal domains, one of the most important data-fusion functions is to eliminate data redundancy, or in a more concrete term, to exploit the correlation structure that exists in sensor data via distributed source coding.

Chou et al. [35] proposed incorporating both distributed source coding and adaptive signal processing in data fusion and exploiting the correlation structure in sensor data to reduce the number of bits transmitted. The system architecture consists of a data-gathering node (sink) and sensors. The data-gathering node sends queries to sensors sequentially to obtain certain information that pertains to the entire field. The design objective here is to devise a computationally inexpensive encoding operation that supports multiple compression rates in the sensors while allowing a more complex decoding procedure in the data-gathering node. The theoretical base used is the Slepian–Wolf theorem [40]: if two discrete random variable **X** and **Y** are correlated, then **X** can be losslessly compressed using $\mathbf{H(X|Y)}$ bits without access to **Y**, where $\mathbf{H(X|Y)}$ is the conditional entropy of **X** given the information of **Y**. Based on the Slepian–Wolf theorem, the authors propose a blind compression method to achieve the theoretical limit given in the theorem. Suppose that the data-gathering node has full information of **Y** (side information) and the difference between **X** and **Y** is less than $2^{i-1}\Delta$, where Δ is the sampling resolution of the sensor readings. With their proposed tree-based codebook, one can encode **X** with only i bits without any knowledge of **Y** and the decoder can fully recover the information of **X**. An example that shows the encoding and decoding operations with the tree-based codebook is depicted in Figure 15.10. The sampling resolution, Δ, in this example is set to 0.1. If the sink has collected the full information of sensor data **Y** (e.g., $\mathbf{Y} = 0.4$) and that the difference between the value of **X** and **Y** is less than 0.2, then sensor **X** only needs to use i (=2) bits to encode its value without the knowledge of **Y**. The deterministic encoding function for **X** is $f(X) = index(X) \bmod 2^i$, where the index function converts the value to the index in the tree-based codebook, for example, $f(0.5) = 5$ and **X** is encoded in 2 bits as **01** because 5 mod

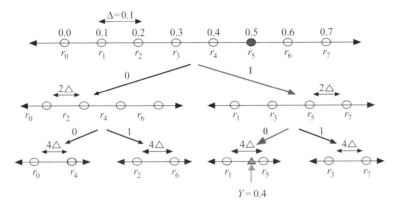

Figure 15.10 An example of the tree-based codebook [35].

4 is equal to 1. After the sink receives the 2-bit (**01**) encoding information of **X**, it traverses the tree and finds the corresponding subcodebook **S**, in which all elements share the same encoding rule. In the given example, $S = \{r_1, r_5\}$. The decoding rule is $\hat{X} = \arg\min_{r_i \in S} \|Y - r_i\|$. Since r_5 is closer to **Y** than r_1, the sink can decode the value of **X** to be 0.5.

The procedures to obtain the correlation structure of sensor data can be divided into two steps. First the data-gathering node collects the uncompressed data from all sensors in the first K rounds to obtain the temporal and spatial redundancy in data. Then in the following round of data collection, an adaptive filtering framework is used at the data-gathering node to learn the correlation structures in the data.

Duarte-Melo et al. [41] address the issue of joint design of data compression and data dissemination. They consider the same system architecture as [35] (i.e., the system consists of a sink and multiple sensors), and formulate the problem as a nonlinear programming problem that maximizes the system lifetime, subject to the constraints on flow conservation, energy, and sampling rates. The last constraint specifies the least sampling rate for Slepian–Wolf type of encoding.

Scaglione and Servetto [36] propose to integrate routing with source coding under a system in which each sensor reports its data to the entire network and all sensors intend to capture certain information that pertains to the entire field within a prescribed distortion value, that is, the joint entropy of all readings. As the data are propagated to a node, they are encoded with the local data, and the compressed data are relayed to the next hop node. It is shown that the aggregation method consumes bandwidth in a scalable way (i.e., below the transport capacity), and thus the problem of vanishing per-node throughput [42] is avoided.

Petrovic et al. [43] propose the *data-funneling* mechanism. In data funneling, multiple reports from different sensors are sent to the sink at approximately the same time. Since these packets have similar headers, they can be merged to a single packet by removing their redundant headers. Considerable saving can be made by a simple concatenation of readings in the packet body. Source coding based on the ordering is further applied to compress the data of concatenated readings. For instance, suppose there are four nodes with ID's 1, 2, 3, and 4 in a region. The readings of the four sensors are integers within the range of [0, 1, ... , 5]. The fourth node that receives the concatenation of readings from the other three sensors can compress the data set with four readings by simply sending three readings. The fourth reading is implicitly encoded by the order of the three readings because there are six (3!) combinations of the ordering relationship. More specifically, the ordering of IDs (1, 2, 3) in the data packet represents that the fourth reading of data are 0, the ordering (1, 3, 2) represents the value of 1, and so on.

15.4.1.3 *Estimation of a System Parameter*   In this category of applications, sensors cooperate to disseminate necessary information to certain nodes, which then proceed to estimate the parameter of interest. The estimation problem can be formulated as an optimization problem, whose objective is to minimize the estimation error. An example of such an optimization problem is to average all the temperature readings of sensors within a room to estimate the temperature of a room. The estimation is optimal with respect to the minimum square error (MSE) criterion. Another example is to track targets to minimize the estimation error of the target's location.

Rabbat and Nowak [34] formulate an optimization problem to estimate a system parameter, θ, given a set of data from n sensors, x. The objective is to minimize the cost function f:

$$\min_{\theta} f(\theta,\ x) = \frac{1}{n} \sum_{i=1}^{n} f_i(\theta,\ x_i) \tag{15.1}$$

The iterative gradient decent method shown in Eq. (15.2) is one of the most popular techniques for solving the nonlinear optimization problem. It can be used at a central server that has the entire set of data:

$$\hat{\theta}^{(k+1)} = \hat{\theta}^{(k)} - \alpha \sum_{i=1}^{n} g_{i,k} \tag{15.2}$$

where $g_{i,k} \in \partial f_i(\hat{\theta}^{(k)})$, α is a small positive step size, and k is the iteration number. In ref. [34], Eq. (15.1) is solved with a decentralized incremental approach by dividing Eq. (15.2) into a cycle of n subiterations. The ith subiteration focuses on optimizing a single component, $f_i(\theta)$, at a node i based on its local data. Just a with PEGASIS, the task of subiterations is rotated hop by hop in a chain. The subiteration starts from a node 1, which inherits the estimation result from the previous iteration, $\varphi_0^{(k)} = \hat{\theta}^{(k-1)}$, where $\varphi_i^{(k)}$ is the result of the ith subiteration within the kth iteration. The sensor i updates the result of the aggregation that it receives from the previous hop, sensor $i - 1$, as follows:

$$\varphi_i^{(k)} = \varphi_{i-1}^{(k)} - \alpha g_{i,k} \quad i = 1, \ldots, n$$

where $g_{i,k} \in \partial f_i(\varphi_{i-1}^{(k)})$. Finally the estimation in the kth iteration ends at the nth subiteration: $\hat{\theta}^{(k)} = \varphi_n^{(k)}$. No more than $O(\varepsilon^{-2})$ iterations are required for the desired accuracy ε. Three applications, robust estimation, localization, and clustering and density estimation, are illustrated that leverage the distributed optimization in ref. [34].

Zhao et al. [2] proposed a leader-based tracking scheme in which samples are collected successively at different time instants and locations to localize the target. A sensor that contains the most information is elected by the previous leader to estimate the current location of the target based on the past belief and the current measurement. Similar to the directed diffusion approach [19], a routing protocol, called *constrained anisotropic diffusion routing* (*CADR*) [44], is used to redirect queries from users to the most qualified leader. The current leader adopts the sequential Bayesian filtering technique to estimate the current location of the target. A sensor, which is estimated to hold the maximum information, is then chosen to be the next leader. This process is termed the *information-driven sensor query* (*IDSQ*).

15.4.2 System Architectures of Data Fusion

As mentioned earlier, the data-fusion mechanisms can also be categorized based on where to place the data aggregators.

Data funneling [43] is intrinsically an energy-efficient routing protocol [45] integrated with data aggregation and compression techniques. The basic idea of data funneling is to build a cost field with the funnel shape to pull the data from sources to the sink. The sink initials directional flooding to send an interest packet toward a target region, as shown in Figure 15.11(a). During the process of forwarding interest packets, a forwarder computes its (energy) cost for communicating back to the sink and updates the cost field in the interest packet.

When a node within the target region receives an interest packet from the nodes outside the region, it designates itself a *border node*. The cost required to reach the sink (i.e., the cost in the interest packet) is recorded, and a field that is used to keep track of the cost to reach the border node is also included. Since there could be multiple "entries" (border nodes) to the target region, a node within the target region might receive multiple interest packets from border nodes. Instead of requesting all the nodes to send individual reports back to the sink, one of the border nodes is responsible for the task of collecting and aggregating all reports in the region and

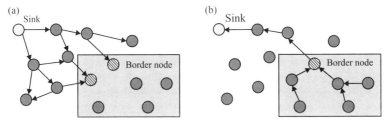

Figure 15.11 Data funneling: (a) directed flooding phase and (b) data communication phase [43].

sending a single packet to the sink. All the sensors within the region share a common schedule of which border node to be the data aggregator during each round of reporting. The schedule is determined by a deterministic function of the costs to reach the sink from all border nodes. Sensors with a longer distance to the designated aggregator send their reports earlier and the readings are concatenated in a single packet to eliminate redundant headers. The data communication process is shown in Figure 15.11(b). After receiving reports from all the sensors within the region, the designated border node further compresses the data by applying a coding technique based on ordering.

DFuse [46] is a distributed data-fusion framework especially designed for video streaming applications. The framework provides the flexibility of data fusion in two aspects. First, a layer of fusion modules provides a set of data-fusion functions for an application to manage video streams. Second, the role of a node (a sink, a relay, or a fusion point) is determined in a distributed way based on the given cost function.

Baek et al. [47] study optimal data-fusion strategies regarding the order of fusion and the organization of fusion devices under two scenarios: networks with a single sink and those with multiple sinks. In the case of a single sink, all sensors send their nonredundant data to the sink. The optimal fusion strategy is to determine the order of compression at each node, so as to minimize the overall energy consumption while faithfully disseminating the data from all sensors to the sink. According to the Slepian–Wolf theorem [40], the sum of rates for any subset of sensors is lower bounded by the conditional entropy, given that the sink has known the data of the set of rest sensors. The optimal solution can be found using a greedy algorithm. The sensor to reach the sink first with the least communication cost transmits its data without compression to the sink. Then, in the increasing order of communication costs to reach the sink, sensors sequentially disseminate their compressed data, given the known side information at the sink. Surprisingly, the optimal solution is independent of the correlation structure of the data, and simply relies on the topology of the network. For instance, the optimal fusion strategy for the sink S in the region consisting of nodes A, B, and C shown in Figure 15.12 is: first S retrieves the full information of node A, $H(A)$, following by $H(B|A)$, and $H(C|A, B)$ sequentially.

In the case of multiple sinks, a three-level hierarchical architecture that includes sensors, compressors, and sinks is considered. Both compressors and sinks can aggregate data from sensors with a compression ratio α ($0 < \alpha < 1$). All the compressed data are destined for any one of the sinks. Therefore, a sensor transmits

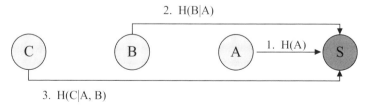

Figure 15.12 The optimal fusion strategy under the case of a single sink.

its raw data either directly to its closest sink or a nearby compressor, which compresses the data and forwards the compressed data to its closest sink. The authors show that under a given α, the optimal organization that minimizes energy consumption is a Johnson–Mehl tessellation, in which the entire sensing field is divided into regions belonging to either a sink or a compressor. In the two extreme cases that α is equal to 0 and 1, representing full and no compression, the optional organization degenerates from Johnson–Mehl tessellation into Voronoi tessellations that constitute the set of sinks and compressors and the set of sinks, respectively.

15.4.3 Trade-Offs in System Design

Depending on the resource constraints, there exist various trade-offs in different data-fusion schemes: energy versus estimation accuracy [34,37], energy versus aggregation latency [38,39], and bandwidth versus aggregation latency [36].

15.4.3.1 Trade-Off Between Energy and Accuracy The requirement of higher accuracy demands more message exchanges and leads to higher energy consumption. Bonnet et al. [5] proposed a distributed periodic aggregation approach to estimate the *maximum* of sensor data in a field, where the maximum of sensor data are modeled to be Gaussian distributed. Compared with multiple "snapshot aggregations," the proposed approach exploits the energy–accuracy trade-off, and provides users with a system-level knob to control the desired accuracy and energy consumption. The distributed optimization approach proposed in ref. [34] shows that $O(\varepsilon^{-2})$ iterations of aggregation are required to achieve the desired accuracy ε.

Similar to a model-based query is supported in ref. [48]. A declarative query-processing engine uses a probabilistic model to answer questions about the current state of the sensor network. The model is based on time-varying multivariate Gaussians, constructed based on the historical data, and updated when new data are available. Given a query specifying the confidence interval, the problem is to choose the best set of new observations such that the cost of collecting new data is minimized.

Ye et al. [49] figure robustness of delivery in the design. Multiple, interleaved paths between the source and the sink enable the system to be more resilient to node or transmission failure. In their proposed *gradient broadcast* (*GRAB*) protocol, the cost field is constructed first [49a]. The cost of a node represents the minimum energy required to forward a packet along a path to the sink. To exploit the redundancy of delivery, a sender (forwarder) broadcasts a packet, and a relay node forwards the packet only if its cost is smaller than the sender's cost. Moreover, the degree of delivery redundancy is controlled, as paths are expanded quickly from the source, maintained within a reasonable width next, and finally shrunk near the sink.

Tilak et al. [50] trade accuracy of information for energy saving in data dissemination. They advocate nonuniform information granularity, that is, the required accuracy of information decreases as the distance from the source becomes longer. Applications in the battlefield or disaster rescue scenarios usually possess such characteristics. Two deterministic and two nondeterministic protocols are designed for nonuniform information dissemination. The proposed protocols trade accuracy of information for energy expenditure by selectively discarding packets from a sensor.

15.4.3.2 *Trade-Off Between Energy and Latency*   Both Schurgers et al. [38] and Yu et al. [39] explore the trade-off between energy consumption and propagation latency from data sources to a sink, but from different perspectives. In ref. [38], energy is saved via directly turning off the radio circuitry when a sensor is not transmitting or receiving data. While the low duty cycle reduces power consumption, it increases the propagation latency from a data source to the sink. The protocol, *sparse topology and energy management* (*STEM*), is proposed to deal with the problem. STEM utilizes dual bands for data transmission and wake-up signaling. The channel for wake-up signaling is operated in a low duty cycle. Each node periodically turns on the radio circuitry for the wake-up channel to hear whether any other node has attempted to communicate with it. Once a node detects such an activity in the wake-up signaling channel, it turns on its radio circuitry for the data channel. The increased latency due to the sleep state is thus bounded by the sleep–listen period in the wake-up channel. STEM is especially well suited for applications with most operations in the monitoring state. For instance, in a fire alarm system, the network only senses the environment in an energy-efficient way and the system stays in the monitoring state most of the time. Once an event takes place, the system quickly changes to the transfer state and reports the event to the data sink in a timely manner.

Ye et al. [39] achieve energy saving via adjusting the modulation scaling factor, that is, the number of bits in a modulated symbol. In general, a smaller modulation scaling factor reduces the power required in the transmission, but increases the transmission time over a link, although the relationship between the power required and the link delay thus increased is not necessarily monotonic. The authors consider a multiple-source single-sink data-aggregation tree, and formulate the problem of finding an optimal schedule of packet transmission to minimize the total transmission energy incurred at all nodes in the aggregation tree, subject to the given propagation latency constraints. A numerical optimal algorithm, a pseudopolynomial, a dynamic programming-based approximation algorithm, and a distributed on-line protocol are developed to solve the problem.

Scaglione and Servetto [36] discuss the trade-off between the bandwidth usage and the decoding delay. They argue that data aggregation along a path leads to better bandwidth usage, but if the aggregation is conducted along multiple parallel paths, the delay incurred in aggregating and sending data to destinations is reduced, but the bandwidth usage (or the corresponding energy consumption) is increased. They then suggest that these two quantities should be linked together by the routing strategy chosen.

15.5 DATA TRANSPORT ISSUES

Most of the data-gathering and -fusion mechanisms reside in or below the network layer, except sensor databases, which reside in the application layer but exploit geographic routing or sink tree for data dissemination and storage. Several researchers have stepped up in the protocol stack and considered the data transport issues. Research along this line can be roughly categorized into two tracks: one treats data transport in sensor networks as an optimization problem, with the objective of maximizing the amount of information (utility) collected at sinks, subject to constraints on flow conservation, energy, latency, and/or channel bandwidth; and the other designs a transport-layer protocol for congestion control and end-to-end reliability in wireless sensor networks.

15.5.1 Utility-Based Data Transport

General approaches along this line of research are to (1) formulate the problem as a nonconvex programming problem and solve it in a centralized manner (so as to provide performance bounds); or (2) formulate the problem (with some approximations) as a convex programming problem that can be solved in a decentralized manner, and devise the corresponding distributed algorithm. Some of the data-gathering and -fusion approaches discussed in Sections 15.3 and 15.4 can be cast in this optimization framework by properly assigning the utility of each packet (e.g., packets that carry updated digests are assigned higher utilities).

In this subsection we first introduce a nonconvex programming problem formulation [51] that is general enough to encompass a wide variety of applications in sensor networks, each with a different objective function and subject to different constraints. Then we discuss various simplified convex or linear programming problems. Note that utility-based approaches have been exploited in conventional wired networks by Kelly et al. [52] and Low and Lapsley [53]. Kelly et al. [52] propose a pricing scheme to achieve weight-proportional fair-rate allocation for users in a wireline network. The same problem considered in ref. [52] is solved by Low and Lapsley [53] differently with the use of the dual optimization problem. A major advantage of Low and Lapsley's solution approach is that the dual optimization problem can be solved in a *distributed* manner via a pricing mechanism.

In the context of wireless sensor networks, there are three types of objective functions in the optimization problem: (1) maximizing the overall utility of the sensor data collected at sinks throughout the system lifetime [51,54,55], (2) maximizing the amount of data extracted from sensors [56,57], and (3) maximizing the system lifetime (or equivalently minimizing the energy expenditure) [57,58]. We formulate the first optimization problem, subject to the flow constraint, the energy constraint, and the channel capacity constraint. As will become clearer, the second and the third problems are special cases of the first one.

For notational convenience, we define the following notion:

$U_s(\cdot)$ the utility function that specifies the commodity generated from a sensor s and sent to a sink (perhaps through multiple routes);

S_n and S_i the set of sensors and sinks in the sensing field;

N_k the set of one-hop neighbors of node k;

$q_{ij}^{(s)}$ the rate of the commodity s that passes from node i to node j;

x_i the source rate originated from node i;

$x_i^{(s)}$ the source rate of commodity s originated from node i; as the commodity s only originates from node s, $x_i^{(s)} = x_i$ if $i = s$; $x_i^{(s)} = 0$, otherwise;

E_i the amount of energy initially equipped with node i;

e_i the energy consumed in the idle state per unit time;

e_s and e_r the additional energy consumed in transmitting and receiving one unit of data rate per unit time;

$\bar{d}_s$ the average end-to-end latency that a packet experiences from a sensor s to a sink;

T the system lifetime defined as the time interval until the first failure of a node due to power depletion;

C_i the channel capacity of node i.

Given this notation, the first problem can be formulated as a nonlinear programming problem as follows:

$$\max_{\mathbf{x,q},T} \left[\sum_{s \in S_n} U_s \left(\sum_{i \in S_i} \sum_{k:i \in N_k} q_{ki}^{(s)}, \bar{d}_s \right) \right] \cdot T \tag{15.3}$$

s.t.

$$\sum_{k:i \in N_k} q_{ki}^{(s)} + x_i \geq \sum_{j:j \in N_i} q_{ij}^{(s)}, \quad \forall\, i \in S_n, \quad s \in S_n \tag{15.4}$$

$$\left\{ e_r \cdot \sum_{k:i \in N_k} \sum_{s \in S_n} q_{ki}^{(s)} + e_s \cdot \sum_{j \in N_i} \sum_{s \in S_n} q_{ij}^{(s)} + e_i \right\} \cdot T \leq E_i, \quad \forall\, i \in S_n \tag{15.5}$$

$$\sum_{j:j \in N_i} \sum_{s \in S_n} q_{ij}^{(s)} \leq C_i, \quad \forall\, i \in S_n \tag{15.6}$$

The objective is to maximize the utility of all received packets throughout the system lifetime over a vector of source rates of commodities $\mathbf{x}$, a vector of link flow $\mathbf{q}$, and the system lifetime (T). As such, the objective function (Eq. (15.3)) is expressed as the product of the system lifetime and the utility of all commodities received at the sinks per unit time. The utility function for the commodity s, $U_s(\cdot)$, is a function of the total rate of the commodity s arriving at sinks and the average end-to-end latency it sustains. The rate of commodity s arriving at sinks is the sum of all the incoming flows of commodity s to any of the sinks. Since flows traveling through different routes to sinks endure different latencies, we express the utility function as a function of the average latency $\bar{d}_s$ to account for the average loss of utility due to the delay. Moreover, with different qualities of data, the different quantized utility functions (such as in ref. [59]) can be used to evaluate the utility of a data packet.

The first constraint (Eq. (15.4)) is the flow constraint. The sum of both the incoming flows of commodity s and the flow of commodity s originated from a node is greater than or equal to the sum of the outgoing flows of commodity s, with the inequality implying that intermediate relay nodes may drop the packets they forward. The second constraint (Eq. (15.5)) is the energy constraint, while the third constraint (Eq. (15.6)) is the capacity constraint, that is, the sum of the outgoing flows of all the commodities from a node i should be less than its channel capacity C_i.

The problem just formulated aims to maximize the total utility received at the sinks, by controlling the parameter vectors x and q (which in turn are related to flow control and routing decisions). As a matter of fact, the preceding problem formulation encompasses a wide variety of requirements and objectives for different applications in sensor networks. In what follows, we discuss six possible design dimensions and their corresponding amendments to the preceding problem formulation:

1. *Flow Conservation*   If intermediate relay nodes are not allowed to discard the packets they forward, the inequality in Eq. (15.4) is changed to an equality. With the flow conservation constraint, for each commodity s, the sum of the incoming flows of commodity s at sinks is equal to the rate x_s originated at node s, and hence the objective function in Eq. (15.3) can be rewritten as

$$\max_{\mathbf{x}, \mathbf{q}, T} \left[\sum_{s \in S_n} U_s(x_s, \bar{d}_s) \right] \cdot T \qquad (15.7)$$

2. *Flow Indivisibility Constraint*   If a commodity from a sensor node s cannot be routed through multiple paths, an additional constraint has to be added in Eq. (15.4) such that for each commodity s, only one incoming and one outgoing flow has positive rate and the others are zero. This makes it more difficult to solve the optimization problem because of its discrete constraint.

3. *Join Design of Flow Control and Routing*   In the problem formulation given in Eqs. (15.3)–(15.6), both the routing and flow-control problems are jointly considered. An alternative approach is to solve the optimization problem in two steps. The routes are determined first by a routing protocol and figured into the optimization problem. The optimization problem then solves the flow-control problem by optimizing the total utility over the vector of source rates, x. Wang et al. [60] show that the optimization problem considering both routing and flow-control decisions together is a NP-hard problem. They also show a trade-off exists between utility maximization and route instability. The routing metric based on pure dynamic pricing information achieves high utility but results in instable routing. Adding static components such as hop count to the routing metric stabilizes routing decisions.

4. *Quality-Driven Utility Function* If the quality of the data are considered, the utility function of each sensor is determined based on the quality of the data sensed; otherwise, the utility functions are the same for all the sensors and then the objective function is reduced to the special case of maximizing the data extraction rate or system lifetime.

5. *Effect of Latency on Utility* Whether the latency affects the utility of the data sensed is application dependent. In general, the utility of data decays with the latency, but the decay function (convex, linear, or concave) varies with the application characteristics. Alternatively, the effect of latency can be figured in as a constraint into the optimization problem.

6. *Energy Awareness* If the energy constraint is not considered, the problem formulation can be simplified as follows: the system lifetime T can be removed from the objective function and the energy constraint (Eq. (15.5)) can be removed.

By either modifying the objective function or relaxing one of the constraint functions, the preceding problem formulation can be reduced to several data-gathering optimization problems considered in the literature. Byers and Nasser [54] consider the optimization problem of maximizing the overall utility of sensor networks during the system lifetime, subject to an energy constraint. The energy constraint is expressed as a high-level cost on sensing, transmission, reception, and aggregation. Chang and Tassiulas [61] devise a routing solution to maximize the system lifetime of sensor networks under the given source rate of nodes and subjected to flow conservation and energy constraints.[1] Without considering the node capacity constraint, the problem reduces to a linear programming problem. Sadagopan and Krishnamachari [56] use an iterative approximation algorithm to solve a similar linear programming problem except by changing the objective function from maximizing the system lifetime to maximizing the data extraction rate. Duarte-Melo et al. [41] formulate a nonlinear programming problem that jointly considers the data-compression and data-dissemination problem to maximize the number of snapshots generated from the network, or equivalently maximize the system lifetime.

Chen et al. [55] first transform the optimization problem (Eqs. (15.3)–(15.6)) to a convex programming problem, by linearizing the energy constraint (Eq. (15.5)). This is done by properly setting the value of the system lifetime in advance and controlling the data rate of a node (and hence its total energy consumption rate) so as to sustain its battery lifetime longer than the specified lifetime. Given the transformed problem, they then derive a distributed solution with the use of the dual optimization problem and the pricing mechanism given in ref. [53]. They devise a simple node-capacity estimation method to measure the node capacity (which changes with the traffic load and nodal distribution, but is needed in the optimization problem) on-line, and incorporate optimization results in selecting routes to further increase the overall utility.

---

[1]Only the transmission power consumption is considered.

The goals of maximizing information extraction and minimizing energy consumption always conflict with each other and could not be satisfied at the same time. Ordonez and Krishnamachari [57] discuss two complementary nonlinear optimization problems under the scenario of a single sink and multiple sensors: one maximizes the total information gathered subject to an overall energy constraint, and the other minimizes the overall energy consumption subject to the minimal information-rate requirements. Other constraints considered include (1) fairness constraints, in which the flow rate from one sensor is restricted to be less than or equal to a portion of the overall rate that arrives at the sink, and (2) channel-capacity constraints in which Shannon's theoretical capacity limit is imposed under an interference-free communication model. These two complementary problems are shown to be equivalent in terms of the correspondence between optimal solutions and constraints.

15.5.2   Data Transport Design for Congestion Control and Reliability

In addition to utility-based approaches, there have been several transport control protocols that are designed for wireless sensor networks for congestion control and end-to-end reliability.

PSFQ [62] focuses on reliable transport, which is necessary under certain scenarios such as transport of debugging messages and critical commands. PSFQ consists of two operations: (1) *pump slowly*, where a packet is generated periodically from the source in a low rate so as not to congest the network; and (2) *fetch quickly*, where once a node detects packet loss (perhaps by detection of a discontinuity in the sequence number), it immediately requests a retransmission from its previous hop. Different from PSFQ, the goal of *CODA* [63] is simply congestion detection and avoidance. It consists of three mechanisms: (1) congestion detection via observing the buffer status and channel loading, (2) open-loop, hop-by-hop backpressure, and (3) closed-loop, multisource regulation.

ESRT [64] intends to provide both reliability and congestion control, although the reliability measure used in ESRT is the ratio of the number of received packets to the number of packets required for reliable event detection. The reliability measure increases when a source increases the data rate, subject to the constraint that the data rate should not exceed the critical threshold of congesting the network. On the other hand, once a forwarder detects congestion, it sets the congestion notification bit in the packet. When the sink receives a packet with the congestion bit set, it broadcasts a congestion notification packet (which is assumed to be able to reach all the nodes in one hop), asking the sources to decrease their data rates.

The objective of the *SPEED* protocol [65] is to disseminate the data from the source to the destination at a steady *speed*, that is, the end-to-end propagation delay is proportional to the distance from the source to the destination. Similar to CODA, Speed employs both open-loop and closed-loop control mechanisms, including back-pressure rerouting and neighborhood feedback loop, to achieve the objective.

Woo and Culler [66] propose a transmission control mechanism at both the medium-access control (MAC) and transport layers. Instead of providing reliability or congestion control services, they focus on the fairness issue for a wireless sensor network with the sink tree as the underlying routing paths.

REFERENCES

1. A. Mainwaring, J. Polastre, R. Szewczyk, D. Culler, and J. Anderson. Wireless sensor networks for habitat monitoring. In *Proceedings of the 1st ACM International Workshop on Wireless Sensor Networks and Applications (WSNA '02)*, Atlanta, Georgia, September 2002.

2. F. Zhao, J. Shin, and J. Reich, Information-driven dynamic sensor collaboration for tracking applications, *IEEE Signal Processing Magazine*, **19**:68–77, March 2002.

3. S. S. Intille. Designing a home of the future. *IEEE Pervasive Computing*, **1**(2):80–81, April–June 2002.

4. I. A. Essa. Ubiquitous sensing for smart and aware environments. *IEEE Personal Communications*, **7**(5):47–49, October 2000.

5. P. Bonnet, J. Gehrke, and P. Seshadri. Querying the physical world. *IEEE Personal Communications*, **7**:10–15, October 2000.

6. D. Estrin, R. Govindan, J. Heidemann, and S. Kumar. Next century challenges: Scalable coordination in sensor networks. In *Proceedings of the 5th Annual International Conference on Mobile Computing and Networks (MobiCom '99)*, August 1999.

7. D. Estrin, L. Girod, G. Pottie, and M. Srivastava. Instrumenting the world with wireless sensor networks. In *Proceedings of the International Conference on Acoustics, Speech and Signal Processing (ICASSP)*, Salt Lake City, Utah, May 2001.

8. B. Krishnamachari, D. Estrin, and S. Wicker. Modelling data-centric routing in wireless sensor networks. In *Proceedings of INFOCOM 2002*, New York, June 2002.

9. S. Madden, M. J. Franklin, and J. M. Hellerstein, and W. Hong. The design of an acquisitional query processor for sensor networks. In *Proceedings of the 2003 ACM SIGMOD International Conference on Management of Data*, pages 491–502, San Diego, California, June 2003.

10. X. Li, Y. J. Kim, R. Govindan, and W. Hong. Multi-dimensional range queries in sensor networks. In *Proceedings of the 1st ACM Conference on Embedded Networked Sensor Systems (Sensys '03)*, Los Angeles, California, November 2003.

11. J. Gehrke and S. Madden. Query processing in sensor networks. *Pervasive Computing*, **3**(1):46–55, January–March 2004.

12. TinyDB, from http://telegraph.cs.berkeley.edu/tinydb/index.htm.

13. Y. Yao and J. Gehrke. Query processing for sensor networks. In *Proceedings of the 1st Biennial Conference on Innovative Data Systems Research (CIDR '03)*, pages 233–244, Asilomar, California, January 2003.

14. TinyOS, from http://www.tinyos.net.

15. S. Ratnasamy, B. Karp, S. Shenker, D. Estrin, R. Govindan, L. Yin, and F. Yu. Data-centric storage in sensornets with GHT, a geographic hash table. *Mobile Networks and Applications*, **8**(4):427–442, August 2003.

16. P. Bose, P. Morin, I. Stojmenovic, and J. Urrutia. Routing with guaranteed delivery in ad hoc wireless networks. In *Proceedings of the 3rd International Workshop on Discrete Algorithms and Methods for Mobile Computing and Communications*, pages 48–55, Seattle, Washington, August 1999.

17. B. Karp and H. T. Kung. Greedy perimeter stateless routing for wireless networks. In *Proceedings of the 6th Annual ACM/IEEE International Conference on Mobile Computing and Networking (MobiCom '00)*, Boston, Massachusetts, August 2000.

18. J. Heidemann, F. Silva, and D. Estrin. Matching data dissemination algorithms to application requirements. In *Proceedings of the 1st International Conference on Embedded Networked Sensor Systems (Sensys '03)*, Los Angeles, California, November 2003.

19. C. Intanagonwiwat, R. Govindan, and D. Estrin. Directed diffusion: A scalable and robust communication paradigm for sensor networks. In *Proceedings of the 6th Annual IEEE/ACM International Conference on Mobile Computing and Networking (MobiCom '00)*, pages 56–57, Boston, Massachusetts, August 2000.

20. W. Heinzelman, A. Chandrakasan, and H. Balakrishnan. Energy-efficient communication protocols for wireless microsensor networks. In *Proceedings of the 33rd Hawaii International Conference on Systems Science (HICSS)*, Volume 8, page 8020, Maui, Hawaii, January 2000.

21. J. Kulik, W. Heinzelman, and H. Balakrishnan. Negotiation-based protocols for disseminating information in wireless sensor networks. *Wireless Networks*, **8**(2/3):169–185, March–May 2002.

22. I. Stojmenovic and X. Lin. Power aware localized routing in wireless networks. *IEEE Transactions on Parallel and Distributed Systems*, **12**(11):1122–1133, November 2001.

23. I. Stojmenovic, M. Russell, and B. Vukojevic. Depth first search and location based localized routing and QoS routing in wireless networks. In *Proceedings of the IEEE International Conference on Parallel Processing*, pages 173–180, Toronto, August 2000.

24. Y. Yu, R. Govindan, and D. Estrin. *Geographic and Energy Aware Routing: A Recursive Data Dissemination Protocol for Wireless Sensor Networks*. Technical Report TR-01-0023, Computer Science, University of California, Los Angeles, California, August 2001.

25. W. Heinzelman, J. Kulik, and H. Balakrishnan. Adaptive protocols for information dissemination in wireless sensor networks. In *Proceedings of the 5th ACM/IEEE International Conference on Mobile Computing and Networking (MobiCom '99)*, Seattle, Washington, August 1999.

26. S. Madden, M. J. Franklin, J. M. Hellerstein, and W. Hong. TAG: A Tiny Aggregation service for ad-hoc sensor networks. In *Proceedings of the 5th Symposium on Operating System Design and Implementation (OSDI 2002)*, Boston, Massachusetts, December 2002.

27. J. Mirkovic, G. P. Venkataramani, S. Lu, and L. Zhang. A self-organizing approach to data forwarding in large-scale sensor networks. In *Proceedings of the IEEE International Conference on Communications (ICC 2001)*, Volume 5, pages 1357–1361, Helsinki, Finland, June 2001.

28. H. S. Kim, T. F. Abdelzaher, and W. H. Kwon. Minimum-energy asynchronous dissemination to mobile sinks in wireless sensor networks. In *Proceedings of the 1st International Conference on Embedded Networked Sensor Systems (Sensys '03)*, Los Angeles, California, November 2003.

29. F. Ye, H. Luo, J. Cheng, S. Lu, and L. Zhang. A two-tier data dissemination model for large-scale wireless sensor networks. In *Proceedings of the 8th Annual International Conference on Mobile Computing and Networking (MobiCom 2002)*, pages 148–159, Atlanta, Georgia, September 2002.

30. S. Bandyopadhyay and E. J. Coyle. An energy efficient hierarchical clustering algorithm for wireless sensor networks. In *Proceedings of INFOCOM 2003*, San Francisco, April 2003.

31. C.-C. Shen, C. Srisathapornphat, and C. Jaikeo. Sensor information networking architecture and applications, *IEEE Personal Communications*, **8**(4):52–59, August 2001.

32. W.-P. Chen, J. C. Hou, and L. Sha. Dynamic clustering for acoustic target tracking in wireless sensor networks. *IEEE Transactions on Mobile Computing*, **3**(3):258–271, July–September 2004.

33. S. Lindsey, C. Raghavendra, and K. M. Sivalingam. Data gathering algorithms in sensor networks using energy metrics. *IEEE Transactions on Parallel and Distributed Systems*, **13**(9):924–935, September 2002.

34. M. Rabbat and R. Nowak. Distributed optimization in sensor networks. In *Proceedings of the 3rd International Symposium on Information Processing in Sensor Networks (IPSN)*, Berkeley, California, April 2004.

35. J. Chou, D. Petrovic, and K. Ramchandran. A distributed and adaptive signal processing approach to reducing energy consumption in sensor networks. In *Proceedings of INFOCOM 2003*, San Francisco, April 2003.

36. A. Scaglione and S. D. Servetto. On the interdependence of routing and data compression in multi-hop sensor networks. In *Proceedings of the 8th Annual ACM/IEEE International Conference on Mobile Computing and Networking (MobiCom '02)*, Atlanta, Georgia, 2002.

37. A. Boulis, S. Ganeriwal, and M. B. Srivastava. Aggregation in sensor networks: An energy-accuracy trade-off. In *Proceedings of the 1st IEEE International Workshop on Sensor Network Protocols and Applications (SNPA 2003)*, Anchorage, Alaska, May 2003.

38. C. Schurgers, V. Tsiatsis, S. Ganeriwal, and M. Srivastava. Optimizing sensor networks in the energy-latency-density design space. *IEEE Transactions on Mobile Computing*, **1**(1):70–80, January 2002.

39. Y. Yu, B. Krishnamachari, and V. K. Prasanna. Energy-latency tradeoffs for data gathering in wireless sensor networks. In *Proceedings of INFOCOM 2004*, Hong Kong, March 2004.

40. D. Slepian and J. K. Wolf. Noiseless encoding of correlated information sources. *IEEE Transactions on Information Theory*, **IT-19**:471–480, July 1973.

41. E. Duarte-Melo, M. Liu, and A. Misra. A computational approach to the joint design of distributed data compression and data dissemination in a data-gathering wireless sensor network. In *Proceedings of the 34th Allerton Conference on Communications Control*, Monticello, Illinois, October 2003.

42. P. Gupta and P. R. Kumar. The capacity of wireless networks. *IEEE Transactions on Information Theory*, **46**(2):388–404, 2000.

43. D. Petrovic, R. C. Shah, K. Ramchandran, and J. Rabaey. Data funneling: Routing with aggregation and compression for wireless sensor networks. In *Proceedings of the 1st IEEE International Workshop on Sensor Network Protocols and Applications (SNPA)*, pages 156–162, Anchorage, Alaska, May 2003.

44. M. Chu, H. Haussecker, and F. Zhao. Scalable information-driven sensor querying and routing for ad hoc heterogeneous sensor networks. *International Journal of High Performance Computing Applications*, **16**(3), Fall 2002.

45. R. C. Shah and J. M. Rabaey. Energy aware routing for low energy ad hoc sensor networks. In *Proceedings of the IEEE Wireless Communications and Networking Conference (WCNC 2002)*, Orlando, Florida, March 2002.

46. R. Kumar, M. Wolenetz, B. Agarwalla, J. Shin, P. Hutto, A. Paul, and U. Ramachandran. DFuse: A framework for distributed data fusion. *In Proceedings of the 1st ACM Conference on Embedded Networked Sensor Systems (Sensys '03)*, Los Angeles, California, November 2003.

47. S.-J. Baek, G. de Veciana, and X. Su. Minimizing energy consumption in large-scale sensor networks through distributed data compression and hierarchical aggregation. *IEEE Journal of Selected Areas in Communications*, **22**(6):1130–1140, August 2004.

48. A. Deshpande, C. Guestrin, S. Madden, J. Hellerstein, and W. Hong. Model driven data acquisition in sensor networks. In *Proceedings of the 30th International Conference on Very Large Data Bases (VLDB 2004)*, pages 588–599, Toronto, Canada, August–September 2004.

49. F. Ye, G. Zhong, S. Lu, and L. Zhang. A robust data delivery protocol for large scale sensor networks. In *Proceedings of the 2nd International Workshop on Information Processing in Sensor Networks (IPSN '03)*, Palo Alto, California, April 2003.

49a. F. Ye, A. Chen, S. Lu, and L. Zhang. A scalable solution to minimum cost forwarding in large scale sensor networks. In *Proceedings of the 10th International Conference on Computer Communications and Networks (ICCCN 2001)*, Scottsdale, Arizona, October 2001.

50. S. Tilak, A. Murphy, and W. Heinzelman. Non-uniform information dissemination for sensor networks. In *Proceedings of the 11th IEEE International Conference on Network Protocols (ICNP '03)*, Atlanta, Georgia, November 2003.

51. W.-P. Chen and L. Sha. An energy-aware data-centric generic utility based approach in wireless sensor networks. In *Proceedings of the 3rd International Symposium on Information Processing in Sensor Networks (IPSN)*, pages 215–224, Berkeley, California, April 2004.

52. F. Kelly, A. Maulloo, and D. Tan. Rate control for communication networks: Shadow prices, proportional fairness and stability. *Journal of the Operational Research Society*, **49**(3):237–252, March 1998.

53. S. H. Low and D. E. Lapsley. Optimization flow control, I: Basic algorithm and convergence. *IEEE/ACM Transactions on Networking*, **7**(6):861–875, December 1999.

54. J. Byers and G. Nasser. Utility-based decision-making in wireless sensor networks (extended abstract). In *Proceedings of the 1st Annual Workshop on Mobile Ad Hoc Networking and Computing (IEEE MobiHoc 2000)*, pages 143–144, Boston, Massachusetts, August 2000.

55. W.-P. Chen, J. C. Hou, L. Sha, and M. Caccamo. *A Distributed, Energy-aware, Utility-based Approach for Data Transport in Wireless Sensor Networks*. Technical Report UIUCDCS-R-2004-2455, Department of Computer Science, University of Illinois at Urbana-Champaign, July 2004.

56. N. Sadagopan and B. Krishnamachari. Maximizing data extraction in energy-limited sensor networks. In *Proceedings of INFOCOM 2004*, Hong Kong, March 2004.

57. F. Ordonez and B. Krishnamachari. Optimal information extraction in energy-limited wireless sensor networks. *IEEE Journal of Selected Areas in Communications*, **22**(6), August 2004.

58. M. Bhardwaj and A. P. Chandrakasan. Bounding the lifetime of sensor networks via optimal role assignments. In *Proceedings of INFOCOM 2002*, New York, June 2002.

59. R.-F. Liao and A. T. Campbell. A utility-based approach to quantitative adaptation in wireless packet networks. *ACM Journal on Wireless Networks (WINET)*, **7**(5):541–557, September 2001.

60. J. Wang, L. Li, S. H. Low, and J. C. Doyle. Can TCP and shortest path routing maximize utility? In *Proceedings of INFOCOM 2003*, San Francisco, April 2003.

61. J.-H. Chang and L. Tassiulas. Energy conserving routing in wireless ad-hoc networks. In *Proceedings of INFOCOM 2000*, Tel Aviv, Israel, March 2000.

62. C. Y. Wan, A. T. Campbell, and L. Krishnamurthy. PSFQ: A reliable transport protocol for wireless sensor networks. In *Proceedings of the 1st ACM International Workshop on Wireless Sensor Networks and Applications (WSNA'02)*, pages 1–11, Atlanta, Georgia, September 2002.

63. C. Y. Wan, S. B. Eisenman, and A. T. Campbell. CODA: Congestion detection and avoidance in sensor networks. In *Proceedings of the 1st International Conference on Embedded Networked Sensor Systems (Sensys '03)*, Los Angeles, California, November 2003.

64. Y. Sankarasubramaniam, O. B. Akan, and I. F. Akyildiz. ESRT: Event-to-sink reliable transport in wireless sensor networks. In *Proceedings of the 4th ACM International Symposium on Mobi Ad Hoc Networking and Computing (MobiHoc '03)*, pages 177–188, Annapolis, Maryland, June 2003.

65. T. He, J. A. Stankovic, C. Lu, and T. Abdelzaher. SPEED: A stateless protocol for real-time communication in sensor networks. In *Proceedings of the 23rd International Conference on Distributed Computing Systems (ICDCS '03)*, Providence, Rhode Island, May 2003.

66. A. Woo and D. Culler. A transmission control scheme for media access in sensor networks, In *Proceedings of the 7th Annual International Conference on Mobile Computing and Networking (MobiCom '01)*, pages 221–235, Rome, Italy, July 2001.

WILEY SERIES ON PARALLEL AND DISTRIBUTED COMPUTING

Series Editor: Albert Y. Zomaya

Parallel and Distributed Simulation Systems/Richard Fujimoto

Mobile Processing in Distributed and Open Environments/Peter Sapaty

Introduction to Parallel Algorithms/C. Xavier and S. S. Iyengar

Solutions to Parallel and Distributed Computing Problems: Lessons from Biological Sciences/Albert Y. Zomaya, Fikret Ercal, and Stephan Olariu (*Editors*)

Parallel and Distributed Computing: A Survey of Models, Paradigms, and Approaches/Claudia Leopold

Fundamentals of Distributed Object Systems: A CORBA Perspective/ Zahir Tari and Omran Bukhres

Pipelined Processor Farms: Structured Design for Embedded Parallel Systems/Martin Fleury and Andrew Downton

Handbook of Wireless Networks and Mobile Computing/ Ivan Stojmenović (*Editor*)

Internet-Based Workflow Management: Toward a Semantic Web/ Dan C. Marinescu

Parallel Computing on Heterogeneous Networks/Alexey L. Lastovetsky

Performance Evaluation and Characterization of Parallel and Distributed Computing Tools/Salim Hariri and Manish Parashar

Distributed Computing: Fundamentals, Simulations and Advanced Topics, *Second Edition*/Hagit Attiya and Jennifer Welch

Smart Environments: Technology, Protocols, and Applications/Diane Cook and Sajal Das

Fundamentals of Computer Organization and Architecture/ Mostafa Abd-El-Barr and Hesham El-Rewini

Advanced Computer Architecture and Parallel Processing/ Hesham El-Rewini and Mostafa Abd-El-Barr

UPC: Distributed Shared Memory Programming/Tarek El-Ghazawi, William Carlson, Thomas Sterling, and Katherine Yelick

Handbook of Sensor Networks: Algorithms and Architectures/ Ivan Stojmenović (*Editor*)